# Handbook of Manufacturing Control

Hermann Lödding

# Handbook of Manufacturing Control

## Fundamentals, Description, Configuration

Translated by: Rett Rossi

Springer

Prof. Dr.-Ing. habil. Hermann Lödding
Koppel 7
20099 Hamburg
Germany
loedding@tu-harburg.de

ISBN 978-3-642-43618-5 ISBN 978-3-642-24458-2 (eBook)
DOI 10.1007/978-3-642-24458-2
Springer Heidelberg New York Dordrecht London



Printed on acid-free paper

Springer is part of Springer Science+Business Media (www.springer.com)

# Foreword to the International Edition

Practitioners lament time and time again that many production engineering theories can rarely be implemented on the shop floor. For decades now, the Institute of Production Management and Technology at the Hamburg University of Technology has taken this as incentive to purse practice based research and to test their findings together with industrial enterprises in day to day operations. The same is true of the Institute of Production Systems and Logistics at the Leibniz University Hannover, which through my education there, significantly influenced my current approach to my profession.

In this tradition, this book is aimed at both the practice oriented theorist, as well as the practitioner who is receptive to a bit of theory. The third target group is students at universities and colleges who wish to extend their understanding of production control.

In professional literature, there is frequently a separation between modeling logistic objectives and approaches to manufacturing control: Modeling is often dominated by queuing theory which is not always easily understandable for practitioners. To a large extent separately from this, consultants and theorists promote methods with which enterprises can control their productions.

This book brings the modeling and methods together: From the more approachable funnel model it derives both the actuating and control variables with which enterprises can control the logistic objectives 'schedule reliability', 'throughput times', 'inventory' and 'utilization'. It then allocates the actuating variables to the tasks and methods regulating them.

This systematic approach facilitates a fundamental understanding of controlling manufacturing in order to attain logistic targets. It has been very well received in German speaking countries both by those working in the industry as well as those involved in research.

With the financial support of the Volkswagen Foundation and the perseverance and diligence of the translator, Rett Rossi, it is now possible to share this knowledge of manufacturing control with the large international community of production management. I am very thankful for this.

The creation of this book has a special history in that it originated, with the aid of a grant from the German Research Foundation (DFG), to a large part in countries other than Germany. During this time, I had the opportunity to discuss theories of manufacturing control with accredited experts and to study their application in factories in very different cultures. I am therefore very grateful to the following professors and their colleagues not only for their professional guidance but also their tremendous hospitality:

- Prof. Paul Schönsleben and Prof. Markus Baertschi from the ETH Centre for Industrial Management (BWI) at the ETH Zurich,
- Prof. João Fernando Gomes de Oliveira from the São Carlos School of Engineering at the University of São Paulo,
- Prof. Mitchell Tseng from the Advanced Manufacturing Institute at Hong Kong University of Science & Technology,
- Prof. Peter Nyhuis from the Institute of Production Systems and Logistics at Leibniz University Hannover.

In addition, a number of colleagues and friends took on the considerable task of correcting the entire book manuscript or large parts thereof and providing valuable suggestions with regards to the content. In particular I would like to thank Prof. Dr.-Ing. Carsten Begemann, Gregor von Cieminski, Daniel Grabe, Dr. Ralf Hieber, Dr.-Ing. Michael Schneider, Erik Thiry, Dr.-Ing. habil. Hans-Hermann Wiendahl as well as my mother Maria Lödding. Oliver Kress provided me with much appreciated information about Kanban, whose practical application I was kindly allowed to study at Daimler's plant in Gaggenau.

The book is based on longstanding research and industrial experience at the Institute of Production Systems and Logistics. With uncommon continuity of content, under the leadership of Professors Kettner, Wiendahl and Nyhuis, research results have been targeted there that have shaped production planning and control in German speaking countries and increasingly also internationally. This applies first and foremost to the Funnel Model with the Throughput Diagram and the Logistic Operating Curves (LOC) as well as the methods developed for them at the institute. In particular, the institute fundamentally contributed to methods for releasing orders and controlling capacities.

My academic advisor Prof. Hans-Peter Wiendahl, deserves a special thanks. He not only encouraged me to write this book and to conduct the project at international universities, but also provided a variety of helpful contacts and carefully examined the manuscript resulting in numerous improvements and impetuses.

I would like to dedicate this book to my incomparable grandmother, an unswerving optimist, to whom I am greatly indebted.

Hamburg Hermann Lödding
July 2011

# Preface

The field of manufacturing control has had an eventful past. It began in the mid-1960s, when the planning and control of continually larger plants and factories with thousands of orders and tens of thousands of operations overwhelmed the experience and imagination of supervisors and managers. Famous rules were seen as last resorts for getting through the eternal target conflict between a higher utilization and punctual delivery. Examples of such rules included "a workshop just has to be provided with a large number of jobs in order to be productive", or "setup times should be avoided as much as possible so combine orders that are the same" and "processing orders according to the shortest processing time (SPT) rule will create short throughput times for orders". What is common about all of these rules and others is that they resolutely support one target, but do nothing to resolve the actual conflict between targets.

Within this context, queuing theory, which had already been successfully implemented in systems for distributing goods and telephone systems, gave rise to hope. Nevertheless, despite great efforts being made both in research and on the shop floor, it never managed to establish itself in discrete manufacturing organized according to the job shop principle. Comprehensive research showed that the mathematical conditions for this model just did not exist in the industry.

Growing computer capacities and the availability of visual displays rang in a new era of production control in the early 1980s. The so-called 'electronic control board' finally promised reliable planning and processing of orders by meticulously planning and pursuing every single operation on every machine. But these hopes too were dashed because the underlying deterministic model was wrong and the effort for planning and maintaining the system was too great.

Parallel to these developments new approaches became known. Kanban, developed in Japan, radically called into question traditional ways of thinking in that orders were drawn from the production based on the supermarket principle (i.e. the pull principle) instead of loading them into the production long before the desired delivery date (push principle). The Load Oriented Order Release (LOOR) – developed for job shop productions and later extended to Load Oriented Production Control and Monitoring – was the first to combine a backlog oriented capacity control with a WIP oriented order release. Other authors emphasized the bottleneck theory, thus the Theory of Constraints (TOC) and Optimized Production Technology (OPT), which was derived from it. Further milestones included CONWIP (Constant Work in Process) and Cumulative Production Figures which was developed in the automobile industry.

In the mid-1980s the lean production wave set-in resulting in the partitioning of factories into autonomous market and product oriented islands, segments and fractals. Each of these was to in turn be controlled as much as possible by the workers. This was accompanied by the increased shifting of parts, components and entire subsystems to suppliers, while production enterprises concentrated on procurement, assembly and distribution. The interest in production control thus greatly diminished and numerous production control systems subsequently disappeared from the market.

In the mid-1990s, the focus became centered on the entire supply sequence from the supplier's supplier up to the customer's customer: The supply chain was born and with that the transparency of the processes stood in the foreground.

Irrespective of these and other developments such as agent control, fuzzy logic and genetic algorithms, for example, the target conflicts and production control tasks naturally continued to exist. Nevertheless, the demands especially with regards to delivery reliability, the increasing forms of production and their connection to production systems as well as the methods available for fulfilling PPC tasks have changed. If one were to ask companies today what the largest problem in logistics is, the main complaint would still be the unreliability of production logistics.

Given this situation, this book is long overdue. With a new systematical approach the author forges the logical correlations between the objectives and tasks of production control. The latter being: order generation, order release, capacity control and sequencing. Each of these tasks is then characterized based on the properties with which they are classified.

The subsequent description of all the methods known in professional literature as well as on the shop floor based on their logic forms the core of this book. With the aid of clear and comprehensive diagrams these methods are made transparent and assessed in regards to their suitability for specific application cases. Nine guiding principles, which continually remind the reader about the correlation between the objectives and actuating/controlled variables, form the foundation for this evaluation.

In conclusion, the author demonstrates how the individual components of manufacturing control are selected and configured together. Seeing some methods are able to fulfill a number of tasks, this is a critical aspect.

This work distinguishes itself not only with its convincing systematic and thorough international research, but also in the way it clearly presents the material and can be directly applied. I hope this book will be well received both in scientific circles as well as on the shop floor and am convinced it will become a standard work in publications on PPC.

Hanover
May 2004

Hans-Peter Wiendahl

# Preface to the International Edition

Since this book was first published it has drawn great interest in the German speaking community of manufacturing control experts. The dissemination of the ideas discussed in it through international journals and presentations at conferences have repeatedly led to requests for an English edition. I am thus very pleased that this has been accomplished with this publication and hope that the Manufacturing Control Handbook will experience an equally positive reception not only in research and education but also in the industrial practice.

Hanover
August 2011

Hans-Peter Wiendahl

# Translator's Notes

I would like to begin by thanking the Volkswagen Stiftung for financially supporting this extensive translation project. Although it has been a considerable undertaking, there are only a few remarks which I would like to make note of here.

The first of the issues I would like to address are the names we have chosen to refer to each of the individual methods presented here. Obviously, many of these methods are known by a number of more or less formal names as well. We have chosen to provide formal names for them, in order to facilitate the comparison of them and to add clarity to the discussions about them. Secondly, as this is a translation of the original German edition, references are made to both German and English sources. For the ease of the reader English references have been marked with an * for easy identification.

Finally, I would just like to add a word with regards to the terms inventory, WIP and stock: Although the differences between these terms are sometimes difficult to define, for the purpose of this book, *inventory* is used as an overall term encompassing both *work-in-process (WIP)* and *stock*. Whereby, *WIP* is the inventory of orders and their work contents between the order release and the order completion, while *stock* is that which is found in a store, either before the order release (e.g., raw goods, procured parts), between processes (e.g., semi-finished goods) or after the order completion.

Berlin  
July 2011

Rett Rossi

# Contents

## Part B Methods for Generating Orders

## Part D Sequencing



## Part E Controlling Capacities

# Index of Abbreviations and Variables

## General Terms (as index or suffix)

| | |
|---|---|
| i,j,k | indices |
| act | actual value |
| m | mean value |
| max | maximum value |
| min | minimum value |
| mw | mean weighted value |
| plan | planned value |
| s | standard deviation |
| v | coefficient of variance |
| (T) | value at time T |

## Variables

| Symbol | Unit | Meaning |
|---|---|---|
| ARW | SCD | advanced release window |
| BiC | units | bin content |
| BL | - | backlog |
| BN | - | position of bottleneck |
| BS | units | base stock |
| c | - | C-Norm parameter |
| CAP | h/SCD | capacity of a workstation |
| CH | % | holding costs for reference period |
| CM | €/unit | material costs/unit |
| COP | - | index of the current operation |
| CP | €/unit | production costs/unit |
| CPF | units | Cumulative Production Figure |

| Symbol | Unit | Meaning |
|---|---|---|
| CR | - | critical ratio |
| CSU | € | setup costs |
| CU | € | cost per unit |
| DDE | SCD | delivery date |
| DEC | % | delivery compliance |
| DEL | SCD | delivery lateness |
| $DEL_{max}^{+}$ | SCD | maximum positive delivery lateness (delay) |
| DEOT | % | on-time delivery |
| DERE | % | delivery reliability |
| DFA | % | discounting factor |
| DMD | units | demand |
| EDO | SCD | end of order processing |
| EDOP | SCD | end of operation processing |
| FR | - | Flow Rate |
| $L_{max}^{+}$ | SCD | maximum positive lateness (delay) |
| LD | h/SCD | load |
| $L_{in}$ | SCD | input lateness |
| $L_{out}$ | SCD | output lateness |
| $L_{rel}$ | SCD | relative lateness |
| LP | % | loading percentage |
| LS | units | lot size |
| NO | - | number of orders |
| NOP | - | number of operations |
| OI | SCD | order interval |
| OUT | h | output in reference period |
| P | SCD | length of the reference period |
| $PRF_i$ | - | production run frequency of Variant i during the planning period |
| RDMD | units/SCD | demand rate |
| ROUT | h/SCD | output rate |
| ROUTO | -/h | output rate (in orders/h) |
| R | SCD | range |
| SB | units | base stock |
| SDO | SCD | start of order processing (order input) |
| SeqC | % | sequencing compliance |
| SERL | % | service level |
| SF | - | stretch factor |
| SL | units | stock level |
| slack | SCD | slack time |
| SOP | units | stock order point |
| SR | % | schedule reliability |
| SSL | units | safety-stock level |
| $t_p$ | min/unit | processing time |

| Symbol | Unit | Meaning |
|---|---|---|
| $t_{su}$ | min | setup time |
| TAD | SCD | administration time |
| TD | SCD | delivery time |
| TDEB | SCD | delivery time buffer |
| TIO | SCD | inter-operation time (operation) |
| TOP | SCD | operation time (operation) |
| $T_{order}$ | SCD | order time point |
| TP | SCD | procurement time |
| TRP | SCD | replenishment time |
| TS | SCD | shipping time |
| TTP | SCD | order throughput time |
| U | % | WIP dependent utilization |
| WC | h | work content |
| WIP | h | WIP |

# Abbreviations

| | |
|---|---|
| CONWIP | Constant Work in Process |
| CORMA | Capacity Oriented Materials Management |
| CPFR | Collaborative Planning Forecasting Replenishment |
| DEWIP | Decentralized WIP Oriented Manufacturing Control |
| EODD | Earliest-Operation-Due-Date |
| ESD | Earliest Planned Start Date |
| FIFO | First in – First out |
| LOOR | Load Oriented Order Release |
| LPT | Longest Processing Time |
| MC | Manufacturing Cell |
| MRP | Material Requirements Planning |
| MRP II | Manufacturing Resource Planning |
| OP | Operation |
| ORLP | Order Release with Linear Programming |
| PAC | Production Authorization Cards |
| Polca | Paired-Cell Overlapping Loops of Cards with Authorization |
| SCD | Shop Calendar Day |
| SCM | Supply Chain Management |
| Slack | Least Slack |
| SPT | Shortest Processing Time |
| WS | Workstation |

# Chapter 1
# Introduction

## 1.1 Intro

Researchers and business managers throughout the world agree that companies which combine products that customers benefit from greatly with superior logistics are especially competitive. Such enterprises often grow more quickly than the market and net particularly high profits. Here, good logistics means consistently achieving logistic targets:

- Successful enterprises require lower inventories of raw, semi-finished and finished goods. As a result, less capital is tied-up in processes for fulfilling orders and can instead be invested in developing innovative products or advanced technologies.
- Companies that are leaders in logistics deliver punctually, therefore gaining their customers' trust and avoiding the high costs resulting from delayed deliveries.
- Enterprises with good logistics deliver quickly, therefore benefiting from sales opportunities and frequently refraining from expensive storage.

Whereas there is an amazing degree of agreement about the goals and the advantages of consistently attaining logistic targets, the actual way to achieve them seems to be difficult. Already back in 1921, Hippler was insisting that the speed with which the workpiece flowed through the production was of utmost importance [Hipp-21]. The advances that have been made since then—largely due to developments in production and information technologies—are immense. Nonetheless, customers continue to complain about delivery times being too long and supplies not arriving punctually. To some extent this can surely be attributed to increased expectations, however, looking at enterprise practices also reveals that there is considerable potential for improvement.

One of the basic challenges lays in the fact that a superior level of target achievement can only be attained by combining various enterprise functions. Successful companies take into consideration logistic demands in areas as diverse as product or technology development, marketing and distribution, factory

H. Lödding, *Handbook of Manufacturing Control*,
DOI 10.1007/978-3-642-24458-2_1, © Springer-Verlag Berlin Heidelberg 2013

planning, strategic and operative production planning as well as in manufacturing control.

This book focuses on the area of manufacturing control as an essential factor for the success of logistics, but also points out connections to other functions as required. I have chosen to concentrate my observations on manufacturing control for four reasons.

### 1 *Manufacturing control is a field that is often neglected both in scholarly research as well as on the shop floor*

The discussion about attaining logistic targets is clearly overshadowed by production planning. There are three main reasons for this: First, production planning offers intellectually challenging tasks such as optimization problems that can be elegantly solved with methods from operation research, artificial intelligence or other scientific disciplines. Second, there is a huge market for enterprise software in which large amounts of money can be made. Correspondingly, conferences are frequently dominated by questions about information technology. And last but not least, a realistic and solid production plan is actually a condition for consistently attaining logistic targets.

However, it is easy to overlook the fact that a good production plan only leads to reliably attaining logistic targets when the manufacturing control also manages to implement this plan. A functioning manufacturing control is therefore imperative for also practically exploiting the theoretical advantages of a progressive production planning. Consequently, considerable attention should be paid to it.

This is all the more critical because the manufacturing control workforce often fails to transfer the discussions that take place on a strategic level about the significance of logistic objectives to the daily operations. Although their personal success is frequently measured based on attaining logistic targets, a large percent of the workforce does not have a clear idea about how they can systematically influence the logistic objectives. If the planning failures of other business functions prevent targets from being attained then there is also no understandable correlation between the employee performance and the key logistic figures.

One of the important goals of this book therefore, is to demonstrate how the logistic objectives are influenced by manufacturing control.

### 2 *The significance of manufacturing control for attaining the logistic targets is increasing*

Even when it cannot be emphasized enough that consistently attaining logistic targets can only be achieved through the coordinated collaboration of numerous corporate functions, manufacturing control is still a particularly important element. On the one hand, poor manufacturing control prevents the attainment of logistic targets even when the general conditions are positive, whereas good manufacturing control can compensate to some extent for mistakes made in planning. On the other hand, not attaining logistic targets creates huge problems in the areas of the supply chain that follow e.g., stock outages in an assembly. Production, therefore, takes on a key logistic position.

In addition, manufacturing control is becoming increasingly vital due to three reasons: First, continually greater expectations are being placed on attaining logistic targets. Second, as a result of the large diversity of variants and uncertain markets it has become quite difficult to reliably predict demand. Accordingly, it is continually more difficult to establish dependable production programs. Manufacturing control therefore has to be able to react quickly to changes in the production plans. And third, the scope of tasks covered by manufacturing control is constantly expanding. Whereas previously manufacturing control was limited to the sequencing on workstations, nowadays it also includes releasing orders into production and regulates capacities. In make-to-stock manufacturing, manufacturing control even takes on to some extent the responsibilities of production planning by generating orders.

***3 New theoretical findings in modeling logistic objectives allow a more effective manufacturing control***

Since the 1980s, the theoretical understanding of the correlations between logistic objectives and how they can be influenced has clearly deepened. New theoretical findings in modeling logistic objectives can trigger new developments in manufacturing control procedures. This is illustrated by a number of examples:

- Both *Little's Law* [Litt-61*] as well as the *Funnel Formula* [Bech-84] formulate a correlation between the production's work-in-process (WIP), its output and the order throughput time. At the beginning of the 1970s, Wight already concluded [Wigh-70*] that regulating the WIP would allow throughput times to be controlled. Subsequently, numerous methods for releasing orders that consequently regulate the production's WIP were developed.
- More recently, considerable advances were made in modeling due date deviations (referred to here also as lateness) as well as schedule reliability [Yu-01]. Accordingly, the mean *lateness* of orders results from the ratio of the production's backlog and output rate. Thus, methods for controlling backlogs will evidently take on greater significance [Brei-01, Pete-96, Remp-03]. Meanwhile, the finding that the *schedule reliability* is a function of the mean and variance of the lateness has lead to the suitability of sequencing rules being reassessed.
- *Logistic Operating Curves* (LOC) describe the conflict between the logistic objectives. Logistic Operating Curves in the production area depict the correlation between the WIP, throughput time and utilization of a workstation [Nyhu-91]. In doing so they allow a Logistic Positioning within the field of conflict created by targeting low WIP levels and short throughput times on the one side and the goal of a high utilization rate on the other hand. In addition, Storage Operating Curves (SOC) quantify the correlation between store levels and the service level of a product [Gläß-95, Lutz-01, Lutz-02, Nyhu-09*]. Although production and Storage Operating Curves have not yet lead to new developments in procedures for manufacturing control, they simplify systematically setting the method parameters. They thus contribute significantly to attaining logistic targets.

***4 The practice of manufacturing control offers considerable potential for improvement***

There are enterprises that have mastered the principles of manufacturing control themselves and attain outstanding results. Generally speaking it has also been a long time since there was a one-sided transfer of knowledge between theory and practice. On the contrary, especially in the area of manufacturing control, many procedures have been developed on the shop floor. Leading scientific institutes therefore cultivate their dialogue with the practice. Nonetheless, there is considerable potential for improvement in the majority of production companies and this potential can frequently be exploited through comparatively simple measures. Thus, there is a remarkable opportunity for enterprises to reduce WIP levels and delivery times and to increase schedule reliability.

On the whole though there is an abundance of manufacturing control methods whose modes of operations and parameters are not commonly known. This book therefore introduces a selection of procedures and explains the tasks these methods fulfill and their parameters. Our focus here is on *methods* because they standardize decisions. In doing so, they help to streamline the decision process and accelerate decision making while at the same time often relieving operators from mindless routine tasks. However, they also involve considerable risks. In particular, they often take away the worker's control in the decision making so frequently, no one verifies whether or not the decision met by the method is still practical with regards to the overall targets. It is also not unusual that expectations linked to implementing a method are disappointed. There are at least five reasons for this:

1. *Lack of potential*: Partly, the production simply does not have the potential required for achieving the logistic targets. In this case, the solution lies not in improving the manufacturing control, but rather in improving the production itself (e.g., through shortening setup times) or in constructing the product to be suitable for manufacturing.
2. *Starting at the wrong point*: The cause of frequently failing to attain targets does not always lie in the area in which the selected manufacturing control method is applied. If for example, the schedule reliability is reduced by a lack of sequencing compliance and a lack of backlog control, applying a method for releasing orders cannot make improvements.
3. *Unsuitable methods*: The method is not suitable for the specific application case e.g., because the diversity of the variants is too great or the methods are not adjusted to the complexity of the material flow.
4. *Incorrectly parameterized method*: The impact of a method is considerably dependent on the parameters set for the method. Nonetheless, only a few enterprises devote enough consideration to this. Many companies wrongly estimate the significance of the parameters. Moreover, they often do not know how they should suitably set the parameters.
5. *Mistakes in implementing*: Failing to attain targets on the shop floor can, at least in part, also be explained by mistakes in implementing methods. A known example of this is the practice of many enterprises of releasing a part of the

orders before the planned start date so that they ensure a punctual completion. As a result the WIP in production increases and urgent orders compete for resources with non-urgent orders. The early completion of a few orders nevertheless inevitably delays the throughput of the remaining orders and can thus cause their lateness. Consequently, the delivery reliability decreases.

With the exception of the first case, in which there is no logistic potential, enterprises can attain their logistic targets and thus in many cases immediately and to a considerable extent increase their competitiveness by improving manufacturing control. In order to do so though, it is necessary to develop a deeper understanding of manufacturing control and the methods implemented. This book aims to promote this understanding and to point out the levers for exploiting existing potential.

## 1.2 A Manufacturing Control Model

Manufacturing control is a component of Production Planning and Control (PPC). It is the job of PPC to pre-plan the continuous production program for a number of planning periods, to derive the required materials and resources from that and to implement the production program despite unavoidable disruptions such as staff shortages, machine malfunctions, supply delays and rejects [Wien-97]. The Aachen PPC Model, which is primarily known in German speaking countries, differentiates between key tasks and cross-functional tasks (see [Lucz-99] as well as Fig. 1.1).

The key tasks of PPC are production program planning, material resource planning as well as the planning and control of external and in-house production. *Production program planning* determines which products in which quantities should be produced during the next planned periods. *Production requirements planning* derives the required materials and resources from the production program. In order to do so it determines the material requirements for components and parts, schedules the production orders and determines the load for the production's capacity groups. Generally, enterprises procure some of the components and parts from suppliers and manufacture the rest themselves. The difference between external procurements and in-house production is ultimately a question of the reference point within the supply chain: The order for a component is an in-house production for the supplier, but an external procurement for the receiving producers. The Aachen PPC Model allocates among other things tasks such as calculating the production lot sizes, finite scheduling, sequence planning and verifying availability to *in-house production planning and control.* Accordingly, tasks such as determining the order lot-sizes, obtaining and evaluating bids, and selecting suppliers belong to *external procurement and control* [Lucz-99]. The cross-functional tasks of PPC coordinate the order development across different production areas (*order coordination*), provide stock (*storage*) and measure the

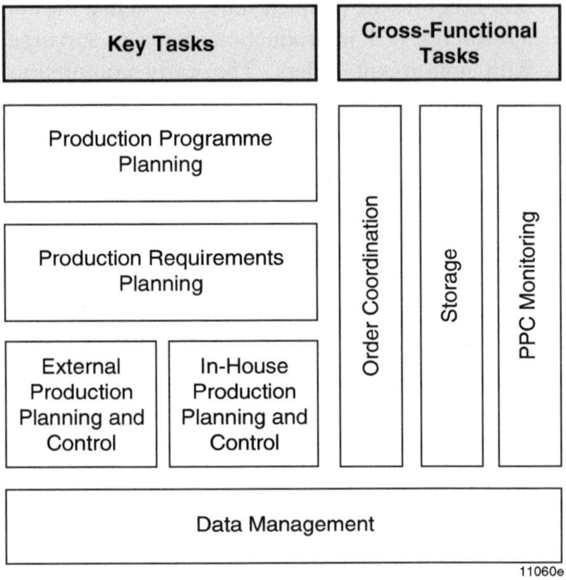


**Fig 1.1** Overview of tasks according to the Aachen PPC model (FIR)

attainment of logistic targets (*PPC Monitoring*). Both the key tasks as well as the cross-sectional tasks are dependent upon diligent *data management* [Lucz-99].

Manufacturing control which is the key task considered the most in this book is primarily referred to in the Aachen PPC Model as part of the in-house production planning and control. It thus forms only a small part of the model. Nonetheless, as already elaborated, it is highly significant for attaining logistic targets. In order to clarify the manufacturing control tasks and how they influence the logistic objectives, I would now like to introduce a manufacturing control model.

According to Wiendahl it is the job of manufacturing control to implement the targets set by production planning even when there are—frequently unavoidable—disruptions [Wien-97b]. This basic understanding of manufacturing control can be transferred to a manufacturing control model (Fig. 1.2). The model consolidates fundamental knowledge from the modeling of logistic objectives, from theories of manufacturing control and from the developments of methods which were gained from research at the University of Hanover's Institute of Production Systems and Logistics. Readers who do not have any previous technical knowledge of this model may have difficulty in immediately grasping the model in its entirety. This is however not necessary as it serves here only as a frame for bringing the manufacturing control tasks logically together with the logistic objectives. The interactions will be extensively discussed in later chapters.

The model consists of four elements: production planning and control tasks, actuating variables, control variables, and logistic objectives. These elements are linked to one another by their interactions:

1. The tasks determine the actuating variables.
2. The control variables result from the deviation of two actuating variables.
3. The control variables determine the logistic objectives.

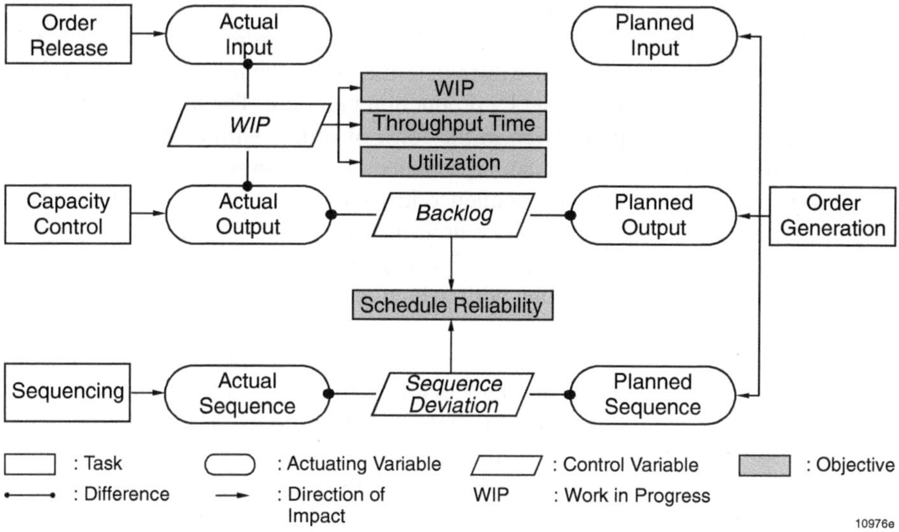


**Fig. 1.2** Manufacturing control model

The elements of the model can be clarified as follows:

1. *Tasks*: Manufacturing control tasks include *releasing orders, controlling capacities* and *sequencing*. The release of orders entails determining the time point and the sequence in which the orders are dispatched for production. It therefore sets the actual input for the production whereas the capacity control ascertains the workers' hours and how long which operator works on which machine. It thus influences the actual output of the production[1]. Sequencing determines the succession in which orders are processed on a workstation. *Generating orders* determines the planned values for the input and output of the production as well as the planned sequence and is therefore a logical component of production planning. It frequently includes many subtasks such as production program planning, material resource planning and scheduling due dates and capacities. However, in make-to-stock productions the order generation is to some extent conducted using very simple procedures, for example, the Kanban Control. These methods—especially because of how quickly decisions have to be made—are frequently classified as manufacturing control methods, although they actually carry out planning tasks. Generally they do not explicitly determine the planned values. However, based on the logic of the procedure and the choice of the parameters it is possible to derive useful planning values.

[1] The production output is essentially influenced by the capacity, however, also by the production's WIP, as well as generally by the market demand for a product and to some degree also by the sequence in which the orders are processed. Representing the output as an actuating variable thus simplifies complex correlations for the sake of greater clarity.

2. *Actuating variables*: The actuating variables for manufacturing control are the *input* and *output* as well as the *sequence* in which the orders are processed. The model contains both the actual values determined by the manufacturing control as well as the planned values set by production planning. The input describes the work received by the production in the form of orders with a specific 'allotted time'. This actuating variable can be described by the amount and time point of the input and also includes the sequence in which the orders enter into production. Accordingly, the output can be described by the amount and time point of the work processed by a production. The sequence of the output is defined as an independent actuating variable with which the orders are processed by the production.
3. *Control variables*: The control variables of a manufacturing control result from the deviation between two actuating variables: The difference between the actual input and actual output of a production defines the production's *work-in-process* or *WIP* for short. Both of these variables are cumulatively measured over time. The WIP control variable impacts the logistic objectives WIP[2] and utilization of the production as well as the throughput times of the orders. The *backlog* is defined as the difference between the planned output and the actual output. Here too, it is necessary to measure these actuating variables cumulatively over time. With a positive backlog the planned output is greater than the actual output. The backlog decisively influences the schedule reliability of the production. The *sequence deviation* describes the deviation of the actual sequence from the planned sequencing. Together with the backlog, it impacts the schedule reliability of the production.
4. *Logistic objectives*: The model includes the basic logistic objectives *WIP, utilization, throughput time* and *schedule reliability*. The work-in-process is defined as the difference between the (cumulated) actual input and the actual output (see above). It influences how much capital is tied-up in the production's WIP and how much floor area is required in the production. The (WIP dependent) utilization describes the probability of not being able to operate a workstation because no order is available. It thus impacts the production costs. The throughput time of the orders is defined as the length of time from the release of the order to its completion. It therefore forms a lower limit for the delivery time of an order. The schedule reliability is defined as a percentage of the orders completed within a specified deviation of a planned completion date, which is smaller than a set due-date tolerance. The schedule reliability directly impacts the delivery reliability of an enterprise from the perspective of the customer.

The model can be employed for a number of purposes. For the practice of manufacturing control the following four seem to be the most significant:

[2] The WIP is both a control variable and logistic objective.

1. *Analyzing the failure to consistently attain targets*: The model makes it possible to analyze the causes for inadequately attaining logistic targets. Too high WIP levels, too long throughput times or a low WIP dependent utilization are the result of the actual input and actual output being insufficiently coordinated. The release of orders, therefore, should be coordinated with the capacity control. Low schedule reliability is caused by a positive backlog or by sequence deviations. If the backlog is positive, the enterprise's planned and actual output need to be better aligned. This concerns the order generation tasks as well as the capacity control. If sequence deviations arise, then the actual sequence needs to be aligned with the planned sequence and where necessary the planning process for determining the planned sequence should also be verified.
2. *Designing a manufacturing control*: The model indicates how a manufacturing control of the logistic objectives can be attained. A *WIP control* uses the order release to link the actual input to the actual output. The WIP, (WIP dependent) utilization and throughput times can therefore be controlled by it. A *backlog control* regulates the capacities so that the actual output constantly follows the planned output and the backlog is as close to zero as possible. This in turn increases the schedule reliability of the production. A *sequence control* impacts the schedule reliability similarly. It sets the processing sequence so that the actual sequence and the planned sequence coincide with another as much as possible.
3. *Developing an understanding of the system used for manufacturing control*: The model helps to create an understanding of the system used for manufacturing control in that it depicts the logical correlations between manufacturing control tasks and logistic objectives. A more comprehensive understanding of the system is further promoted in that the model describes all internal logistic objectives (WIP, throughput time, utilization, schedule reliability) and is not limited to a part of the objectives.
4. *Differentiating between planning and control*: The model illustrates the difference between the production planning and the manufacturing control. The production planning determines the planned values for the actuating variables, and the manufacturing control determines the actual values. This demonstrates the crucial significance of manufacturing control.

Nevertheless, the model also has its limitations, four of these I would like to mention here. All of them are generally the result of trying to keep the model clear and concise:

1. *The model does not depict any external logistic objectives*: The model only describes how the production planning and control influences the internal logistic objectives. The external logistic objectives 'delivery time' and 'delivery reliability', which are important from the customers' perspective, are decoupled from the internal objectives 'throughput time' and 'schedule reliability' where applicable by a planning parameter such as a delivery time buffer or time factor. The correlation between the external and internal logistic objectives is the focus of Chap. 3 (Modeling Logistic Objectives).

2. *The model only depicts one aspect of the interactions*: The model is limited to depicting the strong interactions. In addition to these there are a number of weaker interactions. For example, the sequencing can impact the output rate and therefore the production's output (e.g., with sequence dependent setup times). Furthermore, the production's input impacts the WIP and thus also the possible output. We will closely examine these and other dependencies in the later chapters of this book.
3. *The model does not depict any quantitative correlations*: Although all of the depicted dependencies are based on quantitative correlations, the model itself only describes the qualitative correlations. In the following discussion it will therefore be supplemented by the Hanoverian Funnel Model.
4. *The model does not contain all of the manufacturing control tasks*: The Aachen PPC Model [Lucz-99] allocates additional tasks to manufacturing control. An especially relevant task for attaining logistic targets is the 'distribution of work' [Wien-02]. This task determines on which of a number of alternative work centres an order will be processed. Distributing the work will only be explained indirectly in the following. Generally, however, it can be shown that with identical machines and sequence independent setup times, it is practical not to allocate the orders to a specific machine ahead of time [Wien-02, Wien-97b]. Instead, it is enough that once a machine has completed an order, it always selects the order with the highest priority from a central queue. Even with different machines or sequence dependent setup times there is still a case to be made for not actively dividing work, but rather only fixing a suitable sequencing rule.

## 1.3 Book Layout

The methods described in this book fulfill all of the tasks outlined in the manufacturing control model:

1. generating orders
2. releasing orders
3. sequencing
4. controlling capacities

Although generating orders is a production planning task, it is presented here for different methods of make-to-stock production. These include procedures that generate orders quickly and are event oriented (e.g., Kanban Control) and are thus generally not a component of the periodic planning run in enterprises.

In simple cases, these tasks completely cover the controlling of the order fulfillment process from the point a customers places a demand up to the product delivery. In all cases they influence attaining the logistic targets—as described in the manufacturing control model. Figure 1.3 illustrates the interplay between the individual tasks during the order fulfillment process. Based on the customers' demand, orders are generated which are released according to defined criteria.

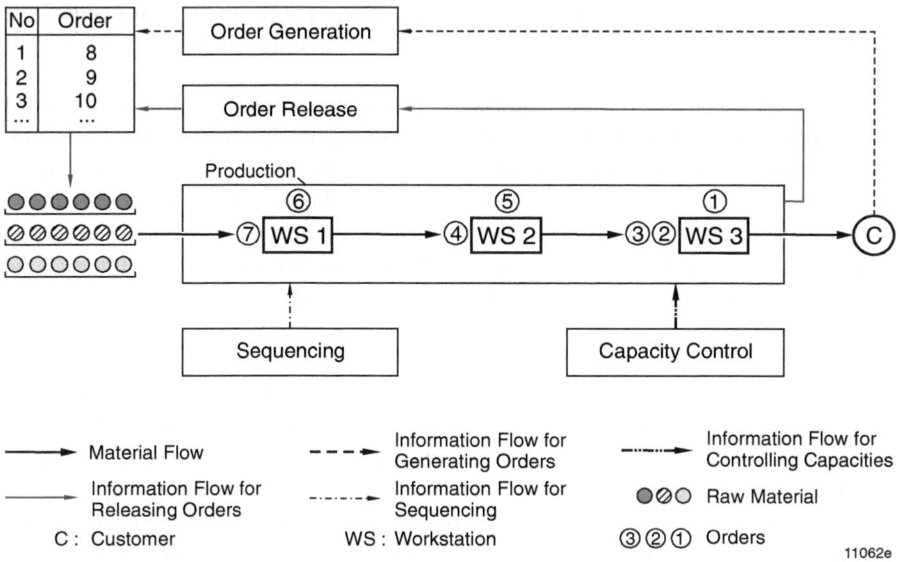


**Fig. 1.3** Flow of information for controlling the order fulfillment process

The capacities are controlled and the sequencing is set at the workstation. Flows of information pertaining to specific tasks then arise from this.

Methods which fulfill tasks other than the four mentioned will be described only briefly, when at all. This applies especially to methods which schedule orders (e.g., finite scheduling), plan capacities, determine lot-sizes, set delivery dates or make decisions about accepting orders. This choice is not intended to signify a devaluation of these methods or to call into question their suitability.

Based on these initial considerations the book is divided into six parts whose contents will be expanded upon in the following (see Table 1.1).

**Part A: Fundamentals of Manufacturing Control** The first part of this book is dedicated to the fundamentals of manufacturing control and consists of five chapters:

**Table 1.1** Layout of the book

| Part | Content | Chapter |
|---|---|---|
| A | Fundamentals of manufacturing control | 2–6 |
| B | Methods for generating orders | 7–15 |
| C | Methods for releasing orders | 16–24 |
| D | Sequencing | 25 |
| E | Controlling capacities | 26–28 |
| F | Configuring the manufacturing control | 29 |

- Chapter 2 defines the logistic objectives and explains their significance for the enterprise's success. In doing so it also accounts for the importance of manufacturing control on the whole.
- Chapter 3 discusses the correlations between the actuating variables, control variables and logistic objectives represented in the manufacturing control model. In addition it depicts the connection between the internal logistic objectives and the external logistic objectives which impact the customers. The modeling is based on the theoretically and practice proven combination of the Hanoverian Funnel Model, the Throughput Diagram and Logistic Operating Curves. This chapter also plays a key role with regards to the designing of manufacturing control methods.
- Chapter 4 describes the production planning tasks. Based on the customers needs these determine the planned values on which manufacturing control orients itself. Production planning is thus tightly interlocked with manufacturing control.
- Chapter 5 discusses the manufacturing characteristics relevant to manufacturing control. On the one hand, this is necessary because the possible applications of one part of the manufacturing control methods are restricted to specific types and principles of manufacturing. On the other hand, it is shown how the logistic objectives can be influenced by different factors which lay outside of manufacturing control.
- Chapter 6 is dedicated to production planning and control across the entire supply chain. Taking up the continuing discussion about managing supply chains and/or networks it focuses on analyzing the causes of increased demand fluctuations in the supply network i.e. the so-called "Bullwhip Effect". From here, requirements especially for generating orders within the supply net can be derived.

**Part B: Generating Orders** An important class of methods generate orders and with that the planned values for the input and output of the production and the order sequencing. The feasibility of these planned values impacts the possible schedule reliability of the production. Furthermore, the methods provide the planned values for the production's WIP and the throughput time of the orders. Customer orders or store outputs usually form the basis for generating orders. Among the techniques applied for generating orders are MRP-II and make-to-stock methods (e.g., Kanban Control, Order Point System or Base Stock).

**Table 1.2** Methods described for generating orders

| Chapter | Method for generating orders |
|---|---|
| 7 | Fundamentals of order generation |
| 8 | Order point system |
| 9 | Kanban |
| 10 | CORMA |
| 11 | Synchro MRP |
| 12 | Hybrid Kanban/CONWIP control |
| 13 | Cumulative production figures |
| 14 | Base stock |
| 15 | Production authorization cards |

Procedures for generating orders are implemented both in production and commercial enterprises. Many companies also control different products using different methods. Therefore, an enterprise can employ a number of methods for generating orders.

Eight different methods for generating orders will be described here, some of which include a number of variations (Table 1.2). The first chapter of Part B describes the fundamentals of order generation. The remaining chapters explain make-to-stock methods which generate orders based on events.

When generating orders, the production's capacity is frequently considered insufficiently or not at all. This is an important reason for combining methods that generate orders along with those for releasing orders (Part C) and controlling backlogs (Part E).

**Part C: Releasing Orders** The order release determines the production's input. Many methods for releasing orders delay or accelerate the release of orders with the goal of controlling the WIP or balancing the load. They are generally implemented in order to align the load with the production's actual WIP situation. The order release can concern an entire order or individual operations of an order. Methods for releasing orders influence the WIP and the (WIP dependent) utilization of a production as well as the order's throughput time.

Eight different methods for releasing orders will be discussed here (Table 1.3). Some of these include in turn a number of variations. The fundamentals of releasing orders will be explained at the beginning of Part C.

**Table 1.3** Methods described for releasing orders

| Chapter | Method for releasing orders |
|---|---|
| 16 | Fundamentals of releasing orders |
| 17 | Due date based order release |
| 18 | CONWIP |
| 19 | Bottleneck control |
| 20 | Workload control |
| 21 | Load oriented order release |
| 22 | Order release with linear programming |
| 23 | POLCA |
| 24 | Decentralized WIP oriented manufacturing control |

**Part D: Sequencing** Sequencing rules determine the sequence in which queued orders are processed on a workstation. They primarily impact the production's schedule reliability and therefore the delivery reliability or service level of an enterprise. In some cases the sequencing can increase the output rate of the production.

Part D introduces sequencing rules relevant to the shop floor and which influence the logistic objectives 'delivery reliability', 'service level', and 'output rate'. Moreover, it explains the significance of a high degree of sequencing compliance.

**Part E: Controlling Capacities** The capacity control ascertains on short notice which measures are to be implemented for adjusting capacities. In particular it determines overtime, shortened work hours and other special measures related to the flexibility of capacities.

It thus primarily impacts the production's backlog and therefore influences the schedule reliability. Part E introduces different methods for controlling the capacities. They differ with regards to the set target (e.g., increasing the schedule reliability/service level, maximizing the output rate) as well as with regards to the area applied (e.g., make-to-stock/make-to-order productions) (Table 1.4).

**Table 1.4** Methods described for controlling capacities

| Chapter | Method described for controlling capacities |
|---|---|
| 26 | Fundamentals of capacity control |
| 27 | Backlog control |
| 28 | Further methods for controlling capacity |
| | Plan oriented capacity control |
| | Due date oriented capacity control |
| | Output rate maximizing capacity control |
| | Inventory based capacity control |

**Part F: Configuring the Manufacturing Control** Every production company has the task of configuring its manufacturing control. In order to do so it has to decide which methods will be implemented for generating orders, releasing orders, sequencing, and controlling capacities. Part F, therefore, explains the criteria which should be taken into consideration when selecting measures and exemplarily presents possible combinations.

**Description and Discussion About Manufacturing Control Methods** The methods described in Parts B, C and E for generating orders, releasing orders and controlling capacities form the main component of this book. Each of the descriptions follow a basic, uniform structure.

- In the first section of each chapter the procedural rules are described and illustrated using an example. A comparatively large amount of space is dedicated to describing the methods. Time and time again discussions with students and practitioners showed that a brief description of the procedure was insufficient for developing a detailed understanding of them.
- The second section discusses the procedural rules from the perspective of various aspects of logistics. In doing so both the strengths and weakness of a method will be considered. As a result important information about the conditions for applying a method arises.
- The third section of a chapter focuses on setting the method's parameters. All too often, this aspect is considered insufficiently. Nonetheless, determining the method's parameters is a decisive factor for successfully applying the method.

For each of the methods it will also be shown how the parameters are connected to the logistic objectives. This allows enterprises to determine the method's parameters in two model based steps: In the first step, the company establishes the planned values for the logistic objectives with a Logistic Positioning. This step is generally the same for all manufacturing control methods and is based on the modeling of the logistic objectives discussed in the section on fundamentals (Part A). In the second step the planned values for the logistic objectives are converted into control variables based specifically on the individual methods.

- The fourth section describes applying the manufacturing control across the supply chain. This is especially relevant for procedures employed in generating orders, which to some degree differ considerably with regards to their suitability for being implemented throughout a supply chain. In contrast, the application of methods for releasing orders and controlling capacities are usually limited to a single manufacturer.
- Finally, the fifth section concentrates on applying the method. This section differs noticeably for each of the methods. Generally, the more frequently a method is used on the shop floor, the more extensive the discussion is. Thus detailed case studies are presented for the Due Date Based Order Release and Order Point Systems. For procedures implemented less frequently however, the section—when at all present—is limited to examples and conditions for applications specified in previous literature. The examples for configuring the manufacturing control, found in Chap. 27, also have a direct relation to companies.

In order to attain logistic targets, a fundamental understanding of manufacturing control is imperative. The logistic objectives are the starting point for the discussion in the next chapter.

## References


[Bech-84] Bechte, W.: Steuerung der Durchlaufzeit durch belastungsorientierte Auftragsfreigabe bei Werkstattfertigung. VDI Progress Reports, Series 2, No.70, Düsseldorf (1984)

[Brei-01] Breithaupt, J.-W.: Rückstandsorientierte Produktionsregelung von Fertigungsbereichen. Grundlagen und Anwendung. VDI Progress Reports, Series 2, No. 571, Düsseldorf (2001)

[Gläß-95] Gläßner, J.: Modellgestütztes Controlling der beschaffungslogistischen Prozeßkette. VDI Progress Reports, Series 2, No. 337, Düsseldorf (1995)

[Hipp-21] Hippler, W.: Arbeitsverteilung und Terminwesen in Maschinenfabriken. Berlin (1921) (quoted from [Wien-97b])

[Litt-61*] Little, J.D.C.: A proof of the queuing formula: L = Lambda W. Oper. Res. **9**(3), 383–387 (1961)

[Lucz-99] Luczak, H., Eversheim, W.: Produktionsplanung und -steuerung. Grundlagen, Gestaltung und Konzepte. Springer, Berlin (1999)

[Lutz-01] Lutz, S., Lödding, H., Wiendahl, H.-P.: Kennliniengestützte logistische Lageranalyse. Ein neuer Ansatz zur Positionierung im Dilemma zwischen Bestand und Servicegrad. ZWF **92**(9), 550–553 (2001)

[Lutz-02] Lutz, S.: Kennliniengestütztes Lagermanagement. VDI Progress Reports, Series 13, No. 53, (2002)

[Nyhu-91] Nyhuis, P.: Durchlauforientierte Losgrößenbestimmung. VDI Progress Reports, Series 2, No. 225, Düsseldorf (1991)

[Nyhu-09*] Nyhuis, P., Wiendahl, H.-P.: Fundamentals of Production Logistics. Theory, Tools and Applications. Springer, Berlin (2009)

[Pete-96] Petermann, D.: Modellbasierte Produktionsregelung. VDI Progress Reports, Series 20, No. 193, Düsseldorf (1996)

[Remp-03] Rempp, B.: Regelungstechnische Untersuchung durchsatzgesteuerter Produktionssysteme. Dissertation, Universität Stuttgart (2003)

[Wien-02] Wiendahl, H.-H.: Situative Konfiguration des Auftragsmanagements im turbulenten Umfeld. Jost-Jetter Verlag, Heimsheim (2002)

[Wien-97] Wiendahl, H.-P.: Betriebsorganisation für Ingenieure. Carl Hanser Verlag, Munich/Vienna (1997)

[Wien-97b] Wiendahl, H.-P.: Fertigungsregelung. Logistische Beherrschung von Fertigungsabläufen auf Basis des Trichtermodells. Carl Hanser Verlag, Munich/Vienna (1997)

[Wigh-70*] Wight, O.: Input/Output control. A real handle on lead time. Prod. Invent. Manag. **11**, 9–31 (1970)

[Yu-01] Yu, K.-W.: Terminkennlinie. Eine Beschreibungsmethodik für die Terminabweichung im Produktionsbereich. VDI Progress Reports, Series 2, No. 576, Düsseldorf (2001)

# Part A
# Fundamentals of Manufacturing Control

# Chapter 2
# Logistic Objectives

Manufacturing control is critical due to the strategic importance of the logistic objectives it influences. Wiendahl differentiates these objectives according to how they influence the logistic performance and logistic costs (see Fig. 2.1 and [Wien-97b]).

From the customers' perspective, the logistic performance (Sect. 2.1) consists of the logistic objectives 'delivery time', 'delivery lateness' and 'delivery reliability' (with make-to-order productions) or 'service level' (with make-to-stock productions). Based on these external objectives, the logistic objectives 'throughput time', 'lateness' and 'schedule reliability', which can be measured within the production, are derived. The external objectives for the logistic performance are then further differentiated as to whether the orders are triggered by a customer contract (make-to-order) or in order to restock a store (make-to-stock). This differentiation is common both in theoretical work and on the shop floor. The service level though, can also be interpreted as a special case of the delivery reliability, in which the planned delivery time (from stores) is by definition zero (immediate delivery, cf. [Wien-02]).

The objectives for the logistics costs (Sect. 2.2) are independent of whether the production is make-to-stock or make-to-order. The internal logistics costs are determined by the costs related to inventory (WIP and stock),[1] utilization and delays. They contribute to the production costs and as such are taken into consideration when calculating the price. Therefore, they also impact the customer.

[1] Translator's Note: I would like to clarify here, that for the purposes of this book the terms "inventory" or "inventories" is used when the author is referring to *both* the work-in-process (WIP) and stores of stock.

H. Lödding, *Handbook of Manufacturing Control*,
DOI 10.1007/978-3-642-24458-2_2, © Springer-Verlag Berlin Heidelberg 2013

| | Logistic Performance | Logistic Costs |
|---|---|---|
| External | *make-to-order production*<br>delivery time<br>delivery date deviation<br>delivery reliability | price |
| | *make-to-stock production*<br>service level | |
| Internal | throughput time<br>due date deviation<br>due date reliability | inventory<br>utilization<br>costs of delays |

11333e

**Fig. 2.1** Logistic performance and logistic costs objectives

## 2.1 Logistic Performance

The logistic performance is made up of the external logistic objectives 'delivery time' (Sect. 2.1.1), 'delivery reliability' (Sect. 2.1.2) and 'service level' (Sect. 2.1.3) as well as the internal logistic objectives 'throughput time' (Sect. 2.1.4), 'lateness', and 'schedule reliability' (Sect. 2.1.5). In the next sections of this chapter each of these objectives will be defined and explained with regards to their significance and how they interact with the other objectives.

### *2.1.1 Delivery Time*

Delivery time is defined as the length of time between the receipt of the order and its shipment [Wien-97b]. Usually it is measured in 'shop calendar days' or 'work days'. Moreover, the length of the delivery time is generally dependent on the stock strategy (see Fig. 2.2).

In a make-to-stock production the delivery time results solely from the shipment of the goods. With orders that are assembled or produced, the time required for the assembly and or production is added to this shipment time. If an article also has to be procured for an order, then the procurement time is also included. Consequently, the delivery time is no longer influenced by just the enterprise.

The strategic significance of the delivery time for the economical success of a company is widely undisputed. Empirical studies have shown that enterprises with short delivery times grow more quickly and achieve higher profits than those with longer delivery times:

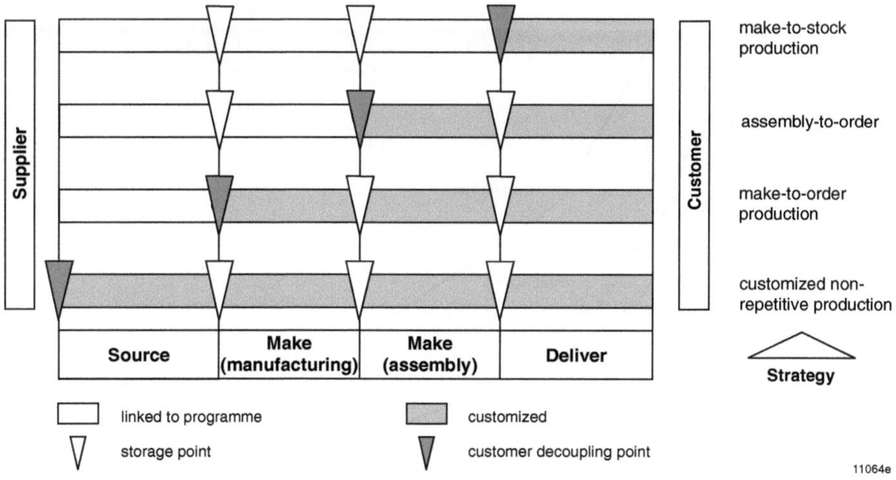


**Fig. 2.2** Stock strategies (Eidenmüller, according to Wiendahl)

- Stalk and Hout suggest that enterprises with considerably shorter delivery times than the competition grow approximately three times faster than the market and yield double the profits. In [Stal-90*] they point out examples for this from a variety of sectors.
- Rommel et al. analyzed the productivity, product utility and speed of successful and less successful enterprises. Successful companies manufacture a product in half the time required by their less successful competitors [Romm-93]. The authors used the production throughput time as a variable which forms the lower limit for the delivery time in a make-to-order production.

This directly raises the questions of why short delivery times are so significant. Obviously, customers attach more utility to a product, the shorter its delivery time is. Thus, when deciding to purchase a product the delivery time is considered just as much as the price and quality. If two offers have a similar price and quality, the decisive factor is the delivery time. If an offer is not superior with regards to all three utility factors then the customer is forced to weigh the factors against one another and trade one off for the other. A 'trade-off' such as this between the delivery time and price is also required especially in the case of supplier dependent price premiums (or so-called 'mark-ups' or 'surcharges') for shorter deliveries. These can be depicted in the form of a ratio between the price and delivery time (see Fig. 2.3).

Figure 2.3 is a reproduction of the price formation for a German circuit board manufacturer which was published on the internet (www.andus.de, November 2003). For decades, they have been charging premiums of up to 150% for especially short delivery times of standard (bi-layer) circuit boards.

The starting point for calculating the price is the base price of a circuit board with the standard delivery time of 15 SCD. The highest premium is charged for the

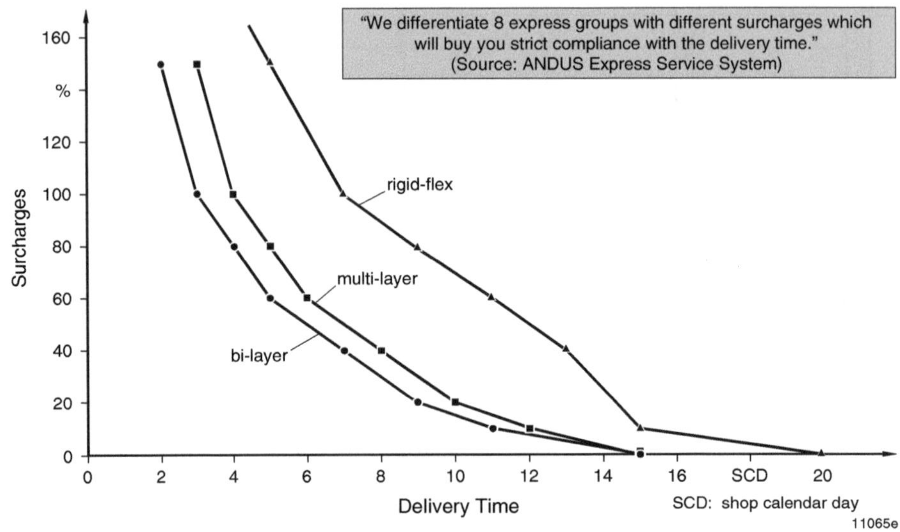


**Fig. 2.3** Ratio of the price and delivery time for an enterprise in the circuit board industry

minimal delivery time of two shop calendar days. Like a large number of circuit board manufacturers, Andus differentiates the price formation according to circuit board technology (bi-layer, multi-layer, rigid-flex). Further examples of price-delivery time relations in other branches are cited by Stalk and Hout [Stal-90*].

Similar to the delivery time and price, the customer also weighs the delivery time and quality. Such is the case when a customer accepts a lower quality product because the higher quality product is not available (or not available in time). A clear example of this is a person who walks into a cafeteria and orders the meal that is prepared the quickest i.e., has the shortest waiting time, instead of the one that tastes the best. In English, accepting a lower quality solution due to the quickness of availability has even become known as a "quick-and-dirty" solution.

In addition to the rather subjective sense of utility, there are also economic reasons for remunerating shorter delivery times:

1. *Follow-up costs of longer delivery times:* If long delivery times cause follow-up costs then accepting price premiums is justified. Examples of this include the threat of a production stand-still due to a lack of available materials in the automobile industry, delaying complex assembly groups in plant construction or default penalties that can be avoided by promptly delivering of procured parts.

The majority of the next reasons can be traced back to follow-up costs due to longer delivery times. Reasons two to five are concerned with cases in which the customer initially feeds goods into a store, from which smaller amounts can then be called-up. A typical example of this are production companies that procure and maintain stores of parts in order to limit their own delivery times and trade. The sixth reason applies solely to order based procurement.

2. *Inventory costs:* In order to ensure the availability of a product the purchaser generally requires higher stock levels of items with long delivery times than those with shorter delivery times (see Sect. 3.1.3). Correspondingly, the capital tie-up costs are higher with longer delivery times than with shorter delivery times [Nyhu-09*].
3. *Inventory risk:* If the demand for a product is less than expected, the excess inventory has to be sold either at a discount or when necessary completely written off. There is a greater risk of this with the higher stock levels of items requiring long delivery times than with lower stock levels.
4. *Responsiveness:* Products with shorter delivery times can be more quickly re-ordered when the demand is higher than expected, thus making it possible to use the non-predicted market opportunity despite low store levels. Moreover, if competitors are dependent on longer delivery times, a (temporary) monopoly can even be attained. An enterprise can use this to skim off the demand and to target an attractive price.

The delivery time for a product determines the necessary forecasting horizon. Since the precision of a forecast decreases with the length of the forecasting horizon, the forecast risk is less with short delivery times [Stal-90*]. Reasons three and four describe the risk of over or underestimating the demand in a forecast horizon. It can thus generally be concluded that a market with uncertain demand cannot only be more quickly served with shorter delivery times, but also more cost-effectively.

5. *Diversity of variants:* Since products with shorter delivery times require lower stock levels to ensure their availability, a greater diversity of products can be provided with the same capital expenditure than with longer delivery times. This allows a greater diversification in meeting the demands of the market.
6. *Delivery time:* With order based procurement, shorter delivery times for the supplier reduce the delivery times of the enterprise from the customer's perspective. This increases the company's ability to compete.

The reasons mentioned above concern the economical benefits of shorter delivery times for the customer (i.e. those requesting the product). Correspondingly, there are also reasons for the supplier to demand higher prices for shorter delivery times.

1. *Number of competitors:* The shorter delivery times are, the fewer suppliers there are and the less competition there is. Those suppliers that remain can therefore enforce higher prices. In extreme cases a supplier attains a monopoly and with that the freedom to fix prices to a very large extent. Accordingly, in evaluating delivery times it is always important to compare them to the typical market standards.
2. *Loss of utilization:* There are two general reasons for an increased loss of utilization with short delivery times: On the one hand, an enterprise can consciously accept a loss of utilization in order to target shorter throughput times in the production. This allows shorter delivery times in a make-to-order

production, however, is linked to higher costs (see Sect. 2.2.2). On the other hand, an enterprise can provide reserve capacities in order to ensure short delivery times also when there is increased demand. With normal or low demands a loss of utilization can result from here as well.

3. *Administration costs:* The administrative effort for a rush order is frequently greater than normal orders. This applies in particular when they are not administered with standard processes.
4. *Delay costs for normal orders:* If the processing of normal orders is deferred for the sake of rush jobs, then the completion of normal orders is delayed. As a result delay costs may arise.
5. *Setup costs:* Rush orders lead to increased setup costs in two cases. First, if the rush job has to be divided in order to be processed simultaneously on a number of similar machines (order splitting). Second, if an enterprise interrupts the processing of other jobs in order to prioritize the processing of a rush job.
6. *Distribution costs:* Rush orders are frequently shipped express resulting in additional transportation costs.
7. *Procurement costs:* With order based procurement the supplier often has to pay an express delivery surcharge in order to attain an especially short procurement time.

The costs mentioned in points two to seven for shorter delivery times diminish the advantages of shorter throughput times and can even outweigh them. It therefore has to be a goal to prevent as many of these costs as possible and to attain short delivery times with minimal costs.

### 2.1.2 Delivery Lateness and Delivery Reliability

The *delivery lateness* refers to the difference between the actual and planned delivery date:

$$DEL = DDE_{act} - DDE_{plan} \tag{2.1}$$

where

| | |
|---|---|
| DEL | delivery lateness [SCD], |
| $DDE_{act}$ | actual delivery date [SCD], |
| $DDE_{plan}$ | planned delivery date [SCD]. |

According to this definition, positive lateness describes a late delivery whereas negative lateness refers to a delivery that is too early.

The delivery lateness can also be measured in relation to the confirmed delivery date or the requested delivery date [Wien-02]. The requested delivery date is the delivery date that was originally specified by the customer. In contrast, the confirmed delivery date is the delivery date promised to the customer and the planned delivery date that which was set in the production planning. The latter is the standard for the

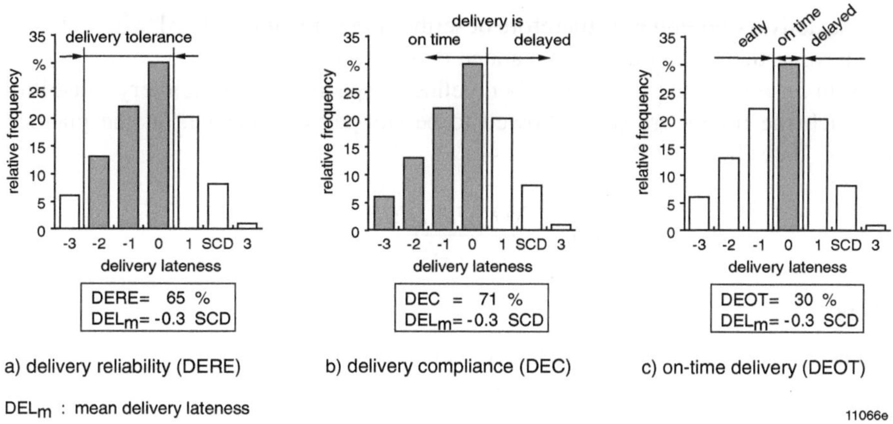


**Fig. 2.4** Definition of delivery reliability, delivery compliance and on-time delivery

production's schedule reliability. In the following it will be assumed that the planned and target due date correspond with one another.

The *delivery reliability* refers to the percentage of orders delivered within a defined delivery reliability tolerance (see Fig. 2.4a).

$$DERE = \frac{NO \text{ with } DEL_{ll} \leq DEL \leq DEL_{ul}}{NO} \cdot 100 \tag{2.2}$$

where

| | |
|---|---|
| DERE | delivery reliability [%], |
| $DEL_{ll}$ | lower limit for permissible delivery lateness [SCD], |
| DEL | delivery lateness [SCD], |
| $DEL_{ul}$ | upper limit for permissible delivery lateness [SCD], |
| NO | number of orders [–]. |

The delivery reliability is always measured for a defined reference period, whereby the absolute values for the upper and lower limits of the lateness can deviate from one another. Two key special types of delivery reliability are delivery compliance (or in-time delivery) and on-time delivery.

*Delivery compliance* is defined as the percentage of orders within a reference period shipped punctually or too early (see Fig. 2.4b):

$$DEC = \frac{NO \text{ with } DEL \leq 0}{NO} \cdot 100 \tag{2.3}$$

where

| | |
|---|---|
| DEC | delivery compliance [%], |
| NO | number of orders [–], |
| DEL | delivery lateness [SCD]. |

The delivery compliance therefore describes the percentage of orders a customer will have at their disposal at the planned delivery date.

With *on-time delivery* Schönsleben defines a special case of delivery reliability in which the orders are only allowed to be shipped on-time i.e., at the planned delivery date [Schö-07*] (see Fig. 2.4c).

$$DEOT = \frac{NO\ with\ DEL = 0}{NO} \cdot 100 \tag{2.4}$$

where

| | |
|---|---|
| DEOT | on-time delivery [%], |
| NO | number of orders [−], |
| DEL | delivery lateness [SCD]. |

Similar to the delivery time the delivery reliability is also strategically significant for the success of a company. The management consultancy company Deloitte & Touche interviewed 900 managers from 35 countries about what the decisive competitive factors are in the twenty-first century. In the survey, managers assigned the most importance to delivery reliability even before factors such as quality, technology and price. At the same time, the managers acknowledged that there was an especially large potential for improving their companies' delivery reliability [Delo-98*].

The validity of such a survey can surely be debated, nonetheless, there are reasons for the vital significance of the delivery reliability:

1. *Customer trust:* Promising a delivery date and then not keeping it creates doubts regarding the suppliers reliability and thus impacts the customer's trust. In the worse case scenario the customer considers a delayed delivery an opportunity to look for a dependable supplier. Conversely, maintaining a high delivery reliability supports a high level of customer loyalty. If the competitors are unreliable, customers have a greater incentive to develop a higher portion of their purchasing volume with the enterprise that is reliable. A high delivery reliability thus positively impacts the company's profits and growth. As with the delivery time, the comparison with competitors is decisive when evaluating delivery reliability.
2. *Follow-up costs of delayed delivery:* A delayed delivery can cause a loss of utilization or even the standstill of a customer's assembly belt. The resulting damage can very easily exceed the value of the product a number of times over. Modern production concepts such as Just-in-time-Management or Supply Chain Management aim to drastically reduce the amount of stock maintained amongst the enterprises in the supply chain. They are thus particularly susceptible to delayed deliveries. Accordingly, only those suppliers who can ensure a high delivery reliability are suitable for these production concepts.
3. *Delivery time:* Especially with order based procurement customers will plan additional delivery time buffers for suppliers who have a low delivery reliability. The customer thus perceives the delivery time as being longer. This in turn leads to

the disadvantages associated with longer delivery times described in the previous section. If the delivery lateness varies, a conflict arises between the 'delivery time' objective and the 'delivery reliability' objective. A short delivery time is then related to a low delivery reliability and vice versa. With a given variance an enterprise has to position itself within this conflict of objectives (see Sect. 3.1.2).
4. *Safety stock:* When stock is supplied through stored goods, the customer is protected from delayed incoming materials by safety stock. The lower the delivery reliability is, the higher the required safety stock level (see Sect. 3.1.3).
5. *Influence on the customer's delivery reliability:* In supply chains, a delayed delivery of a procurement part can endanger the customer's ability to meet their due date. If the customer cannot deliver punctually due to a lack of materials, they too can be subject to the described disadvantages.

The mentioned reasons concern the economic benefits of a high delivery reliability for the company requesting a product. As with the delivery time, there are also reasons for why the supplier demands particularly high prices for strong delivery reliability.

1. *Number of competitors:* The number of possible suppliers of a product decreases, the higher the required delivery reliability is. This improves the negotiation position of the supplier.
2. *Setup costs:* In many enterprises, orders are processed so that as few setup operations are required as possible. This requires correspondingly large stores and causes order sequence interchanges which negatively impact the delivery reliability. If an enterprise opts not to follow such an optimization of the setup times and instead processes orders according to their priority, additional setup operations are required.
3. *Distribution costs:* Orders which are completed late are frequently delivered using express shipping in order to avoid a delayed delivery or at least to reduce the delay. The special costs caused through this can for the most part not be passed on to the customer.
4. *Delivery time and inventory costs:* Many enterprises include a delivery time buffer so that the delayed completion of an order does not directly lead to a delayed delivery. This increases the enterprise's delivery time and causes additional stores of finished goods (see Sect. 3.1.2).

The costs mentioned for delivery reliability in points two to four diminish the advantages and can even outweigh them. The goal therefore has to be avoiding as many of these costs as possible.

### *2.1.3 Service Level*

The most important external logistic objective in make-to-stock productions is the service level. The service level is defined as the percentage of demands within a reference period that were immediately satisfied (see [Wien-97b]).

$$SERL = \frac{NDMD_{immed}}{NDMD} \cdot 100 \tag{2.5}$$

where

| | |
|---|---|
| SERL | service level [%], |
| $NDMD_{immed}$ | number of immediately satisfied demands [−], |
| NDMD | number of demands [−]. |

Alternatively the weighted service level can be calculated. The weighted service level uses the demanded units as a reference value instead of the number of demands. It thus weights larger demands more strongly than demands for smaller delivery amounts. According to [Lutz-01] the weighted service level is calculated as:

$$SERL_w = \frac{NU_{immed}}{NU} \cdot 100 \tag{2.6}$$

where

| | |
|---|---|
| $SERL_w$ | weighted service level [%], |
| $NU_{immed}$ | number of immediately delivered units [−], |
| NU | number of demanded units [−]. |

When a product is not available there is the immediate danger of losing an order. Frequently a customer falls back on a competitor when a product is not directly available. This is related to a loss of sales for the enterprise. Therefore, the service level generally impacts the growth in sales and profits more directly than the delivery reliability in a make-to-order production.

Naturally, an order is not inevitably lost, when an enterprise is unable to deliver a demanded product directly from the stock. Even when the customer agrees to a later delivery date, the negative effects described concerning the delivery reliability are still applicable. There is still the threat of a loss of trust, a decreased customer loyalty and in the worst case the loss of status as a supplier. The customer in turn is subjected to the danger of a loss of utilization and delivery delays in relation to their own customers.

Inventory costs and utilization loss are the two basic reasons that higher costs can emerge for a supplier for a greater service level:

1. *Inventory costs:* There is a conflict of objectives between the service level and stock level of a product. The higher the stock level is, the higher the service level is. If a supplier maintains a large safety stock in order to ensure a high service level, then higher inventory costs arise for them.
2. *Loss of utilization:* In comparison to a make-to-order production, the delivery dates in a make-to-stock production cannot be shifted during periods where there is a higher demand. If an enterprise therefore maintains capacity reserves in order to be able to ensure the service level also when there is a particularly high demand, then there is a loss of utilization during periods with a normal or low demand.

Many enterprises are not able to precisely measure the service level because they do not even compile the non-fulfilled demands. If they instead calculate the service level as the ratio of immediately fulfilled requests to the total number of (immediate or late) fulfilled demands, an unrealistically high service level is indicated.

### 2.1.4 Throughput Time

The throughput time of a (production) order is defined as the length of time between the order's release and the end of its processing [Wien-97b]. Frequently, a simplified day-to-day plan or feedback is assumed.

$$TTP = EDO - SDO \tag{2.7}$$

where

| | |
|---|---|
| TTP | order throughput time [SCD], |
| EDO | end of order processing [SCD], |
| SDO | start of order processing (order input) [SCD]. |

In make-to-order productions the significance of the throughput time results from its influence on the delivery time. It forms the lower limit of the delivery time for an order. Where applicable, additional time segments need to be taken into consideration when calculating the possible delivery time e.g., the time required for procuring necessary raw materials and shipping. The variance of the throughput time also influences the schedule and delivery reliability. If the customer still changes the order specifications after the orders are placed, shorter throughput times are advantageous: With the given delivery date the order can be released for production later than with longer throughput times. Therefore, there is a greater possibility that the customer's desired changes arrive before the order is released.

In *make-to-stock* production the service level sinks with the length and variance of the throughput times, or, the amount of safety stock required to ensure the service level increases (see Sect. 3.1.3).

Especially with sensitive products there is a correlation between the rejects produced and the throughput times. In the electronic industry for example, dust particles can lead to component malfunctions. The throughput time determines how long the components are subjected to dust particles in an unprotected state. Correspondingly, the number of rejects increases with the throughput time.

### 2.1.5 Lateness and Schedule Reliability

*Lateness* describes the deviation of the actual order throughput from the planned order throughput. It can be measured in the output of an order, in its input and in comparison to the planned throughput time (see Fig. 2.5).

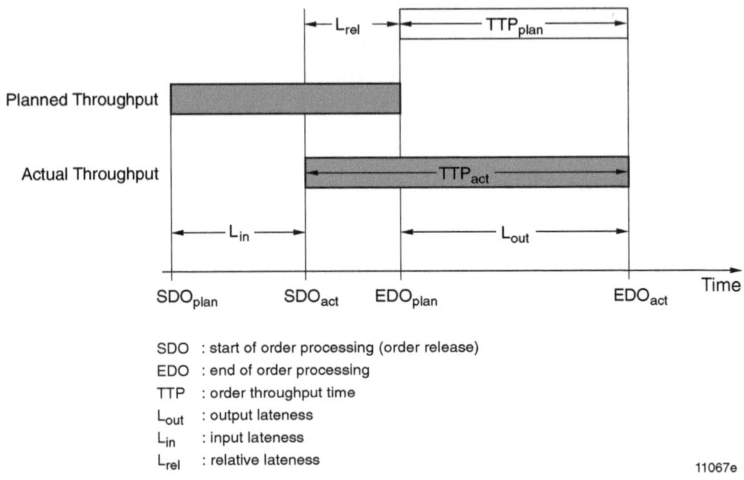


**Fig. 2.5** Definitions of lateness variables (Dombrowski)

The *output lateness* expresses the length of time an order is delayed in its completion in comparison to the planned completion date [Domb-88].

$$L_{out} = EDO_{act} - EDO_{plan} \tag{2.8}$$

where

$L_{out}$ output lateness [SCD],
$EDO_{act}$ actual end of order processing [SCD],
$EDO_{plan}$ planned end of order processing [SCD].

A positive value means the order's completion is delayed; a negative value denotes it is too early. The output lateness directly impacts the delivery lateness and therefore the delivery reliability. This explains its key significance as an internal objective.

The *input lateness* is a measure of how many days late ($L_{in} > 0$) or early ($L_{in} < 0$) an order is released in comparison to the planned start date [Domb-88].

$$L_{in} = SDO_{act} - SDO_{plan} \tag{2.9}$$

where

$L_{in}$ input lateness [SCD],
$SDO_{act}$ actual start of order entry (order input) [SCD],
$SDO_{plan}$ planned start of order entry (order input) [SCD].

These figures are mainly used for purposes of analysis, especially for evaluating the order release. When the start of an order is delayed ($L_{IN} > 0$) it frequently leads to the delayed completion of the order.

The *relative lateness* describes the length of time the throughput of an order through the production is delayed in comparison to the planned throughput time ($L_{rel} > 0$) or expedited ($L_{rel} < 0$). According to [Domb-88] it is calculated as:

$$L_{rel} = TTP_{act} - TTP_{plan} = L_{out} - L_{in} \quad (2.10)$$

where

| | |
|---|---|
| $L_{rel}$ | relative lateness [SCD], |
| $TTP_{act}$ | actual order throughput time [SCD], |
| $TTP_{plan}$ | planned order throughput time [SCD], |
| $L_{out}$ | output lateness [SCD], |
| $L_{in}$ | input lateness [SCD]. |

This figure is also primarily collected for the purpose of analysis. It allows the throughput of an order through the production to be evaluated. If the throughput of an order needs longer ($L_{rel} > 0$), the order is frequently completed too late.

Lateness can be determined for a single order. In contrast, the *schedule reliability* is a measure for evaluating the entirety of all orders within a time period. Similar to lateness, it can be calculated for the output, input and as a relative measure. The output schedule reliability (referred to as 'schedule reliability' in the following) is clearly the most important of these parameters. It is defined as the percent of orders completed within a given due date tolerance [Yu-01] and can be interpreted as the stability of the logistic process [Wien-03].

$$SR = \frac{NO \text{ with } L_{out,ll} \leq L_{out} \leq L_{out,ul}}{NO} \cdot 100 \quad (2.11)$$

where

| | |
|---|---|
| SR | schedule reliability [%], |
| NO | number of orders [−], |
| $L_{out,ll}$ | lower limit for permissible output lateness [SCD], |
| $L_{out}$ | output lateness [SCD], |
| $L_{out,ul}$ | upper limit for permissible output lateness [SCD]. |

The schedule reliability, although sometimes cushioned by a delivery time buffer, directly impacts the delivery reliability and is therefore a very important objective. The schedule reliability itself is influenced by the mean value and variance of the output lateness.

According to Eq. 2.11 and similar to lateness, the definition can be transferred to both the input and relative schedule reliability. And again, these parameters are mainly used for analyses. When the due date tolerance is the same in the input and output, comparing the two values can be particularly informative about the schedule reliability: If the delivery reliability in the input is higher than in the output, the schedule reliability worsens during the orders throughput. Conversely, the production positively impacts the delivery reliability. This comparison is nonetheless only practical, if the mean input lateness and output lateness approximately agree with one another and do not differ greatly from zero.

A critical objective which, nonetheless, is only measured in a minority of businesses is the *planned requested delivery reliability*. It describes the percentage of orders that the enterprise promises for the delivery date requested by the customer. In accordance with a recommendation from Wiendahl, it is also referred to as the capability of the logistic process. The objective uncovers differences between the customers requested date and the target or planned dates of the enterprise. It is calculated as:

$$DERR_{plan} = \frac{NO\ with\ TDE_{plan} = TDE_{req}}{NO} \cdot 100 \qquad (2.12)$$

where

| | |
|---|---|
| $DERR_{plan}$ | planned requested delivery reliability [%], |
| NO | number of orders [−], |
| $TDE_{plan}$ | planned delivery time [SCD], |
| $TDE_{req}$ | requested delivery time [SCD]. |

There are two essential reasons, why an enterprise cannot promise a requested delivery date. In the first case, the enterprise is generally not able to attain the desired delivery time. A typical reason for this would be a production throughput time that exceeds the requested delivery time. In the second case, the company is momentarily unable to confirm the requested delivery date. Although the production throughput time would generally be sufficient for fulfilling the requested delivery date, there are currently insufficient capacities for planning the order in time. In the first case, the key is to increase the requested delivery reliability by shortening the throughput time. In the second case, the enterprise has to increase capacities or the flexibility of the capacities in order to attain a greater requested delivery reliability.

Similar to the planned requested delivery reliability the actual requested delivery reliability can be calculated:

$$DERR_{act} = \frac{NO\ with\ TDE_{act} = TDE_{req}}{NO} \cdot 100 \qquad (2.13)$$

where

| | |
|---|---|
| $DERR_{act}$ | actual requested delivery reliability [%], |
| NO | number of orders [−], |
| $TDE_{act}$ | actual delivery time [SCD], |
| $TDE_{req}$ | requested delivery time [SCD]. |

## 2.2 Logistic Costs

The logistic costs influence the competitiveness of a company just as the logistic performance does. The lower the logistic costs are, the lower the price is with which an enterprise still attains a profit. At a given price, the profits increase as the logistic costs decrease.

In the following sections, the logistic objectives 'inventory' (Sect. 2.2.1), 'utilization' or 'output rate' (Sect. 2.2.2) and 'delay costs' (Sect. 2.2.3) will be discussed.

### 2.2.1 *Inventories*

Generally, inventories can be differentiated between stored stock (stock) and work-in-process (WIP). Stock is comprised of raw materials, semi-finished and finished goods and is measured in quantity of units or evaluated in currency units (e.g., €, $, ¥) for financial analyses. WIP on the other hand is formed by orders that have been released but not yet completed [Wien-97b]. In manufacturing control it is measured in standard hours or as the number of production orders. However, particularly when it comes to balancing, it can also be evaluated in quantity or currency units.

There are four reasons why inventory is critical in production logistics. These include as an objective, as a logistic control variable, as a control variable in a continuous improvement process (CIP) and as an evaluation parameter.

**Inventory as an Objective** Inventory impacts an enterprise's finances due to the capital tied-up in it and the costs which arise from capital being tied-up. The larger the inventory is, the more capital is bound into the company's current assets and the less financial latitude the company has for investing. Conversely, by reducing inventory part of the current assets can be released. This capital can be either reinvested in the company or removed from the enterprise. The latter either reduces debt (borrowed capital) or the company's capital resources.

The interest charges that have to be carried for the capital tied-up in the inventory increase proportionally to the tied-up capital. If a company manages to reduce inventories, then the interest charges decrease and accordingly profits are increased. It is worth taking this measure even if interest charges are a small percentage of the company's total costs. In the majority of enterprises, the costs are not significantly lower than the revenue; the percent rise in earnings is thus basically higher than the percent cost reduction. If a company for example, reduces its costs by 1% with a 5% return on sales, the return on sales increases by 19.5% to 5.95%.

One of the most important financial figures is the return on capital as a ratio of profits to invested capital. It allows the profitability of an enterprise to be compared with other forms of investment. If a company reduces its inventory, then the return on capital increases for two reasons: First, the profits can be increased by the decreased inventory costs and second, the capital tied-up in operations is decreased.

Additional inventory costs are caused by inventory risks. Inventory risks refer to the chance that stock cannot be as economically utilized as planned. The higher the level of stock, the greater the risk. Moreover, stock occupies space in the production and warehouses. Therefore, a plant with less stock requires less area than one with large amounts of stock. Finally, a production facility with less stock is also more comprehensible and thus easier to control.

**Inventory as a Logistic Control Variable** The work-in process of a production influences both the production's utilization (and thus also the output rate) as well as the throughput time of orders. The value of both objectives increases along with the WIP. This conflict of objectives between low WIP levels and short throughput times on the one hand, and a high utilization on the other hand, has been known as the '*dilemma of operation planning*' since the early 1950s [Gute-51]. It can be quantified with the aid of Logistic Operating Curves (see Sect. 3.2.2 as well as [Nyhu-09*]).

In make-to-stock productions the stock level influences the service level. Here too, there is a conflict of objectives. A higher service level requires a high stock level; inversely, an on average low stock level causes a low service level. This *dilemma of inventory management* can be quantified with the Storage Operating Curves (see Sect. 3.1.3 as well as [Gläß-95, Nyhu-09*, Lutz-02]).

**Inventory as a Control Variable in Continuous Improvement Processes** Stores act as a buffer for process disruptions and thus conceal the processes' susceptibilities to disruptions (see Fig. 2.6).

One of the basic ideas of continuous improvement processes in logistics is to gradually decrease stores until problems become obvious. The cause of the problem is then analyzed. After the problem is eliminated, the stores are further decreased until a new problem is revealed and so on.

The continuous reduction of stock leads to controlled processes. As a result, production costs decrease and quality increases [Kraf-88*]. Accordingly, numerous Japanese enterprises consider the attained stock levels and their turnover rate as measures for the process capability and thus for the competency of the corresponding production area [Mond-81*].

The idea to utilize stock as a control variable for continuous improvement processes originated in the Japanese just-in-time philosophy (JIT, cf. [Mond-81*, Mond-98*]). One of the conditions for implementing JIT is that each area of responsibility is able to independently control their inventory. Accordingly, a large number of enterprises who apply this philosophy utilize a Kanban Control system.

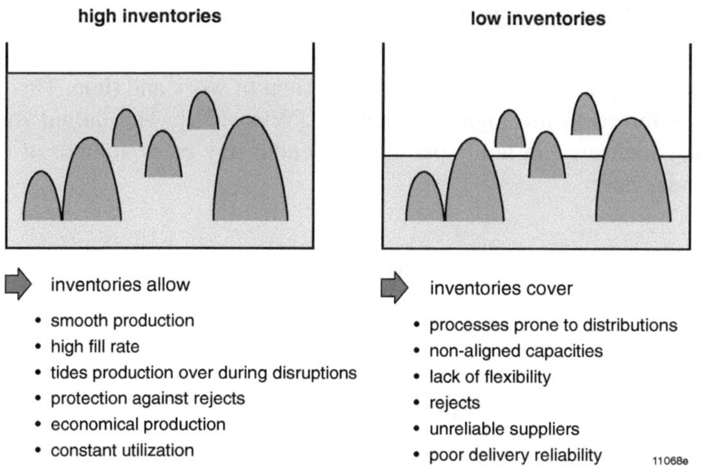


**Fig. 2.6** Sea of inventories (Siemens, cited according to [Wien-97])

With a Kanban Control, the inventory can be controlled decentrally on each value adding level (see Chap. 9 as well as [Mond-81*, Mond-98*]). The same applies for using a POLCA control system to release orders (see Chap. 23 as well as [Suri-98*]) as well as the Decentralized WIP Oriented Production Control (see Chap. 24 as well as [Lödd-01]).

Moreover, the WIP influences the length of time in which quality defects in a process are detected on subsequent workstations. The higher the WIP is, the longer this period of time. Therefore, the higher the WIP is the greater number of rejects produced. Inversely, when the WIP level is low, errors can be quickly discovered and eliminated. This reduces the production costs and increases the productive available capacities.

**Inventory as an Evaluation Parameter** Since inventory can be financially evaluated it is well suited for evaluating other objectives [Lödd-01]. In order to do so it is necessary to convert these objectives into inventories that can be compared. This is especially easy with a make-to-stock production: With the aid of Storage Operating Curves it can be estimated how much an enterprise can reduce the capital tied-up in the finished goods stores, if they manage to improve the throughput time and schedule reliability objectives [Lutz-02, Nyhu-09*]. The company can thus monetarily evaluate the economical utility of the improvement and compare it with the required effort. A comparable procedure is also possible with a make-to-order production in cases where the customer feeds the goods into a store. The cost advantages then emerge on the side of the customer and not in one's own enterprise. In either case, a conversion such as this allows the logistic objectives to be weighed against one another more transparently.

### 2.2.2 Output Rate and Utilization

The physical output rate is defined as the quotient of work and time. This definition can be transferred to the logistic processes [Wien-97b]. The *output rate* is then functionally measured in hours per shop calendar day or as number of orders per shop calendar day:

$$ROUT_m = \frac{OUT}{PE_{ref}} \tag{2.14}$$

where

| | |
|---|---|
| $ROUT_m$ | mean output rate [hrs/SCD], |
| OUT | output in reference period [hrs], |
| $PE_{ref}$ | duration of reference period [SCD]. |

Whereas, the *utilization* describes the ratio of the mean and maximum possible output rate of a workstation [Nyhu-09*], the maximum possible output rate refers to the workstation's available capacities (see Sect. 3.2.1.1):

$$U_m = \frac{ROUT_m}{ROUT_{\max}} \cdot 100 \tag{2.15}$$

where

| | |
|---|---|
| $U_m$ | mean utilization [%], |
| $ROUT_m$ | mean output rate [hrs/SCD], |
| $ROUT_{max}$ | maximum possible output rate [hrs/SCD]. |

Traditionally, the utilization objective is closely monitored. In many enterprises, very expensive machinery in particular is to be utilized as much as possible in order to ensure its refinancing. In other words, from a long-term perspective, it is wise not to excessively dimension the capacities of expensive machines. However, *after* it has been acquired, the cost of financing it are set and can no longer be influenced by the short-term decisions of the manufacturing control. As so-called "sunk costs" they are no longer relevant to decisions and consequently are no longer allowed to be taken into consideration. Thus, one of the tasks of manufacturing control—to utilize expensive machines as much as possible—is not economically based. Quite the contrary, in many cases it leads to high WIP levels and long throughput times.

However, the utilization can influence the revenues and thus increase the gained profit contribution of an enterprise (Fig. 2.7a). If a company with comparable machinery manages to attain a higher output rate than a competing enterprise, then it can distribute the investment costs over more products. It can then either offer a lower price or attain a higher profit as the competitor. Of course in order for this to be true, the extra produced goods must be sold.

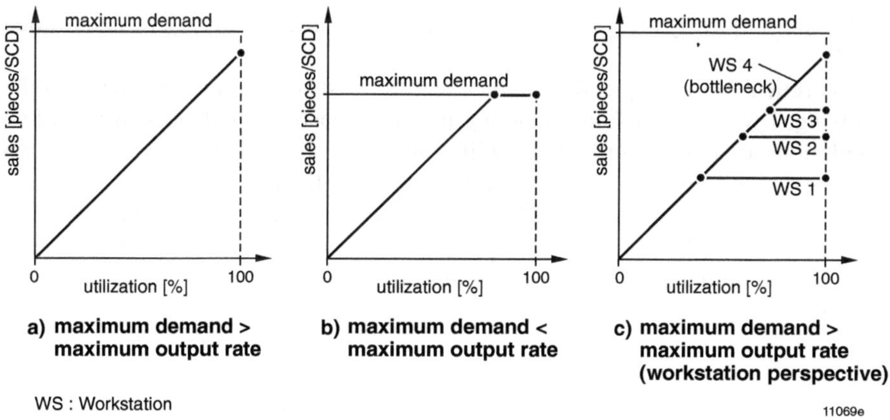


**Fig. 2.7** Impact of utilization on the sales of a company

Otherwise, the demand limits the utilization of a production (see Fig. 2.7b). Due to the overcapacity available in many areas, this is often the case. If in such situations, an enterprise adjusts the number of employees to the demand, the workforce becomes a relevant planning variable.

Even if an enterprise works at the limits of its capacities, it is usually impossible to completely utilize all workstations (see Fig. 2.7c). Generally, the orders are processed on a number of workstations and load them to varying degrees. If the throughput bottleneck of a production is completely utilized, any additional orders that have to be processed on it cannot be produced. If all orders waiting to be released need to be routed through this workstation, then the output rate of the production cannot be further increased. The throughput bottleneck of a production thus determines the output rate of the production and all non-bottleneck workstations (for more information on the bottleneck principle cf. [Gold-84*, Wind-01]). In our example, Workstation 4 is the production's throughput bottleneck. The utilization of the remaining workstations can only be raised by increasing the capacity of the bottleneck. Should this not be possible, many enterprises reduce the time that an operator is assigned to a non-bottleneck workstation. Managing personal thus levels the available capacities on the workstations so that a clearly defined bottleneck is no longer recognizable.

### *2.2.3 Delay Costs*

Delay costs are expenses that result due to shipping an order late. Default penalties or increased costs for required express shipments can, for example, be measured exactly. The costs which take effect over the long-term due to the lack of trust created are frequently more considerable, however, can generally not be determined exactly.

Theoretical work frequently assumes that the costs of delays are known. In the following, I will abstain from modeling the delay costs. Nonetheless, when a few assumptions are made in order to simplify things (e.g., known delay costs) they can easily be derived from the modeling of the delivery reliability and delivery lateness.

The logistic target attainment influences the economical success of the enterprise to a large degree. It thus raises the question of how the logistic objectives themselves can be influenced. The answer to this question is the subject of the following chapter.

## References


[Delo-98*] Deloitte & Touche Consulting: Vision in Manufacturing. Selbstverlag, Düsseldorf (1998)

[Domb-88] Dombrowski, U.: Qualitätssicherung im Terminwesen der Werkstattfertigung. VDI Progress Reports, Series 2, No. 159, Düsseldorf (1988)

[Gläß-95] Gläßner, J.: Modellgestütztes Controlling der beschaffungslogistischen Prozeßkette. VDI Progress Reports, Series 2, No. 337, Düsseldorf (1995)

[Gold-84*] Goldratt, E.M., Cox, J.: The Goal. A Process of Ongoing Improvement. Gower, Aldershot (1984)

[Gute-51] Gutenberg, E.: Grundlagen der Betriebswirtschaftslehre. Die Produktion, vol. 1. Springer, Berlin (1951)

[Kraf-88*] Krafcik, J.F.: Triumph of the lean production system. Sloan Manage.Rev. **30**, 41–51 (1988)

[Lödd-01] Lödding, H.: Dezentrale Bestandsorientierte Fertigungsregelung. VDI Progress Reports, Series 2, No. 587, Düsseldorf (2001)

[Lutz-01] Lutz, S., Lödding, H., Wiendahl, H.-P.: Kennliniengestützte logistische Lageranalyse. Ein neuer Ansatz zur Positionierung im Dilemma zwischen Bestand und Servicegrad. ZWF **92**(9), 550–553 (2001)

[Lutz-02] Lutz, S.: Kennliniengestütztes Lagermanagement. VDI Progress Reports, Series 13, No. 53, (2002)

[Mond-81*] Monden, Y.: Adaptable kanban system helps toyota maintain just-in-time production. Ind. Eng. **13**(5), 28–46 (1981)

[Mond-98*] Monden, Y.: Toyota Production System. An Integrated Approach to Just-In-Time. Engineering & Management Press, Norcross (1998)

[Nyhu-09*] Nyhuis, P., Wiendahl, H.-P.: Fundamentals of Production Logistics. Theory, Tools and Applications. Springer, Berlin (2009)

[Romm-93] Rommel, G., et al.: Einfach überlegen. Das Unternehmenskonzept, das die Schlanken schlank und die Schnellen schnell macht. Schäffer/Poeschel, Stuttgart (1993)

[Schö-07*] Schönsleben, P.: Integral Logistics Management. Operations and Supply Chain Management in Comprehensive Value-Added Networks, 3rd edn. Auerbach, Boca Raton (2007)

[Stal-90*] Stalk, G., Hout, T.M.: Competing Agains Time. How Time-Based Competition is Reshaping Global Markets. The Free Press, New York (1990)

[Suri-98*] Suri, R.: Quick Response Manufacturing. A Companywide Approach to Reducing Lead Times. Productivity Press, Portland (1998)

[Wien-02] Wiendahl, H.-H.: Situative Konfiguration des Auftragsmanagements im turbulenten Umfeld. Jost-Jetter Verlag, Heimsheim (2002)

[Wien-03] Wiendahl, H.-P.: Wie aus der Logistik eine Wissenschaft wurde. In: Wiendahl, H.-P. (ed.) Die wandlungsfähige Fabrik. Integrierte Sicht von Fabrikstruktur, Logistik und Produktionssystemen. Proceedings of the IFA-Fachtagung 2003, pp. 107–142. 20–22 March 2003

[Wien-97] Wiendahl, H.-P.: Betriebsorganisation für Ingenieure. Carl Hanser Verlag, Munich/Vienna (1997)

[Wien-97b] Wiendahl, H.-P.: Fertigungsregelung. Logistische Beherrschung von Fertigungsabläufen auf Basis des Trichtermodells. Carl Hanser Verlag, Munich/Vienna (1997)

[Wind-01] Windt, K.: Engpassorientierte Fremdvergabe in Produktionsnetzen. VDI Progress Reports, Series 2, No. 579, Düsseldorf (2001)

[Yu-01] Yu, K.-W.: Terminkennlinie. Eine Beschreibungsmethodik für die Terminabweichung im Produktionsbereich. VDI Progress Reports, Series 2, No. 576, Düsseldorf (2001)

[illegible]

# Chapter 3
# Modeling the Logistic Objectives

According to a well-known and perhaps already worn out saying from publications on management, one can only control what one can measure. This prerequisite holds true for the targeted control of logistic objectives, however, it alone is not enough. Due to the numerous interactions and conflicts between the logistic objectives and factors which influence them, it is also necessary to understand the intricacies of the system in order to ensure that targets are satisfactorily attained.

Modeling logistic objectives and their mutual dependencies thus takes on a key role. Generally, it serves two purposes: First, it forms the basis for coordinating the parameters for manufacturing control methods with the logistic objectives. Second, fundamental knowledge, which I will refer to here as 'guiding principles for designing manufacturing control methods' can be gained from the models. These guiding principles are utilized for qualitatively evaluating the suitability of manufacturing control methods with regards to logistics.

In the following, this modeling will be presented based on the Hanoverian Funnel Model. This Funnel Model has been continually developed and further refined since the beginning of the 1970s at the University of Hanover's Institute of Production Systems and Logistics. It is also taught at numerous universities and colleges. Since its conception it has proven itself in countless industry and research projects and has been integrated into commercial PPC software. It has thus established itself in the day to day routine of many production companies. Wiendahl and Nyhuis [Nyhu-09*, Wien-97b] provide an in-depth description of the Funnel Model, the Throughput Diagram derived from it and the Logistic Operating Curves (LOC).

This chapter is organized according to the subdivision of the logistic objectives into internal and external objectives (see Chap. 2). The external logistic objectives 'delivery time', 'delivery lateness', 'delivery reliability' and 'service level' are considered first (Sect. 3.1). The aim here is to represent and quantify the dependencies of the internal logistic objectives. The second section of the chapter is dedicated to the internal logistic objectives (Sect. 3.2). Here, two essential elements of the Funnel Model will be explained along with the Throughput Diagram and Logistic Operating Curves. Finally, in the third section, factors that

H. Lödding, *Handbook of Manufacturing Control*,
DOI 10.1007/978-3-642-24458-2_3, © Springer-Verlag Berlin Heidelberg 2013

influence the logistic performance are summarized and generally applicable guiding principles for designing manufacturing control methods are derived from the modeling (Sect. 3.3).

## 3.1 Modeling the External Logistic Objectives

Depicting the dependency of the external logistic objectives on the internal objectives makes it possible to systematically derive targets for the production based on market conditions. Moreover, it indicates the expected degree to which external objectives can be attained when the internal logistic objectives assume specific values. This section is divided into subsections on delivery time (Sect. 3.1.1), delivery reliability (Sect. 3.1.2) and service level (Sect. 3.1.3).

### *3.1.1 Delivery Time*

In the simplest case of a single-stage production, a company's delivery time consists of six components (see for example [Schö-07*]):

$$TD = TTP + TP + TDEB + TS + LDS + TAD \quad (3.1)$$

where

| | |
|---|---|
| TD | delivery time [SCD], |
| TTP | throughput time [SCD], |
| TP | procurement time [SCD], |
| TDEB | delivery time buffer [SCD], |
| TS | shipping time [SCD], |
| LDS | load shift [SCD], |
| TAD | administration time [SCD]. |

*Throughput Time*: The throughput time refers to the length of time from the release of an order until its completion (see Sect. 2.1.4). It is frequently the largest portion of a product's delivery time and can be influenced by the manufacturing control.

*Procurement Time*: When procurement is order based, the time required for procuring purchased parts needs to be taken into consideration. This component is inapplicable when items are procured from in-house stores.

*Shipping Time*: The shipping time refers to the time required for sending completed products. It is dependent on the distance and means of transport. Since it cannot be directly influenced by manufacturing control it is frequently neglected in the modeling.

*Load Shift*: Especially in times when there is an increased demand, capacities are frequently fully utilized for longer periods. Additional orders can then only be planned for periods when capacities are freed and made available again. The orders and thus the load are therefore shifted into the future. The corresponding delayed delivery time is referred to here as a load shift. Its long-term mean is dependent on the fluctuations in the required capacities, the available capacities and the capacities' flexibility. Generally, there is a conflict of objectives between the mean utilization and the delivery time when there are increased capacity demands (*dilemma of planning long-term capacities,* cf. [Pior-84*]). If the capacities are designed based on the maximum required capacity, there is a loss of utilization during lower capacity periods. In contrast, if the capacities are designed based on the mean required capacities, then the supplied capacities will be insufficient for immediately satisfying orders during periods when there is an increased demand. An enterprise must then balance the load and accept delayed delivery times—at least for a portion of the orders.

The required load shift can only be determined exactly with the assistance of a detailed due date and capacity plan. However, they can also be estimated with the aid of the planned backlog. The planned backlog refers to the difference between the requested cumulated output (calculated from the customer's preferred due date) and the planned cumulated output (calculated in consideration of capacity restrictions).

$$BL_{plan} = OUT_{req} - OUT_{plan} \tag{3.2}$$

where

| | |
|---|---|
| $BL_{plan}$ | planned backlog [h], |
| $OUT_{req}$ | requested output [h], |
| $OUT_{plan}$ | planned output [h]. |

The planned backlog therefore refers to the lacking, cumulated capacities required by the enterprise for fulfilling all of the customers' orders by the requested time.

If a few assumptions are made in order to simplify the situation, the required load shift results from the ratio of the planned backlog and the output rate:

$$LDS = \frac{BL_{plan}}{ROUT_m} \tag{3.3}$$

where

| | |
|---|---|
| LDS | load shift [SCD], |
| $BL_{plan}$ | planned backlog [h], |
| $ROUT_m$ | mean output rate [h/SCD]. |

Flexible capacities can ease the conflict of objectives between the capacities' utilization and the delivery time. With regards to personnel, models with flexible work hours, cross-trained employees or the temporary use of part-time staff make it possible to adjust the workforce to the demand (see Chap. 26). Other resources (e.g., machinery) tend to be less flexible. One possibility is cooperating with other enterprises. Either external orders can be taken-on in order to increase utilization or orders can be processed externally when capacities are overloaded. In the ideal case, measures allow the delivery times to be maintained at a low level even with above average demands.

*Delivery Time Buffer*: The delivery time buffer serves to attain high delivery reliability even when throughput times or more specifically the lateness varies strongly. It extends the delivery time and the finished goods store of an enterprise. The greater the required delivery reliability and the more broadly distributed the lateness is, the larger the required delivery time buffer is.

*Administration Time*: The administration time is the time required for generating and closing an order. It is comprised for example, by the verification of availability, the decision about procurement as well as the time required for the production and order management to prepare the order [Schö-07*].

If an enterprise wants to shorten its delivery times, it should first investigate which element is responsible for the largest portion of the delivery time. In many companies, this is often the throughput time and—when there is a high demand—the load shift. The latter can be shortened through higher and/or more flexible capacities whereas shortening the throughput time requires the WIP to be reduced.

In order networks where the manufacturing of a number of components that are to be assembled into an end product needs to be planned, calculating the delivery time is more complex. Figure 3.1 depicts a very simple example of such an order network.

The product P1 is assembled from components C1 and C2. Obviously, the assembly of P1 and the manufacturing of C1 determine the throughput time of the complete order and are thus on the so-called 'critical path'. Orders on the critical path have a higher priority because they determine the throughput time and thus the delivery time of the entire order. Taking into consideration the delivery time buffer on the critical path extends the delivery time of the complete product. The buffer is therefore often sparingly dimensioned. In contrast for orders not on the critical path, the delivery time buffer only extends the total throughput time when exceeding the time buffer (see Fig. 3.1a). It can therefore be dimensioned quite generously to ensure the punctual availability of the relevant components when the actual completion is delayed in comparison to the plan (see Fig. 3.1b). Nonetheless, the conflict with the WIP level needs to be considered. The WIP increases along with the delivery time buffer. An extensive discussion on determining delivery time buffers in order networks can be found in [Wahl-98].

Out of all the delivery time elements, manufacturing control directly influences only the throughput time. However, as previously mentioned the throughput time represents the largest portion of the delivery time in many enterprises. Furthermore, the distribution of the throughput times influences the required delivery time buffer. Modeling the throughput time is therefore a fundamental prerequisite for specifically influencing delivery times.

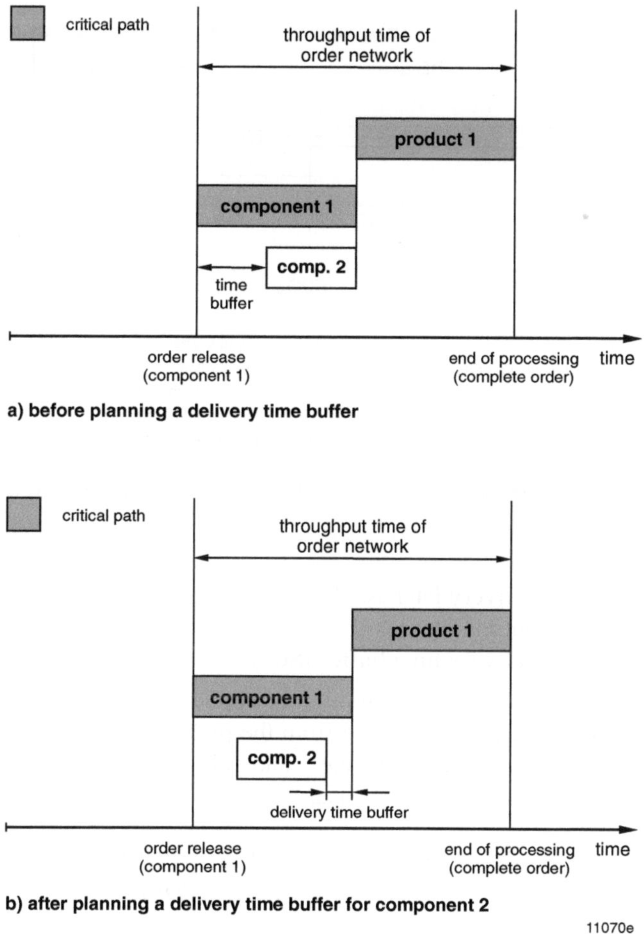


**Fig. 3.1** Determining the throughput time within order networks

### *3.1.2 Delivery Lateness and Delivery Reliability*

The delivery lateness of an order refers to the deviation between the planned and actual delivery date (see Sect. 2.1.2). It is equal to the difference of the order's output lateness and its delivery time buffer (see Fig. 3.2 as well as Eq. 3.4).

$$DEL = L_{out} - TDEB \tag{3.4}$$

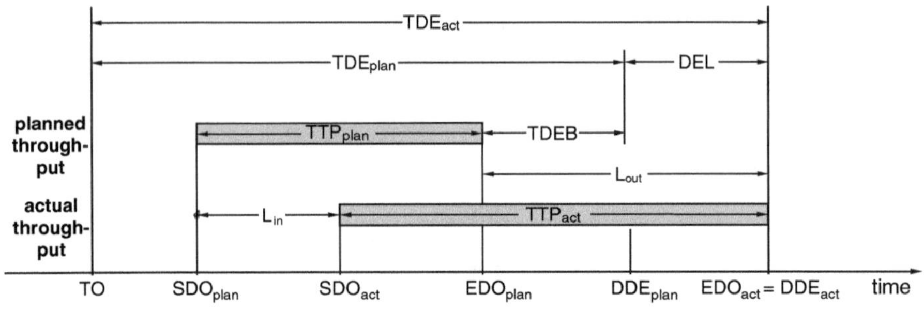


TO : order time point
SDO : start date of order processing (order release)
EDO : end date of order processing
DDE : delivery date
$L_{in}$ : input lateness
$L_{out}$ : output lateness
TDE : delivery time
TTP : order throughput time
TDEB : delivery time buffer
DEL : delivery lateness

**Fig. 3.2** Correlation between the output lateness, delivery time buffer and delivery lateness

where

DEL delivery lateness [SCD],
$L_{out}$ output lateness [SCD],
TDEB delivery time buffer [SCD].

One of the conditions for this equation to be applicable is that the shipping time is negligible or at least does not deviate from the planned value. Many enterprises withhold orders that are completed too early and deliver them only once the agreed upon delivery date is reached. In this case:

$$DEL = \begin{cases} 0 & if\ L_{out} \leq TDEB \\ L_{out}\text{-}TDEB & if\ L_{out} > TDEB \end{cases} \tag{3.5}$$

where

DEL delivery lateness [SCD],
$L_{out}$ output lateness [SCD],
TDEB delivery time buffer [SCD].

The delivery time buffer thus reduces the delivery lateness in comparison to the production's output lateness, accordingly, delivery compliance (also known as 'in-time delivery') increases. Therefore:

$$DEC = \frac{NO\ with\ DEL \leq 0}{NO} \cdot 100 = \frac{NO\ with\ L_{out} \leq TDEB}{NO} \cdot 100 \tag{3.6}$$

where

DEC delivery compliance [%],
NO number of orders [−],
DEL delivery lateness [SCD],

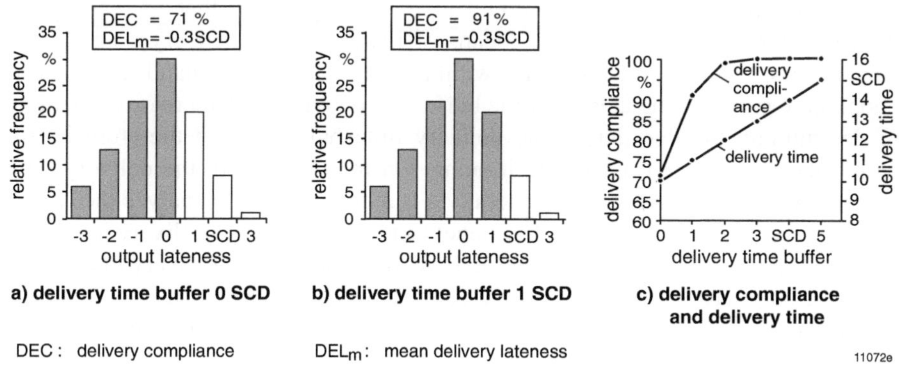


**Fig. 3.3** Influence of the delivery time buffer on the delivery compliance

$L_{out}$ output lateness [SCD],
TDEB delivery time buffer [SCD].

Figure 3.3 depicts the impact of the delivery time buffer on the delivery compliance of a (fictitious) example.

In the initial situation (Fig. 3.3a) the enterprise does not plan any delivery time buffer. Therefore, all orders that are not completed by the planned due date reduce the delivery compliance. With a delivery time buffer of one shop calendar day (Fig. 3.3b) the orders with an output lateness of 1 day can still be punctually delivered. As a result the delivery compliance increases from 71% to 91%. Figure 3.3c illustrates how the delivery compliance develops as a function of the delivery time buffer.

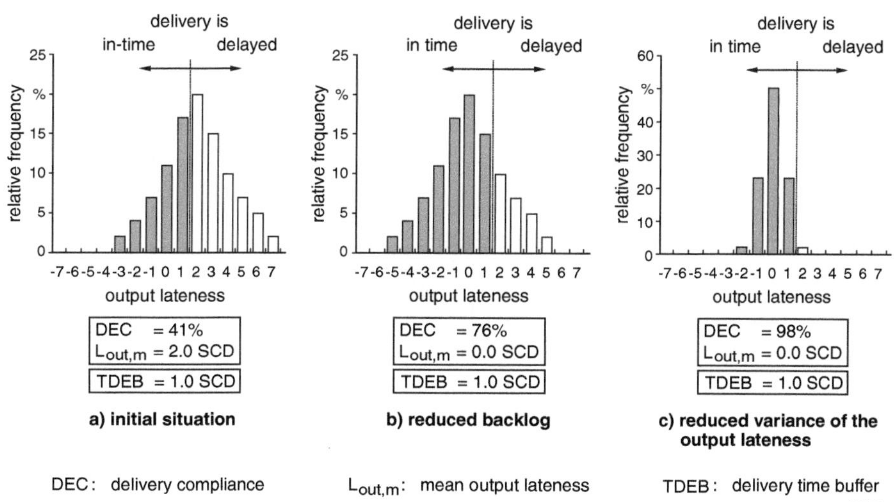


**Fig. 3.4** Measures for increasing the delivery compliance

This graphical representation could give the impression that increasing the delivery time buffer is a suitable means of increasing the delivery compliance. Nonetheless, this is applicable only within limits: On the one hand the finished goods store grows as the delivery time buffer and delivery time increases (dilemma of determining the due date). The delivery time buffer should therefore be set within the frame of a Logistic Positioning. On the other hand, the delivery time buffer only reacts to symptoms. Generally it is more effective to analyze the causes for a large amount of variance in the lateness and to target higher delivery reliability by improving the manufacturing control. Foregoing the use of a delivery time buffer when the output lateness varies however, almost inevitably leads to a lower delivery compliance. It should thus first be reduced once the manufacturing control has been effectively improved.

As mentioned above the variance and mean of the output lateness are factors which decisively influence the delivery compliance (and length of the required delivery time buffer). Figure 3.4 depicts a fictitious example of how the delivery compliance can be improved by reducing the mean output lateness (Fig. 3.4a) and decreasing its variance (Fig. 3.4c). Accordingly, the modeling of the output lateness becomes particularly important.

The basis for modeling the requested delivery reliability is the distribution of the delivery reliability based on the customer's originally requested due date. Figure 3.5 depicts a fictitious distribution for a product group.

The delivery reliability based on the customers original requested due date and calculated according to Eq. 3.1 is comparatively represented for this product group. In the example a standard planned delivery time of 5 days is assumed, thus, all of the orders whose requested delivery time is 5 or more days can be planned for the requested due date.

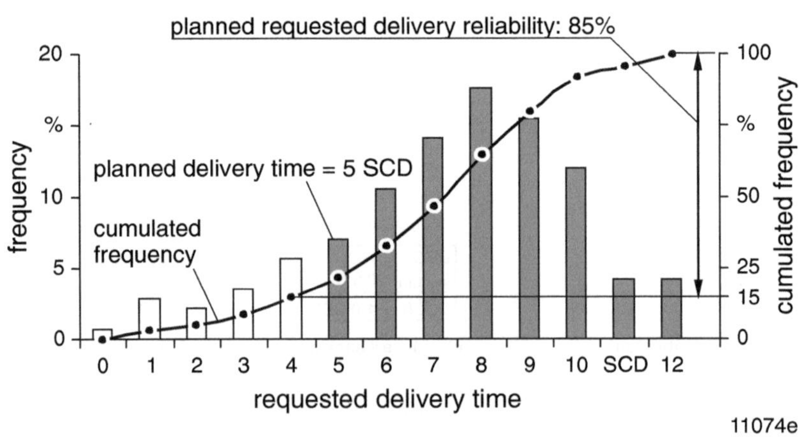


**Fig. 3.5** Calculating the planned requested delivery reliability

If the requested delivery reliability is to be increased, two cases need to be differentiated:

1. Sufficient capacities are available (load shift = 0)
If the calculated requested delivery reliability is too low, then the enterprise should try to reduce the manufacturing throughput time (or another element of the delivery time).

2. Sufficient capacities are unavailable (load shift > 0)
If the load shift is the main contributor to poor requested delivery reliability, then only temporarily increasing the capacities can lead to reducing the delivery time. In both cases the modeling of the planned requested delivery compliance can be traced back to the modeling of the delivery time.

### *3.1.3 Service Level*

The service level of a store (or the stored products) is directly dependent on the mean stock level. The higher the stock level is, the higher the service level (*dilemma of inventory management*). This conflict of objectives can be quantified with the aid of the Storage Operating Curves (SOC). Initially derived by Gläßner and Nyhuis for the delivery delay objective [Gläß-95, Nyhu-09*] they were later transferred by Lutz and Lödding to the weighted service level (see Sect. 2.1.3 as well as [Lutz-01, Lutz-02]).

Figure 3.6 shows the behavior of the weighted Service Level Operating Curve (SLOC). As the mean stock increases, the weighted service level rises. Once the mean stock reaches a certain level, the so-called 'practical minimal stock limit', all demands can be directly serviced from the stores. At this point, the service level

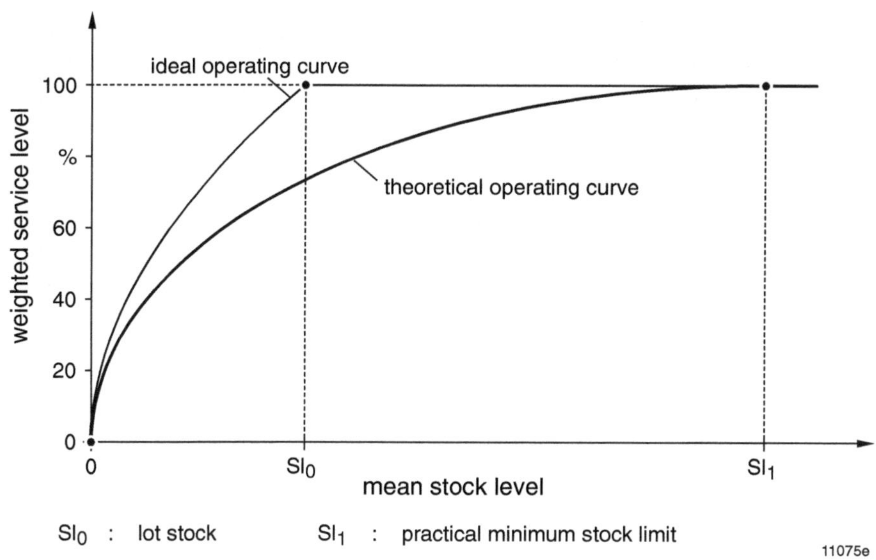


**Fig. 3.6** Ideal and theoretical weighted service level operating curves (Lutz)

reaches 100% and the mean delivery delay is zero. Further increasing the stock no longer increases the logistic performance.

I will not delve into the process for deriving the Storage Operating Curves here and instead encourage those interested in learning more to refer to [Gläß-95, Nyhu-09*] with regards to the Delivery Delay Operating Curve and [Lutz-01, Lutz-02] for the Service Level Operating Curve. According to [Lutz-02] the Service Level Operating Curve can be calculated as:

$$SL_m = SL_0 \cdot \left(\frac{SERL_w}{100}\right)^2 + (SL_1 - SL_0) \cdot \sqrt[c]{1 - \left(1 - \frac{SERL_w}{100}\right)^c} \tag{3.7}$$

where

| | |
|---|---|
| $SL_m$ | mean stock [units], |
| $SL_0$ | lot stock [units], |
| $SERL_w$ | weighted service level [%], |
| $SL_1$ | practical minimal stock limit [units], |
| c | c-norm parameter [−]. |

The parameters for this equation are explained below. The *lot stock* $SL_0$ corresponds to half of the mean store input quantity per store input event (also referred to here as the input lot size) [Gläß-95]:

$$SL_0 = \frac{QIN_m}{2} \tag{3.8}$$

where

| | |
|---|---|
| $SL_0$ | lot stock [units], |
| $QIN_m$ | mean store input quantity (input lot size) per store input event [units]. |

The *c-norm parameter* influences the curvature of the Service Level Operating Curve. Low c-values are favorable and are attained with narrowly distributed plan deviations. For moderately distributed plan deviations a c-value of 0.33–0.37 can be assumed [Nyhu-09*]. When employing Logistic Operating Curves in the industry, the stock level and service levels can be measured during a representative time period and the value of the c-norm value set so that the Service Level Operating Curve runs through the measured value. In order to do so it is necessary that the service level does not decrease as a result of organizational problems [Lutz-02].

The *practical minimal stock limit* $SL_1$ corresponds to the stock level at which the service level is 100%. According to [Nyhu-09*] it is calculated as:

$$SL_1 = \frac{QIN_m - QOUT_m}{2} + \sqrt{\left(L^{+}_{\max} \cdot RDMD_m\right)^2 + \left(Q^{-}_{\max}\right)^2 + \left((RDMD_{\max} - RDMD_m) \cdot TRP\right)^2} \tag{3.9}$$

where

| | |
|---|---|
| SL1 | practical minimal stock level [units], |
| $QIN_m$ | mean store input quantity (lot size) per input event [units], |

| | |
|---|---|
| $QOUT_m$ | mean store output quantity (lot size) per output event [units], |
| $L_{max}^{+}$ | maximum positive lateness (delay) [SCD], |
| $Q_{max}^{-}$ | maximum negative quantity deviation (partial delivery) [units], |
| $RDMD_{max}$ | maximal rate of demand [units/SCD], |
| $RDMD_m$ | mean rate of demand [units/SCD], |
| TRP | replenishment time [SCD]. |

The first summand describes the lot stock that results solely from the different lot sizes in the input and output. The lot stock increases with the store input quantity, which frequently corresponds to the size of the manufacturing lots. The second summand describes the safety-stock required to ensure a high service level even when there are disruptions. Three types of disruptions are considered (see Fig. 3.7):

1. *Delivery Lateness*: If a procured order enters the store later than planned, then the demand has to be serviced from the safety-stock during the delay. The safety-stock required for this increases with the delivery lateness and the mean demand rate.
2. *Delivered Quantity Deviation*: The fill rate can be endangered if the actual delivered quantity falls below the planned delivery quantity. In order to avoid this, the maintained safety-stock level should be equal to the maximum negative quantity deviation.
3. *Demand Deviation*: The demand is generally not constant, but rather fluctuates around a mean value. If the demand is higher than planned during the replenishment time, a part of the demand has to be serviced from the safety-stock level. The safety-stock required for this increases with the duration of the replenishment time and the level of the maximum demand rate.

The safety-stock required for balancing the plan deviations is statistically aggregated in the second summand of Eq. 3.9. This equation is conditional on the statistic independence of the individual plan deviations. As a result of the squaring,

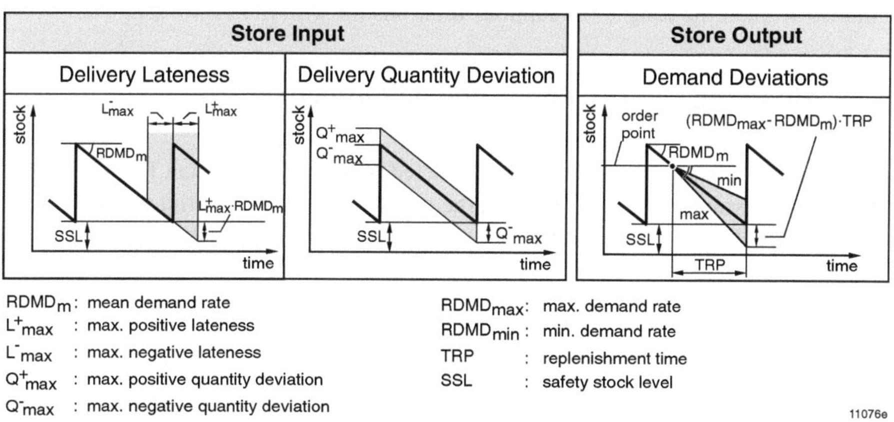


**Fig. 3.7** Plan deviations in storage processes (according to Nyhuis)

plan deviations that can be buffered through a low safety-stock are weighted less. Experience has shown that in many enterprises demand fluctuations and/or delivery lateness dominate the required safety-stock.

The manufacturing control influences the required safety-stock in two different respects: First, the mean and variance of the output lateness determine the maximal delivery lateness. Second, the throughput time is a fundamental element of the delivery and replenishment time. If it is possible to improve the throughput time the required safety-stock will be reduced. With the throughput time and output lateness, the external target attainment is determined by the same internal logistic objectives in make-to-stock production as in make-to-order productions.

The capacities' flexibility is frequently more significant in make-to-stock productions than in make-to-order productions. When the available capacities insufficiently satisfy the capacity requirements, it is usually not possible to shift the delivery date without breeching the promised delivery. Generally, make-to-stock productions are connected with the promise of satisfying demands directly from stock. As soon as the entire increased demand (and not only for a single variant) consumes the safety-stock, the service level is reduced. A similar effect emerges in a make-to-order production if an enterprise promises fixed delivery times and thus limits their load flexibility.

Incidentally, in this situation it would be a mistake to increase the safety-stock as this would only trigger additional production orders, further increasing the load and thus the backlog of the production. In addition, orders in production for which there is still sufficient stock in the store would compete for limited capacities. Generally, only three alternatives remain for the enterprise to eliminate the cause of the problem: extending their own capacities, placing orders externally, or increasing the price. The goal of the latter is to curb the demand to a value which can be attained with the available capacities. Increasing the price has two advantages: First, it improves the company's profits so that capital required for extending the capacities can be more easily financed. Second, the products reach the customers who can achieve the greatest utility with them. Nonetheless, increasing the price can also worsen the long-term supplier relationship with the customer.

## 3.2 Modeling the Internal Logistic Objectives

The internal logistic objectives can be modeled with the Hanoverian Funnel Model. This encompasses the Funnel Model in the narrower sense, the Throughput Diagram derived from it (Sect. 3.2.1) and the Logistic Operating Curves (Sect. 3.2.2).

### *3.2.1 Funnel Model and Throughput Diagram*

The Funnel Model and Throughput Diagram allow the chronological trend of the logistic objectives to be visualized. A prerequisite for this is defining the work content and operation time (Sect. 3.2.1.1). The smallest modeling unit of the Funnel

Model are the components of the throughput element of an operation. The throughput element describes how an order passes through a workstation (Sect. 3.2.1.2). The Throughput Diagram can be derived from the modeling of a workstation (Sect. 3.2.1.3) and allows quantitative relationships between the logistic objectives to be determined.

#### 3.2.1.1 Work Content and Operation Time

The initial parameter for the Funnel Model is the work content. It corresponds to the standard time allotted for the processing of an operation on a workstation. The work content results from the setup time, the lot size and the processing time per unit [Wien-97b]. It is measured in hours and taken from the production schedule.

$$WC = \frac{LS \cdot t_p + t_{su}}{60} \tag{3.10}$$

where

| | |
|---|---|
| WC | work content [h], |
| LS | lot size [units], |
| $t_p$ | processing time [min/unit], |
| $t_{su}$ | setup time [min]. |

The statistical distribution of the work content on a workstation impacts the attainable throughput time of a workstation. The mean and standard deviation of the work content are thus key logistic figures:

$$WC_m = \frac{\sum_{i=1}^{n} WC_i}{n} \tag{3.11}$$

where

| | |
|---|---|
| $WC_m$ | mean work content [h], |
| $WC_i$ | work content of operation *i* [h], |
| n | number of operations [−]. |

The standard deviation describes the dispersion of a distribution [Nyhu-09*].

$$WC_s = \sqrt{\frac{\sum_{i=1}^{n} (WC_m - WC_i)^2}{n}} \tag{3.12}$$

where

| | |
|---|---|
| $WC_s$ | standard deviation of the work content [h], |
| $WC_m$ | mean work content [h], |
| $WC_i$ | work content of operation *i* [h], |
| n | number of operations [−]. |

The work content's coefficient of variance puts the standard deviation of the work content in relation to its mean and allows the variance of different work content distributions to be compared with one another [Nyhu-09*].

$$WC_v = \frac{WC_s}{WC_m} \tag{3.13}$$

where

| | |
|---|---|
| $WC_v$ | coefficient of variance for the work content [−], |
| $WC_s$ | standard deviation of the work content [h], |
| $WC_m$ | mean work content [h]. |

For scheduling, it is necessary to indicate the time a workstation is occupied by an order in the unit 'shop calendar days' (SCD). This figure, referred to as the operation time, is calculated as the quotient of the work content and maximum possible output rate [Nyhu-09*]:

$$TOP = \frac{WC}{ROUT_{\max}} \tag{3.14}$$

where

| | |
|---|---|
| TOP | operation time [SCD], |
| WC | work content [h], |
| $ROUT_{max}$ | maximum possible output rate [h/SCD]. |

The maximum possible output rate corresponds to the respective minimum of the available resource capacities, including personnel [Nyhu-09*]. Disruptions which decrease capacities (e.g., due to machine failure) should be taken into consideration in the calculation.

#### 3.2.1.2 Throughput Element

The throughput element describes the throughput of an order on a workstation. A production order is considered first (see Fig. 3.8).

The production order is comprised of an assembly order and two manufacturing orders (Fig. 3.8a). The critical path for this basic order network is formed by the manufacturing order II and the assembly order. Manufacturing order II consists of four operations OP1–OP4 (Fig. 3.8b). Depicted here is the throughput element for

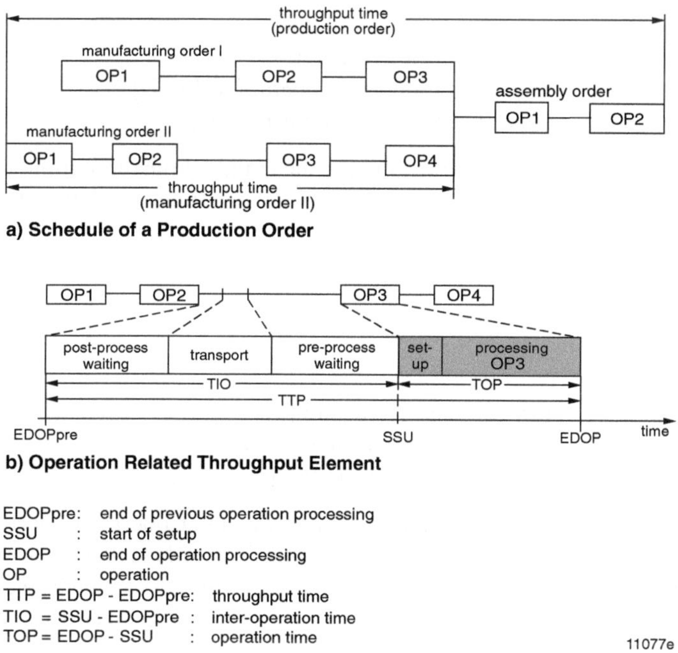


**Fig. 3.8** Throughput time components and throughput time element (Nyhuis/Wiendahl)

the third operation. It begins directly after the previous operation has been completely processed (or when in the case where it is the first operation, directly following the start of the order). It thus contains the waiting time after processing as well as the time required to transport the lot to the next workstation. Once there, the order usually lands in a queue and has to wait until the processing of the orders preceding it are completed (waiting time before processing). Afterwards, the setup on the machine is changed and the processing of the order begins. The throughput time of an operation includes all of the throughput element's components and according to [Wien-97b] is calculated as:

$$TTP = EDOP - EDOP_{pre} \tag{3.15}$$

where

| | |
|---|---|
| TTP | throughput time (operation) [SCD], |
| EDOP | end of operation processing [SCD], |
| $EDOP_{pre}$ | end of previous operation processing [SCD]. |

The periods for waiting after processing, transportation and waiting before processing are summarized as the inter-operation time; the setup and processing times form the operation time. Correspondingly, according to [Wien-97b]:

$$TTP = TIO + TOP \tag{3.16}$$

where

| | |
|---|---|
| TTP | throughput time (operation) [SCD], |
| TIO | inter-operation time (operation) [SCD], |
| TOP | operation time (operation) [SCD]. |

A highly significant figure for evaluating the length of the throughput times is the (unweighted) flow rate. It sets the mean throughput time in relation to the mean operation time [Wien-97b]:

$$FR = \frac{TTP_m}{TOP_m} \tag{3.17}$$

where

| | |
|---|---|
| FR | (unweighted) flow rate [−], |
| $TTP_m$ | mean throughput time [SCD], |
| $TOP_m$ | mean operation time [SCD]. |

The greater the flow rate is, the greater the inter-operation portion of the throughput time is. A higher flow rate therefore generally indicates a higher WIP level. For numerous applications a suitable flow rate of 3–5 can be derived from the Logistic Operating Curves Theory (see comments on normalized LOC Fig. 3.15). The flow rate can be calculated for a workstation as well as for an entire production.

A related parameter is the weighted flow rate. When calculating the mean, it weights both the throughput time and operation time with the work content of the related orders (see [Wien-97] for an extensive discussion on this).

#### 3.2.1.3 Throughput Diagram

According to a suggestion by Bechte [Bech-84] the Funnel Model, as its name suggests, models the workstation of a production as a funnel: Orders enter it thus forming the WIP on the workstation up until they leave the funnel or workstation after being processed (Fig. 3.9).

The size of the orders symbolizes how large the work content is. Accordingly, the size of the funnel opening represents the output rate that can be increased up to the maximum capacity.

The Throughput Diagram is created by measuring the input and output of the funnel and cumulatively plotting them over the time with their work content in standard hours [Bech-84]. If at the start of an investigation period there are already orders at the workstation, the start of the initial WIP of the input curve will be vertically displaced by the work content of these orders. The vertical distance between the input and output curves therefore, always corresponds to the actual WIP at the workstation. The horizontal distance between the input and output curve describes the range of the workstation. This parameter is closely connected to the workstations' throughput time and expresses how long the WIP will be sufficient for supplying the station with work. If the orders are processed in the sequence

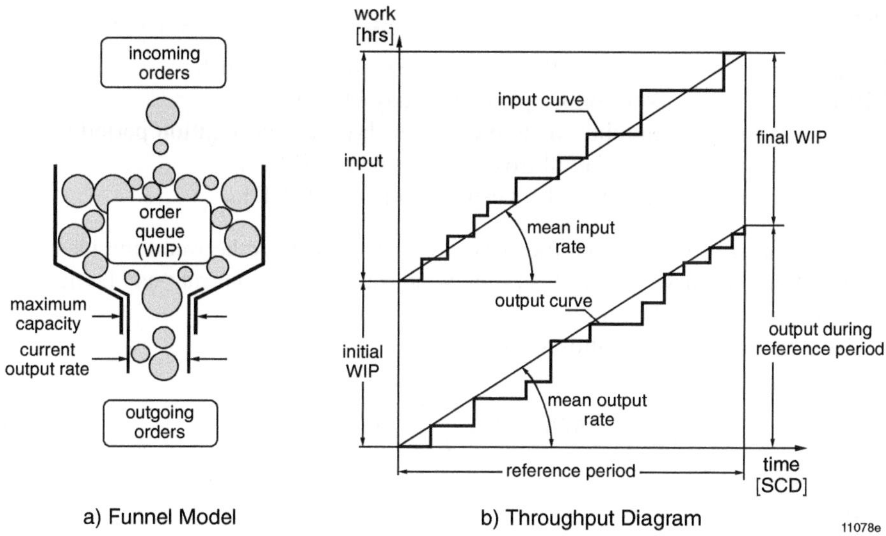


**Fig. 3.9** Funnel model and throughput diagram (Bechte cited according to Wiendahl)

of their input (FIFO principle), the range corresponds to the throughput time of an order entering the workstation [Wien-97b].

The WIP at the end of a specific day $D_n$ is calculated according to the following equation:

$$WIP(D_n) = \sum_{D=D_1}^{D_n} IN(D) - \sum_{D=D_1}^{D_n} OUT(D) + WIP_{ini}(D_1) \tag{3.18}$$

where

| | |
|---|---|
| $WIP(D_n)$ | WIP on shop calendar day $D_n$ [h], |
| $IN(D)$ | work content of incoming orders on shop calendar day D [h], |
| $OUT(D)$ | work content of outgoing orders on shop calendar day D [h], |
| $WIP_{ini}(D_1)$ | initial WIP at beginning of shop calendar day $D_1$ [h], |
| $D_1$ | first day of investigation period [SCD], |
| $D_n$ | day for which the WIP is being calculated [SCD]. |

For the mean WIP during the investigation period (see [Nyhu-09*]):

$$WIP_m = \frac{\sum_{D=D_1}^{D_n} WIP(D)}{n} \tag{3.19}$$

where

| | |
|---|---|
| $WIP_m$ | mean WIP during investigation period [h], |
| WIP(D) | WIP on shop calendar day $D_n$ [h], |
| n | number of shop calendar days in investigation period [−], |
| $D_1$ | first day of investigation period [SCD], |
| $D_n$ | last day of investigation period [SCD]. |

A relationship between the mean WIP (vertical distance between the idealized input and output curves), the mean range and the mean output rate can be derived from the Throughput Diagram.

Using trigonometry it follows that:

$$\frac{OUT}{PE_{ref}} = ROUT_m = \frac{WIP_m}{R_m} \text{ or } R_m = \frac{WIP_m}{ROUT_m} \tag{3.20}$$

where

| | |
|---|---|
| OUT | output [h], |
| $PE_{ref}$ | duration of reference period [SCD], |
| $ROUT_m$ | mean output rate [h/SCD], |
| $WIP_m$ | mean WIP [h], |
| $R_m$ | mean range [SCD]. |

This correlation is referred to as the 'Funnel Formula', accordingly, the range (and thus the throughput time) of a workstation results from the ratio of the WIP and output rate [Bech-84, Nyhu-09*, Wien-97b]. In order to reduce the throughput time, a production can either lower the WIP or increase the output rate. The latter is only possible in certain exceptional cases, since the output rate is derived from the market demand or the capacity restrictions and is thus not suitable as a control variable for the throughput time. Inversely, an enterprise should use an increase in demand to reduce the throughput time of orders. *The WIP thus remains the most important control variable for the throughput time* (see Sect. 2.2.1). The WIP can either be regulated through a WIP control or indirectly through the planned throughput times of the orders.

The WIP cannot be arbitrarily reduced as doing so could lead to a loss of utilization with regards to personnel and other resources (*dilemma of operations planning*). Therefore, a company general has three tasks:

- *WIP control*: An enterprise should have the ability to closely control the WIP.
- *WIP evaluation*: Furthermore it should know which WIP level is economically attainable. This is, for example, possible by conducting a Logistic Positioning with the aid of Logistic Operating Curves (see Sect. 3.2.2).
- *Reducing the required WIP level*: An enterprise should strive to gradually reduce the required WIP level. To do so, it can first reduce the setup times in order to be able to reduce the mean work content. Second, it should employ a lot-sizing method that reduces the variance of the work content. Third it can improve how

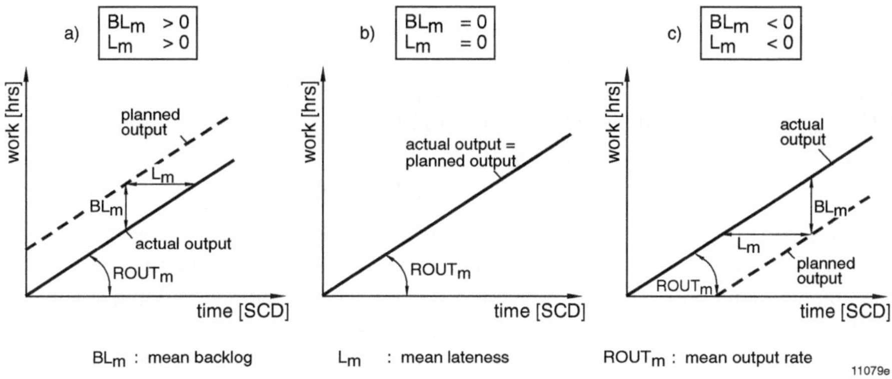


**Fig. 3.10** Representation of the backlog and lateness in an idealized throughput diagram

the parts flow (see Sect. 5.3). Finally, the company should increase the reliability of the technical processes.

In addition to the throughput time, the lateness can also be represented in a Throughput Diagram. In order to do so the planned output, instead of the input, is plotted cumulatively over the time (see Fig. 3.10).

The start of the planned output curve is displaced by the backlog which is present at the beginning of the investigation period. The backlog is calculated at every point in time as the difference between the planned output and the actual output [Pete-96]:

$$BL(t) = OUT_{plan}(t) - OUT_{act}(t) \tag{3.21}$$

where

BL backlog [h],
$OUT_{plan}$ planned output [h],
$OUT_{act}$ actual output [h].

Whereas a positive backlog means the production is behind schedule, a negative backlog indicates that the observed workstation has accomplished more than planned. A situation such as the latter could also be referred to as a lead, however, for the benefit of a unified concept, this is abstained from here.

In the Throughput Diagram the vertical distance between the planned and actual output curve corresponds at every point to the *backlog* of the production. When the orders are processed according to their priority, the horizontal distance corresponds to the orders' *lateness*.

Based on trigonometry, the following relation between the output lateness, backlog and output rate can be derived [Yu-01]:

$$\frac{OUT_{act}}{PE_{ref}} = ROUT_m = \frac{BL_m}{L_{out,m}} \tag{3.22}$$

where

| | |
|---|---|
| $OUT_{act}$ | actual output [h], |
| $PE_{ref}$ | duration of reference period [SCD], |
| $ROUT_m$ | mean output rate [h/SCD], |
| $BL_m$ | mean backlog [h], |
| $L_{out,m}$ | mean (calculated) output lateness [SCD]. |

Consequently the mean lateness results from the ratio of the mean backlog to the mean output rate. With a positive backlog the orders are completed on average later than planned and with a negative backlog, earlier. The lateness calculated according to this equation is a parameter that is weighted with the work content. It can thus deviate from the mean lateness measured from the feedback data if the extent of the lateness is correlated with the orders' work content. This is, for example, the case when orders with the shortest (or longest) work content are processed first i.e., when the SPT (or LPT) rule is applied. With the SPT rule, the measured mean lateness has a tendency to be lower than the calculated lateness, whereas with the LPT rule it tends to be higher. Both priority rules, however, increase the variance of the output lateness which in turn reduces the schedule reliability of the production (see the extensive discussion on the SPT rule in Sect. 25.4).

If the processing sequence and thus the extent of the lateness are *not* correlated with the work content, then the mean lateness measured from the backlog and the orders' feedback data generally agree well with one another. A positive backlog however arises before the orders are completed and the lateness can be measured. Therefore when there is a positive backlog the lateness calculated from the backlog is ahead of the measured lateness (a quite similar effect occurs with the range and (weighted) throughput time, cf. [Nyhu-09*, Wien-97b]).

The prerequisites for a high level of schedule reliability can be derived from this relationship between the backlog, lateness and output rate. On the one hand a backlog free production (backlog = 0) is required, and on the other hand the orders need to be completed in the planned sequence.

**Backlog Free Production** The backlog is calculated as the difference between the planned and actual output. Accordingly, it is the job of production planning to set a planned output in the schedule and capacity planning that can be realistically attained by the production. If necessary, capacity increases requiring a longer amount of time to be implemented have to be introduced in time. Furthermore, close coordination with the sales department is necessary especially when the possible demand requires capacities, which, even considering the available flexibility of the capacities, cannot be met. In a situation such as this, an honest enterprise would not promise any due date that it knows cannot be met (or can only be met at the cost of other orders). It is also the job of manufacturing control to actually attain the planned output. If a production backlog arises, manufacturing control has to try to catch up by temporarily increasing capacities (e.g., with overtime).

It is not enough to reduce the backlog of an observation period on average to zero, rather the backlog needs to remain at zero during the entire investigation

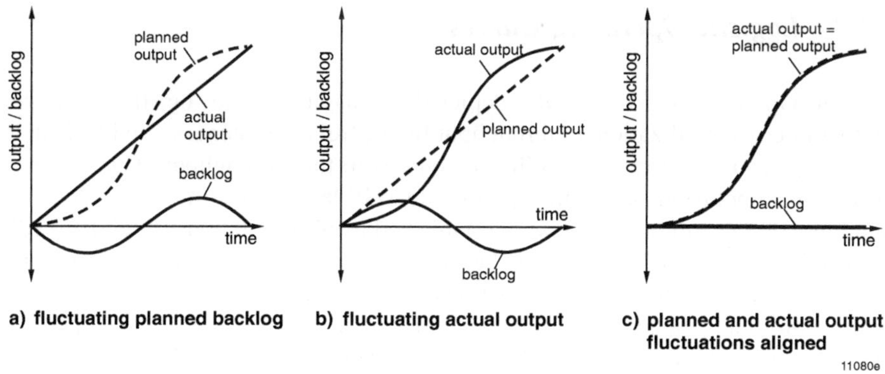


**Fig. 3.11** Generating backlog fluctuations

period. Otherwise there will inevitably be a positive mean lateness during some period intervals and a negative mean lateness during others. This type of oscillating backlog can be caused by fluctuations in the planned output (Fig. 3.11a) or the actual output (Fig. 3.11b). When such fluctuations cannot be avoided, the behavior of the planned and actual output needs to be coordinated (Fig. 3.11c).

**In-sequence Processing** Even when a production does not have a backlog its schedule compliance can be unsatisfactory. This is the case when orders are not processed in sequence according to their priority i.e. the processing of urgent orders is delayed while other orders that do not yet need to be processed are pulled forward. Sometimes the reasons for this are inescapable, for example, a lack of available materials. Sequencing inter-changes however are often not effective when it comes to the capability of the logistic process. In many enterprises, there seems to be two particularly frequent motives for sequencing inter-changes:

- *Early release and processing of particularly important orders*. Frequently, orders that are considered particularly important are released and processed early in order to absolutely ensure their punctual completion. If the orders are thus completed early, it is inevitably connected with other orders being completed later than planned. This applies for a backlog that is greater than or equal to zero. Therefore, even particularly important orders should not be processed before the planned date.
- *Inadequate target communication*. In many enterprises the significance of processing orders in the planned sequence is insufficiently communicated. To some extent no completion dates are set for operations thus an operator can hardly know which orders are the most urgent. Moreover, sequencing inter-changes only need to be justified in rare cases.

Companies can measure the sequencing compliance in order to localize a possible cause for poor delivery reliability and to eliminate the problem. Section 25.5 introduces figures suitable for doing so.

### 3.2.2 *Logistic Operating Curves*

Logistic Operating Curves (LOC) depict the behavior of the logistic objectives 'output rate' (or 'utilization'), 'throughput time', 'relative lateness', and 'schedule reliability' as a function of the WIP. In contrast to the Throughput Diagram, they thus allow conclusions about the appropriate WIP level to be drawn directly. The first Logistic Operating Curves were established using simulations [Bech-84, Erdl-84, vonW-89]. By implementing deductive-experimental approaches, Nyhuis successfully derived approximation equations for the logistic objectives 'output rate', 'utilization', and 'range' [Nyhu-91]. Together with Ludwig he also developed the Inter-Operation Time Operating Curve and the Throughput Time Operating Curve [Ludw-92]. Yu completed the Logistic Operating Curves for the relative lateness and relative schedule reliability [Yu-01].

In the following section the behavior (Sect. 3.2.2.1) as well as the deduction and calculation (Sect. 3.2.2.2) of the Logistic Operating Curves will be described. Section 3.2.2.3 then presents how the Logistic Operating Curves can be implemented in scheduling orders.

#### 3.2.2.1 Behavior of the Logistic Operating Curves

Logistic Operating Curves quantify the mutual dependencies between the logistic objectives. They generally represent means, from which the individual values more or less strongly deviate. The reference variable for the Logistic Operating Curves is always the mean WIP (see Fig. 3.12).

The *mean output rate* initially increases proportional to the WIP (underload zone). In this WIP zone, the workstation is only minimally utilized. All arriving orders are processed directly after input (no queue).

As the WIP continues to increase, orders begin to compete for resource; queues develop and with that a WIP buffer. The mean output rate only increases degressively in the transition zone. Increasing the WIP further beyond the transition zone only increases the mean output rate insignificantly. In such cases the output rate slowly approaches the maximum possible output rate (overload zone).

The mean *Range Operating Curve* is derived directly from the Funnel Formula as the ratio of the WIP to the output rate. In the underload zone, the orders are processed directly after their input into the workstation. The mean range corresponds to the minimal range and results from the operation time of the orders. As the WIP climbs, the orders increasingly have to wait until other orders are processed (transition zone). The mean range however initially increases disproportionately smaller to the WIP. Only in the overload zone with an almost constant output rate does the mean range grow almost proportionally to the WIP.

The mean *Throughput Time Operating Curve* runs parallel to the range operating curve and with varying work content is shifted lower to smaller values. This is applicable assuming that the processing sequence of the orders is independent of

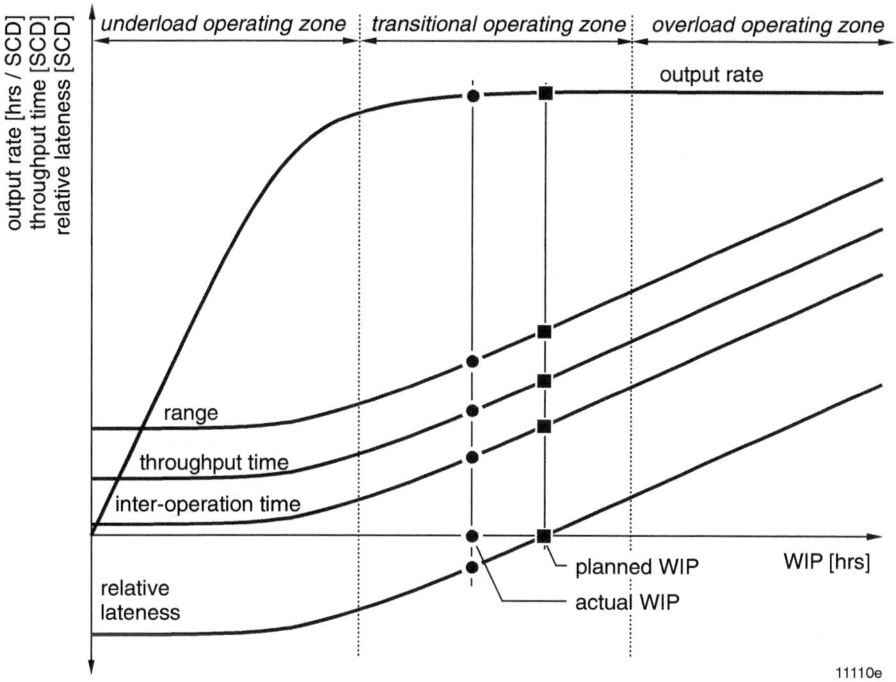


**Fig. 3.12** Behavior of logistic operating curves for selected logistic objectives (Nyhuis/Ludwig/Yu)

the work content (see [Nyhu-09*]). Moreover, the calculated throughput time runs ahead of the measured throughput time (see [Nyhu-09*, Wien-97b]). The distance between the Range Operating Curve and the Throughput Time Operating Curve illustrates that the range is a weighted (with the work content) variable. Orders with large work content are weighted more than those with smaller orders, at the same time due to the longer operation times they also remain part of the WIP longer. Accordingly, the distance between the Range Operating Curve and the Throughput Time Operating Curve is especially emphasized when the work content varies greatly.

The mean *inter-operation time operating curve* in turn, runs parallel to the Throughput Time Operating Curve and is shifted downwards to smaller values, whereby the shift is equal to the value of the mean operation time. In the underload zone, the inter-operation time corresponds to the minimum inter-operation time which is determined from the transport time and other process dependent requirements (e.g., when for technical reasons a product needs to be stored while it cools down or dries). It can frequently be ignored in which case the inter-operation time in the underload zone is equal to zero. In the transition and overload zones, the mean inter-operation time increases approximately proportional to the mean WIP and continually longer waiting times arise.

The mean *relative lateness operating curve* also runs parallel to the Range Operating Curve. It intersects the WIP axis at the planned WIP, which results from the planned throughput times of the operations. Below the planned WIP, the

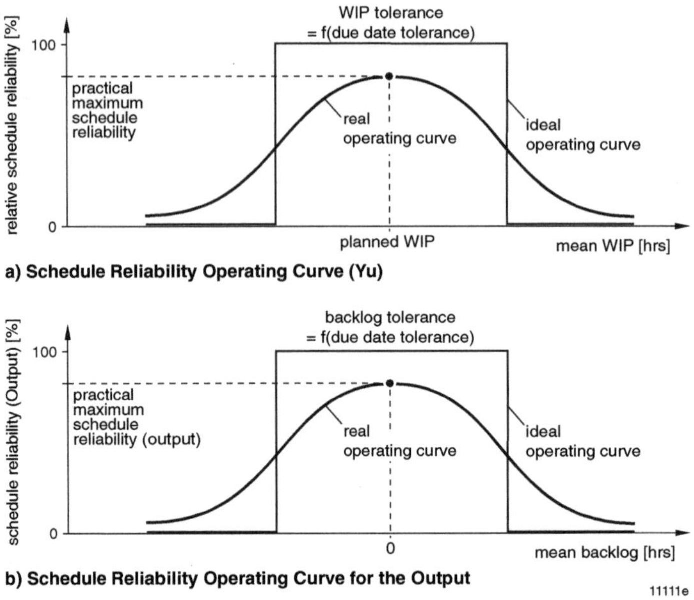


**Fig. 3.13** Logistic operating curves for the relative schedule reliability and for the output schedule reliability

relative lateness is negative i.e., the orders are processed quicker than planned. Correspondingly, the order throughput time above the planned WIP is delayed ($L_{rel,m} > 0$).

The relative schedule reliability is closely connected to the relative lateness. It is calculated as a percentage of the orders whose throughput time deviates at most by a value defined by the due date tolerance from the planned throughput time [Yu-01]. If all of the orders enter the workstation at the planned time, then the relative schedule reliability also corresponds to the schedule reliability of the workstation's output. As the relative schedule reliability approaches the planned WIP it also approximates the practical maximal relative schedule reliability (Fig. 3.13a).

The practical maximal schedule reliability describes the relative schedule reliability which, given the due date tolerance and throughput time variance, cannot be exceeded. As the WIP continues to increase, the schedule reliability further decreases.

The Logistic Operating Curve for the relative schedule reliability (otherwise known as Schedule Reliability Operating Curve) is calculated based on the distribution of the throughput times for a workstation. If the variance of the throughput time is zero, then the ideal Logistic Operating Curve depicted in Fig. 3.13 is applicable. If the span of the throughput times exceeds the due date tolerance, then the due date tolerance cannot reach 100% when the actual WIP equals the planned WIP. The maximum value of the actual Schedule Reliability Operating

Curve is therefore usually lower than 100. When interpreting the relative Schedule Reliability Operating Curve it should be taken into consideration that the variance of the throughput time can also be increased by applying the slack time rule. If the processing of an order which entered the workstation late has to be accelerated in order to allow an order to be completed when planned, the schedule reliability is increased for the output. The relative schedule reliability nonetheless sinks.

Thus, in addition to the relative schedule reliability, it is possible to represent the schedule reliability for the output of the manufacturing (or a workstation) as a Logistic Operating Curve. Figure 3.13b illustrates a recommendation for doing so based on theoretical considerations and which still has to be validated by simulation trials. The procedure and form of representation are as similar as possible to those for the relative schedule reliability. The Schedule Reliability Operating Curve shows the trend of the delivery reliability over the mean backlog. With a mean backlog of zero hours, the mean output lateness is zero shop calendar days. Under the condition that there is a symmetric due date tolerance and an equally symmetric distribution of the output lateness the schedule reliability therefore attains its highest value. If the span of the output lateness exceeds the due date tolerance a schedule reliability of 100% cannot be attained. The practical maximal schedule reliability then takes on a value smaller than 100%. The ideal Schedule Reliability Operating Curve results—also in analogy to the relative Schedule Reliability Operating Curve—from the assumption that the variance of the output lateness is zero. In particular, the conditions for the symmetric due date tolerance and distribution of the output lateness are frequently not given in practice. In both cases, given a backlog of zero hours, the maximum value of the schedule reliability is not attained.

The course of the Output Rate Operating Curve and the Throughput Time Operating Curve apply under the condition that the workstation's entire WIP could actually be processed. If a part of the workstation's WIP is blocked i.e., cut-off from the processing, the resulting Logistic Operating Curve is flatter (see Fig. 3.14). A workstation then attains a specific output rate only with a higher mean WIP. Reasons for blocked WIP include, for example, a lack of materials or a lack of tools. The blocked WIP which results from the manufacturing control are significant for evaluating manufacturing control methods.

Decentralized manufacturing control methods block orders in order to prevent a workstation from producing more than the capacities of the following workstation permits. Blocked WIP is thus an inherent disadvantage of a decentralized WIP control in comparison to a centralized WIP control. Moreover WIP can also be blocked when a workstation is only allowed to process orders after a release date has been reached.

The general behavior of the Logistic Operating Curves is independent of a specific workstation; only the scaling of the axis differs. Accordingly, the Logistic Operating Curves can be normalized in order to gain generally applicable Logistic Operating Curves. In this case, the logistic objectives 'WIP', 'output rate', and 'throughput time' refer to a reference value. The reference value for the output rate of a

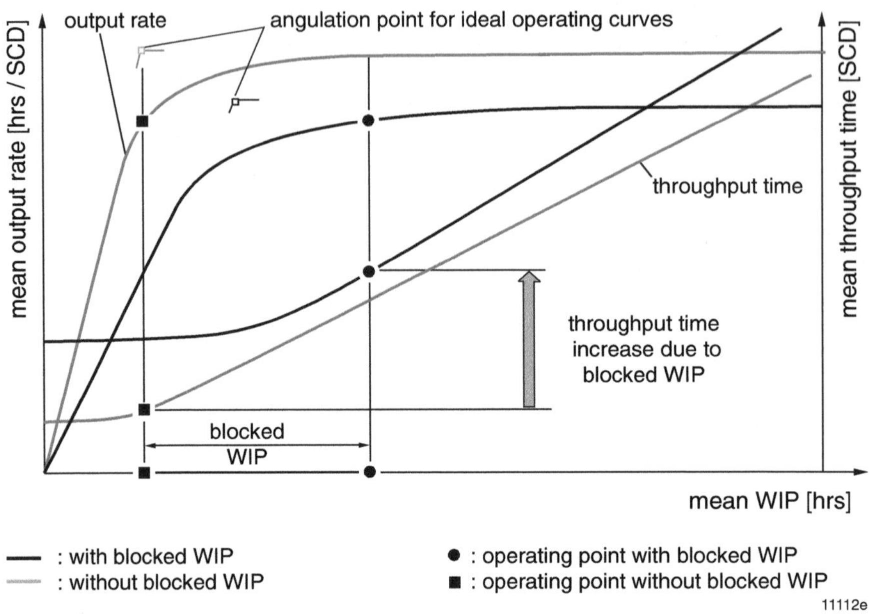


**Fig. 3.14** Impact of blocked WIP on the course of the logistic operating curves

workstation is its maximum possible output rate; the utilization of the workstation results as a normalized objective. Moreover, the mean WIP of a workstation is derived from its ideal minimum WIP which is calculated from the Logistic Operating Curves that correspond to the mean WIP of an ideal production process (see Sect. 3.2.2.2). The resulting normalized WIP variable is the relative WIP [Nyhu-09*]:

$$WIP_{rel} = \frac{WIP_m}{WIPI_{\min}} \cdot 100 \tag{3.23}$$

where

$WIP_{rel}$ relative WIP [%],
$WIP_m$ mean WIP [h],
$WIPI_{min}$ ideal minimum WIP [h].

The unweighted flow rate is the normalized throughput time variable (see Eq. 3.17). It sets the mean throughput time in relation to the mean operation time.

The normalized Logistic Operating Curves represent the course of the utilization and flow rate over the relative WIP (see Fig. 3.15).

The basic behavior of the Logistic Operating Curve does not differ from that of the variables of a specific workstation. Similar to the output rate, the utilization initially increases proportionally and then only degressively with the relative WIP. In the overload zone the curve approaches the maximum utilization of 100%.

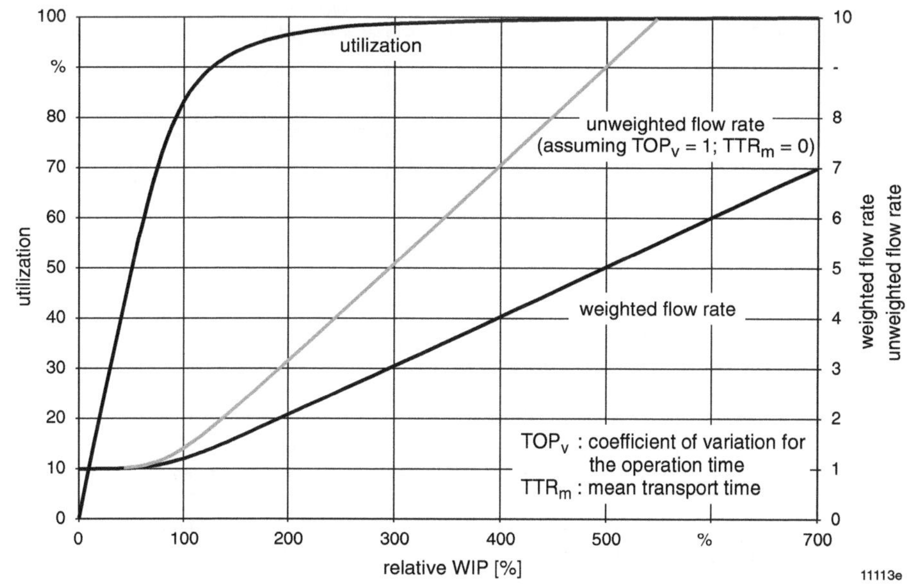


**Fig. 3.15** Normalized logistic operating curves (Nyhuis)

The unweighted Flow Rate Operating Curve runs parallel to the WIP axis when the relative WIP is extremely low and takes on the value of one (throughput time = operation time or minimum throughput time). With higher WIP levels the unweighted flow rate increases almost linearly to the relative WIP. The greater the coefficient of variance for the work content, the steeper the slope of the operating curve. In contrast, the course of the weighted Flow Rate Operating Curve is independent of the variance of the work content. If the work content is the same for all orders, then the Logistic Operating Curves for the weighted and unweighted flow rate coincide with one another [Nyhu-09*]. It can be very informative to compare the Utilization Operating Curve with the Flow Rate Operating Curve. In order to attain a high degree of utilization, it is necessary to plan in throughput times corresponding to 3–5 times the mean operation time.

#### 3.2.2.2 Deriving and Calculating the Logistic Operating Curves

The calculation of Logistic Operating Curves can be traced back to Nyhuis [Nyhu-91]. The basis for the computational equation is the definition of an ideal production process. Ideal Logistic Operating Curves can then be derived for this ideal process and together with the so-called c-norm parameter can in turn serve as the foundation for approximating actual Logistic Operating Curves.

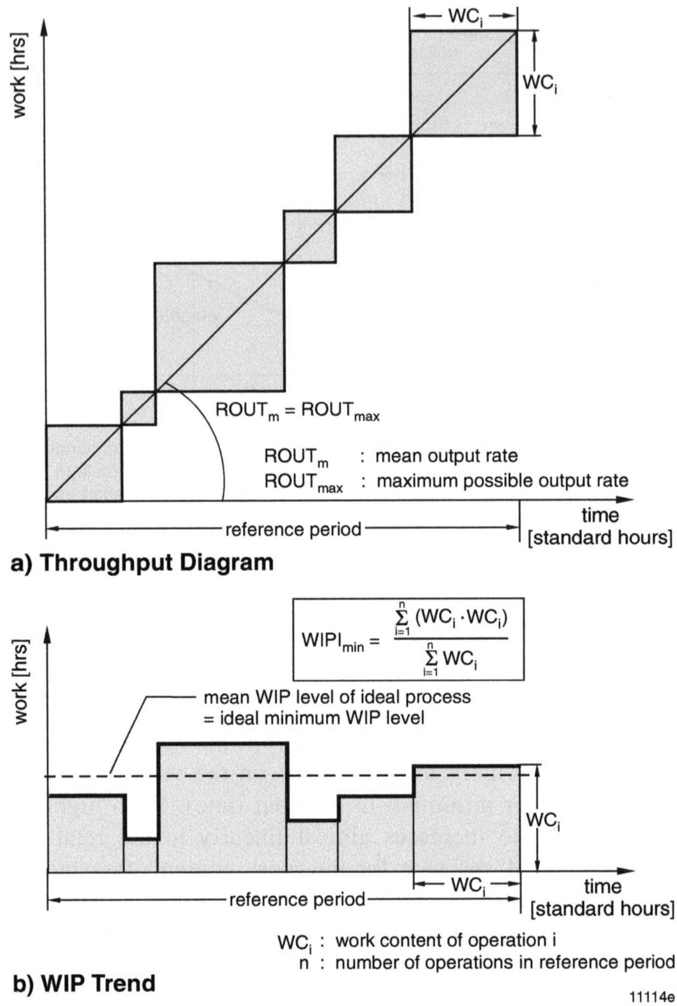


**Fig. 3.16** Representation of an ideal production process in a throughput diagram (Nyhuis)

## Deriving Ideal Logistic Operating Curves

Von Wedemeyer defined an ideal production process characterized by the following conditions [Nyhu-09*, vonW-89]:

- There is exactly one order at every workstation at any point in time.
- An order is processed immediately after it enters the workstation.

Figure 3.16 depicts the Throughput Diagram for this ideal process.

In an ideal production process a workstation neither has to wait for an order (utilization = 100%) nor does an order have to wait for another order to be processed (inter-operation time = 0 or throughput time = operation time).

For the sake of simplification, the time axis is given in standard hours. When the axes are equally scaled, the throughput elements of the order are thus represented as squares. The mean WIP of the ideal production process is denoted as the ideal minimum WIP of a workstation. It is the Logistic Operating Curves' most important parameter and is calculated according to [Nyhu-91] as:

$$WIPI_{\min} = \frac{\sum_{i=1}^{n}(WC_i \cdot WC_i)}{\sum_{i=1}^{n} WC_i} = WC_m + \frac{WC_s^2}{WC_m} \tag{3.24}$$

where

| | |
|---|---|
| $WIPI_{min}$ | ideal minimum WIP [h], |
| $WC_i$ | work content of an operation i [h], |
| n | number of operations [−], |
| $WC_m$ | mean work content [h], |
| $WC_s$ | standard deviation of the work content [h]. |

If a workstation consists of a number of workplaces, the ideal minimum WIPs are added [Nyhu-09*]:

$$WIPI_{\min} = \sum_{i=1}^{nwp} WIPI_{\min,i} \tag{3.25}$$

where

| | |
|---|---|
| $WIPI_{min}$ | ideal minimum WIP [h], |
| $WIPI_{min,i}$ | ideal minimum WIP of the individual workplace *i* [h], |
| nwp | number of work places [−]. |

If the minimum inter-operation times (e.g., for the transport to a workstation) cannot be ignored, then the ideal minimum WIP is extended by the minimum inter-operation time. In order to facilitate this, the inter-operation time is therefore measured as an exception here in hours (see [Nyhu-09*]).

$$WIPI_{\min} = WC_m + \frac{WC_s^2}{WC_m} + TIO_{\min} \tag{3.26}$$

where

| | |
|---|---|
| $WIPI_{min}$ | ideal minimum WIP [h], |
| $WC_m$ | mean work content [h], |
| $WC_s$ | standard deviation of the work content [h], |
| $TIO_{min}$ | minimum inter-operation time [h]. |

Ideal Logistic Operating Curves can be derived from the ideal production process (see Fig. 3.17, [Nyhu-91, vonW-89]). To do so the WIP of the ideal production process is varied in theory. When a workstation is operating with its ideal minimum WIP the workstation attains its maximum possible output rate and is fully utilized.

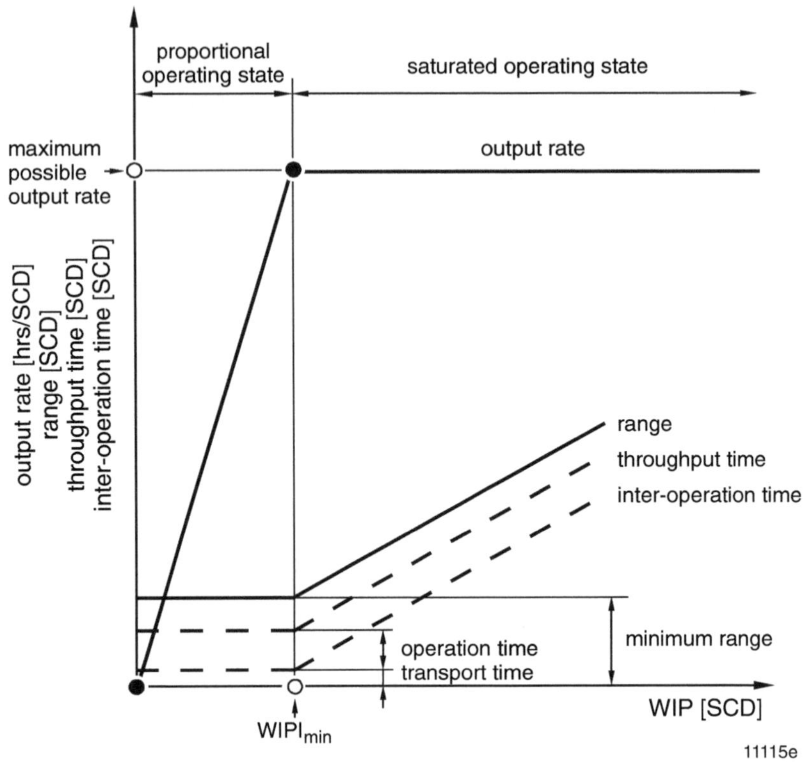


**Fig. 3.17** Construction of the ideal operating curves for the output rate and range (Nyhuis/Wiendahl)

Increasing the WIP above the ideal minimum WIP does not increase the output rate. The ideal Output Rate Operating Curve runs parallel to the WIP axis (saturated operating state). In contrast lowering the WIP below the WIPI is inevitably connected with a loss of utilization. The output rate of a workstation decreases proportionally to the WIP (proportional operating state of the ideal OROC).

The Range Operating Curve results directly from applying the Funnel Formula. In the proportional operating state the ratio of the WIP to the output rate is constant, the Range Operating Curve therefore runs parallel to the WIP axis. In contrast, the output rate in the saturated operating state no longer increases. Accordingly, the slope of the ideal ROC increases proportional to the WIP. Nyhuis derives further ideal Logistic Operating Curves for the logistic objectives 'throughput time' and 'inter-operation time' [Nyhu-09*].

The ideal Logistic Operating Curves already set the general behavior of the actual Logistic Operating Curves.

## Approximating Actual Logistic Operating Curves

The assumptions of the ideal production process are of course not practical. In a production where orders are processed on a number of workstations one after the other, the variance of the work content alone prevents orders from arriving at exactly the same time as the processing of the previous orders is completed. Furthermore, experience has taught that planned tasks cannot be implemented exactly on time by the production. Nonetheless, this is one of the requirements for such an ideal production process.

Thus, on the shop floor, in addition to the active WIP, a WIP buffer is required to ensure the utilization of the workstations. The actual Logistic Operating Curves deviate from the ideal Logistic Operating Curves especially in the area around the inflection point (see Fig. 3.18). In contrast, there is a strong agreement between the actual and ideal Logistic Operating Curves when the WIP is either extremely low or high.

Nyhuis developed an approximation equation for the actual Output Rate Operating Curve based on the c-norm function. It does not apply exactly, but depicts the reality with sufficient enough approximation. It has been successfully applied both by enterprises in the industry as well as for developing methods. Up until now it can only be stated implicitly [Nyhu-09*].

$$WIP_m(ROUT_m) = WIPI_{\min} \cdot \frac{ROUT_m}{ROUT_{\max}} + WIPI_{\min} \cdot \alpha_1 \cdot \left(1 - \sqrt[4]{1 - \frac{ROUT_m}{ROUT_{\max}}}\right)^4 \tag{3.27}$$

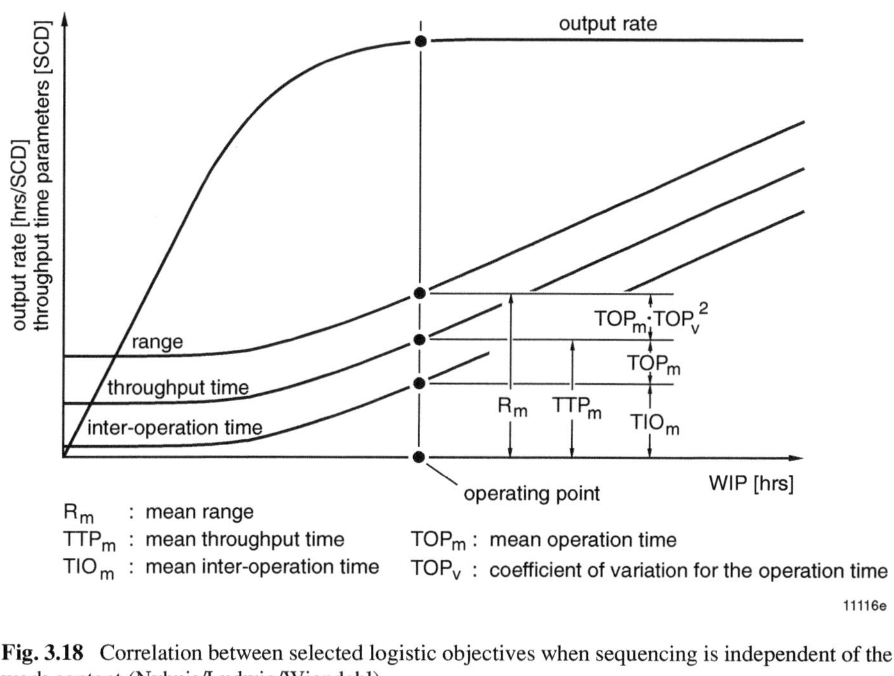


**Fig. 3.18** Correlation between selected logistic objectives when sequencing is independent of the work content (Nyhuis/Ludwig/Wiendahl)

where

| | |
|---|---|
| $WIP_m$ | mean WIP [h], |
| $WIPI_{min}$ | ideal minimum WIP [h], |
| $ROUT_m$ | mean output rate [h/SCD], |
| $ROUT_{max}$ | maximum possible output rate [h/SCD], |
| $\alpha_1$ | stretch factor [−]. |

The first summand of the approximation equation corresponds to the active WIP and agrees with the ideal Output Rate Operating Curve. It increases proportionally with the utilization as a ratio of the mean and maximum possible output rate.

The second summand calculates the WIP buffer and is determined by three factors. First, the required WIP buffer increases proportional to the ideal minimum WIP. The ideal minimum WIP thus generally influences both components of a workstation's WIP. Second, the required WIP buffer grows disproportionally along with the utilization. Accepting a minimal loss of utilization thus allows the WIP to be reduced considerably [Nyhu-09*]. Third, the behavior of the approximated Output Rate Operating Curve is determined by the empirical stretch factor $\alpha_1$. A stretch factor of $\alpha_1 = 10$ has proven to be reliable for job shop productions [Nyhu-09*].

The stretch factor (and thus the required WIP buffer) can be influenced by the manufacturing control. The better aligned the input and output of a workstation is, the smaller the stretch factor is. Therefore, highly flexible capacities and the possibility of shifting orders (load flexibility) positively impact the logistic potential of a production [Nyhu-09*]. Moreover, the required WIP buffer is reduced when a number of individual workstations access a common queue [Nyhu-09*].

Here too, the Range Operating Curve results directly from the Output Rate Operating Curve and Funnel Formula. In order to do so, the ratio between the mean WIP and the mean output rate is formed for each point on the Output Rate Operating Curve and plotted over the mean WIP.

The Throughput Time Operating Curve can then be derived from the Range Operating Curve [Nyhu-09*]:

$$TTP_m = R_m - TOP_m \cdot TOP_v^2 \tag{3.28}$$

where

| | |
|---|---|
| $TTP_m$ | mean throughput time [SCD], |
| $R_m$ | mean range [SCD], |
| $TOP_m$ | mean operation time [SCD], |
| $TOP_v$ | coefficient of variance for the operation time [−]. |

Consequently, the difference between the range and the throughput time increases with the mean and variance of the operation time; when the operation times are the same ($TOP_v = 0$) the mean values of the range and throughput time are the same.

The mean Inter-Operation Operating Curve results directly from the Throughput Time Operating Curve and Eq. 3.16:

$$TIO_m = TTP_m - TOP_m \tag{3.29}$$

where

| | |
|---|---|
| $TIO_m$ | mean inter-operation time [SCD], |
| $TTP_m$ | mean throughput time [SCD], |
| $TOP_m$ | mean operation time [SCD]. |

Equation 3.28 applies under the condition that the processing sequence is not dependent on the work content i.e., that orders with large work content (or small) are not prioritized for processing.

Furthermore, the Relative Lateness Operating Curve can be derived from the Throughput Time Operating Curve. In order to do so, according to Eq. 2.10 the mean planned throughput time is only subtracted from the actual throughput time [Yu-01]:

$$L_{rel,m} = TTP_m - TTP_{m,plan} \tag{3.30}$$

where

| | |
|---|---|
| $L_{rel,\,m}$ | mean relative lateness [SCD], |
| $TTP_m$ | mean (actual) throughput time [SCD], |
| $TTP_{m,plan}$ | mean planned throughput time [SCD]. |

#### 3.2.2.3 Applying Logistic Operating Curves for Determining the Planned Throughput Times

One of the more important parameters for many manufacturing control methods is the planned throughput time of orders. Due to the outstanding significance of the planned throughput times and because the Logistic Operating Curves play a key role in calculating them, I would now like to discuss Flow Rate Oriented Scheduling as developed by Ludwig und Nyhuis [Ludw-92, Ludw-95]. This scheduling technique is used in the following chapters for setting the method parameters. It determines the order throughput time within the frame of a Logistic Positioning with the aid of Logistic Operating Curves. Figure 3.19 illustrates how Flow Rate Oriented Scheduling functions.

Flow Rate Oriented Scheduling determines the order throughput times as the sum of the operations' throughput times. Each of the operation's throughput times is in turn calculated from a workstation specific operation and inter-operation time. The main component of the method is deriving the planned-inter-operation time from a Logistic Positioning with the aid of the workstation's Logistic Operating Curves. The exact procedure is subsequently explained based on a universal example.

The example is divided into calculating the Logistic Operating Curves, the Logistic Positioning and finally the calculating the order throughput time.

**Calculating the Logistic Operating Curves**

The calculation of the Logistic Operating Curves will be explained based on a fictitious example of a lathe. During the planned period, it should complete the following 30 orders (Table 3.1).

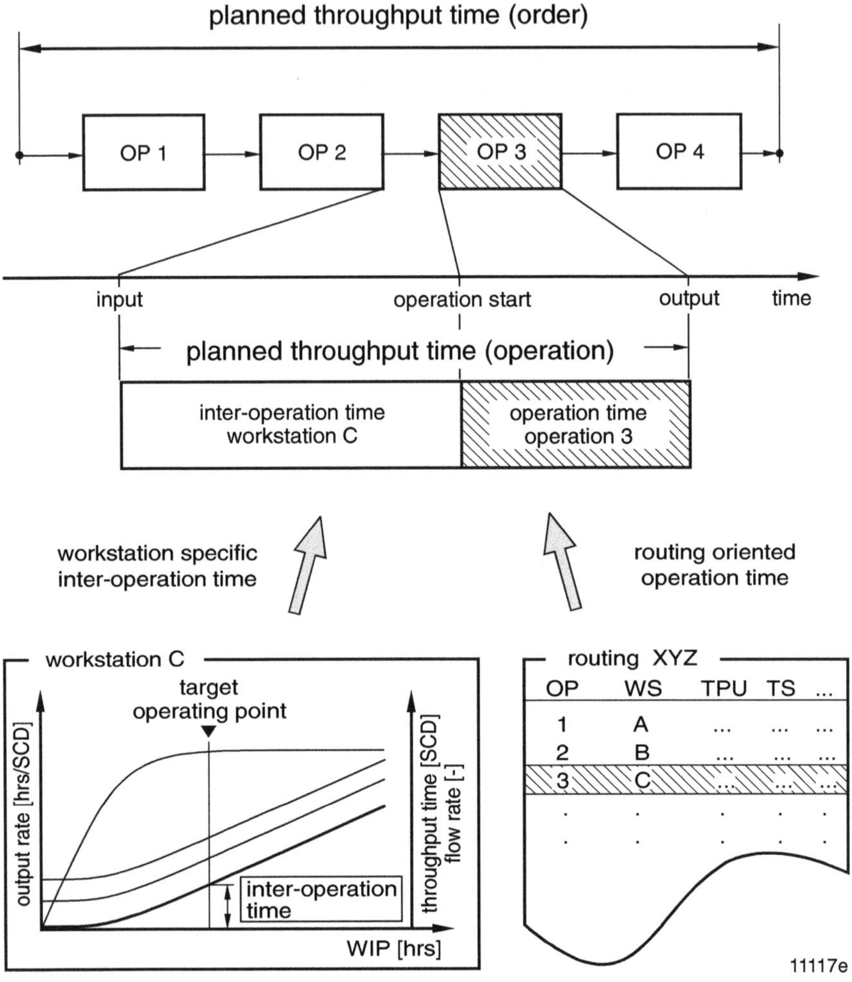


**Fig. 3.19** Flow rate oriented scheduling (Ludwig/Nyhuis)

According to Eq. 3.24 the ideal minimum WIP is calculated as:

$$WIPI_{\min} = \frac{\sum_{i=1}^{30} WC_i^2}{\sum_{i=1}^{30} WC_i} = \frac{4 + 16 + ... + 1}{2 + 4 + ... + 1} = 7.5hrs \tag{3.31}$$

where

$WIPI_{min}$ ideal minimum WIP [h],
$WC_i$ work content of an operation i [h].

**Table 3.1** List of orders to be completed during the planning period

| Nr. | WC | Nr. | WC | Nr. | WC | Nr. | WC | Nr. | WC |
|---|---|---|---|---|---|---|---|---|---|
| 1 | 2 | 7 | 4 | 13 | 7 | 19 | 12 | 25 | 4 |
| 2 | 4 | 8 | 8 | 14 | 11 | 20 | 3 | 26 | 2 |
| 3 | 1 | 9 | 11 | 15 | 3 | 21 | 1 | 27 | 11 |
| 4 | 6 | 10 | 1 | 16 | 5 | 22 | 4 | 28 | 9 |
| 5 | 5 | 11 | 9 | 17 | 5 | 23 | 1 | 29 | 2 |
| 6 | 3 | 12 | 5 | 18 | 8 | 24 | 10 | 30 | 1 |

WC: Work Content

The lathe is operated for one shift (8 h/SCD). Disruptions are negligible. Accordingly, the maximum possible output rate is 8 h/SCD. This parameter is sufficient for determining the course of the Output Rate Operating Curve using Eq. 3.27 (Fig. 3.20).

The output rate (or utilization) is varied in order to gain data points for mapping the operating curve. Using Eqs. 3.20 (range), 3.28 (throughput time), and 3.29 (inter-operation time) the Logistic Operating Curves for the throughput time parameters can be derived from the Output Rate Operating Curve.

Nyhuis and Wiendahl [Nyhu-09*] provide an extensive discussion on calculating the Logistic Operating Curves. Field investigations indicate that the WIPI on a workstation only changes negligibly over time since, although the order spectrum changes in its composition, the mean and variance of the work content barely changes [Nyhu-09*]. If this is the case, then the WIPI of a workstation can be stored as a set value which should be updated in suitable intervals. This procedure is advantageous in that the statistical figures can be determined on the basis of a greater number of individual values.

## Logistic Positioning

Logistic Positionings are used by enterprises to set the planned WIP for a workstation. The concept of a Logistic Positioning was developed by Wiendahl and solves what Gutenberg referred to as 'the dilemma of operation planning'. It takes into consideration the fact that it is impossible both in theory and in practice to define just one practical optimization parameter. Based on a strategically important primary logistic objective (e.g., throughput time), the goal is to find a consistent value for the remaining objectives. Figure 3.21 depicts three possible Logistic Positionings for the workstation, which in this case is the observed lathe.

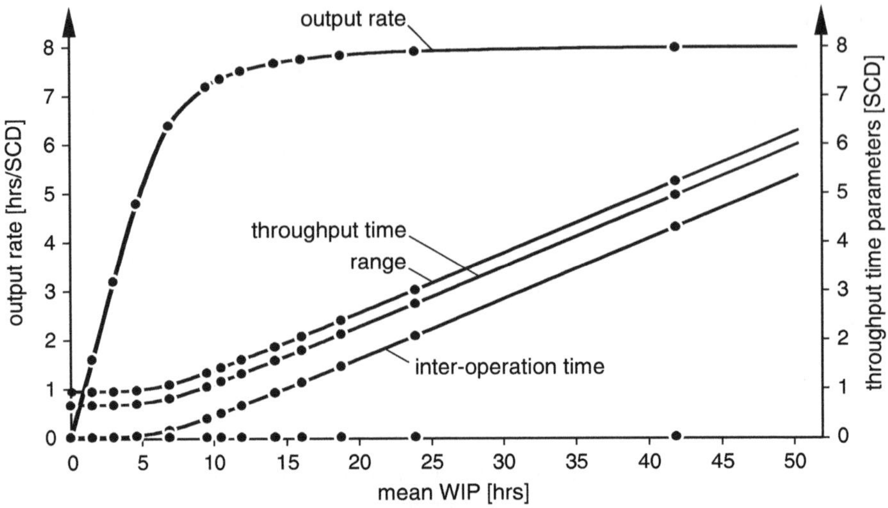


| utilization [%] | relative WIP [%] | output rate [hrs/SCD] | WIP [hrs] | range [SCD] | throughput time [SCD] | inter-operation time [SCD] |
|---|---|---|---|---|---|---|
| 0 | 0 | 0.0 | 0.0 | 0.9 | 0.7 | 0.0 |
| 20 | 20 | 1.6 | 1.5 | 0.9 | 0.7 | 0.0 |
| 40 | 40 | 3.2 | 3.0 | 0.9 | 0.7 | 0.0 |
| 60 | 62 | 4.8 | 4.7 | 1.0 | 0.7 | 0.0 |
| 80 | 92 | 6.4 | 6.9 | 1.1 | 0.8 | 0.1 |
| 90 | 127 | 7.2 | 9.5 | 1.3 | 1.0 | 0.4 |
| 92 | 140 | 7.4 | 10.5 | 1.4 | 1.1 | 0.5 |
| 94 | 159 | 7.5 | 12.0 | 1.6 | 1.3 | 0.7 |
| 96 | 189 | 7.7 | 14.3 | 1.9 | 1.6 | 0.9 |
| 97 | 213 | 7.8 | 16.1 | 2.1 | 1.8 | 1.1 |
| 98 | 250 | 7.8 | 18.8 | 2.4 | 2.1 | 1.5 |
| 99 | 318 | 7.9 | 23.9 | 3.0 | 2.7 | 2.1 |
| 99.9 | 557 | 8.0 | 41.9 | 5.2 | 5.0 | 4.3 |

11118e

**Fig. 3.20** Calculating the logistic operating curves

*Position 1* is in the underload zone of the Output Rate Operating Curve. Correspondingly, the machine is only minimally utilized (approx. 50%). However, as a result the inter-operation and throughput times for the workstation are very short. Such a Logistic Positioning can be selected if:

1. The workstation is not critical with regards to capacities, and
2. Costs related to the loss of utilization, in particular for personnel waiting for work are low in comparison to the advantages resulting from the shorter throughput times.

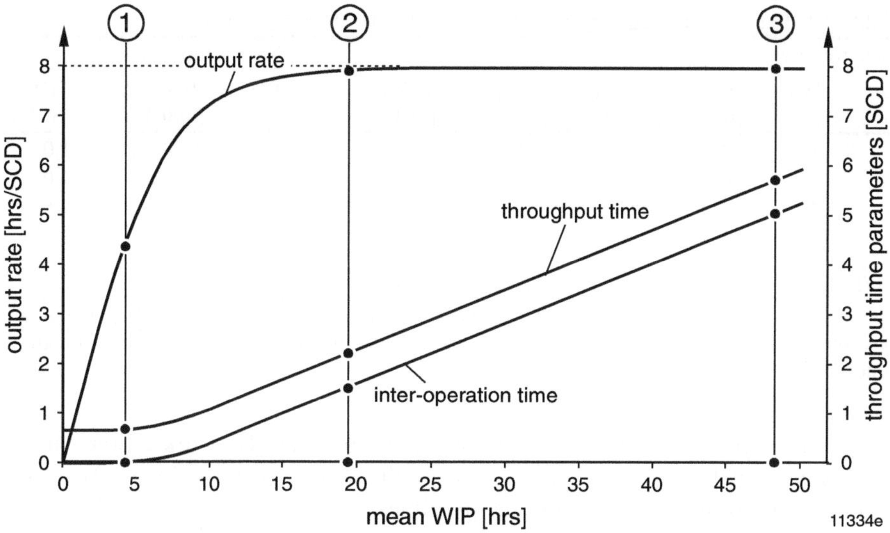


**Fig. 3.21** Alternative logistic positionings for a workstation (example)

In practice, a Logistic Positioning in the underload zone is in part selected for workstations used only sporadically to which flexible personnel can be allocated as needed.

*Position 2* is in the Output Rate Operating Curve's transition zone. It is characterized by a high utilization (98%) and moderate throughput or inter-operation times ($TTP_m = 2.25$ SCD; $TIO_m = 1.5$ SCD). Such a Logistic Positioning is typical for workstations that have fixed personnel allocated to them. It should be selected for the lathe in the example.

*Position 3* is characteristically fully utilized and has long throughput and inter-operation times. A Logistic Positioning of this sort is only consciously selected if the workstation is a throughput bottleneck for the production and the production is generally working at the limits of its capacities. Nonetheless, many enterprises position themselves in the overload zone unconsciously: Not knowing the behavior of the Logistic Operating Curves they thus attach a strong security factor to the planned throughput times in order to avoid a loss of utilization. As a result excessive throughput times and WIP levels arise, frequently also on non-bottleneck workstations.

### Calculating the Order Throughput Times

If the planned inter-operation times and the planned output rate are determined for all workstations, then the orders can be scheduled. The procedure for doing so is shown in Table 3.2, based on the example of Order 4 which is to be processed on the lathe machine in the second operation.

The operation times according to Eq. 3.14 are calculated as the quotient of work content (for Operation 1: 2 h) and the maximum possible output rate of the workstation (for Workstation 1: 8 h/SCD). The inter-operation times are taken from

**Table 3.2** Example of calculating the operation and order throughput times

| Data for order 4 | | | | | | Workstation data | | |
|---|---|---|---|---|---|---|---|---|
| OP [–] | WS [–] | WC [h] | $TOP_{plan}$ [SCD] | $TIO_{plan}$ [SCD] | $TTP_{plan}$ [SCD] | WS [–] | $ROUT_{max}$ [h/SCD] | $TIO_{plan}$ [SCD] |
| 1 | 1 | 2 | 0.25 | 1.0 | 1.25 | 1 | 8 | 1.0 |
| 2 | 3 | 6 | 0.75 | 1.5 | 2.25 | 2 | 8 | 1.0 |
| 3 | 2 | 4 | 0.50 | 1.0 | 1.50 | 3 | 8 | 1.5 |
| 4 | 4 | 2 | 0.25 | 1.0 | 1.25 | 4 | 8 | 1.0 |
| TOTAL | | 14 | 1.75 | 4.5 | 6.25 | | | |

*OP* operation, *WC* work content, $TIO_{plan}$ planned inter-operation time, $ROUT_{max}$ maximum possible output rate, *WS* workstation, $TOP_{plan}$ planned operation time, $TTP_{plan}$ planned throughput time, $TIO_m$ mean inter-operation time

the workstation data, so that the throughput times of the workstation can be calculated. Finally, the throughput time of the orders (here: 6.25 SCD) results from the sum of the operation throughput times.

## 3.3 Conclusions from the Modeling

In the previous section the logistic objectives were quantitatively modeled. This next section aims at giving an overview of the factors which influence the external logistic objectives (Sect. 3.3.1). Moreover, guiding principles for designing manufacturing control methods (Sect. 3.3.2) will be derived from the modeling.

### *3.3.1 Correlations Between the Logistic Objectives*

In the following, the correlations between the logistic objectives 'delivery reliability' (as the most comprehensive schedule objective), 'delivery time', and 'service level' will be briefly considered. These logistic objectives represent the external logistic performance (see Fig. 2.1). Due to the variety of factors which influence pricing, it is almost impossible to represent the correlations for the external logistic costs in a comparative form.

#### Delivery Reliability

Figure 3.22 depicts the parameters which influence the delivery reliability. The most important of these are the mean and variance of the delivery lateness. In addition, the delivery tolerance agreed upon with the customer determines the calculated delivery reliability.

The mean and variance of the delivery lateness (which when applicable may be absorbed by an enterprise's internal delivery time buffer) result from the output lateness' mean and variance.

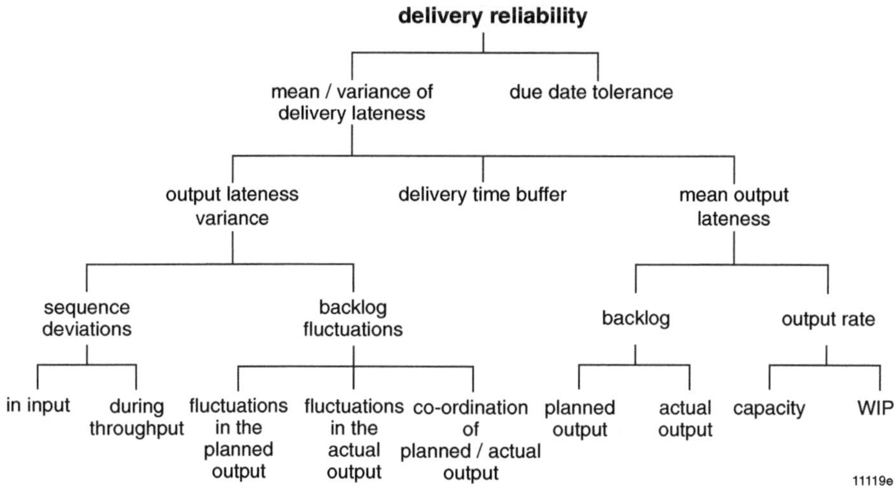


**Fig. 3.22** Parameters that influence the delivery reliability

The *mean output lateness* is calculated as the quotient of the backlog and output rate. The backlog results in turn from the difference between the planned and actual output. It is the job of production planning therefore to provide a realistic planned output. The manufacturing control, on the other hand, strives to adjust the actual output rate to the planned output.

When there is a backlog, increasing the output rate positively impacts the output lateness in two ways: First of all, the backlog is directly reduced and secondly, the increased output rate shortens the lateness resulting from a specific backlog. In order to be able to attain the planned output rate, the required personnel and resource capacities have to be available. Moreover, there has to be sufficient WIP planned so that breaks in the material flow on the (bottleneck) workstation are avoided.

The *output lateness variance* results from sequence deviations, fluctuating backlogs or a combination of the two. The influence of the backlog fluctuations demonstrates that it is not enough to reduce the mean backlog of a planning period. Rather, the backlog cannot be allowed to take on a positive value for the entire period, since otherwise delays will be caused.

Sequence deviations can either arise in the input or during the orders' throughput. Input related interchanges can be analyzed with the use of a distribution table for the input lateness. The delayed input of an order can be caused for example by a lack of materials or tools. Frequently, it is then practical to release another order early which can use the planned capacities. Moreover, attempts to balance the load can result in sequence deviations. The sequencing of orders can change during their throughput when an operator does not process them according to their due date priority, but rather according to another criterion, e.g., saving on setup times. This is a very common practice in the industry. For a high delivery compliance it is thus critical to set targets for sequencing and to measure these (see Chap. 25).

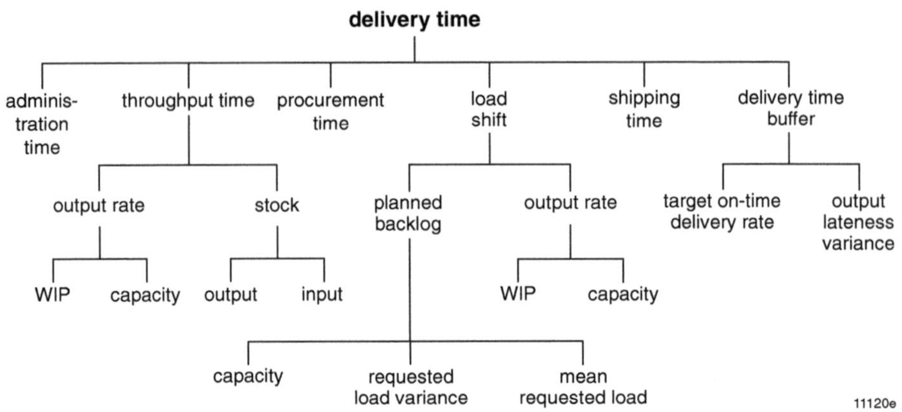


**Fig. 3.23** Parameters that influence the delivery time

### Delivery Time

Figure 3.23 depicts the parameters that influence the logistic objective 'delivery time'. Due to their close relation to the manufacturing control, the throughput time and load shifts are explained in detail.

The *mean throughput time* results from the ratio of the WIP to the output rate. The WIP is calculated as the difference between the cumulated input and output. The output rate is determined by the capacities and the WIP (without WIP, the workstation cannot productively implement the capacities). Since the output rate sets the system output, the output rate and WIP mutually influence one another.

The *load shift* is determined by the relation between the peak required capacities and the flexibility of the capacities as well as by how they are aligned. Depending on whether the throughput time or the load shift dominates the delivery time, the key to shorter delivery times is in reducing throughput times or increasing the capacities and/or flexibility of the capacities.

### Service Level

Figure 3.24 depicts the parameters that influence the service level. These include both the positive demand deviations in a replenishment cycle that can arise due to a delayed delivery or increased demand as well as the maximum under-delivery and safety stock.

The maximum positive lateness (delay) results from the output lateness' mean and variance (see Fig. 3.22 for a more extensive breakdown of these factors). Together with the mean values of the demand rate, it determines the maximum additional demand due to an extended replenishment time.

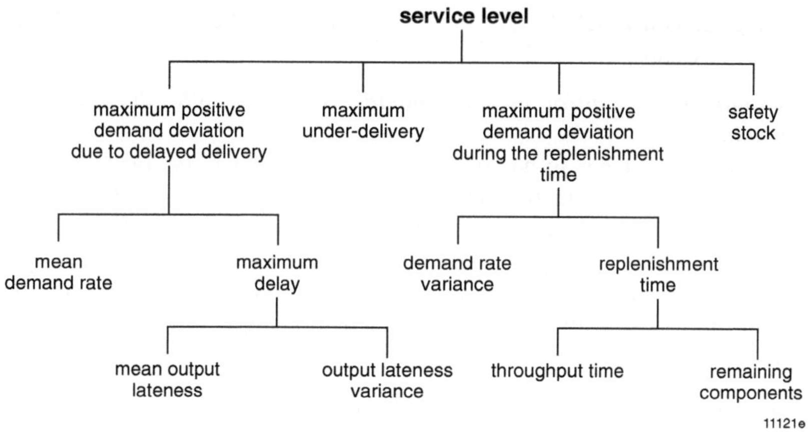


**Fig. 3.24** Parameters that influence the service level

During the replenishment time, the maximum positive demand deviation increases along with the variance of the demand rate and the length of the replenishment time. Similar to the delivery time, the replenishment time can be divided into various components of which, from the perspective of manufacturing control, the throughput time is the most relevant.

### *3.3.2 Guiding Principles for Designing Manufacturing Control Methods*

Based on the modeling of the internal logistic objectives, guidelines for evaluating and designing manufacturing control methods can be derived (see [Lödd-01]). To a large extent these are oriented on the basic laws of production logistics, formulated by Wiendahl and Nyhuis and transfer these to problems related to manufacturing control [Nyhu-09*, Wien-97b].

1. *A manufacturing control method should allow the WIP of a production or workstation to be regulated at a defined level.*

This ability is connected to two important advantages for a production. First, it allows an enterprise to actively position itself within the dilemma of operation planning and to consistently align the target attainments with regards to the logistic objectives 'WIP', 'output rate' (or utilization), and 'throughput time'. Second, regulating the WIP increases the predictability of the throughput times. It thus supports the scheduling precision and with that a high delivery schedule reliability.

With regards to the order release, controlling the WIP can be disadvantageous in that the planning tasks cannot be exactly implemented. In some cases, this disadvantage can outweigh the advantages of controlling the WIP. This is discussed extensively in Chaps. 16 and 29.

2. *A manufacturing control method should reduce the WIP fluctuations to a minimum.*

There are two advantages to having minimal WIP fluctuations: First, it results in minimal throughput time fluctuations. This simplifies scheduling and thus supports the schedule reliability. Second, it also leads to the actual Output Rate Operating Curve closely approximating the ideal Output Rate Operating Curve. A specific output rate can therefore be more easily attained with lower WIP levels and shorter throughput times than with stronger WIP fluctuations (see also the fourth guiding principle for designing manufacturing controls).

3. *A manufacturing control method should cause as little blocked WIP as possible.*

The most important reason for this, results from the unfavorable behavior of the operating curves when there is blocked WIP (see Fig. 3.14). A specific output rate can be attained only with higher WIP levels than without blocked WIP. Moreover, blocked WIP causes sequencing interchanges and thus negatively impacts the schedule reliability of a production (see also the fifth guiding principle for designing manufacturing controls and the ninth basic law of production logistics).

4. *A manufacturing control method should contribute to closely aligning the load to the capacities.*

Load balancing accelerates or delays the release and/or throughput of orders with the aim of avoiding breaks in the material flow (accelerating) or preventing a build-up of WIP (delaying). As a result the Output Rate Operating Curve behaves more favorably [Nyhu-09*] and a specific output rate can then be attained with less WIP. At the same time balancing the load can require breeching the third and fifth guiding principles for designing production methods. The more carefully scheduling and capacity planning are conducted, the less manufacturing control has to balance the load and the less impact balancing the load has.

5. *A manufacturing control method should cause as few sequencing interchanges as possible.*

In order to ensure a high degree of schedule reliability, workstations should process orders according to their urgency. This reduces the variance of the output lateness and thus supports a high schedule reliability. In particular, order release methods that force sequence interchanges could under certain circumstances decrease the productions schedule reliability. Sequence interchanges should thus be reduced to a minimum. This guiding principle can also conflict with the fourth guiding principle for designing manufacturing controls (see Chap. 25 for a detailed assessment of sequencing rules).

6. *A manufacturing control method should be able to balance schedule deviations within the context of backlog control.*

If there is a threat of a backlog developing, it is the job of manufacturing control to prevent this or to decrease it once it has arisen. In order to do so capacities should be temporarily increased. The goal of regulating the backlog is when possible not to allow the mean lateness to become positive. Together with the fifth guiding principle, this should ensure the production's schedule reliability. Alternatives to controlling backlogs are discussed in Chap. 28.

7. *A manufacturing control method should take into consideration the principle of bottlenecks.*

This guiding principle leans toward Goldratt, who clearly illustrates the particular importance of a bottleneck in his novel "The Goal" [Gold-84*]. The bottleneck principle, according to which bottleneck workstations determine the manufacturer's output rate, contains two basic consequences for manufacturing control. First, the manufacturing control should ensure that there are no material flow disruptions on the bottleneck workstation of a production, that is, they should ensure that there is always sufficient WIP. The mean WIP level required to do so is a measure of the efficiency of a manufacturing control method. Second, capacities should be increased primarily on bottleneck workstations; whereas increasing capacities on non-bottleneck workstations, which do not increase the output rate of the entire production, should be avoided.

8. *A manufacturing control method should be as simple as possible.*

Not only do complicated manufacturing control methods create more opportunities for errors than simple methods, they also require explanations which in turn lowers their acceptance. Simple manufacturing control methods can therefore also be implemented more easily in the industry than complicated ones. When the quality of the solutions is comparable or better, the simpler method is always preferred. If the more complicated method leads to better results in theory, it should be considered whether or not the gained utility can actually be attained on the shop floor and whether the extra effort for introducing it and using it is justifiable.

9. *A manufacturing control method should communicate information about demands as quickly and clearly as possible.*

The delayed or distorted communication of information about demands leads to increased demand fluctuations on the preceding stages of a supply chain. As a result they frequently cause capacities to be over-dimensioned, inventory fluctuations and a general build-up of inventory.

The theoretical basis for this ninth guiding principle is extensively discussed in Chap. 6. It applies exclusively to methods for generating orders.

In the following, the guiding principles are used to evaluate methods for generating and releasing orders as well as for controlling capacities. Before doing so though, the next chapter focuses on the responsibilities of production planning. Production planning generates the planned values which should be implemented by the manufacturing control and is therefore critical to attaining logistic targets.

## References


[Bech-84] Bechte, W.: Steuerung der Durchlaufzeit durch belastungsorientierte Auftragsfreigabe bei Werkstattfertigung. VDI Progress Reports, Series 2, No.70, Düsseldorf (1984)

[Erdl-84] Erdlenbruch, B.: Grundlagen neuer Auftragssteuerungsverfahren für die Werkstattfertigung. VDI Progress Reports, Series 2, No. 155, Düsseldorf (1984)

[Gläß-95] Gläßner, J.: Modellgestütztes Controlling der beschaffungslogistischen Prozeßkette. VDI Progress Reports, Series 2, No. 337, Düsseldorf (1995)

[Gold-84*] Goldratt, E.M., Cox, J.: The Goal. A Process of Ongoing Improvement. Gower, Aldershot (1984)

[Lödd-01] Lödding, H.: Dezentrale Bestandsorientierte Fertigungsregelung. VDI Progress Reports, Series 2, No. 587, Düsseldorf (2001)

[Ludw-92] Ludwig, E., Nyhuis, P.: Verbesserung der Termineinhaltung in komplexen Fertigungsbereichen durch einen neuen Ansatz zur Plandurchlaufzeitermittlung. In: Görke, W., Rininsland, H., Syrbe, M. (eds.) Information als Produktionsfaktor, pp. 473–483. Springer, Berlin (1992)

[Ludw-95] Ludwig, E.: Modellgestützte Diagnose logistischer Produktionsabläufe. VDI Progress Reports, Series 2, No. 362, Düsseldorf (1995)

[Lutz-01] Lutz, S., Lödding, H., Wiendahl, H.-P.: Kennliniengestützte Logistische Lageranalyse. Ein neuer Ansatz zur Positionierung im Dilemma zwischen Bestand und Servicegrad. ZWF **92**(9), 550–553 (2001)

[Lutz-02] Lutz, S.: Kennliniengestütztes Lagermanagement. VDI Progress Reports, Series 13, No. 53, (2002)

[Nyhu-91] Nyhuis, P.: Durchlauforientierte Losgrößenbestimmung. VDI Progress Reports, Series 2, No. 225, Düsseldorf (1991)

[Nyhu-09*] Nyhuis, P., Wiendahl, H.-P.: Fundamentals of Production Logistics. Theory, Tools and Applications. Springer, Berlin (2009)

[Pete-96] Petermann, D.: Modellbasierte Produktionsregelung. VDI Progress Reports, Series 20, No. 193, Düsseldorf (1996)

[Pior-84*] Piore, M. J., Sabel, C. F.: The second industrial divide. Possibilities for Prosperity, Basic Books, New York (1984)

[Schö-07*] Schönsleben, P.: Integral Logistics Management. Operations and Supply Chain Management in Comprehensive Value-Added Networks, 3rd edn. Auerbach, Boca Raton (2007)

[vonW-89] von Wedemeyer, H.-G.: Entscheidungsunterstützung in der Fertigungssteuerung mit Hilfe der Simulation. VDI Progress Reports, Series 2, No. 176, Düsseldorf (1989)

[Wahl-98] Wahlers, T.: Modellgestützte Analyse und Verbesserung der logistischen Merkmale komplexer Serienprodukte. VDI Progress Reports, Series 2, No. 475, Düsseldorf (1998)

[Wien-97] Wiendahl, H.-P.: Betriebsorganisation für Ingenieure. Carl Hanser Verlag, Munich/Vienna (1997)

[Wien-97b] Wiendahl, H.-P.: Fertigungsregelung. Logistische Beherrschung von Fertigungsabläufen auf Basis des Trichtermodells. Carl Hanser Verlag, Munich/Vienna (1997)

[Yu-01] Yu, K.-W.: Terminkennlinie. Eine Beschreibungsmethodik für die Terminabweichung im Produktionsbereich. VDI Progress Reports, Series 2, No. 576, Düsseldorf (2001)

# Chapter 4
# Production Planning Tasks

Production planning schedules the production for the next planning periods in advance. From the perspective of manufacturing control, the relevant results of the production planning process are the values planned for the input and output as well as the chosen sequence (see the manufacturing control model in Sect. 1.2). Since production planning therefore directly impacts manufacturing control, a brief overview of it is provided here.

The depiction of production planning presented here is oriented on the Aachen PPC Model which is based on longstanding research and consulting projects at the at RWTH Aachen University's Research Institute for Operations Management (FIR). The Aachen PPC Model differentiates between four reference views of production planning and control (PPC), the view from the task level is explained here (see Fig. 1.1). The task view subdivides the responsibilities of production planning and control into key tasks (Sect. 4.1) and cross-sectional tasks (Sect. 4.2).

The Aachen PPC Model contains both planning and control tasks. The latter are extensively discussed in the following main chapters of this book, thus we will focus exclusively on planning tasks in this chapter.

## 4.1 Key Tasks of Production Planning

The key tasks of production planning include: planning the production program (Sect. 4.1.1), planning the production requirements (Sect. 4.1.2), planning in-house production (Sect. 4.1.3) and planning external procurements. Since the last task is not a subject of this book, it will not be covered here (see however Chap. 6 on production planning in supply chains).

H. Lödding, *Handbook of Manufacturing Control*,
DOI 10.1007/978-3-642-24458-2_4, © Springer-Verlag Berlin Heidelberg 2013

### 4.1.1 Production Program Planning

It is the task of production planning to develop the enterprise's production program which includes the quantity to be produced for every article and for every planning period. It thus determines the net primary requirements of the enterprise. The starting point for planning the production is the sales plan which contains the quantity to be delivered for every product and every planning period (gross primary requirements). In particular, customer orders and projected sales flow into the sales plan. The latter are usually set by the marketing and sales department. Beyond that, the sales plan also includes all additional requirements such as spare parts, demo models, or prototypes. It is critical here that all sources of demands are covered (see [Voll-91*]). The gross primary requirements are compared to the available inventory in order to determine the net primary requirements and with that the production program. The safety-stock is also considered here (see Fig. 4.1).

American publications in particular, recommend aggregating the production program to a production plan (see [Voll-91*]) which serves as an input parameter for an enterprise's business plan. It can be formulated either as a cumulated production output (e.g., production of 100,000 chairs) or as a sales target (e.g., ten million Euro). Consistency between the production plan and the production program also has to be ensured when changes are made to the production program. If the production of an article is to be increased, then either the production of another article needs to be reduced (production plan remains the same) or the production plan has to be changed.

With seasonal sales, the enterprise has to decide if it follows the sales course (chase strategy) or if it levels the production program (Fig. 4.2).

Leveling the production program has the advantage of a constant utilization; however, one of the prerequisites for leveling it is a high inventory. It is thus tied to committing a large amount of capital and an increased inventory risk. The chase strategy, which follows the fluctuations in sales, can only be realized with flexible personnel resources. In addition, it usually requires the capacities of other resources (e.g., machinery) to also be generously dimensioned. In the industry, a compromise between the extremes of a leveling production program plan and chasing the fluctuations is frequently selected.

| Planning Period | 1 | 2 | 3 | 4 | 5 |
|---|---|---|---|---|---|
| Projected Sales (Gross Primary Requirement) | 20 | 25 | 22 | 24 | 26 |
| - Inventory | 30 | 20 | 20 | 20 | 20 |
| + Safety-Stock | 20 | 20 | 20 | 20 | 20 |
| = Planned Production (Net Primary Requirements) | 10 | 25 | 22 | 24 | 26 |

11122e

**Fig. 4.1** Tabular production program plan for a product

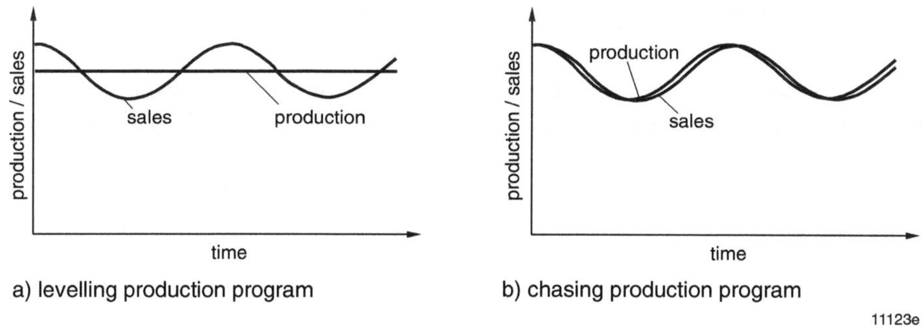


**Fig. 4.2** Production planning with seasonal sales course

**Table 4.1** Bills of capacities with the capacities required for products

| Product A | | Product B | |
|---|---|---|---|
| Workstation [–] | Capacity requirement per unit [min/unit] | Workstation [–] | Capacity requirement per unit [min/unit] |
| 1 | 2 | 1 | 3 |
| 2 | 3 | 2 | 3 |
| 3 | 1 | 5 | 2 |
| 4 | 5 | 6 | 5 |

The production program sets the planned output of a workstation and with that the required capacities. In order to estimate the capacity requirements, the capacity requirements for each article on all workstations (or at least on the critical ones) are calculated per unit and multiplied by the planned number of pieces. So-called capacity bills can be used for doing so. These contain the required capacities for an article on a workstation (Table 4.1).

In order to generate the capacity bills, the capacity requirements for the parts and components of the end product are taken into consideration. Data here includes the part list for the products and the work plans for the products, components and parts. If the required capacities are not available from the start, enterprises are dependent on planning with so-called reference profiles. This method is based on deriving projected capacity requirements for an order from known capacity requirements for comparable orders [Hegi-74].

Depending on the precision of the calculation, lot sizes and lead times are also considered [Voll-91*]. The aim here is to indicate the required capacities as early as possible in order to create a basis for making decisions about investments. If the production program cannot be realized with the available capacities, it needs to be modified. A realistic production program is a fundamental prerequisite for a high degree of delivery reliability. From the perspective of manufacturing control, the feasibility of the production program is the main requirement of the production program planning. Moreover, in make-to-order productions the capacity load should ideally already be considered when determining the delivery date.

## 4.1.2 *Production Requirements Planning*

It is the responsibility of those planning the production requirements to derive the necessary materials and resources from the production program. The first step consists of determining the materials required for the parts and components, and then generating the manufacturing orders (Sect. 4.1.2.1). In the next step, the manufacturing orders are scheduled and the capacity requirements for the manufacturing resources are calculated (Sect. 4.1.2.2).

### 4.1.2.1 Material Requirements Planning

Generally, a product consists of various components and parts which are assembled together in a number of steps. It is the responsibility of material requirements planning, to determine the required components and parts from the product requirements and to appropriately allocate these for the different periods. Four factors need to be taken into consideration here:

1. *Number of parts for each end product:* The number of parts for each end product is stored in a product's bill of materials. In order to calculate the gross material requirements from the production program, all products in which a part goes into need to be considered.
2. *Stock:* The quantity that actually has to be manufactured is reduced when there is a store of the parts that exceeds the required safety-stock. By taking this into consideration the gross material requirements are converted into net material requirements.
3. *Lead times:* The demand for a part arises earlier than the demand for a product that it belongs to. The reason for this is the throughput time that is required to produce a product from a part. This throughput time is frequently taken into consideration by a so-called 'lead time offset' with which the demand for a part is shifted chronologically towards the present in comparison to that of the product. If the lead times are stored as a fixed planning parameter, it does not necessarily ensure that the required capacities are actually available. Furthermore, in make-to-stock productions there is the danger that demand fluctuations are strengthened as a result of a lead time offset (see Chap. 6). Alternatives to a lead time offset include a WIP offset (see Chap. 13) or finite scheduling.
4. *Consolidating requirements and lot sizing:* In order to save setup times and to increase available capacities, the requirements for consecutive planning periods are often consolidated. In order to avoid delays, all of the requirements are then allocated to the earliest planning period. Consolidating the requirements thus increases the WIP. If requirements are rounded up to a (optimal) manufacturing lot size, then the demand increases. Both the consolidated requirements and increased demand impact dependent materials, where they can blur both the time at which materials are required as well as the level of demand.

Material requirements planning results in manufacturing orders with end dates. These form the input parameters for the throughput scheduling, the capacity requirements planning and the balancing of the load and capacities. Determining the material requirements is automated by PPC software in many enterprises. It is particularly important to dedicate special attention to maintaining and constantly updating the lead times and the safety stock.

For parts that are made-to-stock, procured-for-stock or are inexpensive, enterprises refrain to some degree from a deterministic assessment of requirements via a 'bill explosion' and instead use either statistic or heuristic methods. In cases such as this, consumption oriented manufacturing control methods such as Kanban controls or order point systems are frequently employed.

#### 4.1.2.2 Schedule and Capacity Planning

It is the responsibility of schedule and capacity planning to determine the start date of the orders manufactured in-house (throughput scheduling) as well as the required capacities (capacity requirements planning) and to align the capacities when there is a capacity overload. This results in the planned start dates for the orders and planned processing dates for the operations on the one hand, and the capacity requirements profile of the workstations on the other hand. Furthermore, planning the schedule also encompasses scheduling the procurement orders.

**Throughput Scheduling** The purpose of the throughput scheduling is to determine the planned start date and the planned end dates of the operations based on the order's end date (backwards scheduling). Occasionally scheduling is conducted according to a forwards or mid-point principle. Forward scheduling calculates an end date from a given start date. It is mainly applied when the capacities of an enterprise are highly utilized and/or the customer cannot name a preferred (realizable) delivery date. It is frequently conducted counter to the limited capacities and in this case requires that the capacities are already given. The same applies to mid-point scheduling, which first plans the operation of the order that is conducted on the bottleneck workstation and then the order end date using forwards scheduling of the operations after the bottleneck. Correspondingly, the planned start date results from a backwards scheduling of the operations before the bottleneck. Mid-point scheduling thus emphasizes the throughput bottleneck.

The key parameter for throughput scheduling is the order throughput time. Standard PPC software usually calculates the order throughput time as the sum of the operation throughput times. The latter is in turn sub-divided into an inter-operation time and operation time (Sect. 3.2.1). If finite scheduling is not required, the inter-operation time can be determined using Flow Rate Oriented Scheduling (Sect. 3.2.2.3).

Unfortunately most PPC systems fail to offer a possibility for setting the inter-operation times based on a model and to (partially) automate their maintenance. This is one of the reasons that enterprises struggle to attain their logistic targets.

Moreover, throughput scheduling is frequently conducted without verifying capacities. Capacity bottlenecks are revealed in the capacity requirements planning and removed by the capacity alignment.

**Capacity Requirements Planning** It is the responsibility of capacity requirements planning to determine the required capacities for each resource and each planning period. The input parameters for planning the capacity requirements are the scheduled orders with the work content for each operation and the schedule deviations which result from the production. If existing schedule deviations are not taken into consideration, further scheduling cannot be realistically completed: On the one hand, additional capacity requirements cannot be recognized when the production's output is lower than planned (backlog $> 0$), and, on the other hand, an enterprise cannot use the additional free capacities that arise when the production's output is higher than the planned (backlog $< 0$). The required capacities per resource and planning period are frequently represented as a bar chart (Fig. 4.3).

In the example, the required capacities exceed the normal capacities in planning period one as well as in periods four to nine. The maximum capacity is surpassed in period seven. In planning periods two, three and ten onward the normal capacities are not fully utilized.

**Capacity Alignment** Capacity alignment is thus responsible for coordinating or balancing the available capacities and the required capacities. In order to do so, an enterprise can either adjust the capacities to the demand (utilizing the capacities' flexibility) or adjust the demand to the capacities (utilizing the load's flexibility). In the latter case, it shifts the processing of orders or allocates operations to external enterprises. When doing so, the impact on the remaining operations of the orders as well as in particular the start and end dates needs to be considered. With complex products, this can also impact dependent manufacturing and assembly orders. Shifting the end date of an order should be avoided as much as possible, since this means that the production program cannot be implemented as planned and/or that a delivery date cannot be met. If this cannot be avoided, the production program should be changed. Figure 4.3b depicts how required capacities and available capacities can be aligned with one another. The capacities are adjusted as much as possible to the required capacities. Only part of the peak demand in period seven was pulled into period six. The planned output curve of the Throughput diagram can be directly derived from the capacity requirements profile (Fig. 4.3c).

The duration of the planning periods can usually be provided by the user in most PPC systems. The degree of detail that is required decreases with the length of the planning horizon. Planning orders minute-to minute can be practical especially for short-range plans as this ensures that the planned tasks are feasible.

Carefully planning capacities is a fundamental prerequisite for consistently attaining logistic targets. Even a well controlled production cannot compensate for the deficits in the production planning especially when the capacity requirements for the planned orders exceed the available capacities. In particular, correct planning

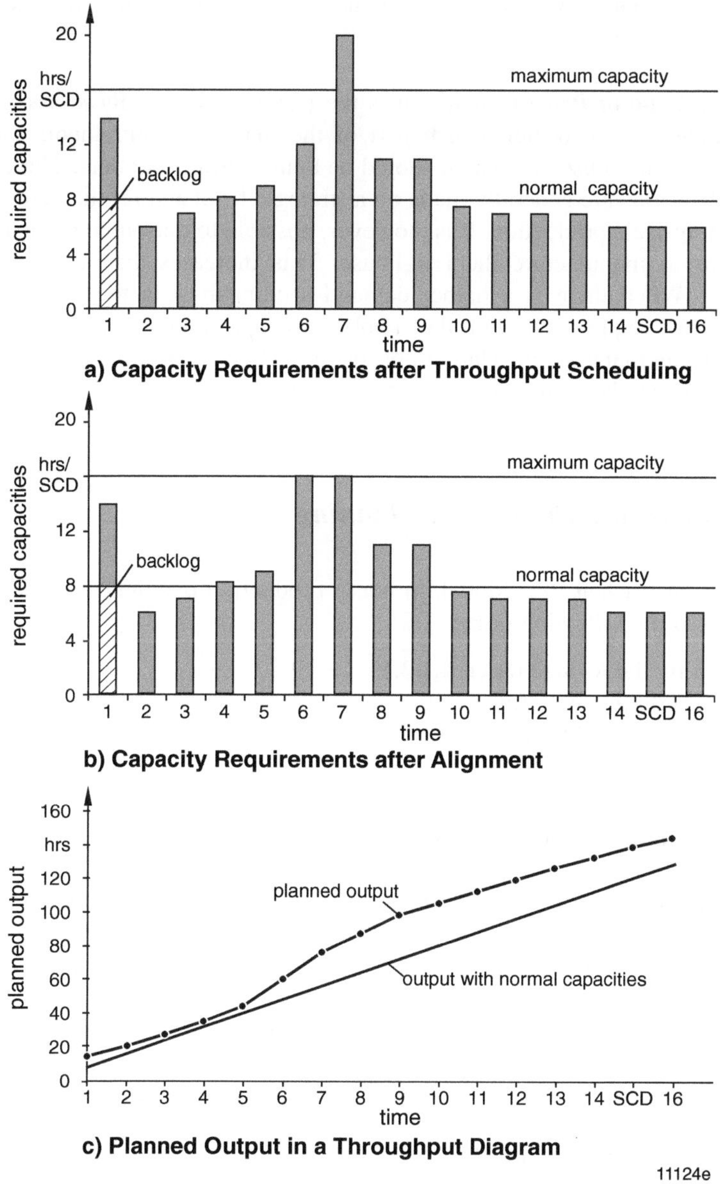


**Fig. 4.3** Determining and aligning the capacity requirements (example)

values for the standard hours and available capacities are critical for realistically planning capacities.

**Assigning Type of Procurement** It is the responsibility of those assigning the procurement type to decide which part of the material requirements should be manufactured in-house and which should be bought from suppliers. These 'make-or-buy' decisions are generally made on a strategic level and set the vertical range of the enterprise's production. It is, however, possible to outsource particular parts as well as to manufacture them in-house. This increases the flexibility of the enterprise. When there is a higher demand a company can then place a larger portion of the requirement with the supplier. Correspondingly, when there is less demand, it can manufacture a larger portion itself. In this special case, the make-or-buy decision can be made on the operative level.

## *4.1.3 In-House Production Planning*

Based on the Aachen PPC Model, in-house production planning is classified here according to the following tasks:

- calculating the lot size (Sect. 4.1.3.1)
- finite scheduling and sequencing (Sect. 4.1.3.2)
- verifying availability (Sect. 4.1.3.3).

### 4.1.3.1 Calculating the Lot Size

Determining the lot size of a specific product is a responsibility of the procurement department as well as the in-house production. For the in-house production it is a task that is pertinent especially for the products that are made-to-stock. However, material requirements can also frequently be economically integrated for various customer orders. Calculating the lot size plays an important role in that it influences a number of entrepreneurial targets. The majority of lot sizing methods aim to minimize the lot size dependent costs, particularly the setup and inventory costs.

*Setup costs:* Frequently, the setup on a machine has to be converted when a variant different from the previous one has to be manufactured. Costs arise as a result of these setup changes, for example, for personnel or materials used for the conversion. Moreover, it reduces the productively available capacities. If the production's output is in turn decreased, then opportunity costs arise which are equal to the product's lost profit contribution that could otherwise have been produced and sold. The setup costs per manufactured part decreases with the lot size.

*Finished goods stock:* The stock of finished goods and with it the tied-up capital, interest costs, storage costs and inventory risk increase with the size of the lot.

*Work-in-process:* Nyhuis was able to prove that the lot size influences the WIP required to fully utilize a workstation in the production [Nyhu-91]. As with the finished goods stock, the tied-up capital, interest costs, and inventory risk increase with the size of the lot.

The basic method for calculating the lot size according to Andler (Europe) or Harris (USA) calculates the lots size at which the sum of the setup costs and storage costs reach a minimum. Other methods such as the Throughput Oriented Lot Sizing Method [Nyhu-91] also take into consideration WIP costs in the production.

Generally speaking though, it is very difficult to consider all of the influences for smaller lot sizes in calculation methods. The reasons for this include:

1. *Cost fixation:* Since most lot sizing methods are fixated on the costs, they neglect important advantages of smaller lot sizes. In particular, the calculation of lot sizes influences the work content of the workstation and thus determines the behavior of the Logistic Operating Curves. With that, the lot size calculation determines the economically attainable throughput time of a production. The shorter the throughput times, the shorter the attainable delivery times (make-to-order productions) or the lower the required safety stock levels are (make-to-stock productions) are. In a borderline case, smaller lots even make it possible to forego a make-to-stock production and to convert to a make-to-order production. Generally, as the throughput times are reduced the ability for a production to react increases and the necessity of rush orders decreases. Furthermore, their fixation on costs falls short: The lot sizing methods influence the tied-up capital and with that the enterprise's return on capital, whereas the maximum return on capital is attained with smaller lots—not with a minimal of costs.
2. *Setup costs are over-valued:* Independent of whether a workstation limits a production or works at the capacity limit, many enterprises assess the setup times at the machine's hourly rate. Generally, in addition to the personnel costs this includes the depreciation of the investment. Since the investment costs are not relevant to the decision when calculating the lot size this causes excessively large lot sizes. If an enterprise is operating at the capacity limit and the setup times influence the production throughput, then the profit contribution the company loses due to the setup times should be considered when evaluating them.
3. *Setup times longer than necessary:* Setup times can often be significantly reduced through technical and organizational measures. Evidence of this has been given both in theoretical observations as well as practical applications [Früh-90, Shin-85*, Take-99].

The lot size calculation determines the structure of the work content or production and with that the behavior of the Logistic Operating Curves. It thus provides the frame within which the manufacturing control can influence the WIP and throughput times with the order release.

The lot size calculation is relevant not only for the in-house production but also for the parts ordered from suppliers (see Chap. 8).

### 4.1.3.2 Finite Scheduling and Sequence Planning

It is the responsibility of finite scheduling to precisely plan the timing of the operations on the resources. With that, the finite scheduling also plans the sequencing in which the orders should be processed. It goes beyond planning the sequencing though because it also sets the exact point in time at which an operation should be processed.

Many critiques generally question the practicality of finite scheduling because the start and completion dates are usually planned to the exact minute and are therefore frequently not met on the shop floor. In particular, when an enterprise has sufficient capacities, it can consistently attain logistic targets without the effort required for finite scheduling.

Nonetheless, the finite scheduling at least ensures that the plan for the orders can theoretically be implemented. This is not ensured without finite scheduling—especially when the due date and capacity planning is poor. Generally, finite scheduling therefore makes particular sense when the capacities are tightly dimensioned as it requires the allowed times and information provided about the capacities to be extremely precise.

In the meantime, there are a number of methods for conducting finite scheduling. A selection of these is briefly outlined below:

- *Interactive finite scheduling with control centers:* (Electronic) control centers visualize the allocation of the operations to the resources, usually in the form of Gantt charts (Fig. 4.4) [Dang-86]. The production planners can re-design

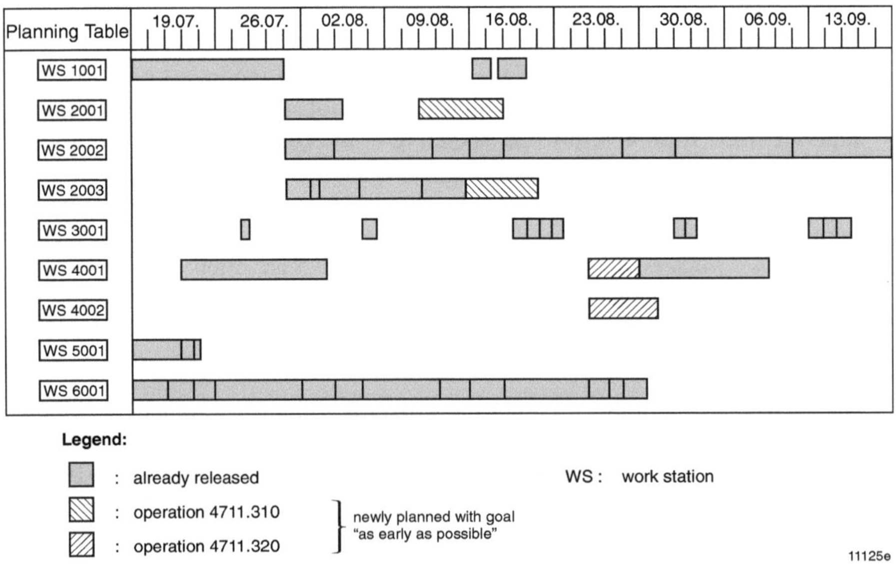


**Fig. 4.4** Employing a control center for resource finite scheduling (Schönsleben)

individual operations so that the finite scheduling is generally conducted interactively between the control center and the planner.
- *Finite scheduling with sequencing:* Sequencing rules determine the row in which the processing of orders is planned on the resources. The sequencing rules allocate orders different priorities, taking into consideration criteria such as the delivery date of an order or its work content. One of the advantages of this method is the minimal effort involved in it. The resulting planned sequence is not inevitably sufficient for attaining the logistic targets.
- *Optimizing algorithms:* A large amount of research is aimed at finding the planned sequence best suited for attaining a defined target. The users of these algorithms formulate a target function and the selected algorithms then determines a planned sequence with which the target function takes on an (approximate) optimal value.

The formulation of the target function also generally influences the quality of the planned sequence and is in no way trivial. Thus, for example, the aim to maximize the planned delivery reliability frequently leads to absurd sequence plans: Since the extent of the delay is not taken into consideration in the target function, the algorithms generate production plans with which many orders are planned within the due date tolerance. However, in order to do so, extremely long delays for a few orders are accepted. Moreover, orders can be planned according to the due date without optimizing algorithms when there are sufficient capacities.

Various methods of operations research have been applied to the problem of sequence planning. Among these are:

- Branch and Bound Techniques [Hill-90*]
- Tabu Search [Pine-99*]
- Simulated Annealing [Kirk-83*]
- Genetic Algorithms [Niss-97].

#### 4.1.3.3 Verifying Availability

It is the responsibility of those verifying availability to confirm that the necessary materials, resources and documents (e.g. NC Program) are on hand. Otherwise the order is not released.

## 4.2 Cross-Sectional Tasks of Production Planning and Control

The cross-sectional tasks of PPC serve to facilitate the inter-departmental integration and optimization of the PPC [Lucz-99]. The Aachen PPC Model contains three cross-sectional tasks:

- Order Coordination (Sect. 4.2.1)
- Storage (Sect. 4.2.2)
- Monitoring PPC (Sect. 4.2.3)

### 4.2.1 Order Coordination

The cross-sectional task referred to as 'order coordination' organizes the order fulfillment across various departments within the enterprise. It plays an especially significant roll for complex orders such as when building plants and often takes over the customer contact. As a result there is a set contact person instead of having to communicate with different employees about the construction, preparations, procurement, manufacturing and assembly. According to the Aachen PPC Model, order coordination consists of the following sub-tasks [Lucz-99]: processing offer, clarifying the order, approximately scheduling order, approximately planning resources and maintaining the order records.

### 4.2.2 Storage

It is the responsibility of storage management, to quickly and cost-efficiently supply stored goods when requested as well as to be able to constantly provide information about current stock levels. Within the context of *controlling the stock movement* all of the storage's physical input and outputs are compiled and compared with the PPC system. In addition, the *inventory control* includes the management of reserved stores that are physically located in the warehouse, but are only allowed to be picked for reserved purposes. The *store location and bin management* allocates the stock to each of the storage sites. It thus allows the targeted access to various stored goods. Where necessary, storage management also conducts the *batch management* and traces back each part of the inventory.

Moreover, due to accounting regulations, all enterprises have to *take inventory* and compare the actual stock levels with those recorded in the books. Typical causes for deviations are false entries or inputs and outputs that have not been booked (theft, breakdowns in the storage management system, etc.). *Storage controlling* is frequently part of PPC monitoring. It indicates, for example, which articles are responsible for a large portion of the stock, which articles are slow-sellers or with which articles the service level is unsatisfactory. Along with Storage Throughput Diagrams and Storage Operating Curves, analyzing the demand behavior (UVW or XYZ Analyses [Gläß-95]) is an important instrument for controlling stores.

### 4.2.3 PPC Monitoring

The task referred to here as 'PPC monitoring' is responsible for measuring the logistic target attainment, uncovering the causes for plan deviations and recommending measures for resolving problems. Both the planned data from the PPC system as well as the feedback data from the production and/or from the enterprise's individual departments form the data base for monitoring the PPC. Information gained from the PPC monitoring can flow directly back into the production planning. For example, a production backlog should be taken into consideration when planning new production orders. Monitoring the PPC is thus an important link between the production planning and the operational reality. The PPC can be monitored based on actual data as well as from plan data.

Plan Data Monitoring, as it is referred to here, usually applies to the production planning. It can be used for example to evaluate the target fulfillment of the production plan or to indicate discrepancies such as a capacity overload. The Plan Data Monitoring is usually neglected by the majority of enterprises. Consequently, production is occasionally called to account for an insufficient target attainment which it has not even caused.

What I refer to here as 'Actual Data Monitoring' is quite common and focuses on the actual target attainment. The aim here is to compile the objectives together with the factors that influence them. Thus for example, not only the schedule reliability should be measured, but also the backlog and the sequencing compliance. Only then is it possible to recognize why targets are not sufficiently attained. In addition, the quality of the data is decisive. In particular, monitoring the PPC on the shop floor is hampered by feedback data, and to some degree plan data, which has poor quality.

Furthermore, the PPC monitoring should be differentiated according to the enterprise's departments. A procurement monitoring primarily serves to logistically evaluate suppliers, whereas, a production monitoring judges the logistic performance of the production, the assembly monitoring the logistic performance of the assembly and so on. The focus of the PPC monitoring is frequently the manufacturing and/or manufacturing control. It can either draw upon orders (order monitoring) or resources (resources or workstation monitoring). Throughput Diagrams, Logistic Operating Curves, and distributions for the logistic objectives are fundamental aids for monitoring the production. Moreover, the production monitoring should not only indicate plan deviations, but also enumerate possible improvements and thus provide practical, attainable targets.

The logistic target attainment measured in the production monitoring is generally influenced by the characteristics of the observed production. The following chapter introduces important properties of a production that play a critical role in controlling it.

## References

[Dang-86] Dangelmaier, W., Aldinger, L.: Kurzfristige Fertigungssteuerung mit Leitständen. Werkstattstechnik **76**(2), 101–104 (1986)

[Früh-90] Frühwald, C.: Analyse und Planung produktionstechnischer Rüstabläufe. VDI Progress Reports, (1990)

[Gläß-95] Gläßner, J.: Modellgestütztes Controlling der beschaffungslogistischen Prozeßkette. VDI Progress Reports, Series 2, No. 337, Düsseldorf (1995)

[Hegi-74] Hegi, O., Wiendahl, H.-P.: Produktplanung eines Maschinenbaukonzerns. Congena-Texte **1/2**, 3–14 (1974)

[Hill-90*] Hillier, F.S., Lieberman, G.J.: Introduction to Operations Research, 5th edn. McGraw-Hill, New York (1990)

[Kirk-83*] Kirkpatrick, S., Gelatt, C.D., Vecchi, M.P.: Optimization by simulated annealing. Science **220**(4598), 671–680 (1983)

[Lucz-99] Luczak, H., Eversheim, W.: Produktionsplanung und -steuerung. Grundlagen, Gestaltung und Konzepte. Springer, Berlin (1999)

[Niss-97] Nissen, V.: Einführung in evolutionäre Algorithmen. Vieweg, Braunschweig (1997)

[Nyhu-91] Nyhuis, P.: Durchlauforientierte Losgrößenbestimmung. VDI Progress Reports, Series 2, No. 225, Düsseldorf (1991)

[Pine-99*] Pinedo, M., Chao, X.: Operations Scheduling with Applications in Manufacturing and Services. Irwin/McGraw-Hill, New York (1999)

[Shin-85*] Shingo, S.: A Revolution in Manufacturing. The SMED System. Productivity Press, Portland (1985)

[Take-99] Takeda, H.: Das synchrone Produktionssystem. Just-in-time für das ganze Unternehmen. Verlag Moderne Industrie, Landsberg (1999)

[Voll-91*] Vollmann, T.E., Berry, W.L., Whybark, D.C.: Manufacturing Planning and Control Systems. Irwin McGraw-Hill, Homewood/Boston (1991)

# Chapter 5
# Key Manufacturing Characteristics

In the previous chapters, the focus has primarily been on differentiating between make-to-order and make-to-stock productions. However, there are a number of additional characteristics that influence the manufacturing control. Two proven classification criteria that we will examine in this next chapter are the manufacturing principle (Sect. 5.1) and the type of manufacturing (Sect. 5.2). Moreover, a close connection to the logistic objectives allows productions to be classified according to the type of part flow (Sect. 5.3), whereas, the number of variants (Sect. 5.4), the complexity of the material flow (Sect. 5.5) as well as the fluctuations in both the customers' demands and the required capacities (Sect. 5.6) impact the applicability of various manufacturing control methods.

## 5.1 Manufacturing Principles

Wiendahl classifies industrial manufacturing principles according to their spatial structure [Wien-97]. He outlines five different manufacturing principles: the function or workshop principle, the flow principle, the cellular principle, the on-site principle and the workbench principle (Fig. 5.1).

*Workshop productions* arrange workstations according to the function principle i.e., similar types of workstations are integrated spatially into a workshop. The parts are usually transported from workshop to workshop in lots. Workpieces that have different processing sequences can thus be produced without a problem. Moreover, the system supports employees in specializing in a specific function. Within a workshop employees can frequently operate a number of machines; to some degree even simultaneously. The main disadvantage of workshop productions is their long throughput times [Wien-97].

*Flow productions* are the opposite of workshop productions. Workstations are arranged according to the material flow of variants being manufactured. In contrast to workshop productions the parts are transported to the next workstation directly after their completion. Very short throughput times are thus targeted. The cycle

H. Lödding, *Handbook of Manufacturing Control*, 
DOI 10.1007/978-3-642-24458-2_5, 

| Organizational Criterion | Manufacturing Principle | Spatial Structure | Example |
|---|---|---|---|
| people | workbench principle | 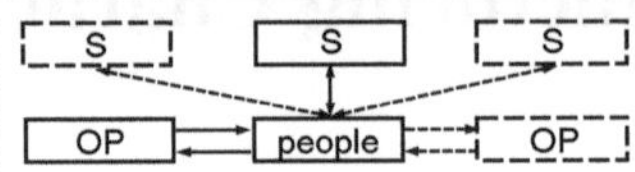 | manual work centers<br>tool shop |
| product | on-site principle | 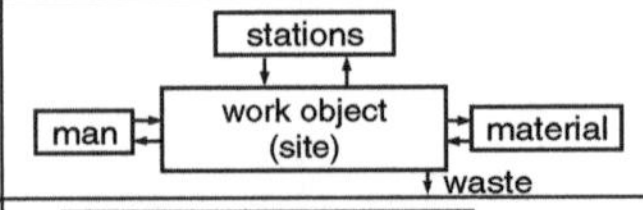 | heavy machine construction<br>shipyard |
| technology | function principle or job shop principle | 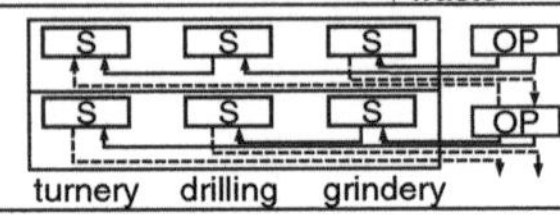 | turnery<br>drilling<br>grindery<br>welding |
| material flow of part family | cellular principle<br>group principle | 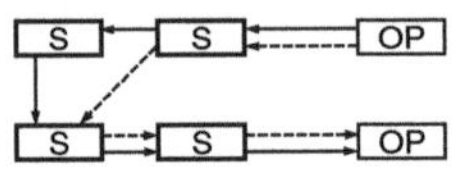 | manufacturing cell<br>assembly cell<br>production segment |
| material flow of defined variants | flow principle | 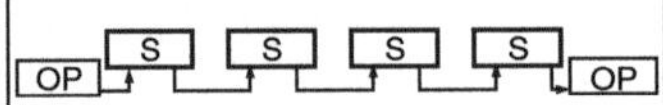 | production line<br>assembly line |

11126e

OP : operation S : station

**Fig. 5.1** Layout criteria for the spatial structure of industrial manufacturing principles (Kettner, Schmidt, Greim according to Janisch)

times on the individual workstations are generally coordinated with one another i.e., a standard production cycle is fixed. Usually only a tightly limited number of variants can be produced in flow productions. They are frequently highly automated and capacities tend to be inflexible. Due to the generally small WIP buffers, disruptions tend to quickly impact downstream workstations (workstations thus run idle) and upstream workstations (workstations are blocked).

*Cellular* or *group productions* are between workshop and flow productions. Similar to flow productions, the workstations are arranged according to the material flow of the variants produced, usually in the form of a U. However, less demands are placed on the cycle of the individual workstations. They are less automated and deviations from the standard material flow are to some extent permissible. With cellular productions, employees take on a highly significant role as they can often operate and setup all of the workstations in a manufacturing cell (multiple machine assignment). Furthermore, the capacities of the manufacturing cell can be controlled by the number of operators and the operating times. Generally, the parts are transported to the next workstation, directly following their processing (one-piece flow principle, see Sect. 5.3.2). The throughput times are thus very short. Similar to flow productions the demands on the process reliability are high. However, due to the minimal degree of automation and the concentration on personnel it is simpler to attain.

*On-site productions* hold a special position: with this manufacturing principle, workstations (machines) are usually brought to very large workpieces. The production lot-size is however generally very small. Possible applications include building large machinery or ships. The production planning and control in on-site productions tends to be structured similar to projects.

The *workbench principle* centers the production on the employees who process the part on a workbench and access the necessary machines as required. The operator then usually executes a number of operations one after the other or very complex individual operations. Frequently, the individual parts are completely processed so that the throughput times are minimal. The workbench principle is applied in the industry primarily for assembly tasks. It lends itself to situations where the required machines are small and reasonably priced, and, when dividing up the operations is inefficient.

## 5.2 Types of Production

Schomburg classifies productions according to the average size of the production run (lot size) and how frequently the product runs are repeated. In doing so he distinguishes four types of productions: one-time productions, single or small lot-wise productions, serial productions and mass productions [Scho-80] (Fig. 5.2).

A *one-time production* is characterized by a small run i.e., only one small lot is produced and the product is not manufactured again (repetitions = 0). Moreover, one-time productions tend to be related to the workshop or on-site principle. Cellular or flow productions can only be considered when the product can clearly be allocated to one product family.

A *single* or *small series production* is characterized by a small run and a low number of repetitions. Similar to one-time productions, single or small series productions seem to be predominantly organized according to the workshop principle. Make-to-stock productions, however, are an exception; generally items are produced according to customer orders.

*Serial productions* attain larger runs and are repeated often. With serial productions, the proportion organized according to the cellular or group principle

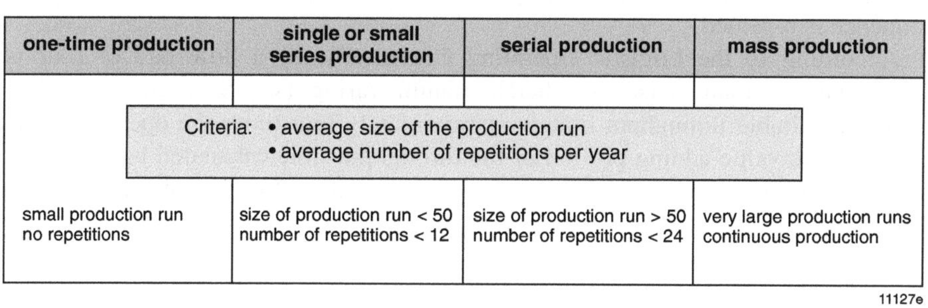


**Fig. 5.2** Differentiation of the various production types (Schomburg 1980, cited according to [Lucz-99])

increases while the proportion of those related to the workshop principle decreases. Furthermore, combined productions in which a part of the product is made-to-stock and another part made-to-order are typical for serial productions.

*Mass productions* attain very large runs. Usually the production is organized according to the flow principle and make-to-stock productions are dominant. The applicability of mass productions decreases as the number of variants increases.

Concepts such as mass customization or flexible final production levels aim to combine the cost benefits of mass production with the flexibility of workshop productions [Pill-98*, Pine-93, Röhr-02, Wien-04].

## 5.3 Part Flow

The type of part flow has an outstanding influence on the WIP levels and throughput times of a production. The standard case depicted in the Logistic Operating Curves is transportation by lots (Sect. 5.3.1). However, the ideal logistic case is the direct transportation of a part to the following workstation (one-piece flow, Sect. 5.3.2). Between lot-wise transportation and one-piece flow is overlapping production (Sect. 5.3.3). From the perspective of logistics, batch production is the least efficient type of part flow (Sect. 5.3.4) because it leads to long throughput times and high WIP levels.

### *5.3.1 Lot-Wise Transportation*

With lot-wise transportation a complete lot is processed on a workstation and only transported to the next workstation once the last part of the lot is processed. Before being processed a part first waits for more urgent orders to be completed, then for the workstation to be setup and then for the earlier pieces in the same lot to be processed. Once it is processed the part has to wait for the remaining pieces in its lot to be processed. The lot is then transported to the next workstation where the same sequence is repeated.

According to the Logistic Operating Curves Theory a flow rate of four is appropriate in many cases for batch manufacturing (see Sect. 3.2.2.1). That means, a suitable throughput time is approximately four times the operation time of a lot. The value-adding portion of the throughput time calculated based on the part is much lower and dependent on the lot size. With a flow rate of four and a lot size of one-hundred, the flow rate for the part is four-hundred. The part is thus only actually processed a quarter percent of the throughput time.

A key place to start shortening the throughput times is thus either reducing the lot size or getting rid of the restrictions of the lot-wise transports. As mentioned above, in the ideal case each part is transported to the next workstation directly after it is processed (so-called 'one-piece flow').

### 5.3.2 One-Piece Flow

With one-piece flow a part is transported from one workstation to the next workstation directly after it has been processed (Fig. 5.3).

Before a one-piece flow line, an order usually waits for the completion of more urgent lots, after which the workstation is setup for the processing of the lot. Within a one-piece flow line each of the parts are transported to the next workstation directly after they have been processed. They only have to wait for the processing of other parts from the same order—and these only if the processing times for the individual units are different thus creating a build-up of WIP. After the one-piece flow line the manufacturing is again completed in lots.

Introducing the one-piece flow principle is therefore not related to abolishing batches. It only eliminates the correlation between the manufacturing and transport lot. Nor does a one-piece flow require a specific manufacturing principle. Whether the production is organized as a manufacturing cell or according to the flow principle only influences the logistic modeling of the production negligibly. It is therefore conceivable to introduce one-piece flow sections within a workshop production. This seems obvious especially in areas where orders pass through specific workstations in the same sequence. Nonetheless, this generally requires changes in the layout of the production.

Similar to lot-wise transportation a one-piece flow can be modeled with the aid of Logistic Operating Curves [Lödd-01b]. In order to do so the entire one-piece flow line is understood as a single workstation. Every order loads this workstation with a work content corresponding to the time from the start of the setup up until the completion of the order's last part. The length of the work content is fundamentally influenced by three factors: the time for setting up the workstation, the cycle-time

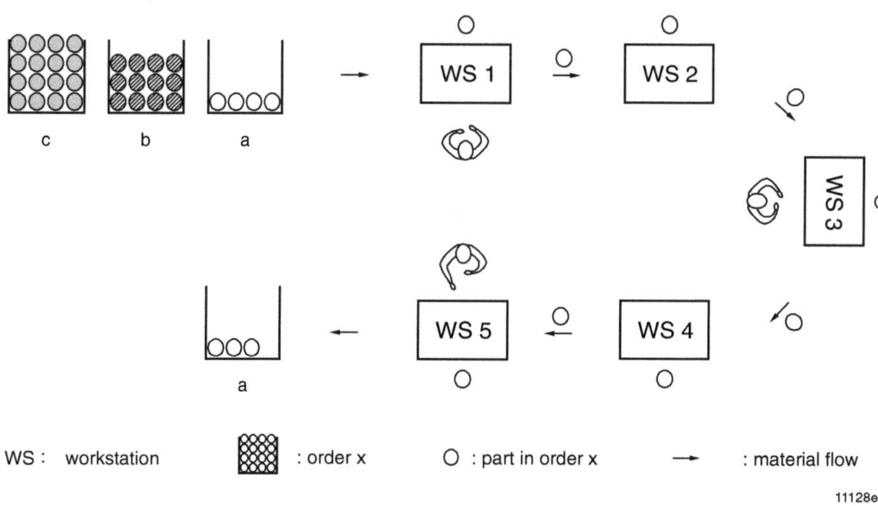


**Fig. 5.3** Manufacturing cell with one-piece part flow

and the lot size. The cycle-time corresponds to the time that passes between the completions of two parts. It is calculated as the maximum processing time of a unit on the workstation.

Similar to lot-wise transportation an ideal production process can be described based on this definition of the work content. In it, the one-piece flow line is continually fully utilized. At the same time however, every order can be processed directly after its arrival. The equations for determining the Logistic Operating Curves can be directly transferred due to the identity of the ideal production process. Nonetheless, when the initial workstations on a one-piece flow production have already been setup for the next order and begun processing it while the end workstations are still processing the previous order the equations for the ideal minimum WIP and the maximum possible output rate have to be modified (see [Lödd-01b]). This increases the maximum attainable output rate.

From the perspective of logistics a one-piece flow system thus replaces a series of workstations with a single higher level workstation. Figure 5.4 compares the material flow, Logistic Operating Curves and the attainable throughput times for batch transport and one-piece flow systems.

The exact behavior of the Logistic Operating Curves for the one-piece flow line results from the allowed times on the individual workstations and the lot size of the orders. Roughly speaking, the throughput time of a one-piece flow line is slightly longer than the throughput time of the longest operation of an equivalent production employing lot-wise transports. The throughput time potential thus corresponds approximately to the sum of the operation throughput times of the remaining workstations.

As is typical for the one-piece flow principle, employees are assigned to operate multiple machines. It is frequently more important to highly utilize employees than

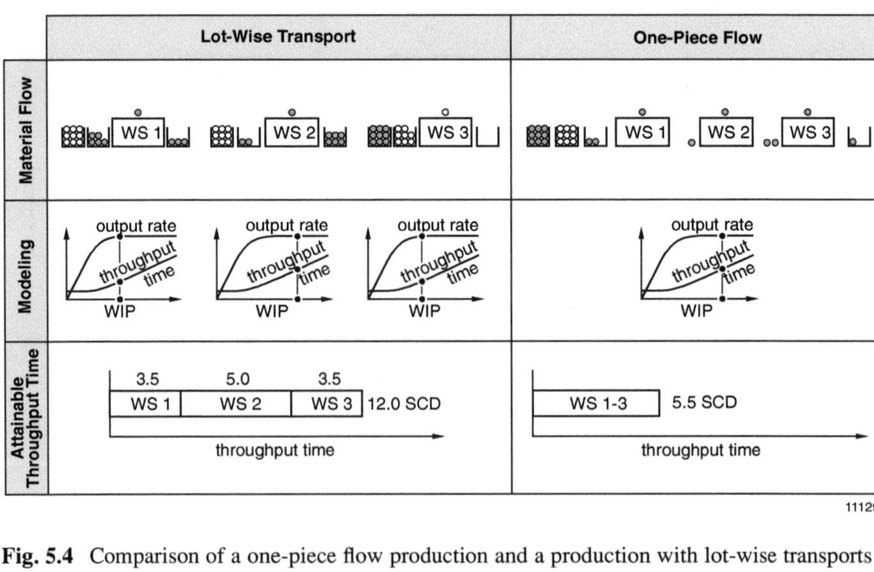


**Fig. 5.4** Comparison of a one-piece flow production and a production with lot-wise transports

machinery. In comparison to lot-wise transports, the productivity of the employees in a one-piece flow line is exceptionally high in many cases. However, multiple machine assignment requires employees to be cross-trained.

In contrast to lot-wise productions, productions in one-piece flow lines are almost inevitably connected to blockages and starvations: If the workstations' have different unit processing times, the slowest workstation provides the cycle time for the entire one-piece flow line. Accordingly, the capacity of the quicker workstations can not be fully utilized. Similarly, losses of utilization can arise due to varying setup times or as a result of sequential setups at the workstations. Moreover, disruptions on individual workstations quickly impact the remaining workstations. However, even with lot-wise transports it is usually not possible to balance the load on all of the workstations. This should also be kept in mind when evaluating the loss of utilization.

### *5.3.3 Overlapped Manufacturing*

In between lot-wise transportation and one-piece flow productions is overlapped production. In overlapped manufacturing an already processed portion of a lot is transported to the following workstation and processed there (Fig. 5.5).

There are two fundamentally different possibilities for implementing overlapped manufacturing. On the one hand, it can serve to specifically accelerate individual orders. This method is quite commonly employed in the industry in order to complete rush orders as quickly as possible. Quite often it is connected with postponing and delaying other orders.

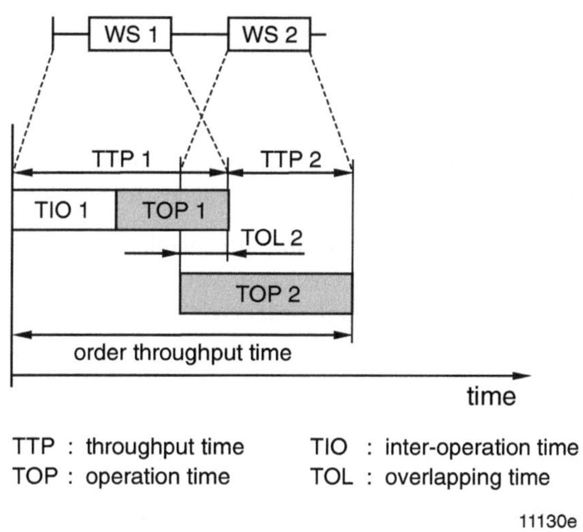


**Fig. 5.5** Overlapped manufacturing of orders

On the other hand, overlapped manufacturing can be implemented for reducing the attainable throughput times and WIP levels. In this case, the following workstation only tries to manufacture an 'overlap order' when there is no other order in the workstation's WIP. The overlapped manufacturing in this case serves to avoid a loss of utilization on the workstation. The additional effort required for transportation is limited to situations where there is a threat of breaks in the material flow on a workstation. Overlapped productions thus pose fewer demands on the layout and/or transport system of a production than one-piece flow does which generally requires the workstations to be closely located to one another.

From a logistic perspective, the overlapped production changes the behavior of the Logistic Operating Curves for the following workstation (see [Lödd-01, Nyhu-09*]). In particular the ideal minimum WIP of the workstation is reduced. A specific utilization rate is therefore already attained at a lower WIP level and with shorter throughput times than a non-overlapped production in a comparative state. This effect is especially pronounced when the individual processing times are similar, the transportation times short and the minimum amount transported small [Lödd-01].

In comparison to lot-wise transports, overlapped production poses high demands on the in-house transport. Should these demands be met, the potential of an overlapped production can, however, be exploited quite easily.

### *5.3.4 Batch Production*

In a batch production a number of lots are processed together. Generally the lots have to belong to the same type of batch i.e., have to be able to be processed under the same conditions. Oven processes are a common example of batch-wise productions. Usually for these, orders that are processed for the same duration with the same oven temperature are collected until there are enough to completely load the oven.

Batch productions cause high WIP levels and long throughput times. From a logistics perspective, they are thus not advantageous at all. Enterprises that use batch production find that a large portion of an order's throughput time is caused by these processes. There are a four fundamental reasons for this:

1. *Collection time:* Generally after arriving at a batch station an order cannot be immediately processed because there is not yet a complete batch. This means that the order has to wait until enough orders of the same batch type have arrived at the workstation. This time is referred to here as 'collection time'. The more orders there are in a batch and the more infrequent the batch type of the order is, the longer the collection time is.
2. *Preparation time:* After a batch is completed, it frequently has to be prepared for the actual processing. For oven processes for example, it is frequently necessary to mount the parts on an oven belt. Although this preparation time does not usually tie-up capacities on the batch workstation, it does extend the order throughput time and increase the WIP on the workstation.

3. *Parallel processing:* In a normal (single) workstation only one order is processed at a time, correspondingly, only one batch is processed on a batch workstation at a time. Since it contains more orders though, a generally higher WIP is required to fully utilize the batch workstation.
4. *Load peaks on downstream workstations:* Batch productions negatively impact downstream workstations. The more constant the load on a workstation is, the less WIP is required to ensure its utilization. With lot-wise productions however, a number of orders are completed at the same time. This leads to a very irregular load on the downstream workstations. Accordingly, the throughput times are longer than average not only for the batch processes themselves, but also, for the downstream workstations.

Where possible, batch processes should thus be avoided. This can frequently be attained by changing the technology (e.g., replacing heat treatments in an oven with inductive surface hardening). If an enterprise is unable to refrain from a batch process, it should be designed and controlled as optimally as possible:

1. *Coordinated input of similar batch types:* In order to reduce the collection time, the release of orders of the same batch type should be coordinated by the manufacturing control. Their arrival times at the batch workstation should be as close to one another as possible. This applies especially to batches that are run less frequently.
2. *Reducing preparation time:* Similar to the setup times of normal workstations, the preparation time for batch workstations can be reduced through technical and organizational measures.
3. *Reducing the batch size:* Reducing the batch size is a very effective means of minimizing the disadvantage of lot-wise productions. In view of the logistic objectives for example, two small ovens are preferred over one large oven. The advantages here are shorter collection times, frequently shorter preparation times and a more constant load on downstream workstations. If workstations access a common queue, the required WIP buffer is also reduced.

It should also be noted here that batch processes can also be modeled with the aid of Logistic Operating Curves [Schn-02b, Stüh-99].

## 5.4 Number of Variants

With regards to manufacturing control, another outstanding characteristic of a production is the number of variants. The more variants have to be controlled, the more complex the manufacturing control tasks are.

If only a few variants are manufactured in a production, then generally simple and self-regulating manufacturing control methods can be successfully implemented (see [Schö-07*]). Implementing methods with buffers for specific variants allows a part of the production requirement planning and in-house planning to be shifted to manufacturing control.

With an increasing number of variants, implementing more complex PPC software can usually no longer be circumvented. Furthermore, the number of variants influences which production principle can be implemented and automating the manufacturing process becomes more difficult. Accordingly, the range of application for a synchronized flow production is reduced. The diversity of the variants has increased considerably in the last decades. A customized product can be considered the endpoint of this development as is propagated for example by the vision of a customized mass production [Pill-98*, Pine-93].

For the manufacturing control, the extent to which the diversity of the variants on the production level impacts the production is critical. The goal should be to keep the number of variants as small as possible and to form the variants as late in the production as possible. Modular systems, platform concepts, same part strategies or shifting the variant forming production process further along in the assembly according to the concept of the flexible final production stage, all target this goal [Röhr-02, Wien-04].

In the ideal case, diversity of the product variants does not impact the production at all. This allows components to be pre-manufactured independent of the customers and reduces the delivery time to the customer. In the operational practice this ideal case is rarely attained, therefore, the ability to finely control the logistics of variant rich productions becomes all the more significant.

## 5.5 Complexity of the Material Flow

Closely related to the number of variants is the complexity of the material flow which is influenced by two factors: The number of workstations that possibly precede or follow one another and the number of backflows in the material flow. Backflows arise when the workstations in a production can no longer be arranged in a sequence that allows all orders to flow through in the same direction. The more backflows there are, the more complex the material flow is.

Accordingly, a production line has the least complex material flow (Fig. 5.6a). Every workstation has at the most one predecessor and one successor, moreover, there are no backflows. The complexity of the material flow increases with the number of possible preceding or subsequent workstations (Fig. 5.6b) as well as the number of backflows (Fig. 5.6c). Büchel and Sainis developed key variables with which the complexity of the material flow could be measured and evaluated [Büch-68, Sain-75].

The complexity of the material flow directly impacts the suitability of various manufacturing control methods. A simple material flow is also easier to control than a complex material flow. Whereas backflows can generally call into question the applicability of decentralized production methods in particular [Lödd-01], a number of methods exploit the complexity of the material flow in order to target a more constant utilization by changing the sequence when releasing orders. This type of order release loses its effectiveness with very simple material flows.

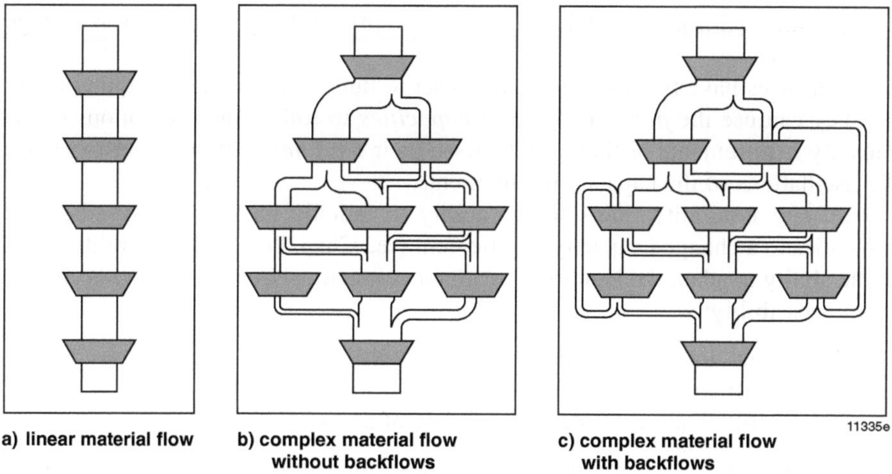


**Fig. 5.6** Forms of material flow complexity

Furthermore, the complexity of the material flow plays a role when selecting the type of part flow. The one-piece flow principle can be realized with a simple material flow as well as to some extent with a large number of variants, however, due to the layout alone, this is often not possible with complex material flows. Thus with a large number of variants, the goal should be to keep the complexity of the material flow at a minimal and to strive for a uniform throughput of the variants through the production.

## 5.6 Fluctuations in Customer Demands and Capacity Requirements

Generally, the customer demand for an enterprise's product is irregular thus the demand rate fluctuates. In a make-to-stock production the fluctuations in the demand have to be buffered by the safety stock to ensure a constantly high service level.

In the production, demand rate fluctuations cause the capacity requirements to oscillate independent of whether the productions are make-to-order or make-to-stock. For the most part these oscillations are generally less than the demand fluctuations since the increased demand for one product often balances the decreased demand for another product. The resulting demand rate fluctuations are, however, usually still so considerable that an enterprise has to react to them. This also applies in particular, when the bottleneck system of the production changes over time due to the demand fluctuations. This is the case, for example, when the increased customer demand strains a workstation that normally has sufficient capacities with an average product mix.

Enterprises have two possibilities to react to fluctuating capacity requirements: They can use the *flexibility of their capacities* to follow the oscillations of the capacity requirements or they can exercise their *load flexibility* in order to reduce the oscillations of the capacity requirements.

Capacity flexibility refers to the ability to quickly and efficiently adapt the capacities to a changed capacity requirement (see Chap. 26). Load flexibility refers to the ability to adapt the load to the available capacities. Load flexibility is created by the possibility:

- of declining customer orders when the capacities are fully booked.
- of placing orders or operations externally.
- of being able to shift the start or end-date of an order.

The start date of an order can be shifted—within limits—when the throughput time of an order is smaller than the delivery time. The end date of an order is generally less flexible.

It is critical for the manufacturing control that on the whole, the load and capacity flexibility is sufficient for balancing out the fluctuations in the demand rate. Otherwise, it is theoretically impossible to ensure that logistic targets are consistently attained. Incidentally, this is often enough the case on the shop floor: If through a favorable economic situation or through measures that promote sales, the sales department acquires more orders than the production can complete, then a high due date reliability cannot be targeted even with good manufacturing control. In contrast if the load and capacity flexibility is sufficient to compensate for oscillations in the capacity requirements, at least a high due date reliability can be attained. Insufficiently attaining targets can then frequently be traced back to errors in the manufacturing control.

Incidentally, strong demand fluctuations are not inevitable—in many cases, enterprises can attain a steady demand by cooperating with customers. This co-operation is one of the key focuses of supply chain management which is discussed in the following chapter.

## References

[Büch-68] Büchel, A.: Aufbau eines Simulationsmodells der Werkstättenfertigung auf der Basis eines Markov-Prozesses. Juris Druck + Verlag, Zurich (1968)

[Lödd-01] Lödding, H.: Dezentrale Bestandsorientierte Fertigungsregelung. VDI Progress Reports, Series 2, No. 587, Düsseldorf (2001)

[Lödd-01b] Lödding, H., Hernández, R., Wiendahl, H.-P.: Dimensionierung und Bewertung von Fertigungsinseln mit logistischen Kennlinien. Io Manage. **70**(1–2), 25–32 (2001)

[Lucz-99] Luczak, H., Eversheim, W.: Produktionsplanung und -steuerung. Grundlagen, Gestaltung und Konzepte. Springer, Berlin (1999)

[Nyhu-09*] Nyhuis, P., Wiendahl, H.-P.: Fundamentals of Production Logistics. Theory, Tools and Applications. Springer, Berlin (2009)

[Pill-98*] Piller, F.T.: Kundenindividuelle Massenproduktion. Die Wettbewerbsstrategie der Zukunft. Carl Hanser Verlag, Munich/Vienna (1998)

[Pine-93] Pine, B.J.: Mass Customization. Harvard Business School Press, Boston (1993)

[Röhr-02] Röhrig, M.: Variantenbeherrschung mit hochflexiblen Produktionsendstufen. Dissertation, Universität Hannover 2002, veröffentlicht in: Fortschritt-Berichte VDI, Reihe 2, Nr. 622, Düsseldorf (2002)

[Sain-75] Sainis, P.: Ermittlung von Durchlaufzeiten in der Werkstattfertigung aus Daten des Fertigungsprogrammes mit Hilfe der Warteschlangentheorie. Dissertation, University of Hanover, Hannover, (1975)

[Schn-02b] Schneider, M.: Manufacturing system operating curves. In: Nyhuis, P., Wiendahl, H.-P. (eds.) Fundamentals of Production Logistics, pp. 112–117. Springer, Berlin (2002)

[Scho-80] Schomburg, E.: Entwicklung eines betriebstypologischen Instrumentariums zur systematischen Ermittlung der Anforderungen an EDV-gestützte Produktionsplanungs- und -steuerungssysteme im Maschinenbau, Dissertation, RWTH Aachen (1980)

[Schö-07*] Schönsleben, P.: Integral Logistics Management. Operations and Supply Chain Management in Comprehensive Value-Added Networks, 3rd edn. Auerbach, Boca Raton (2007)

[Stüh-99] Stühler, L., Schmidt, M., Wiendahl, H.-H.: Produzieren im turbulenten Umfeld der Halbleiterindustrie Teil 2: Einsatz der Betriebskennlinie im turbulenten Umfeld. PPS-Management **4**(3), 14–20 (1999)

[Wien-97] Wiendahl, H.-P.: Betriebsorganisation für Ingenieure. Carl Hanser Verlag, Munich/Vienna (1997)

[Wien-04] Wiendahl, H.-P., Gerst, D., Keunecke, L. (eds.): Variantenbeherrschung in der Montage. Konzept und Praxis der flexiblen Produktionsendstufe. Springer, Berlin (2004)

# Chapter 6
# Supply Chain Management (SCM)

Production enterprises are integrated into logistic networks. They obtain raw materials and components from suppliers and deliver products to customers who either further process these, sell them unchanged, or, use and/or consume them. The customer utility and the cost of the end product are dependent on the performance of the entire logistic network. Frictional losses frequently arise at the interfaces between companies in a logistic network. Accordingly, it is here that costs can be decreased and conditions for a strong logistic performance (from the end customer's perspective) can be created.

Few management concepts have received as much attention since the 1990s as supply chain management. Supply chain management (SCM) is dedicated to the layout and operation of logistic networks. Section 6.1 introduces the concept of SCM and explains its objectives and tasks. One of the fundamental reasons for a coordinated supply chain management is the so-called "bullwhip effect" which describes how the demand fluctuations increase along the supply chain. Causes and effects of the 'bullwhip' as well as suitable counter-measures are discussed in Sect. 6.2.

A number of methods which generally developed independent of one another are summarized under the term supply chain management. Section 6.3 briefly describes a selection of these methods.

## 6.1 Definition

Figure 6.1 depicts the simplified SCOR model for a simple supply chain (also known as demand or value adding chains) from the supplier of the supplier up to the customer of the customer. The process of order fulfillment is described by three basic processes: deliver, source and make. These in turn are coordinated by a higher level planning process. Possible storage and distribution centers as well as carriers are not explicitly represented in the figure.

The actors in the supply chain procure raw materials, transform them into interim or end products and distribute these to the end customer [Lee-92*, Zäpf-03].

H. Lödding, *Handbook of Manufacturing Control*,
DOI 10.1007/978-3-642-24458-2_6, 

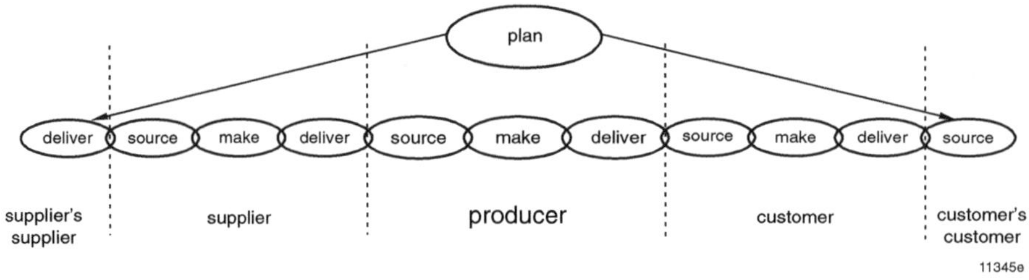


**Fig. 6.1** Modeling a supply chain (supply chain council)

The frequently used and highly illustrative term 'supply chain' suggests linear relations between the companies involved. However, the operational practice is characterized by a network of enterprises. On the one hand, companies cooperate with a variety of customers and suppliers. One the other hand, they frequently produce a variety of products and are thus part of different supply chains. Accordingly, the concept of logistic networks (or supply networks) is also becoming increasingly common [Hieb-02*].

Numerous definitions for supply chain management can be found in publications (for an overview see [Bech-97*]). Simchi-Levi, Kaminsky and Simchi-Levi define supply chain management as follows [Simc-00*].

"Supply chain management is a set of approaches used to efficiently integrate suppliers, manufacturers, warehouses and stores so that merchandise is produced and distributed at the right quantities, to the right locations, and at the right time in order to minimize system-wide costs while satisfying service level requirements."

Supply chain management, in contrast to production planning and control, considers logistic networks or supply chains, not individual enterprises. Whereas, the focus is on the cooperation between the partners within the logistic network, the goal is to orient all of the partners on the greatest possible overall utility and to prevent or minimize the frictional loss. In the ideal case, a competitive advantage results from this thanks to the reduced costs and/or increased customer utility (e.g., shorter delivery times). The supply chain management profits from the advances in the information and communication technology. These reduce the costs for the production, preparation and distribution of information.

One of the essential objectives of supply chain management is to better control the flow of goods and information in a supply chain. Should this objective not be met, there is a threat that the demand information will be distorted by the so-called bullwhip effect.

## 6.2 Bullwhip Effect

The following section explains three factors:

- what the bullwhip effect is and what consequences it has for an enterprise in a logistic network (Sect. 6.2.1).
- what the causes of a bullwhip effect are (Sect. 6.2.2).
- how the bullwhip can be prevented or at least reduced (Sect. 6.2.3).

## 6.2.1 Clarification and Consequences

Forrester discovered the bullwhip effect already at the end of the 1950s when he was investigating industrial demand fluctuations [Forr-61*]. Since then, the effect is also known as the 'Forrester effect'. He described the phenomena whereby fluctuations in demands increase within the supply chain. The farther the partner is from the end customer, the greater this increase is. In a supply chain consisting of a customer, an end producer, a supplier and a sub-supplier, the end producer experiences the least demand fluctuations and the sub-supplier the most (Fig. 6.2).

The bullwhip effect lowers the logistic performance of a network and increases the logistic costs, thus damaging the competitiveness of a network. These negative effects, apply to three aspects in particular:

- *Dimensioning and/or utilizing capacities:* The erratic demands cause fluctuations in the capacity requirements and thus pose a problem for enterprises: If a company dimensions capacities according to the mean demand, then it will regularly cause supply bottlenecks during periods in which the demand is higher. If it adjusts the capacities to the maximum demand, then it risks high losses of utilization. The bullwhip effect thus promotes the development of overcapacities and negatively impacts the enterprise's delivery reliability.
- *WIP fluctuations:* The different amplitude of the fluctuations in the capacity requirements in a supply chain lead to WIP fluctuations in the individual stages of the supply chain. If one of the partners in the supply chain delivers more than the next stage in the chain forwards on, a stockpile develops. In the reverse case, the store is reduced. A larger store involves more tied-up capital, whereas a smaller store endangers the delivery reliability.

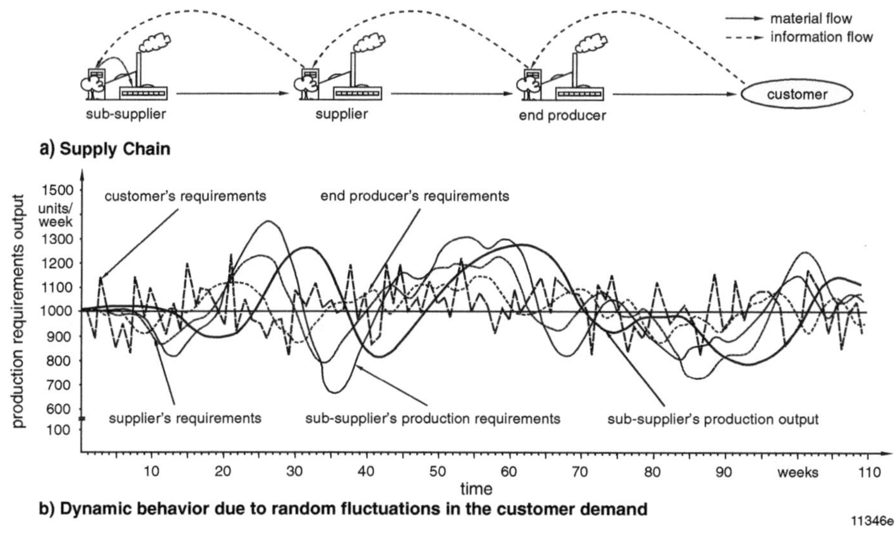


**Fig. 6.2** Visualization of the bullwhip effect (Forrester, cited according to Breithaupt)

- *WIP increase:* The level of safety-stock required to ensure a sufficient service level increases with the demand fluctuation (for a more extensive discussion see Sect. 3.1.3). Since the bullwhip effect increases the demand fluctuations, it increases the necessary safety-stock.

Similar to production planning and control, one of the essential aims of supply chain management should therefore be to reduce or completely prevent the bull-whip effect. In order to do so, it is necessary to analyze the causes of the increased demand fluctuations.

### 6.2.2 Causes

There are at least seven causes that create the bullwhip effect and one which limits it:

1. constant throughput times.
2. decentralized sales planning [Lee-97*, Lee-97b*].
3. history based forecasting methods [Lee-97*, Lee-97b*].
4. lot sizing [Lee-97*, Lee-97b*].
5. price fluctuations [Lee-97*, Lee-97b*].
6. ordering excessively when there are supply bottlenecks [Lee-97*, Lee-97b*].
7. turn-around-times for information.
8. capacity restrictions.

***1. Constant throughput times increase demand fluctuations because they cause WIP fluctuations within the different stages of a supply chain.***

Constant throughput times are one of the fundamental causes of the bullwhip effect. They increase demand fluctuations because they cause WIP fluctuations. According to Little's Law [Litt-61*, Nyhu-09*]:

$$WIP_m = TTP_m \cdot ROUT_m \tag{6.1}$$

where

| | |
|---|---|
| $WIP_m$ | mean WIP [units], |
| $TTP_m$ | mean throughput time [SCD], |
| $ROUT_m$ | mean output rate [units/SCD]. |

If an enterprise plans with constant throughput times, the planned WIP will increase when the demand grows and decrease when the demand shrinks. This change in the planned WIP directly impacts the (cumulated) planned output of the preceding stage in the supply chain. The relation between the cumulated output of the subsequent stage in the supply chain, its WIP and the cumulated output of the preceding stage can be described as:

$$OUT_{plan,i} = OUT_{plan,i+1} + WIP_{plan,i+1} \tag{6.2}$$

where

| | |
|---|---|
| $OUT_{plan,i}$ | cumulated planned output from stage i of the supply chain (preceding stage) [units], |
| $OUT_{plan,i+1}$ | cumulated planned output from stage i +1 of the supply chain (subsequent stage) [units], |
| $WIP_{plan,i+1}$ | planned WIP from stage i + 1 of the supply chain (subsequent stage) [units]. |

The preceding supply stage therefore has to increase their output rate above the customer's increased demand in order to allow the WIP to increase on the subsequent supply stage. Based on Little's Law, the longer the throughput times are, the higher the WIP's increase (see the following discussion on decentralized sales planning).

Figure 6.3 depicts an overview of the correlations based on a supply chain with one producer, one retailer and one customer (Fig. 6.3a). The trend of the demand is shown in the second part of the figure (Fig. 6.3b) while the diagram of the Cumulative Production Figures is shown in the third (Fig. 6.3c).

The cumulative production figure of a variant corresponds to the cumulated output of the variant at a specific measuring point. In this case the producer's cumulative production figure corresponds to the cumulated output of the producer (and this the cumulated input of the retailer). Accordingly the cumulative production figure of the retailer is defined as the cumulated output of the retailer and marks the transfer of the product to the end customer. In order to exclude other influences, the retailer is constantly informed about the exact current demand and conveys his orders directly on to the producer. Moreover, the lot sizing is neglected.

The retailer's strategy is to consistently maintain a store of goods equivalent to that required for one period (he thus strives for a constant throughput time of one period). At time t1 the demand increases from four to eight units per period. The retailer thus increases the store from four to eight units, in order to in turn ensure a stock range of one period.

In this case, at time t1 the producer perceives a four unit jump in demand. This one time jump in the demand is necessary in order for the retailer to double his inventory and also has to be managed from the producer side in addition to the double demand. A reverse figure would result when the demand decreased. The retailer would then decrease their store and thus not place any orders with the producer during a transition time.

Generally, constant throughput times lead to a procyclical increase in demand fluctuations. Nevertheless, constant planned throughput times and fluctuating planned WIP levels are however not uncharacteristic for all three of the basic logistic reference processes defined by Kuhn i.e., production/testing, storage and transportation [Kuhn-95].

*Production/testing:* Most PPC systems use constant planned throughput times not only for both components and parts but also for workstations.

*Storage:* Applying a period batch control leads to (approximately) constant storage throughput times. Moreover, dimensioning the safety-stock plays a critical

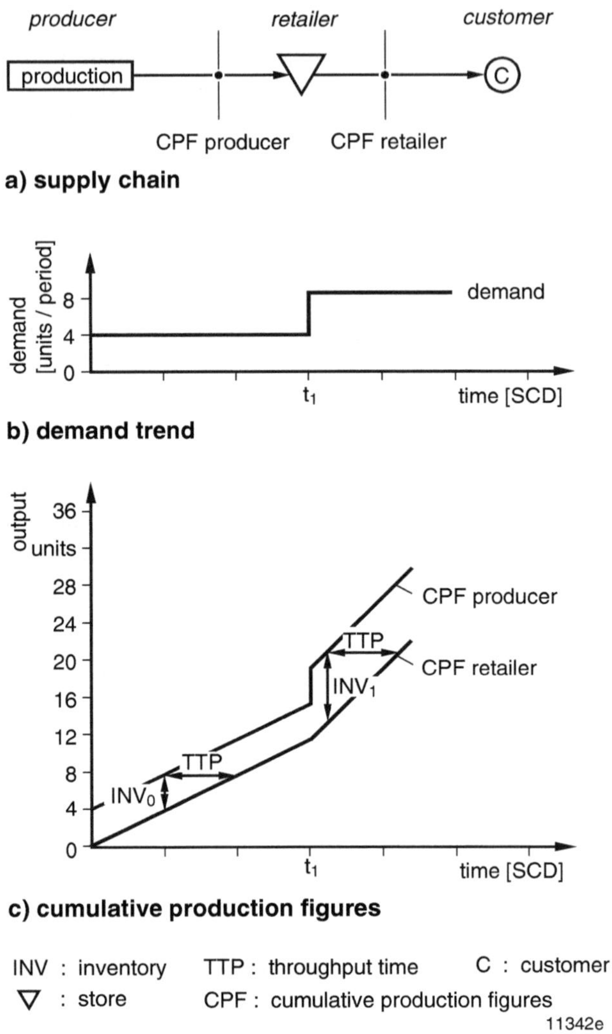


**Fig. 6.3** Depiction of the bullwhip effect with a sudden increase in demand

role for all methods of make-to-stock productions. If the safety-stock grows along with the customer demand, it then also leads to an increased load for the preceding supply stages when there is greater customer demand. Thus, if we consider the bullwhip effect, the widespread practice of setting the safety-stock in the form of a safety range as a multiple of the (average) daily/weekly/monthly demand is particularly unsuitable.

*Transportation:* Constant throughput times are also not unusual in transportation and result for example, when there are set transport rhythms (e.g., cyclical schedules or supplying on specific weekdays).

2. ***Forecasting demand based on orders in the subsequent stages of the supply chain increases the demand fluctuations in the supply chain.***

An enterprise has two possibilities for estimating the demand for a product. It can either use the end customers' demands for predicting sales (in the following: centralized sales planning) or the orders of the subsequent stage in the supply chain (in the following: decentralized sales planning).

Assuming a constant throughput time (see above), a decentralized sales planning considerably strengthens the bullwhip effect. In order to further discuss this I would like to consider a supply chain with an end customer, a producer and a supplier (Fig. 6.4a).

The trend for the end customer's demand oscillates similar to a sinus curve around a mean. Similar to the planned throughput time between the supplier and producer, the planned throughput time between the producer and the retailer is constant. If the supplier tries to maintain the throughput time to the end customer as a constant (centralized sales planning), then the increased fluctuations in the planned output rate in comparison to the producer can be explained by the longer throughput times to the end customer (according to Eq. 6.1 the WIP fluctuations increase the length of the throughput time).

This situation changes fundamentally if the supplier plans the sales based on the planned output rate of the producer. In this case, the supplier takes into consideration only the throughput time to the producer and neglects the throughput time from the producer to the customer. In comparison to centralized sales planning, the demand fluctuations increase (Fig. 6.4b).

This is caused by the supplier misconstruing the producer's increased planned output rate as a long term increase in demand which only serves to build-up the amount of WIP in the supply chain. Consequently, the supplier tries to raise its WIP to correspond with this imagined demand. Accordingly, the supplier's increase in the planned output rate turns out to be much higher than that of the producer. The impact of a decentralized sales planning can also be mathematically described with the assistance of Little's Law [Schö-03].

Among other factors, Simchi-Levi, Kaminsky and Simchi-Levi trace the increase in demand fluctuations back to the forecasting method [Simc-00*]. Accordingly, demand fluctuations increase significantly more with decentralized sales planning than with centralized sales planning (see also the following section).

3. ***Forecasting based on historical demands increase the demand fluctuations in the supply chain.***

Simchi-Levi, Kaminski and Simchi-Levi also investigate how forecasting methods based on historical demands influence the bullwhip effect [Simc-00*]. One of their prerequisites is that the partners in the logistic network utilize an order point system (for an extensive discussion see Chap. 8). With an order point system, whenever the store of a specific variant reaches or falls below an order point the enterprise places a repeat order with the supplier. The order quantity is frequently dependent on the mean and standard deviation of the expected demand [Simc-00*]. If the enterprise specifies the mean and standard deviation of the demand based on historical data,

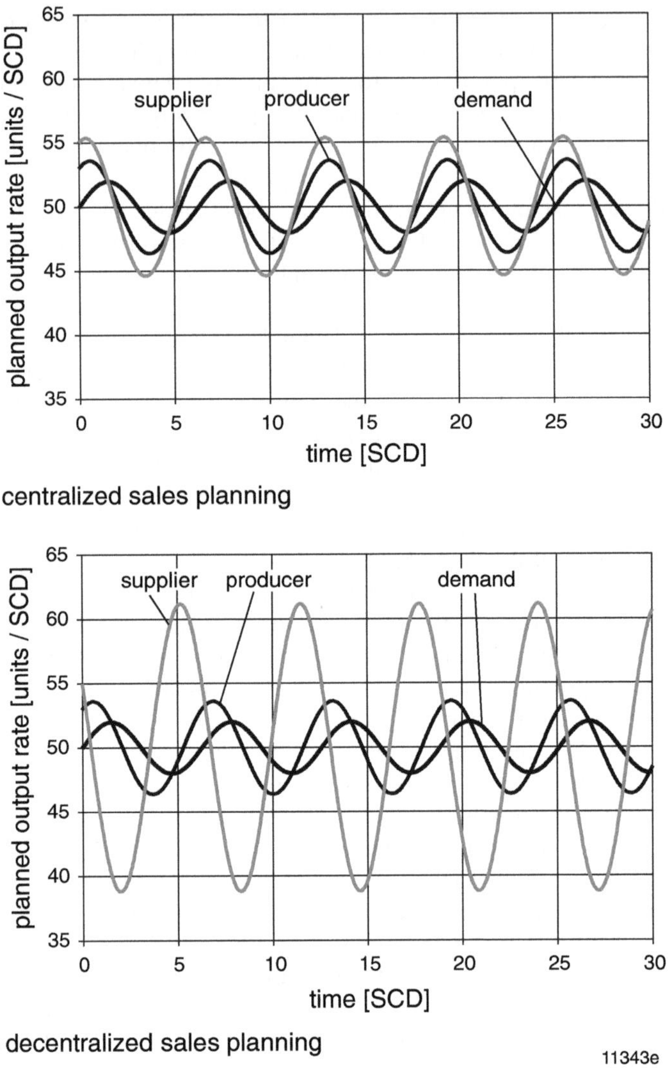


**Fig. 6.4** Bullwhip effect with centralized and decentralized sales planning

the mean and standard deviation of the measured demand fluctuates over time. As a result different safety-stocks are calculated. This leads to changed order points and order quantities and thus causes additional demand rate fluctuations in the preceding stages of the supply chain. As a result, from the perspective of this stage, the demand fluctuations are increased. The amplification increases with the length of the throughput time. This effect can however be reduced through a large number of measured values [Simc-00*], which are taken into consideration when calculating the mean.

***4. Lot sizing distorts the information about demands and causes demand fluctuations in the supply chain.***

As mentioned, in order to reduce setup costs or (lot size independent) ordering costs, enterprises combine demands into lots. Even with a constant demand from the end customer this leads to the supplier perceiving a distorted demand.

The constant demand from the end customers is transformed into demand peaks equal to the lot size at some time points and periods of time where there is no demand (Fig. 6.5). Lee, Padhmanaban and Wang [Lee-97b*] examined the bullwhip effect that results when a number of retailers place orders in a set order rhythm. In order to do so they differentiated three cases:

- *Positively correlated order point:* The retailers place orders in a set rhythm. The order points are the same (e.g., always Monday).
- *Non-correlated order point:* The retailers place orders independent of one another in a set rhythm and do not coordinate the order points.
- *Negatively correlated order point:* The retailers order in a set rhythm. The order points of the individual retailers are nonetheless different (e.g., Retailer A on Monday, Retailer B on Tuesday etc.).

Positively correlated order points distort the demand the most. The producer perceives a very high demand for a brief time period. Subsequently, there are long time periods in which there is a low demand, if at all. Non-correlated, random order points also distort the demand. There are random times at which numerous retailers place orders and other times at which only a few orders are placed. The information is distorted less though than when the order points are positively correlated. In the ideal case, where order points are negatively correlated, the distortion of the demand that results from building lots can be completely nullified [Lee-97b*].

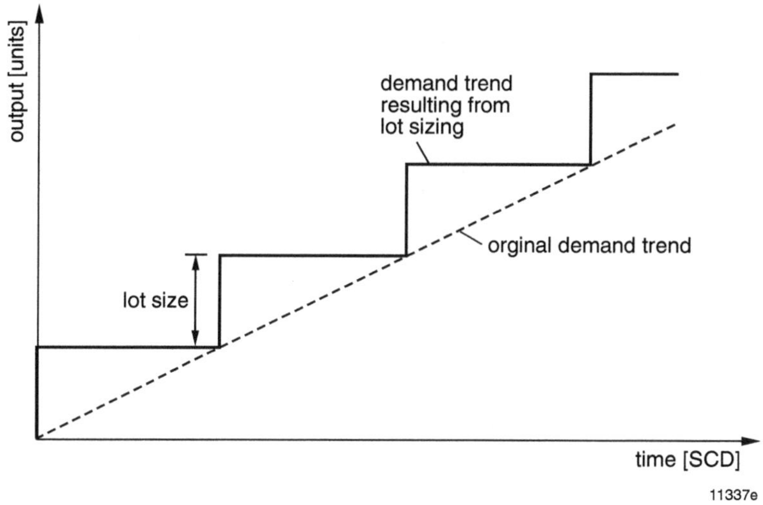


**Fig. 6.5** Distorting the demand by forming lots

If and how the order points are correlated in the practice, depends on the individual cases. Lee, Padhmanaban and Wang mention two reasons for a positively correlated order point [Lee-97*]:

- *Periodic MRP runs:* Numerous enterprises conduct their MRP runs periodically e.g., at the beginning of the week or month. Accordingly, the orders are also triggered at the beginning of the week or month.
- *Quarterly or year-end peaks:* Many enterprises are evaluated quarterly or yearly. They thus strive particularly near the end of an evaluation period to attain the projected sales and pull demands forward where necessary. As a result the orders increase although there is no corresponding demand from the end customer.

***5. Price fluctuations can contribute to increased demand fluctuations.***
Many enterprises vary the price of their product and offer their end customers or retailers temporarily reduced prices through marketing actions. As a result the customers begin to speculate with the goods. They pull demands forward when prices are low and put them off when the prices are high. This purchasing behaviour increases the demand fluctuations and thus contributes to the bullwhip effect [Lee-97*, Lee-97b*].

***6. Supply bottlenecks can cause customers to place orders larger than required and thus increase demand fluctuations.***
Should the demand for a product exceed that which is available, a supply bottleneck results. Suppliers are then forced to ration the product and to divide the available quantity among customers. If in such situations the supplier fulfils the same percentage of each order, it can lead the customer to order more as they actually require. A customer will therefore order twice as much as required when the expected order fulfillment rate is 50%. The orders then exceed the actual demand and this distorts the demand information [Lee-97*, Lee-97b*].

***7. Long turn-around-times for information increase demand fluctuations in the supply chain.***
Turn-around-times for information can distort the demand information. If there is a delay in communicating an increase in demand to a producer (Fig. 6.6 at t2 instead of t1), the producer cannot increase their output rate in time. The WIP in the supply chain is reduced, difficulties with deliveries can arise and the producer ends up developing a backlog. In order to catch-up on this backlog, the producer has to temporarily increase their output rate more strongly than the increased customer demand requires (in Fig. 6.6 from $t_2$ to $t_3$).

Long turn-around-times for information arise for example, when one of the partners in a logistic network plans based on the orders of the subsequent partner and not on the basis of the end customers' demands. The partners' order rhythms and their synchronization then provide the turn-around-time for the information.

***8. Capacity restrictions limit the bullwhip effect.***
Limited capacities can lead to supply bottlenecks as well as tempt retailers to place excessive orders, thus, strengthening the bullwhip effect. Conversely, limited

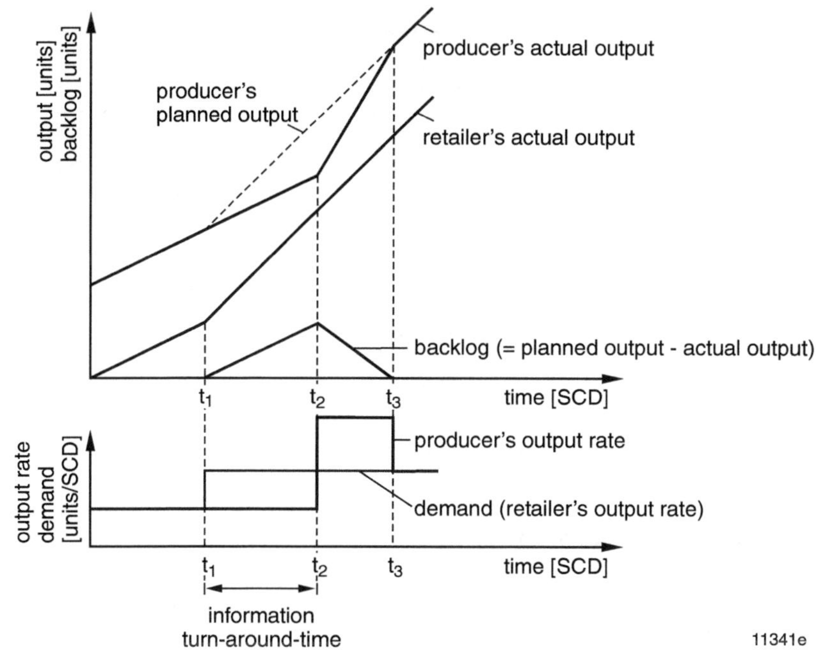


**Fig. 6.6** Influence of information turn-around-times on the bullwhip effect

capacities can also contribute to suppressing the bullwhip effect. The arguments put forward in a number of the previous sections implicitly require unlimited capacities in that they refer to the planned values for the WIP and output rate. They thus only apply if the actual process corresponds to the planned process and this would be unrealistic at the latest when the planned output rate exceeds the maximum capacities. In this case, the WIP in the supply chain cannot be as strongly increased as planned when there is a heightened demand. Accordingly, it is not necessary to reduce the WIP as much when the demand is lower. Thus the demand fluctuations that need to be considered in the practice are less than the theoretical considerations would allow to be expected.

This throws a surprising light on the efforts of many enterprises to increase the flexibility of their capacities when they are unable to (punctually) meet the orders of the next partner in the supply chain. From the perspective of the entire supply chain, increasing capacities only removes the possibility of limiting the bullwhip effect. Consequently, the demand fluctuations continue to increase, so that, where applicable, supply bottlenecks arise for other partners in the supply chain. The vicious cycle of capacity planning in supply chains thus closes (Fig. 6.7).

It should be pointed out here, that the vicious circle is not verified by empirical data; rather it is only based on the analysis of the causal relationships of the bullwhip effect. In any case, it seems more practical to reduce the bullwhip effect than to increase the flexibility of the capacities, thus, we will turn our attention to measures suitable for doing so in the next section.

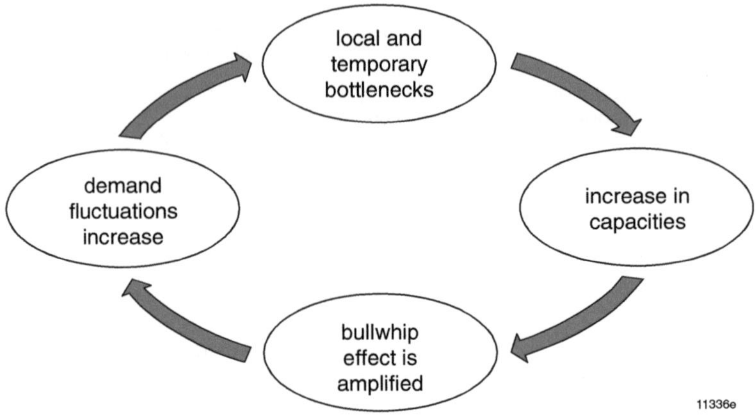


**Fig. 6.7** Vicious cycle of capacity planning in supply chains

## *6.2.3 Counter-Measures*

The goal of supply chain management, similar to production planning and control should be to prevent or at least reduce the bullwhip effect as much as possible. Measures suitable can be derived directly from the causes of the bullwhip effect. In the following eight measures will be explained:

1. controlling the WIP.
2. reducing throughput times.
3. centralized sales planning.
4. delayed forecasting methods.
5. small lot sizes.
6. reducing price fluctuations.
7. cooperation of retailers and producers when supply bottlenecks arise.
8. preventing long turn-around-times for information.

Measures 1, 3 and 8 are directly related to generating orders and are summarized into one guiding principle, which will serve to evaluate methods for generating orders in the following chapters:

***Guiding Principle:*** *Methods for generating orders should communicate information about demands as quickly and clearly as possible.*

With this in mind, using delayed forecasting methods (fourth measure) makes particular sense when setting method parameters since safety-stock levels should only be adjusted very carefully in order to avoid distorting information about demands.

Moreover, the task of reducing throughput times (second measure) is influenced by a number of factors. The most important of these are: the type of material flow, the determined lot sizes, the planned throughput times and the order release.

Furthermore, whereas reducing the lot size (fifth measure) impacts the production, it often requires decreasing setup times. Finally, the last two measures 'minimizing fluctuations in the price' (sixth measure) and 'cooperating with retailers' (seventh measure), frequently cannot be influenced by either the production planning or control. I would now like to examine each of these measures a bit more closely.

***1. Controlling the WIP prevents demand fluctuations from being amplified.*** One very effective way of preventing the bullwhip effect is to provide constant WIP levels instead of constant throughput times. This is first explained for the case in which the demand suddenly doubles in comparison to the original value (Fig. 6.8).

With the task of maintaining constant planned throughput times (Fig. 6.8a) the increased demand causes—as shown in Fig. 6.3—a sudden jump in the planned output and with that discontinuity in the output rate curve (Curve 1 see the above comments). This can only be prevented if the output rate is raised before the increase in the demand (Curve 2). The required offset time corresponds directly to the throughput time. In order to increase the output rate in time the raise in the demand has to be predicted.

In comparison, maintaining a constant WIP (Fig. 6.8b) prevents discontinuity in the planned output. The producer's output rate constantly corresponds with the demand. Contrary to the case with the constant planned throughput times, the increase in demands is not amplified and it is generally not necessary to predict customers' demands. Since it is usually very difficult to precisely forecast the demand this is an important advantage of a constant WIP compared to the constant throughput times.

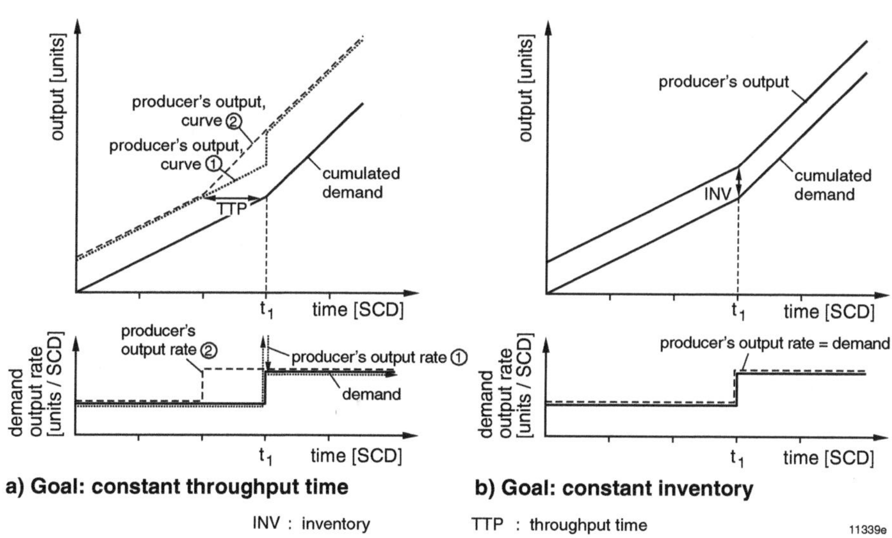


**Fig. 6.8** Comparison of the tasks 'constant throughput times' and 'constant WIP' when there is a sudden change in demand

Quite similar conclusions can be drawn for a sinus shaped demand curve (Fig. 6.9).

When the task is to maintain constant planned throughput times, the demand fluctuations are amplified (Fig. 6.9a) and the bullwhip effect can become quite extensive. In contrast when the task is to maintain a constant planned WIP, the curve for the producer's planned output results from a parallel shift of the retailer's planned output. The output rate and the demand are thus in agreement at all times and the bullwhip effect is prevented.

The question then arises as to what extent the WIP can be regulated on the shop floor. In turn, we need to consider the three logistic reference processes production/testing, storage and transportation.

### *(a) Production/testing*

The WIP level required to utilize a workstation is independent of the workstation's output rate. Accordingly, a constant WIP can at least in principal be attained even when the demand is fluctuating. Increasing the WIP level, however, is necessary if additional machinery is procured or the lot size enlarged when there is a higher demand.

### *(b) Storage*

The mean stock level is influenced by the order size and the safety-stock. Both factors should be decoupled as much as possible from temporary demand fluctuations. Consequently, when the aim is to avoid WIP fluctuations an order point system (with a constant order quantity) is superior to a period batch control (with variable order quantity).

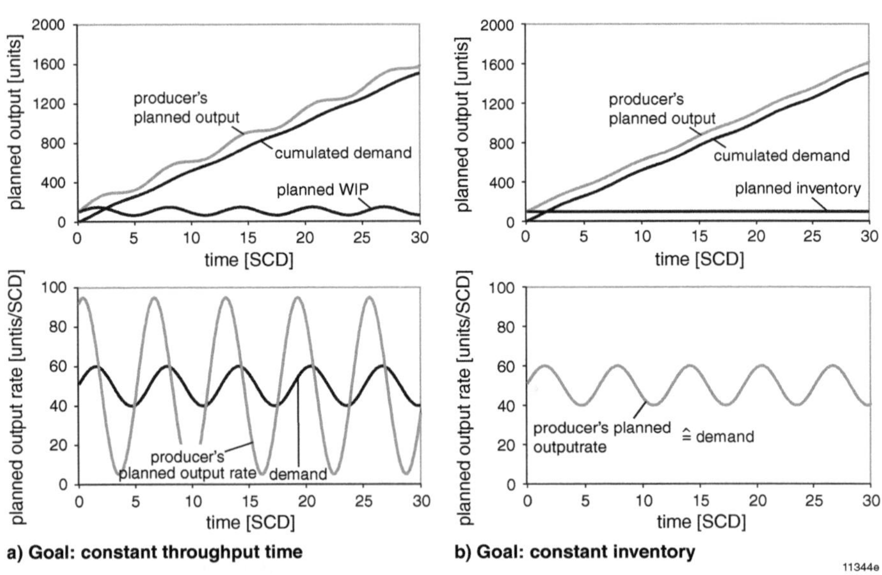


**Fig. 6.9** Comparison of the tasks 'constant throughput times' and 'constant WIP'

Order lot sizes and safety-stock levels should not be changed when there are temporary demand fluctuations. Nevertheless, changes in demand that are effective over the long-term can certainly influence the necessary safety-stock and lot size. I would like take a closer look at this by considering the safety-stock in an example where there is a long-term increase in demand.

If the short-term demand fluctuations also increase with the average demand, then the safety-stock should be increased when the replenishment time remains the same. In contrast, if it is possible to reduce the replenishment time when there is increased demand, increasing the stock level can be avoided. Higher short-term market fluctuations can then be compensated for by the lower replenishment times.

The manufacturing control has a number of possibilities for attaining shorter replenishment times when there is an increased demand. The most important of these is regulating the WIP in the production (see above discussion).

Beyond this correlation, there is a very pragmatic reason for not increasing the stock level when there is greater demand: In this case, the producers are already operating with increased capacities in order to manage the higher demand. In order to develop the stock levels, the capacities have to be further increased. Many methods for calculating safety-stock have a distinct procyclical effect. This is particularly disadvantageous if the capacity requirements exceed the available capacities. In this case, the orders that essentially serve to build the safety-stock compete for limited resources with orders that are required to fulfill customers' concrete demands.

A similar argument can be made for a lower demand. If the safety-stock is reduced when there is a low demand, the decreased demand has a twofold impact on the producer: Since the retailer first exhausts its existing stores, the retailer places even fewer orders than those required to fulfill the already low demand.

***(c) Transportation***

The transportation throughput time consists of two components: The time spent waiting for the transport and the time in which the goods are actually transported. If the carriers increase their hours when there is a higher demand, lower throughput times result independent of how the transport and waiting time are weighted. In this case, a constant WIP level can be realized. Otherwise, the relation of the waiting time to the transport time needs to be taken into consideration.

If the waiting time dominates, the transport rhythm can be shortened when there is greater demand. Transports are then more frequent. Although the quantity moved per transport remains the same, the transport throughput times are shorter and a constant WIP can at least be approximately attained.

If the transport time causes the main portion of the transport throughput time it is frequently more difficult to reduce the transport throughout time. An example of this is a cargo ship that transports goods from Asia to Europe. Here, the transport throughput time can generally not be reduced.

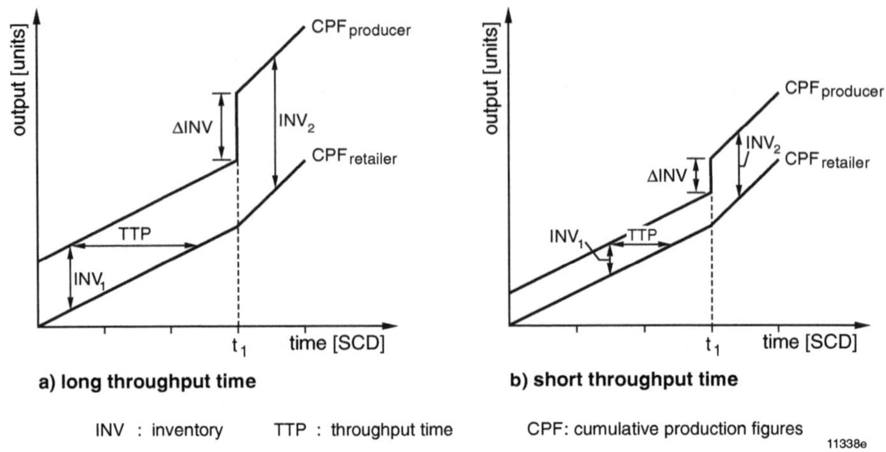


**Fig. 6.10** Impact of the throughput time on the bullwhip effect

***2. Shortening the throughput time minimizes the amplification of demand fluctuations.***

Shorter throughput times minimize the increase in demand fluctuations, however unlike controlling the WIP, it does not prevent them (Fig. 6.10).

Both the bullwhip effect caused by constant throughput times as well as that caused by forecasting methods based on historical data increase along with the throughput time. Figure 6.10 clarifies how the bullwhip effect caused by constant throughput times is reduced with shorter throughput times. In both cases the customer demand doubles at time $t_1$, as can be seen in the retailer's Cumulative Production Figures. Consequently, the retailer's WIP also doubles when the throughput time is constant. Since the initial WIP level ($WIP_1$) is higher with longer throughput times (Fig. 6.10a) than with shorter throughput times (Fig. 6.10b), when the demand is doubled the jump in the Cumulative Production Figures is all the more greater with the longer throughput times. The same applies for the distortion of the demand.

Shortening the throughput times should always be strived for then, when the WIP cannot be controlled.

***3. Centralized sales planning minimizes the amplification of demand fluctuations in the supply chain.***

Both the research on the impact of constant throughput times on the bullwhip effect as well as the already discussed analysis of forecasting methods by Simchi-Levi, Kaminski and Simchi-Levi lead to the same conclusion: The direct relationship to the end customers' demands minimizes the bullwhip effect. This requires, however, very quick turn-around-times for information.

When WIP levels are constant, the demand fluctuations do not build-up over the supply chain; controlling the WIP ensures the coupling with the end customers' demands.

***4. Delayed forecasting methods prevent demand fluctuations from increasing by using statistical effects.***

Adjusting the mean demand too quickly to current measured values strengthens the bullwhip effect; accordingly, a delayed adjustment is preferable. When calculating moving averages, many periods should be taken into consideration in order to avoid amplifying demand fluctuations [Simc-00*].

***5. Small lot sizes minimize the amplification of demand fluctuations in the supply chain.***

Combining orders into lots distorts the demand. The larger the lots are the more the demand is distorted. Correspondingly, reducing the lot size contributes to minimizing the bullwhip effect. This measure should be endorsed all the more since small lot sizes—as already shown—are connected to a number of other logistic advantages.

Furthermore, it is possible to convey the exact customer demand independent of the orders, thus ensuring that the information is not distorted.

***6. Avoiding uncoordinated price fluctuations prevents demand fluctuations from being amplified by speculative purchasing.***

Since price fluctuations can cause demand fluctuations, implementing them should be coordinated. Especially in the consumer goods industry the trend is to discuss and implement the concept of every day low prices [Lee-97*, Lee-97b*]. This concept avoids short-term discounts for marketing purposes in favor of a low price strategy that is effective over the long-term.

In contrast a counter-cyclical price policy, which increases the price when there is a high demand and reduces it when there is a low demand, can contribute to smoothing demand fluctuations. This is a standard strategy especially in industries with a high proportion of fixed costs.

***7. Cooperating with the retailer and distributing goods based on previous sales can avoid excessive orders when there are supply bottlenecks.***

The practice of ordering more than necessary in order to attain sufficient delivery quantities even when there are supply bottlenecks distorts the information about demands and contributes to the bullwhip effect. This practice can be avoided by cooperating together with the retailer. Moreover, when there is a supply bottleneck it is also helpful to distribute the product to the retailers according to previous sales and not according to the order quantity. This removes the incentive for a retailer to order more than what is actually required [Lee-97*, Lee-97b*].

***8. Decreasing or avoiding information turn-around-times prevents demand fluctuations from being further amplified.***

Modern information and communication technology allow turn-around-times for information to be reduced to a minimum. As a consequence the bullwhip effect resulting from long turn-around-times can be avoided (Fig. 6.11).

Generally speaking, there are thus effective possibilities for preventing or at least reducing the bullwhip effect which is frequently an important goal of supply chain management methods.

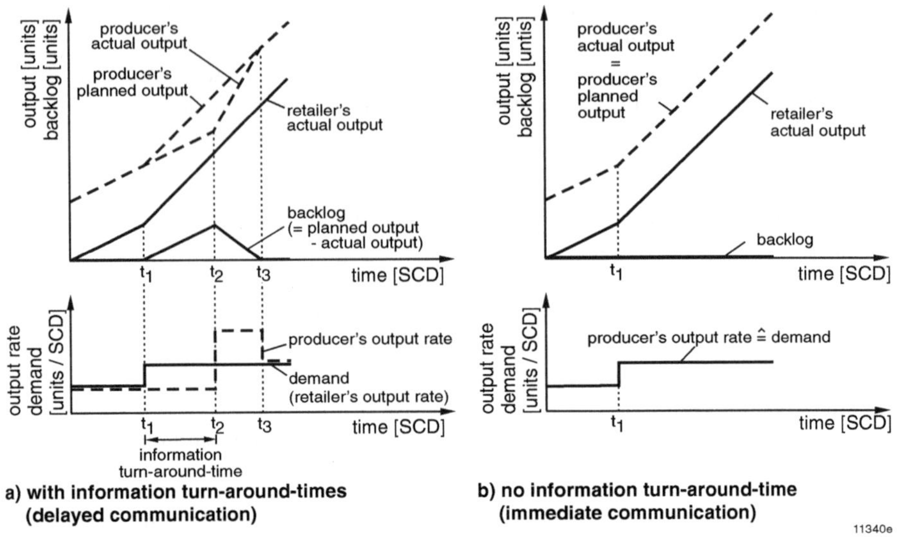


**Fig. 6.11** Preventing the bullwhip effect by minimizing information turn-around-times

## 6.3 Methods

The following section provides an overview of supply chain management methods:

- Section 6.3.1 explains two important fundaments of SCM methods: information and cooperation.
- Section 6.3.2 presents the concept of vendor managed inventories and mentions comparable SCM methods.
- Section 6.3.3 introduces the idea of cooperative production planning for logistic networks.

There are numerous other methods that PPC however does not directly focus on and are thus not dealt with here.

### *6.3.1 Fundamental Principles*

Methods for managing supply chains are characterized by how they deal with the two basic principles: information (Sect. 6.3.1.1) and cooperation (Sect. 6.3.1.2). Let us turn our attention to these now.

#### 6.3.1.1 Information

Traditionally, an enterprise primarily has information which can be collected within the enterprise itself. In comparison, the information collected by partners within the logistic network is lacking. To overstate it, one could say that a producer knows its customers' orders, but not the demand of the end customer. It measures the stores in the finished goods warehouse, but cannot draw any conclusions about the stores in the preceding or subsequent stages of the supply chain. The enterprise knows which delivery reliability it has attained from the perspective of its customers, but is uninformed about the delivery reliability of the supply chain from the perspective of the end customer.

Accordingly, without this information it is quite difficult for an enterprise to adjust its decisions to the needs of the entire supply chain. One of the basic tasks of supply chain management then is to compile relevant information for the entire supply chain and to supply it to all of the partners in the logistic network. Essential information includes:

*Information about the demand:* The end customers' demands should be provided to partners within the logistic network as quickly as possible. This prevents the part of the bullwhip effect that results from decentralized sales planning or long turn-around-times for information.

*Information about the WIP:* Many enterprises do not know the WIP levels of the following stage in the logistic supply chain. This makes decisions about which product variants should be produced (operative production program planning) and how these should be allocated to the individual distribution channels more difficult.

Both the information about the demand and about the WIP can be clearly represented in the form of a cumulative production figure diagram (also known as a progress number diagram). A cumulative production figure diagram is the minimum amount of information required for an effective operative supply chain management. The following information can further increase the effectiveness and efficiency of the supply chain management:

*Information about capacities:* Partners within a logistic network provide information about the load and uilization of their capacities. This allows the impact of capacity restrictions to be recognized ahead of time and in particular for bottlenecks which arise in other enterprises to be taken into consideration. In the long-term capacity planning, the capacities and the flexibility of them can be coordinated with one another in the logistic network.

*Early warning systems:* If it is foreseeable that orders will not be completed in time, early warning systems notify the partner(s) that follow in the supply chain. The early warning system provides partners with the possibility to react to the plan deviations in time.

*Information about costs:* Particularly in highly advanced cooperations, partners within a logistic network exchange information about costs with one another. This makes it possible to minimize the costs of the complete network (in contrast to minimizing costs locally in the individual partners). Some partners in the logistic network can consciously accept higher costs if through that large cost savings can be attained by other partners. Fairly compensating for such local cost increases and distributing the profits resulting from the cooperation to the partners in the logistic network requires a high degree of transparency regarding the costs. This transparency is often a bone of contention and an obstacle within cooperations. Essential parts of the supply chain management can, however, be attained without such extensive transparency about costs.

*Information about attaining targets:* The partners in a supply chain should have access to information about the logistic performance of the supply chain from the end customers' perspective. The performance criteria result directly from applying the logistic objectives (see Chap. 2) to the supply chain. According to that the fundamental performance parameters include:

- Delivery time and throughput time.
- Delivery reliability and schedule reliability.
- Service level (in case the end customer is served from stores).

Correspondingly, the following objectives apply to the logistic costs of a supply chain:

- WIP and/or inventory costs.
- Utilization and/or output rate of the supply chain.

Both the SCOR model from the Supply Chain Council as well as Hieber's dissertation [Hieb-02*] contain comprehensive systems for the key logistic figures of supply chains. In comparison, the exchange of information in the operational practice is frequently under developed. Even in the distribution systems of in-house productions current inventory information are to some extent unavailable for the distribution centers [Hieb-02*]. Furthermore, industry analyses reveal distorted perceptions: Enterprises generally believe that they pass on more information to suppliers as they think they obtain from their own customers [Hieb-02*].

#### 6.3.1.2 Cooperation

Cooperation is the second fundamental principle of supply chain management. The cooperation is aimed at avoiding local optimizations and instead considering the needs of preceding and subsequent partners when making decisions in order to attain the best solution for the entire network. Thus, the retailer should not build-up its stores at the same point in time that the producer is already working at its capacity limits due to increased demand. Conversely, the producer can select a manufacturing lot size which is smaller than the most economical size for that producer alone, if the resulting advantages for the subsequent (e.g. shorter throughput times) and preceding partners (e.g., leveled demand) outweigh the disadvantages for the producer.

One of the conditions for the cooperation between enterprises is mutual trust. Moreover, the cooperation has to be clearly beneficial for all partners in the cooperation (i.e., it has to be a win-win situation). If an enterprise accepts higher costs in order to reduce the costs of the remaining partners, the cost savings should consequently be appropriately shared.

Hieber and Schönsleben developed a morphology for describing the degree of cooperation between enterprises [Hieb-02*, Schö-02b]. They differentiate three main criteria:

*Collaboration in a logistic network:* The pertinent characteristics describe the type and intensity of the collaboration between the network partners on a strategic level. Hieber and Schönsleben mention among others, the following characteristics: orientation (of the partners) on a network strategy and network interests, mutual dependence within the network, mutual trust and openness, and, the balance of power between partners.

*Coordination in logistic networks:* This criterion characterizes the operation of the logistic network. Here the authors note the following factors for example: intensity of information exchange, the links between or interlocking of logistic processes, the formality of the coordination, the degree of communication, and, the implementation of IT systems.

*Composition of logistic networks:* This criterion describes the logistic network and the business relationships between the partners. Characteristics here include among others: number of value-adding levels (network depth), number of products (network range), geographical distribution of the network, time horizon of the business relationships, economical and legal business interests.

In the following two sections, two specific methods of SCM will be discussed. They are both based on the two basic principles of SCM—information and cooperation.

### 6.3.2 Vendor Managed Inventories

With the concept of Vendor Managed Inventories (VMI) the producer plans the amount of stock for all levels of the distribution net. In order to facilitate this the retailers provide the producer with current information about the end customers' demand (point-of-sales data). Consequently, the producer no longer receives orders from the retailer and is thus no longer evaluated based on the delivery reliability or the service level from the perspective of the retailer, but rather directly by the service level from the customer's perspective.

The Vendor Managed Inventories concept strives to:

*Prevent the bullwhip effect:* The producer can plan the stock for the entire distribution net based on current demand information. Accordingly, the producer has extremely good conditions for preventing demand fluctuations from being amplified. This allows many of the following advantages of VMI.

*Reducing WIP:* By utilizing VMI, enterprises can reduce the WIP within the supply chain without endangering the level of service perceived by the end customer. This has been confirmed in cases where VMI has been used in the consumer goods industry.

*Increasing service levels from the perspective of the end customer:* The service level from the end customers' perspective should be boosted without having to increase the WIP in the supply chain. This has been successful in industrial applications. As a consequence of the increased service level, enterprises report considerable growth in sales [Simc-00*].

*Reducing the administrative effort:* The administrative effort is decreased especially for the retailer. Its operative responsibility is reduced to the task of compiling the end customers' demands and passing it on. In addition, it measures the service level of the supply chain. Both tasks can be extensively automated. In contrast the effort required from the producer has the tendency to grow since its responsibilities increases. In particular, the required qualifications change. Generally though, the administration effort should clearly decrease. The operative communication between the producer and the retailer is reduced to conveying the end customers' demands. Neither the retailer nor the producer has to administer the orders.

*Consistent utilization of the producer:* If the bullwhip effect is successfully reduced, the producer can utilize its capacities more consistently. With a given capacity, the producer can serve a higher demand in the long term. This in turn reduces the capital costs.

The concept of Vendor Managed Inventories fundamentally changes the cooperation between the producer and the retailer; the retailer hands over a large percentage of its responsibility to the producer. In doing so it increases its dependence on the producer. The success of the retailer's business is decisively influenced by the producer's ability to plan the distribution network's WIP. In contrast the producer's responsibility for its own economical success increases.

The question of how long the WIP in the supply chain belongs to whom can cause conflicts. In a number of industry applications, the WIP remains the producer's until the product is sold to the customer. With this solution, the retailer is not responsible for any of the capital tied-up in the current assets for products included in the VMI concept. As a result the retailer can directly realize the profits of the concept. This is worth it for the producer in the case where the advantages of the VMI exceed the disadvantages of owning the WIP longer.

Vendor Managed Inventories are a very extensive form of cooperation between the retailer and producer. Other concepts are aimed in the same direction without attaining the same degree of cooperation (see [Simc-00*]):

*Quick response manufacturing:* The producer receives current data regarding the end customers' demands, which should allow the producer to better synchronize the production and demand. The retailer, however, continues to generate orders. An important goal of Quick Response Manufacturing is to reduce delivery times.

*Continuous replenishment:* The producer and retailer agree upon constant delivery periods and a target WIP. The producer receives current data about the end customers' demands and determines the delivery quantity from that. The goal is to continually decrease the WIP but still attain a high service level from the perspective of the customer.

### 6.3.3 Collaborative Planning, Forecasting and Replenishment

The concept of Collaborative Planning, Forecasting and Replenishment (CPFR) further develops the concept of Vendor Managed Inventories and especially emphasizes the cooperation aspect. The partners agree upon a joint business plan and sales forecast as well as on the production program. Related data are saved in a central databank which all partners can access.

The procedure for developing the sales forecast and production program generally does not differ from the previously described production planning and control process (see Chap. 4). However, it combines the planning activities of both the producer and the retailer. One of the difficulties in implementing the CPFR concept can be coordinating the production program for a number of customers. Another concern of CPFR is standardizing the communication and data management required for the joint planning and forecasting process.

This discussion about supply chain management completes our look at the fundamentals of manufacturing control. The next part of the book is dedicated to generating orders. The main focus in doing so is make-to-stock productions. If they prefer, readers who are only interested in make-to-order productions can continue with Part C and the release of orders.

## References

[Bech-97*] Bechtel, C., Jayaram, J.: Supply chain management: a strategic perspective. Int. J. Logist. Manag. **8**(1), 15–34 (1997)

[Forr-61*] Forrester, J.: Industrial Dynamics. MIT Press, Cambridge, MA (1961)

[Hieb-02*] Hieber, R.: Supply Chain Management. A Collaborative Performance Measurement Approach. Vdf-Verlag, Zurich (2002)

[Kuhn-95] Kuhn, A.: Prozeßketten in der Logistik: Entwicklungstrends und Umsetzungsstrategien. Praxiswissen Dortmund, (1995)

[Lee-92*] Lee, L.H., Billington, C.: Managing supply chain inventory: pitfalls and opportunities. Sloan Manage.Rev. **33**(3), 65–73 (1992)

[Lee-97*] Lee, L.H., Padmanabhan, V., Whang, S.: The Bullwhip effect in supply chains. Sloan Manage.Rev. **38**, 93–102 (1997)

[Lee-97b*] Lee, L.H., Padmanabhan, V., Whang, S.: Information distortion in a supply chain: the Bullwhip effect. Manag. Sci. **43**(4), 546–558 (1997)

[Litt-61*] Little, J.D.C.: A proof of the queuing formula: L = Lambda W. Oper. Res. **9**(3), 383–387 (1961)

[Nyhu-09*] Nyhuis, P., Wiendahl, H.-P.: Fundamentals of Production Logistics. Theory, Tools and Applications. Springer, Berlin (2009)

[Schö-02b] Schönsleben, P., Hieber, R.: Gestaltung von effizienten Wertschöpfungspartnerschaften im Supply Chain Management. In: Busch, A., Dangelmaier, W. (eds.) Integriertes Supply Chain Management. Theorie und Praxis effektiver unternehmensübergreifender Geschäftsprozesse, pp. 45–62. Gabler, Wiesbaden (2002)

[Schö-03] Schönsleben, P., Lödding, H., Nienhaus, J.: Verstärkung des Bullwhip-Effekts durch konstante Plan-Durchlaufzeiten. PPS-Management **8**(1), 41–44 (2003)

[Simc-00*] Simchi-Levi, D., Kaminsky, P., Simchi-Levi, E.: Designing and Managing the Supply Chain. Concepts, Strategies, and Case Studies. Irwin McGraw-Hill, Boston (2000)

[Zäpf-03] Zäpfel, G.: Supply Chain Management. In: Baumgarten, H., Wiendahl, H.-P., Zentes, J. (eds.) Springer Experten System Logistik-Management. Strategien, Konzepte, Praxisbeispiele, volume 2, part 7, article 7/02/03/01, pp. 1–32

# Part B
# Methods for Generating Orders

# Chapter 7
# Fundamentals of Generating Orders

The order generation process is responsible for creating manufacturing orders from customer orders, material withdrawals or production programs. It thus determines the planned input and planned output of the production as well as the planned sequence in which the orders should be completed. At the same time, it sets the planned values for the enterprises internal WIP, throughput time and utilization. Figure 7.1 depicts the role of generating orders in the manufacturing control model. The task and accompanying actuating variables are outlined in bold.

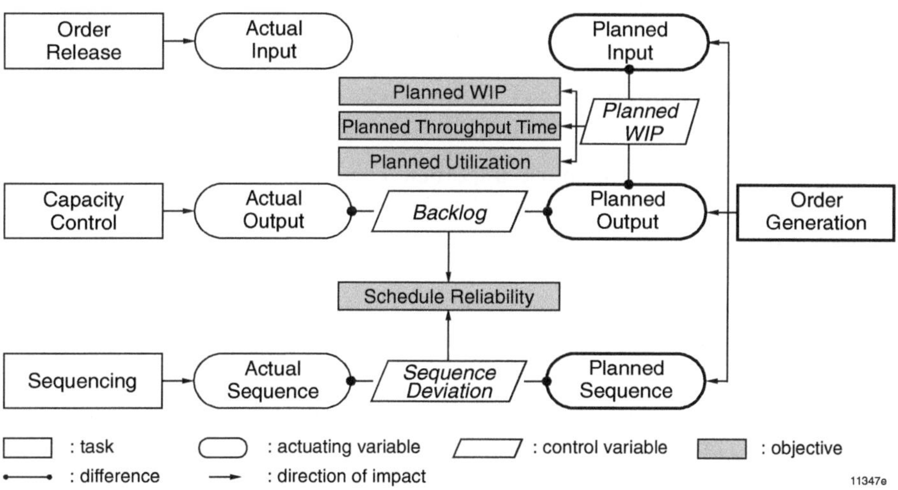


**Fig. 7.1** The role of generating orders in the manufacturing control model

The feasibility of the planned output and planned sequence is also decisive for the schedule reliability. Generating orders is therefore a very important planning task generally. Nonetheless, it is partially conducted in the production which is why it will be extensively described in the second part of this book—although not as

H. Lödding, *Handbook of Manufacturing Control*,
DOI 10.1007/978-3-642-24458-2_7, © Springer-Verlag Berlin Heidelberg 2013

fully as the process of releasing orders. The methods considered here will be those exclusively used in make-to-stock productions. Since the task of generating orders is clearly allocated to production planning in make-to-order productions it will not be dealt with further.

Three characteristics for classifying order generations can be identified:

- The *type of trigger* for an order generation in make-to-order productions is different from that in make-to-stock productions (Sect. 7.1). In make-to-order productions a customer's order triggers one or more manufacturing orders which cover the requirements for the customer's order. In make-to-stock productions the requirements for a customer's order are covered by a manufacturing order that was already generated before the customer order was.
- Accordingly, the *scope* of the order generation depends on whether a method can simultaneously create orders for one or more levels of a product's part list (Sect. 7.2).
- The *trigger logic* of the order generation determines the logic according to which the orders are created (Sect. 7.3). Decisions about generating new orders can either be made at specific times (periodic order generations) or according to defined events (event oriented order generation).

Section 7.4 discusses further criteria for classifying order generations and presents an overview of the methods we will be examining. Section 7.5 clarifies problems that can arise from using fixed lot sizes, whereas, Sect. 7.6 outlines how limited capacities can be taken into consideration when generating orders.

## 7.1 Type of Trigger

The type of trigger is the most important characteristic in classifying order generations. It differs between make-to-order productions (Sect. 7.1.1) and make-to-stock productions (Sect. 7.1.2).

### *7.1.1 Make-to-Order Productions*

In a make-to-order production, the generation of a manufacturing order is triggered by a customer's order. The manufacturing order serves to fulfil the customer's order that triggered it. In the simplest case of a one level product the customer's order is directly converted into a manufacturing order (Fig. 7.2a).

This is the case especially with customized products, for example, when manufacturing circuit boards. With customer neutral products, enterprises combine orders from time to time in order to save on setup costs (Fig. 7.2b). Usually this involves capital intensive goods for which make-to-stock productions would be too expensive or risky. Conversely, it is also possible to generate a number of manufacturing orders from one customer order (Fig. 7.2c [Wien-02]). This can be practical especially when the customer prefers different delivery dates for subsets

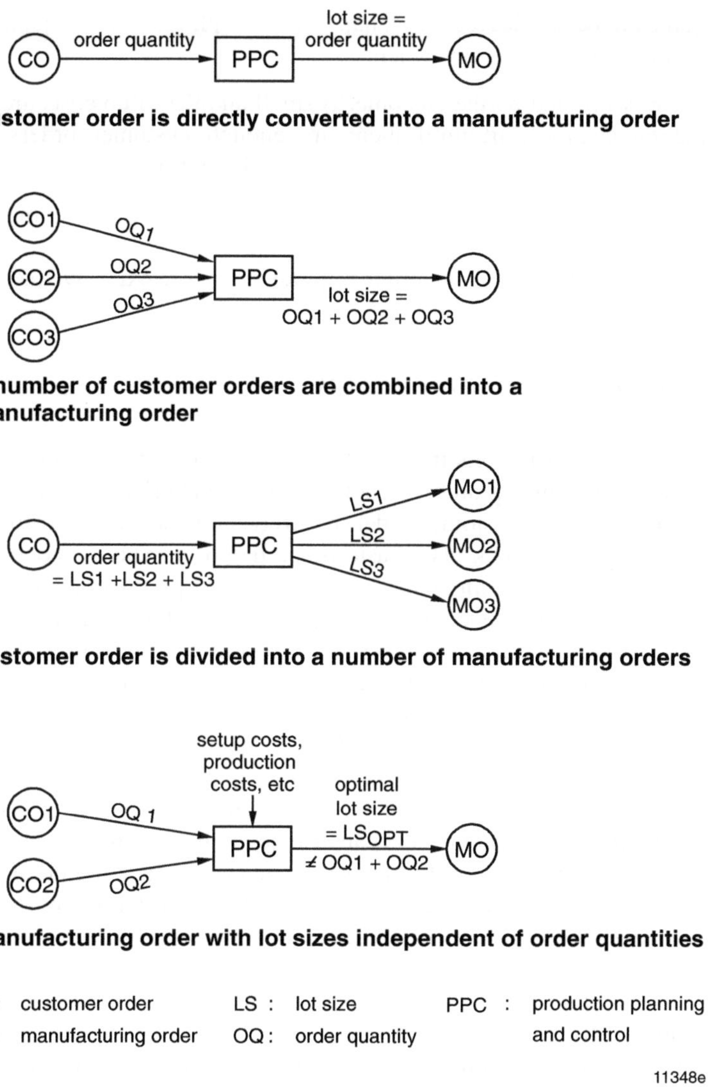


**Fig. 7.2** Generating orders with one level products and make-to-order productions

of the total ordered quantity (e.g., with frame orders) or the ordered quantity is very large and would cause large work contents.

If the manufacturing lot size is completely decoupled from the customer's order quantity (e.g., in order to continually use a fixed lot size that is optimal with regard to costs), a customer's order then generally triggers a larger quantity for production (Fig. 7.2d). This type of policy inevitably causes stores [Wien-02] and requires existing stock to be considered (see Chap. 4). Decoupling the order quantity and the production lot size is necessary when:

- the quantity ordered by the customer is small and the setup costs are high.
- production cannot wait until there are enough customer orders to start a manufacturing order (e.g., with very short delivery times).

Order releases such as this combine make-to-order and make-to-stock productions.

With multi-level products the order generation is more extensive (see Sect. 7.2).

### *7.1.2 Make-to-Stock Productions*

In make-to-stock productions there is no direct connection between a customer's order and a manufacturing order. Generally a manufacturing order does *not* cover the requirements of the customer order that triggers it; rather the customer's order is filled from existing stock. The basic properties of a make-to-stock production result from this fact: In comparison to make-to-order productions it allows much shorter delivery times, however, it requires stores to be maintained. Moreover, a make-to-stock production is generally a repetitive production. Products are frequently customer neutral and the generation of an order is usually triggered by an event e.g., by withdrawing or reserving material.

Theoretically, with multi-level products a requirement explosion with time offsets coordinates the production better than a make-to-stock production: In the ideal case the material is delivered at the exact moment it is required. With frequent repetition, though, very good results can also be attained with multi-level products in a make-to-stock production. Particular advantages include the program flexibility, the ability to self-regulate and the speed with which the many methods used in make-to-stock productions can react to unplanned events in the production or demand.

The make-to-stock production methods described later in this book are based on the assumption that generated orders enter into the store at the end of a replenishment time. In doing so, they *do not* take into consideration any capacity restrictions. The methods generate orders for one variant without considering other variants.

In contrast to periodic production planning with finite scheduling these methods neither set the planned values for the input and output of the production nor the sequencing rules for the orders. The *planned output* can only be determined from the generated orders and the replenishment time required for the orders generated. The same applies for the *planned sequence* in which the orders are to be produced, whereas, a *planned input* can be calculated based on the assumption of a planned throughput time. It is however questionable, whether or not the actual input should be oriented on this planned input since the latter is not explicitly set.

The obvious lack of quality in the planning is one of the fundamental reasons why our discussion about manufacturing control is focused on methods from make-to-stock productions: With all methods that thoroughly plan the schedule and capacities, the manufacturing control benefits from the clearly defined planned values. However, with the make-to-stock production methods presented here, these values first have to be derived to some extent. Thus, there is a danger of neglecting the manufacturing control tasks on the shop floor which remain crucial for attaining logistic targets in make-to-stock productions. This is a direct consequence of the fact that the manufacturing control has to compensate for the inadequacies of the production planning in order to ensure that logistic targets can be consistently attained.

At this point I would like to briefly discuss the selection of methods for generating orders. Due to the above mentioned reasons only methods from make-to-stock productions that do not apply finite scheduling are presented. Within this group the two methods that are by far the most common and widespread methods are first described, namely the Order Point System (Chap. 8) and the Kanban system (Chap. 9). Capacity Oriented Material Management (CORMA, Chap. 10) extends the Order Point System in two important areas: First, it generates the orders already before the order point is reached and thus allows the load to be balanced better. Secondly, it takes into consideration the demand that appears during the replenishment period of a product. Both allow logistic targets to be better attained. Synchro MRP (Chap. 11) and hybrid Kanban/CONWIP control (Chap. 12) are extensions of the Kanban system. Synchro MRP emphasizes the integration of the Kanban control in a higher level production planning. Furthermore, it addressed the key situation in which a greater number of variants is to be controlled with a Kanban system. The hybrid Kanban/CONWIP control is employed for generating multi-level orders and is based on a hybrid control loop structure.

The Cumulative Production Figures method which is also sometimes referred to as Progress Numbers (Chap. 13) is described for two important reasons: On the one hand it is very well suited for cross-plant order generation in supply chains. On the other hand, it can be understood as a generic method since it can be used to emulate many other methods for generating orders just by selecting the appropriate method parameters.

Although developed earlier than Cumulative Production Figures, the Base Stock technique can be understood as a special case of it which clearly and directly communicates demand information to all levels in a supply chain (Chap. 14). In Chap. 15, another generic method—the Production Authorization Cards system (PAC)—is described which, nonetheless, is more complicated to work with than Cumulative Production Figures.

## 7.2 Scope of the Order Generation

Single level methods for generating orders, create orders for every part number in separate control loops (Sect. 7.2.1) whereas multi-level methods generate orders also for dependent requirements (Sect. 7.2.2).

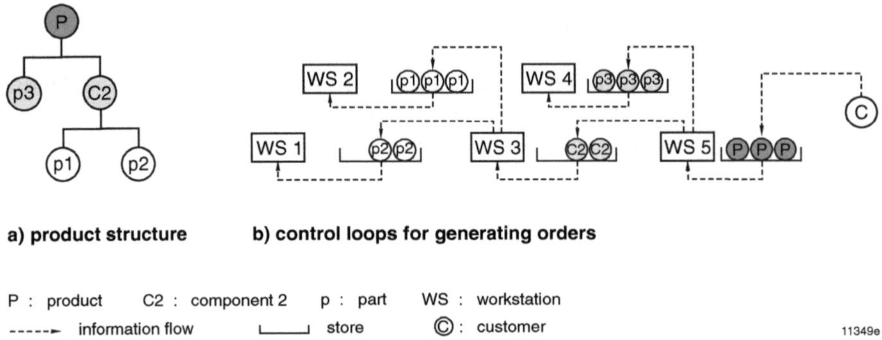


**Fig. 7.3** Controlling multi-level products with single level methods

**Table 7.1** Methods for single level order generation

| Chapter | Production control method |
|---|---|
| 8 | Order point system |
| 9 | Kanban |
| 10 | CORMA |

## *7.2.1 Single-Level Order Generation*

A large percentage of methods generate orders for only one product or part number. Nevertheless, complex product structures can also be controlled with these methods. Generating orders for dependent requirements requires the control loops of the parts to be connected with the higher level assembly units (Fig. 7.3).

Single level methods refrain from the possibility of directly conveying information about the actual customer demand across a number of the product structure's levels. As a result information is frequently delayed. Single level methods for generating orders include (Table 7.1).

## *7.2.2 Multi-level Order Generation*

Methods for multi-level order generation create orders across a number of part list levels. With that they take into consideration dependent requirements and can more quickly communicate information about the customer's requirements than single level methods. At the same time, they require more from the information processing. Generating orders for multi-level products is more complicated for a number of reasons:

First, a customer's order triggers a number of manufacturing orders (one manufacturing order for every part number). Second, it has to be taken into consideration what quantity of a part is incorporated into the end product. Third, certain parts are integrated into a number of end products (so-called 'multiple-use'). Accordingly, a number of possible combinations of requirements arise. Fourth, there is the necessity of coordinating the scheduling of the manufacturing orders for the individual parts.

Three basically different mechanisms can be used to coordinate the manufacturing orders: a lead time offset with a fixed lead time, a WIP offset (see Chap. 13) or a finite scheduling of the orders.

A *lead time offset with a fixed lead time* (as is conducted for example in MRP) shifts the end date of a part in comparison to the end date of the component in which it is included by the lead time of the component. The result is so-called 'time schedules' (Fig. 7.4).

This arranges the corresponding throughput elements for all the procurement, manufacturing and assembly orders based on backwards scheduling. In the case where no delivery time buffer is planned, the throughput element of an order then ends at the point in which the (assembly) order on the higher level begins. The entire delivery time can be read from the time plan as the length of time from the earliest start until the latest end of all the throughput elements. The lead times

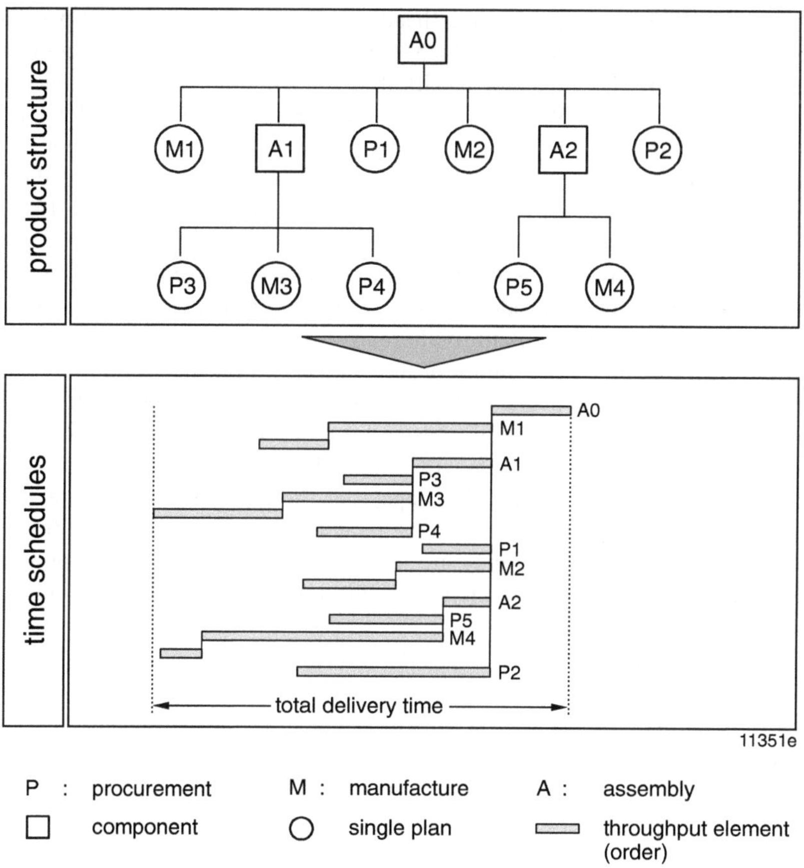


**Fig. 7.4** Deriving the time schedule from the product structure overview (according to Wahlers)

are stored as planning parameters in the PPC system. For practical purposes they should be determined from the workstation specific inter-operation times and order specific operation times and should be updated regularly (see Chap. 3).

Compared to a non-coordinated method, one of the advantages here is the later start date for a number of manufacturing orders which reduces the capital tied-up in the WIP. However, one of the disadvantages is the increased vulnerability to throughput time fluctuations in the manufacturing orders for components. The fluctuations though can be counteracted where necessary by an additional delivery time buffer (see Chap. 3).

Furthermore, it is not uncommon for the entire throughput time of the order network to be longer than the delivery time. In these cases, the order generation is dependent on the demand forecast. If this forecast is not accurate the—to some extent considerable—changes in the demands for parts frequently lead to supply bottlenecks or to larger stores of semi-finished products. One alternative is to simplify the order network by storing one component or part and to shorten the entire delivery time via this customer decoupling point.

With a *WIP offset* the need for stores of semi-finished products is accepted. The WIP offset is defined as a quantitative 'head-start' in a production that a part has in comparison to the component in which it is integrated (e.g., with a cumulative production figure of 200 units and a WIP offset of 100 units a target production of 300 units results). With a WIP offset, it is not necessary to explicitly forecast the demand although the size of the safety-stock of course includes an assumption about the variance of the customers' demand. WIP offsets lead to more constant load trends, but nonetheless can only be implemented in repetitive productions. WIP offsets are supported for example by Base Stock or Cumulative Production Figures.

Both WIP offsets as well as the lead time offsets with fixed lead times do not consider whether or not the production has sufficient capacities for completing the production orders punctually. A third alternative therefore is a *finite scheduling of the orders* with careful consideration of the available capacities. Finite scheduling plans every operation of every order on every workstation minute for minute. Where necessary the availability of tools and devices are also verified. It is usually unrealistic to assume that the orders will actually be completed at the planned time, however, by considering the available capacities the finite scheduling does ensure that no unrealistic production plans are generated. Finite scheduling methods will not be handled here though as they are a subject of production planning and not manufacturing control. An introduction to this topic is provided for example by Pinedo and Chao [Pine-99*].

Methods for generating multi-level orders include (Table 7.2).

**Table 7.2** Methods for generating multi-level orders

| Chapter | Production control method |
|---|---|
| 11 | Synchro MRP |
| 12 | Kanban/CONWIP hybrid |
| 13 | Cumulative production figures |
| 14 | Base stock |
| 15 | Production authorization cards (PAC) |

## 7.3 Trigger Logic

Methods can either generate orders at a specific point in time (periodic order generation, Sect. 7.3.1) or according to defined events (event oriented order generation, Sect. 7.3.2).

### 7.3.1 *Periodic Order Generation*

Periodic order generation decides at pre-set times about the generation of new orders. Between these pre-set times, no orders are generated. Order generation time points are usually at the beginning of a week or a work day. Periodic order generation is typical of methods that generate orders in production planning. Of all the methods presented here only Period Batch Control (Sect. 8.1) generates orders periodically. Other methods can at least in principle be implemented at set order generation time points, however, their procedural logic suggests an event orientation.

An important disadvantage of periodic order generation is its inertness. It can only take into consideration changes in demands at the next order generation time point. Furthermore, it generates a number of orders at the same time and thus hinders a constant order generation. Whereas this characteristic can be a clear advantage for coordinating transports, it leads to an intermittent order input in the production. The disadvantages can be prevented by shortening the periods between the planning processes. When required, periodic order generation can also limit the effort needed for controlling the production since it does not require that information relevant to generating orders be continuously compiled.

### 7.3.2 *Event Oriented Order Generation*

Event oriented order generation makes decisions about generating orders according to specific events. Generally these events are:

- the input of a customer's order.
- the withdrawal of materials from a store (or when materials are reserved).
- changes in the method's parameters (e.g., increasing the safety-stock).

Generating orders based on events allows a quick reaction to these events and thus prevents delays normally associated with periodic order generation. Nonetheless, it generally poses higher demands on the compilation and processing of information. The methods are thus generally designed very simply and take into consideration minimal information when deciding about generating an order. This can at least in principle negatively influence the quality of the decision. Table 7.3

**Table 7.3** Methods with event oriented order generation

| Chapter | Production control method |
|---|---|
| 8 | Order point system |
| 9 | Kanban |
| 10 | CORMA |
| 11 | Synchro MRP |
| 12 | Kanban/CONWIP hybrid |
| 13 | Cumulative production figures |
| 14 | Base stock |
| 15 | Production authorization cards (PAC) |

lists methods with event oriented generation that are frequently applied on the shop floor or cited in professional publications.

## 7.4 Classifying Methods for Generating Orders

In order to provide an overview of the methods dealt with here, they will be classified in the following according to four criteria. These criteria are:

- the scope of order generation.
- the trigger logic.
- the character of the method.
- the primary criterion for generating orders.

The first two criteria used for classification i.e., the scope of the order generation and the trigger logic have already been dealt with. The overall character of the method is referred to as 'specific' when the rules for the method are clearly set and the user can only adjust defined method parameters. By adapting the parameters, however, the general effect of the method remains unchanged. With generic methods, the user can also influence their principle effects either by defining open procedural rules or by setting configuration parameters. Generic methods are usually able to depict a number of different specific methods.

The *primary criterion for generating orders describes* when and/or according to which events an order generation is typically triggered. It is thus a further particularization of the trigger logic.

Based on these four criteria, Table 7.4 depicts the classification of the methods dealt with here for generating orders.

*Scope of order generation:* Period Batch Control, Order Point System, Kanban control and Capacity Oriented Material Management (CORMA) can clearly be classified as single level order generators. Synchro MRP combines the Kanban control, a single level order generator, with a usually multi-level order generation from the production program planning. The hybrid Kanban/CONWIP control also combines elements of both single and multi-level order generators in the first production stage. According to the general logic of Base Stock, it can be clearly classified as a multi-level order generator, whereas with the generic systems for

**Table 7.4** Classification of order generation methods

| Method | Scope of order generation | Trigger logic | Character of method | Primary criterion for generating |
|---|---|---|---|---|
| Period batch control | Single level | Periodic | Specific | Time of order |
| Order point system | Single level | Event oriented | Specific | Order point |
| Kanban | Single level | Event oriented | Specific | Withdrawal of materials |
| CORMA | Single level | Event oriented | Specific | Order release |
| Synchro MRP | Single/multi-level | Event oriented | Specific | Withdrawal of materials/ production program |
| Hybrid Kanban/ CONWIP control | Single/multi-level | Event oriented | Specific | Withdrawal of materials |
| Base stock | Multi-level | Event oriented | Specific | Demand |
| Cumulative production figures | To be defined | Event oriented | Generic | To be defined |
| PAC | To be defined | Event oriented | Generic | To be defined |

generating orders (Cumulative Production Figures, Production Authorization Cards), the user defines the scope of the order generation.

*Trigger logic:* Only Period Batch Control method generates orders at intervals, the remaining methods are event oriented. It is however, at least considerable that the event oriented methods could be activated only at specific points in time in order to periodically generate orders.

*Character of the method:* Generating orders using Cumulative Production Figures and with Production Authorization Cards is generic in nature whereas the remaining methods have specific procedure rules.

*Primary criterion for generating orders:* Period Batch Control generates orders at a defined point in time; the Order Point System when the stock level attains or falls below a defined order point. With the Kanban control, withdrawing materials triggers a replenishment order as is also the case with the hybrid Kanban/CONWIP control and Synchro MRP. With a so-called daily production program the latter, however, also takes into consideration a further criterion for generating orders. At first glance, the point in time at which orders are generated with the Capacity Oriented Material Management is unusual. It generates a replenishment order for a variant every time a previously generated replenishment order is released for production. Base Stock is oriented on the customer demand when generating orders, whereas the user has to first define the criterion for generating orders with the generic methods (Cumulative Production Figures and PAC).

## 7.5 Considering Lot Sizes

A number of enterprises generally produce specific parts and components only in defined lot sizes. If these are taken into consideration in the requirement explosion the demand is frequently distorted. This can easily be clarified with an example:

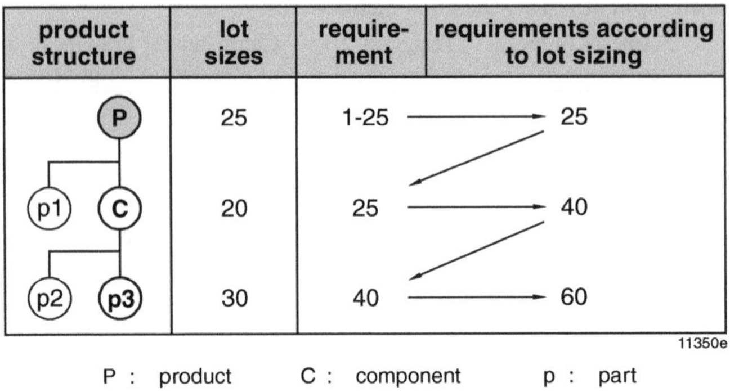


**Fig. 7.5** Impact of lot sizing on requirement explosion

A product P consists of a component C and a part P1 (Fig. 7.5); the component in turn consists of two parts P2 and P3. The parts P1 and P2 will not be considered in the following discussion.

An MRP planning run would generate an order for 25 units when there is a demand for 1–25 products. This order would cause a material requirement of 25 components. In order to cover this demand, two lots of component C would have to be triggered. This creates a material requirement of 40–P3 parts, which could be covered by two runs of two lots each consisting of 30—P3 parts. As a result of the lot sizing in this example 60 P3 parts are required to cover a demand of 1–25 products.

This effect contributes to so-called 'MRP nervousness'. This refers to the phenomena whereby even small changes in the demand (in the example from 25 to 26 product units) can cause significant chances in the demand for parts [Voll-91*]. Naturally, the effect can also arise with methods other than the MRP.

The demand distortion can be eased by coordinating the lot sizes (in the example through a standard lot size of 25 pieces). If safety or base stock are planned for the individual stages, the lot size does not inevitably have to be considered in the requirement explosion. The part requirement is directly derived from the primary requirements. The lots are then first formed after the requirement explosion. An example of this is the Cumulative Production Figures principle (Chap. 13).

## 7.6 Considering Capacity Restrictions

One thing is common among the make-to-stock production methods described in Part B of this book: The procedure rules do *not* take into consideration the production's capacity restrictions. This is grounds for constant criticism in evaluating the MRP-II concept (not presented here) even though the method explicitly considers capacity restrictions. In contrast, the methods described here for make-to-stock productions are often spared this criticism although they actually do not consider the capacities.

The reason for ignoring the available capacities is simple: Whether directly or indirectly, it is always the customer who is responsible for generating orders. However, up until now the customer has only considered the supplier's capacity restrictions in exceptional cases.

If the capacities required to fulfil customer demands exceeds the available capacities, it is generally impossible to punctually comply with the customer's wishes (see Chaps. 2 and 3). In make-to-order productions it is possible to either decline customer orders or to negotiate a delivery date that can be complied with. In make-to-stock productions there is less of a possibility to negotiate. However, one possibility here is the price formation (see Chap. 4). In the following we will consider how the problem of limited capacities can generally be dealt with when generating orders.

**Verifying the Feasibility of Production Programs** After a production program is created the feasibility of it is evaluated. If it is not feasible it is adjusted until it is. There are a number of possibilities for doing so which differ with regards to their level of detail and the extent to which they can be automated:

- *Approximate capacity planning using capacity bills:* Using capacity bills it is possible to very roughly verify whether or not the production program is generally feasible (see Chap. 4 as well as [Voll-91*]).
- *Workstation's load profile:* Based on the orders that are temporarily generated, a load profile can be created for the workstation. It is the responsibility of the planner (manual re-planning) or the planning algorithm (automated re-planning) to re-plan the orders so that capacity overloads can be avoided.
- *Finite scheduling with limited capacities:* With finite scheduling every order is planned up to the minute on the workstations (and where applicable also for the workforce and tools). The production program for a period is then derived from the resulting load plan.

**Variant Specific WIP Control** A variant specific WIP control can prevent the build-up of WIP that results when the capacities are not balanced because the demand exceeds the available capacities. One of the prerequisites for this are tight control loops that do not comprise an entire production but rather at most sectors of it. This is frequently the case with Kanban controls which couple the order generation with the material withdrawal of the subsequent workstation. Generally as a result, only as many orders are generated as that which the following system can complete. By linking the workstations this control mechanism is transferred to the entire production. The effectiveness of this mechanism though is limited by a large number of variants.

**Combinations with Methods for Releasing Orders** Methods for generating orders are combined with order release techniques that control the WIP in the production (see Part C). This type of method prevents orders from being released into the production if the WIP exceeds a defined level. As a result, a significant part of the negative impact of not coordinating the order generation with the capacities can be avoided (see the extensive description in Chap. 16).

Having described the fundamentals of generating orders in this chapter, we will now turn to the most well known method for generating orders—the Order Point System.

## References

[Pine-99*] Pinedo, M., Chao, X.: Operations Scheduling with Applications in Manufacturing and Services. Irwin/McGraw-Hill, New York (1999)

[Voll-91*] Vollmann, T.E., Berry, W.L., Whybark, D.C.: Manufacturing Planning and Control Systems. Irwin McGraw-Hill, Homewood/Boston (1991)

[Wien-02] Wiendahl, H.-H.: Situative Konfiguration des Auftragsmanagements im turbulenten Umfeld. Jost-Jetter Verlag, Heimsheim (2002)

# Chapter 8
# Order Point System

The Order Point System is one of the oldest and most commonly used methods in procurement and production logistics. Throughout the world it is applied to control storage levels in production and commercial enterprises. Moreover, the system's parameters influence the service level of the controlled variants and thus of the enterprise.

The Order Point System generates an order for a variant when the stock level for it (including expected store inputs from open stock orders) reaches or falls below a defined order point. Enterprises employ this system in procurement, semi-finished and finished goods warehouses. Through the order lot size and the time at which the orders are generated it directly impacts a preceding production.

Many enterprises would be able to reduce their inventories and with that their tied-up capital if they appropriately set the system's three parameters 'order lot size', 'safety-stock' and 'replenishment time' and took into consideration the interaction with the production.

In the following, the procedural rules for the Order Point System will be presented (Sect. 8.1) and discussed with regards to the logistic objectives (Sect. 8.2). In doing so variations of the method such as the Period Batch System or reserving stock will be considered.

Section 8.3 shows how the parameters for the Order Point System can be set, while Sect. 8.4 discusses its frequent implementation across supply chains in the industry (Sect. 8.4). Finally in Sect. 8.5 we will take a look at a typical example.

## 8.1 Process Description

Four variations including the basic Order Point System will be presented here:

- Basic system (Sect. 8.1.1).
- Period Batch System (Sect. 8.1.2).
- Order Point System with variable lot sizes (Sect. 8.1.3).
- Reserving stock in the Order Point System (Sect. 8.1.4).

H. Lödding, *Handbook of Manufacturing Control*,
DOI 10.1007/978-3-642-24458-2_8,  

### 8.1.1 Basic System

The Order Point System controls the variants of an enterprise independently from one another. It generates a replenishment order with a set lot size when the stock reaches or falls below the order point. In doing so, expected store inputs are added to the stock. Expected store inputs include the delivery quantities of all replenishment orders that have been generated but not yet entered into the store. Accordingly, the condition for generating an order is:

$$SL + \sum SO_{open} \leq SOP \tag{8.1}$$

where

| | |
|---|---|
| SL | stock level [units], |
| $SO_{open}$ | delivery quantity of open stock orders [units], |
| SOP | stock order point [units]. |

Enough orders are generated until the sum of the stored stock and the delivery quantity of open stock orders once again exceeds the order point. The replenishment order(s) enter into the store after the replenishment time and increase the stock of the variant. The Order Point System leads to a saw-tooth shaped trend of the stock level when the store output quantities are small in comparison to the input quantity (Fig. 8.1).

If the store is supplied from a production that manufactures the stock variants, then the replenishment orders for the different variants compete in the production for the available capacities (Fig. 8.2).

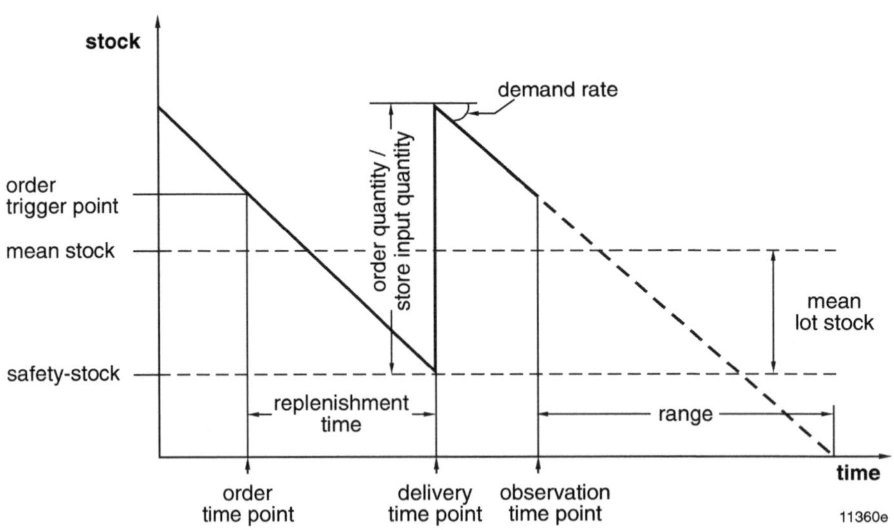


**Fig. 8.1** General storage model when applying the order point (Wiendahl)

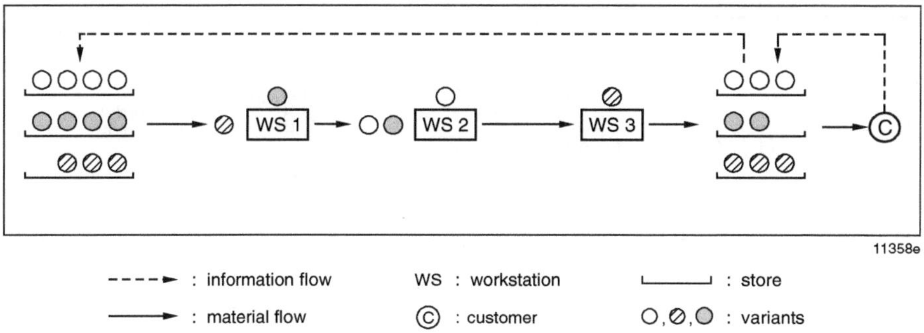


**Fig. 8.2** General depiction of the order point system

Additionally, in many enterprises make-to-order systems are implemented for producing a portion of the orders (hybrid production). The customer orders then have to be coordinated with those generated from the Order Point System.[1] Based on an example, I would now like to discuss the procedural rules for the Order Point System.

An enterprise generates orders for variants A, B and C using the Order Point System. The orders for further variants (D—F) are generated specifically for customers, however, are not considered in the following. All of the variants are processed on workstations WS 1 to WS 3 (Fig. 8.3).

In particular, we will consider Variant A with a stock level of 100 units. The enterprise set the order point at 60 units. In the initial situation there are no replenishment orders for Variant A in the production (Fig. 8.3a).

If a customer demands 30 units of Variant A (Step 1 in Fig. 8.3b), the enterprise delivers the variants to the customer (Step 2) so that the stock level is then reduced to 70 units (Step 3); the stock therefore does not fall below the order point. Later a customer calls up 30 units of Variant A (Step 1 in Fig. 8.3c). Due to the delivery (Step 2) the stock sinks below the order point to 40 units (Step 3). The Order Point System generates a replenishment order for the order lot size of 50 units (Step 4). The order is immediately released for production and the base material is supplied. From the perspective of the inventory control this order is considered an open input (Step 5). The sum of the stock level and the open stock orders thus increases to 90 units so that the order point is once again clearly exceeded. The generated order enters into the queue for Workstation 1 and waits to be processed. Until it is completed and enters the finished goods store all of the customer demands have to be supplied from the 40 units which are currently in stock.

The first demand is soon entered (Fig. 8.3d): A customer calls-up 10 units of Variant A (Step 1). Due to the delivery, the store is reduced to 30 units

[1] One method that dedicates considerable attention to this aspect is Capacity Oriented Material Management (Chap. 10).

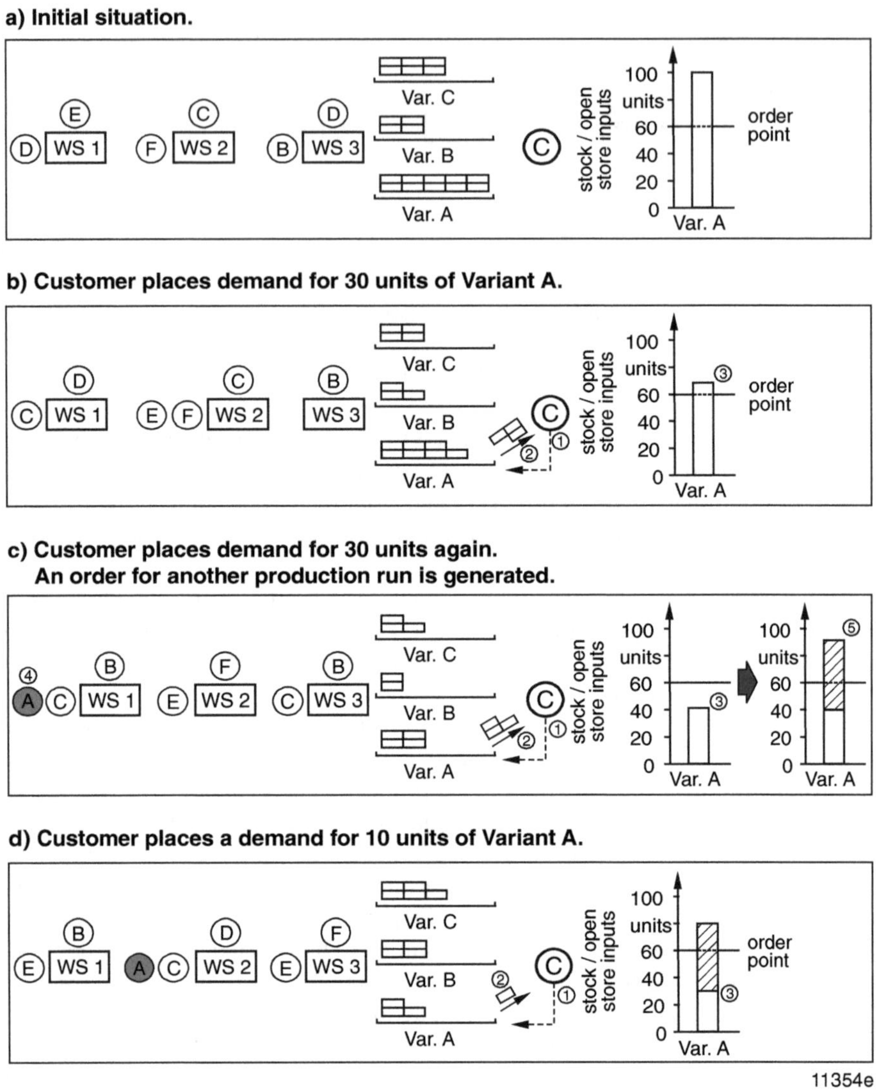


**Fig. 8.3** How the order point system functions

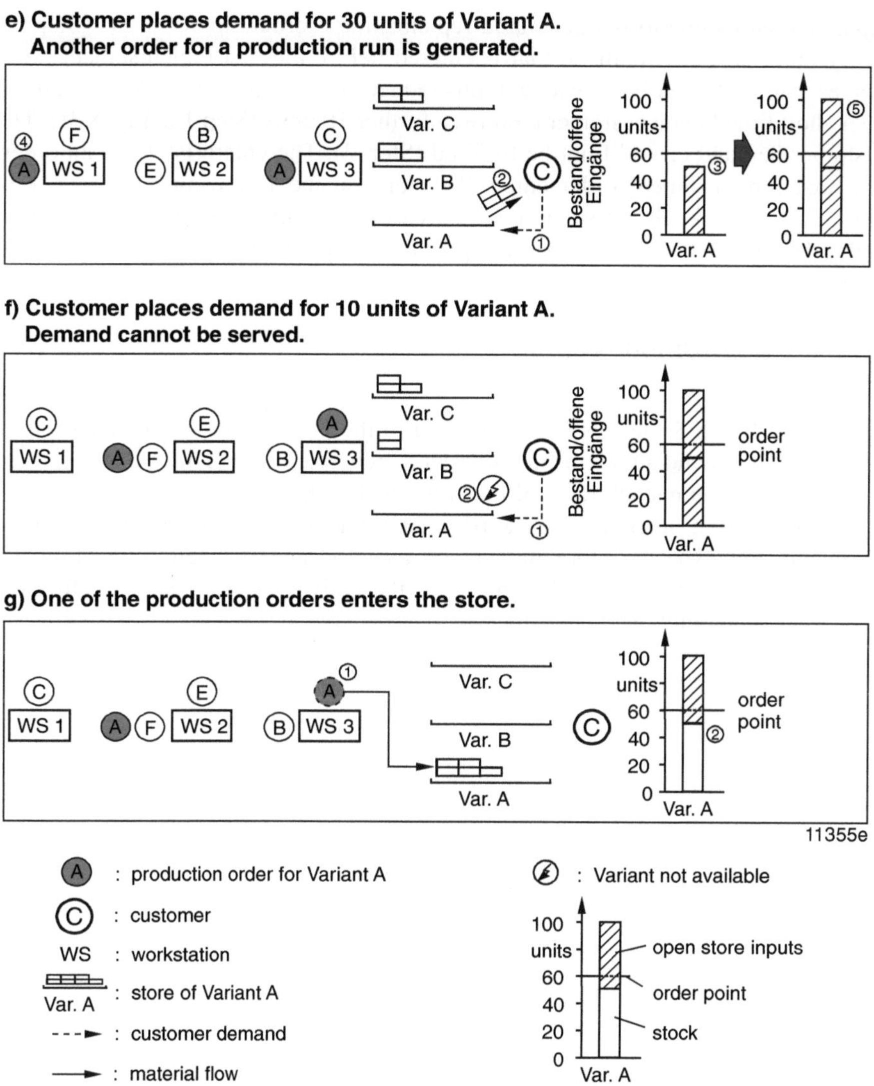


**Fig. 8.3** (continued)

(Steps 2 and 3). The replenishment order for Variant A has now reached Workstation 2. In the meantime additional orders are generated for other variants.

Some time later there is a demand for 30 units (Step 1 in Fig. 8.3e). The stock is completely depleted by this delivery (Step 2). The sum of the stock and expected store inputs falls below the order point of 60 units (Step 3). The Order Point System generates an additional replenishment order for Variant A for 50 units (Step 4). This illustrates why the open inputs are taken into consideration: If the order generation was solely oriented on the stock level, no orders would be generated and a high service level would also not be ensured for the future. The sum of the expected store inputs increases due to the generated replenishment orders to 100 units (Step 5). Every other customer demand that is entered before the first replenishment order arrives, can no longer be served and thus reduces the service level for that variant.

A short time later a customer requires a further 10 units (Step 1 in Fig. 8.3f). The customer demand can still not be fulfilled (Step 2). The enterprise loses an order. The first replenishment order enters the store on the following day (Step 1 in Fig. 8.3g). As a result the stock level increases to 50 units (Step 2). Correspondingly, the sum of the expected store inputs is reduced to 50 units.

### 8.1.2 Period Batch System

The Period Batch System is closely related to the Order Point System. The stock level is recorded and a replenishment order is generated for the difference between it and the maximum stock level of that variant (Fig. 8.4).

The maximum stock level is a parameter which is set by the user. Where applicable the expected store inputs are also considered when calculating the order quantity. The latter can be ignored if the delivery always occurs before the next order time point. The order lot size is calculated as:

$$LS_{ord}(T_{ord}) = SL_{\max} - SL(T_{ord}) - \sum SO_{open}(T_{ord}) \tag{8.2}$$

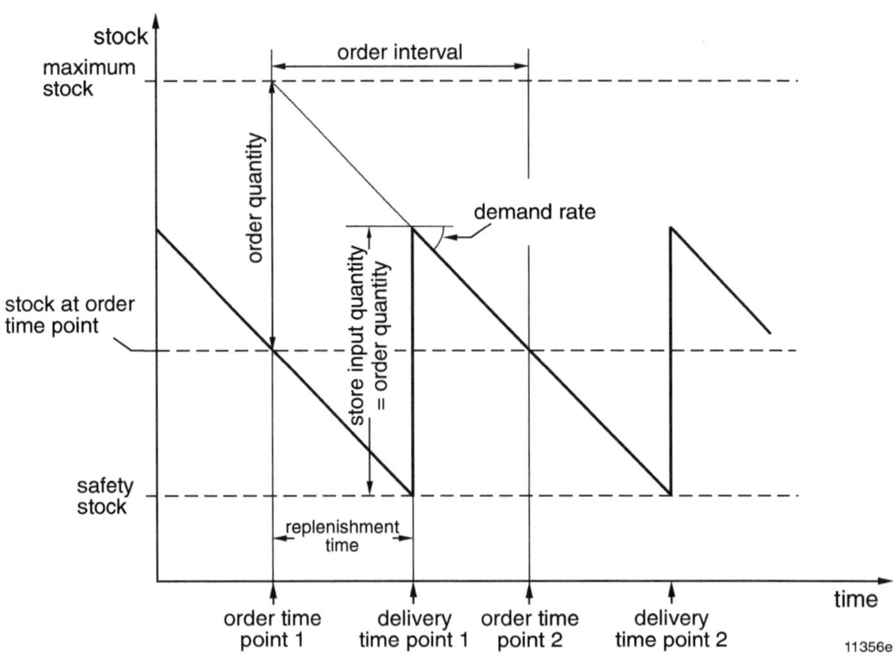


**Fig. 8.4** Ideal storage curve with a period batch system

where

| | |
|---|---|
| $LS_{ord}$ | order lot size [units], |
| $SL_{max}$ | maximum stock level [units], |
| $SL$ | stock level [units], |
| $SO_{open}$ | quantity of open stock orders [units], |
| $T_{order}$ | order time point [SCD]. |

With the Period Batch System, the order lot size varies with the store input between two order points. This can be disadvantageous if the Period Batch System is supposed to control a preceding production. In this case, it is usually practical to set the manufacturing lot size as a function of the setup costs. The Period Batch System is thus generally implemented when a specific delivery rhythm has been agreed upon with the supplier. It then synchronizes the delivery point of different variants and thus allows deliveries to be bundled.

### *8.1.3 Order Point System with Variable Lot Sizes for Orders*

The Order Point System with variable lot sizes for orders combines elements of both the conventional Order Point System and the Period Batch System: As with the conventional Order Point System an order for replenishment is generated when the stock level, excluding the expected store inputs, reaches or falls below the order point. As with the Period Batch System, the order lot size corresponds to the difference between the defined maximum stock and the current stored stock.

Unfortunately, the method combines more of the disadvantages of the methods than the advantages. As with the conventional Order Point System it is not particularly well suited for coordinating the order date with the defined delivery date. Moreover, as with the Period Batch System the size of the lots for the orders is variable.

Unlike the Period Batch System, the order lot size cannot fall below a minimum quantity equal to the difference between the maximum stock level and the order point. Thus in many cases, in the end manufacturing control will not differ fundamentally from conventional Order Point Systems with constant lot sizes.

Finally, I would like to consider a simple example in order to clarify the Order Point System with variable order lot sizes. The object of consideration is a product, which has fallen below the order point due to a customer's demand (Table 8.1).

**Table 8.1** Applying the order point system with variable lot sizes for calculating the order lot size

| Product P1 | Order point = 60 units<br>Maximal stock level = 100 units | | |
|---|---|---|---|
| Time point | Stock level [units] | Expected store inputs [units] | Order lot size [units] |
| Before orders are generated | 30 | 0 | |
| After orders are generated | 30 | 70 | 70 |
| Before delivery | 20 | 70 | |
| After delivery | 90 | 0 | |

At the time observed there are 30 units in store with an order point of 60 units. The method generates a replenishment order for 70 units (maximum stock level minus actual stock level). As a result an expected store input, equal to that of the order lot size e.g., 70 units, arises. During the delivery time, the stock level is reduced to 20 units due to a customer demands. With the delivery of the replenishment order the stock level increases to 90 units, the expected store input returns to zero.

### 8.1.4 Order Point System with Reserved Stock

Many enterprises reserve stock, when there are existing orders that first have to be supplied at a later point in time. Usually stock is reserved with the intention to ensure that a promised delivery is met. The reserved stock can no longer be implemented to serve other customer demands and is defined as the difference between the physical and the available stock. Reserved stock reduces the available stock. At every point in time:

$$SL_{avail}(T) = SL(T) - SL_{reserv}(T) \tag{8.3}$$

where

| | |
|---|---|
| $SL_{avail}$ | available stock [units], |
| SL | stock [units], |
| $SL_{res}$ | reserved stock [units]. |

The system generates a replenishment order as soon as the available stock reaches or falls below the order point:

$$SL_{avail} + \sum SO_{open} \leq SOP \tag{8.4}$$

where

| | |
|---|---|
| $SL_{avail}$ | available stock [units], |
| $SO_{open}$ | quantity of open stock orders [units], |
| SOP | stock order point [units]. |

Accordingly, reserving stock can prematurely trigger replenishment orders. If the order heralds a future increased demand, then generating a replenishment order early is justified. When this is not the case, it leads to a higher stock level, without increasing the service level. On the contrary, the reserved stock can reduce the service level. This is the case, if, before the point in time in which it is required:

- A customer demand first arrives that cannot be serviced due to the reservation.
- A store input subsequently arrives that would have been sufficient to service the customer demand for which the stock had been reserved.

Particularly with large purchase quantities, customers often place orders with enough lead time for the supplier to adjust to the increased demand. In this case, it is frequently practical to generate a special production order for these orders, which is planned using backwards scheduling starting with the date the product is required. This has the additional advantage that the demand trend, which has to be managed by the Order Point System, is not distorted by single demands.

## 8.2 Discussion of the Procedural Rules

Although the Order Point System is very common, it frequently fails to sufficiently attain logistic targets both in stores (high stock levels, insufficient service levels) as well as in the production (high WIP levels, lateness). Based on the guiding principles for designing manufacturing control methods developed in Sect. 3.3.2, I would now like to discuss the weaknesses inherent to this system.

When its parameters are not skilfully set, the method's efficiency decreases (see Sect. 8.3). Moreover, the control task itself, requires a great deal of effort. In many enterprises, the make-to-stock production contains de facto the promise to be constantly ready to supply goods regardless of the time or quantity of the demand. Even with extremely good manufacturing control, this requires either high stock levels or very flexible capacities. It is thus important to first reduce the demand's variance and therefore to simplify the control task itself.

1. *A manufacturing control method should allow the WIP of a production or workstation to be regulated at a defined level.*

A WIP control releases an order as soon as the WIP falls below the planned WIP. As a result, orders can enter into production both earlier or later than the planned release date. That is, the Order Point System limits the maximum stock level for each of the variants and thus indirectly also the production's maximum WIP for all of the variants. With a large number of variants though, this WIP limit has practically no effect; the Order Point System is no longer able to regulate the production's WIP. It neither retains orders when the WIP level is too high, nor does it release orders early when the WIP level is too low. The latter is not at all possible since the orders are released immediately after they are generated. Consequently, there is no pool of known but not yet released orders. Thus, one of the basic conditions for regulating the WIP is lacking.

It is thus frequently practical to supplement the Order Point System with a WIP control so to be able to profit from the advantages it provides. To do so the replenishment orders have to be generated *before* the stock level reaches the order point. The orders can then be sorted according to the sequence of their look-ahead-time. The look-ahead-time indicates how much time is anticipated to pass before the order point is reached (see the Capacity Oriented Material Management developed by Schönsleben and discussed here in Chap. 10). The Order Point System can then be combined with whichever order release method is preferred.

2. *A manufacturing control method should reduce the WIP fluctuations to a minimum.*

Due to the lack of a method for controlling the WIP, the Order Point System cannot prevent WIP fluctuations. Quite the contrary, the production is extremely susceptible to WIP fluctuations: The replenishment orders are generated at randomly distributed points in time. The production planning thus has no possibility of influencing the production's load and coordinating it with capacities. Instead, WIP fluctuations can only be prevented by adjusting capacities to the load fluctuations.

3. *A manufacturing control method should cause as little blocked WIP as possible.*
The Order Point System does not cause any blocked WIP.

4. *A manufacturing control method should contribute to closely aligning the load to the capacities*
One of the more prominent features of the Order Point System is that it does *not* align the load with the capacities. Orders are automatically generated when the stock level falls below the order point and thus define the production's load. To ensure that logistic objectives are consistently attained, an enterprise must be able to coordinate the capacities with the load. As a result, planning the due dates and capacities takes on a critical role. The primary task for planners is to determine the required capacities and to orient the capacities on this demand. Planning the required capacities is made more difficult by the generation of replenishment orders which entail short forecasting periods. Thus, it can also be practical here (i.e., with the capacity requirement planning) to estimate when the Order Point System might generate replenishment orders based on the look-ahead-times.

In addition, an enterprise can supplement the Order Point System with a load balance. In turn, it is necessary to generate the orders before the stock level reaches the order point. The orders can then be released early, if, by doing so the capacities are used more consistently. This is the basic idea behind Capacity Oriented Material Management (Chap. 10).

Moreover, an enterprise can strive to reduce the fluctuations in the demand trend e.g., set an upper limit for demands that can be served from stock. Larger demands are then generally supplied with a sufficient delivery time according to the specific orders. This measure can also be flanked by different pricing schemes (more favorable prices per unit for longer delivery times). Further measures for reducing demand fluctuations result from the supply chain management (see especially Chap. 6).

5. *A manufacturing control method should cause as few sequencing interchanges as possible.*
The Order Point System does not cause any sequence interchanges. It should, nonetheless, be combined with a practical method for setting the processing sequence (see Chap. 25). If the sequencing takes into consideration the demand during the replenishment time, the safety-stock in the finished goods store can be reduced.

6. *A manufacturing control method should be able to compensate for schedule deviations within the context of backlog control.*
It is one of the essential responsibilities of the manufacturing control to prevent backlogs and to reduce backlogs once they have developed. To do so, capacities are temporarily increased. The backlog control, however, is not a component of the Order Point System. Due to the usually fluctuating and difficult to forecast load, a backlog control is a prerequisite for being able to ensure that logistic targets are

consistently attained. The Order Point System should thus be combined with a backlog control. This requires a backlog measure that is adjusted to the system (see Chap. 27).

7. *A manufacturing control method should take into consideration the principle of bottlenecks.*

The Order Point System does not take into consideration the bottleneck principle.

8. *A manufacturing control method should be as simple as possible.*

The Order Point System is very simple. This is an essential reason for why it is so common in the industry.

9. *Methods for generating orders should communicate information about the demands as quickly and clearly as possible.*

One of the fundamental weaknesses of the Order Point System is its failure to adequately communicate information about demands. It transfers demand information only at the time of ordering and then with a set order lot size. It does not directly convey changes in the demand rate, but rather does so indirectly via shorter intervals between order times. As a result, the turn-around-time for information is longer. An avoidable distortion of the demand information can also occur when the safety-stock is inappropriately adjusted to the demand rate. Where possible, the system should be supplemented with a separate method for communicating information about the demand.

## 8.3 Determining the System Parameters

The most important parameters for the Order Point System are the order point and the order lot size. The order point can be divided into three components: the safety-stock, the mean demand during the replenishment time and a portion dependent on the store output lot size. The latter takes into consideration that the stock level is frequently lower than the order point at the time the order is placed.

$$SOP = SSL + TRP \cdot RDMD_m + P(Q_{out}) \tag{8.5}$$

where

| | |
|---|---|
| SOP | stock order point [units], |
| SSL | safety-stock level [units], |
| TRP | replenishment time [SCD], |
| $RDMD_m$ | mean demand rate [units/SCD], |
| $P(Q_{out})$ | output lot size dependent parameter [units]. |

Thus, those using the Order Point System have to set the following system parameters for each variant:

1. Safety-stock (Sect. 8.3.1).
2. Replenishment time (Sect. 8.3.2).
3. Output lot size dependent parameter (Sect. 8.3.3).
4. Order lot size (see Sect. 8.3.4).

Finally, Sect. 8.3.5 discusses the parameters for the Period Batch System which can basically be traced back to the Order Point System.

## *8.3.1 Safety-Stock*

In this next section, I would like to introduce an equation for calculating the safety-stock (Sect. 8.3.1.1). The plan deviations that the safety-stock should buffer are integrated into this equation and include the following (for a more extensive discussion of these see Sect. 3.1.3):

1. deviation from the maximum positive due date (Sect. 8.3.1.2).
2. deviation from the maximum negative delivery quantity (Sect. 8.3.1.3).
3. deviation from the maximum positive demand (Sect. 8.3.1.4).

### 8.3.1.1 Equation for Calculating the Safety-Stock Level

The safety-stock level should be determined based on a Logistic Positioning between the conflicting objectives for the stock level and service level. To do so, the service level for the article should first be calculated and a target service level set (see Fig. 3.6). For a specific weighted target service level the necessary safety-stock is calculated as [Lutz-02]:

$$SSL\left(SERL_{w,targ}\right) = SL_0 \cdot \left(\left(\frac{SERL_{w,targ}}{100}\right)^2 - 1\right) + (SL_1 - SL_0) \cdot \sqrt[c]{1 - \left(1 - \frac{SERL_{w,targ}}{100}\right)^c} \tag{8.6}$$

where

| | |
|---|---|
| SSL | safety-stock level [units], |
| $SERL_{w,targ}$ | weighted target service level [%], |
| $SL_0$ | lot stock level [units], |
| $SL_1$ | practical minimum stock limit [units], |
| c | C-Norm-Parameter [−]. |

The equation for calculating the formula's parameters $SL_1$ (Eq. 3.8) and $SL_0$ (Eq. 3.7) are provided in Sect. 3.1.3. According to [Nyhu-09*], for a service level of 100% the following safety-stock is therefore required:

$$SSL_{SERL=100\%} = \sqrt{\begin{array}{c}\left(DEL_{\max}^{+} \cdot RDMD_m\right)^2 + \left(Q_{\max}^{-}\right)^2 + \\ \left(\left(RDMD_{\max} - RDMD_m\right) \cdot TRP\right)^2\end{array}} \tag{8.7}$$

where

| | |
|---|---|
| $SSL_{SERL=100\%}$ | safety-stock level (for a service level of 100%) [units], |
| $DEL^{+}_{max}$ | maximum positive delivery lateness (delay) [SCD], |
| $RDMD_m$ | mean demand rate [units/SCD], |
| $Q^{-}_{max}$ | maximum negative delivery quantity deviation (under-delivery) [units], |
| $RDMD_{max}$ | maximum demand rate [units/SCD], |
| TRP | replenishment time [SCD]. |

The delivery lateness and the delivery quantity deviation are influenced by the production. Accordingly, the interactions between changes in the safety-stock level and the production in particular, should be taken into consideration. This applies especially to the delivery lateness. I would now like to discuss the individual parameters of the equation more closely.

#### 8.3.1.2 Maximum Positive Delivery Lateness

The maximum positive delivery lateness expresses the maximum amount of time by which the replenishment time is exceeded. The longer the delivery is delayed, the higher the required safety-stock level is. The maximum positive delivery lateness is influenced by the mean and variance of the production's lateness.

The *mean lateness* results directly from the production's backlog. It is the responsibility of schedule and capacity planning as well as the backlog control to prevent a backlog or to reduce a backlog that has already developed as quickly as a possible by increasing capacities. That is, it is possible to define a maximum production backlog that can be buffered by the safety-stock, however, it would be problematic to dynamically adjust the safety-stock level to the relevant current mean lateness. Doing so, could cause the enterprise to then fall into a vicious cycle, in which the safety-stock is continually increased without improving the service level (Fig. 8.5).

In the initial situation, the production's backlog increases due to a temporary higher demand. As a result the production's lateness grows and, consequently, the due date lateness. In order to protect itself from the unreliability of the supply, the enterprise increases its safety-stock. As a consequence, the Order Point System temporarily generates more replenishment orders. The production's load thus continues to increase and as a result the backlog and production lateness also continue to grow, creating the vicious circle.

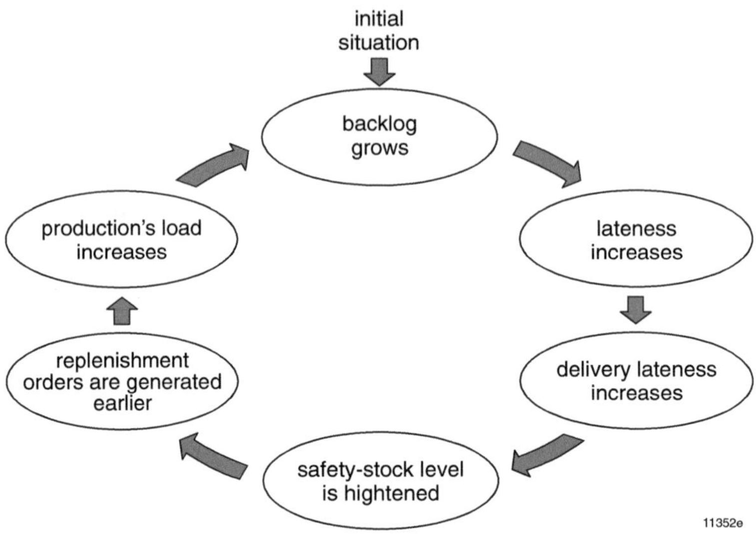


**Fig. 8.5** Vicious cycle of adjusting the safety-stock level

The *variance of the due date lateness* is especially dependent on the sequence interchanges during the throughput as well as, where applicable, the disruptions resulting from an insufficiently reliable process and availability of initial materials. The goal should be to continually reduce the variance of the due date lateness in order to keep the required safety-stock as low as possible. Unlike with the mean due date lateness, an enterprise should measure the variance of the due date lateness and adjust the safety-stock accordingly. It can thus use a lower variance to reduce stores. In order to determine the variance that is to be counteracted by the safety-stock, the distribution of the due date lateness can be analyzed (Fig. 8.6).

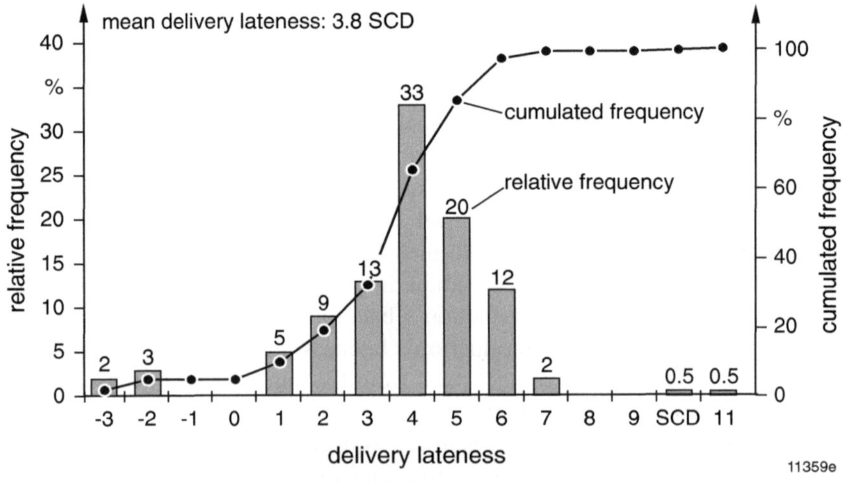


**Fig. 8.6** Distribution of the due date lateness (example)

The mean due date lateness in the (fictitious) example is just under four shop calendar days and is caused by a production backlog. Almost 99% of the orders are delivered not more than 7 days late. Accordingly, the enterprise could set a safety-stock level that covers a variance of the due date lateness of three shop calendar days. It then also classifies delays above and beyond that as extreme cases which should not be covered by the safety-stock.

The mean due date lateness is too high. If the enterprise accepts a maximum production backlog that causes a 2 day delay, then the maximum positive delivery lateness is set at five shop calendar days (mean of 2 days + variance of 3 days). When there is a larger backlog, then the enterprise will not attain its desired service level. This, however, is not the result of the safety-stock being too low, but rather is due to the production's backlog being too large. Accordingly, the capacities should be increased and not the safety-stock level.

#### 8.3.1.3 Maximum Negative Delivery Quantity

The maximum negative delivery quantity (also known as the 'maximum under-delivery') expresses the maximum amount the actual delivered quantity can fall short of the planned delivery quantity. Usually technological factors determine the maximum under-delivery. The better an enterprise controls the production process, the lower the maximum under-delivery is. In many cases, it can be ignored when determining the safety-stock level. Enterprises can estimate the maximum negative delivery quantity based on the historical data of the product or process.

#### 8.3.1.4 Maximum Demand During Replenishment Time

The safety-stock should buffer an increased demand during the replenishment time. The maximal demand during the replenishment time is frequently estimated based on previous demand data. Table 8.2 depicts an example of a list with the demand data for 30 weeks.

Based on this list, two methods for estimating the maximum demand will now be explained. The replenishment time for Alternative A is 2 weeks and for Alternative B 4 weeks. The weekly demands listed here have a normal distribution (mean: 100 units; standard deviation: 30 units).

The *first possibility* is to find the two or four consecutive weeks with the highest demand in the list. In order to facilitate this, the weekly demands are added together in the table according to the replenishment time (column 3 and 4). In the example the highest demands arise in weeks 5 and 6 (Alternative A: 289 units) and weeks 5–8 (Alternative B: 530 units). With a mean demand of 100 units per week, the safety-stock then has to be set at 89 units (Alternative A: 289 units—2 weeks · 100 units/week) or 130 units (Alternative B: 530 units—4 weeks · 100 units/week). The simplicity of use is one of the most important advantages of this procedure. Furthermore, the method can be applied regardless of the shape of the demand's distribution. One of the disadvantages though is that the maximum demand

**Table 8.2** Demand trend for a product (example)

| Week | Weekly demand | Mean demand within | |
|---|---|---|---|
| | | 2 weeks (A) | 4 weeks (B) |
| [–] | [units] | [units] | [units] |
| (1) | (2) | (3) | (4) |
| 1 | 82 | | |
| 2 | 23 | 105 | |
| 3 | 74 | 97 | |
| 4 | 119 | 193 | 298 |
| 5 | 131 | 250 | 347 |
| 6 | 158 | **289** | 482 |
| 7 | 102 | 260 | 510 |
| 8 | 139 | 241 | **530** |
| 9 | 106 | 245 | 505 |
| 10 | 71 | 177 | 418 |
| 11 | 56 | 127 | 372 |
| 12 | 117 | 173 | 350 |
| 13 | 59 | 176 | 303 |
| 14 | 114 | 173 | 346 |
| 15 | 100 | 214 | 390 |
| 16 | 124 | 224 | 397 |
| 17 | 79 | 203 | 417 |
| 18 | 88 | 167 | 391 |
| 19 | 97 | 185 | 388 |
| 20 | 110 | 207 | 374 |
| 21 | 146 | 256 | 441 |
| 22 | 54 | 200 | 407 |
| 23 | 114 | 168 | 424 |
| 24 | 83 | 197 | 397 |
| 25 | 100 | 183 | 351 |
| 26 | 79 | 179 | 376 |
| 27 | 121 | 200 | 383 |
| 28 | 75 | 196 | 375 |
| 29 | 125 | 200 | 400 |
| 30 | 114 | 239 | 435 |

determined using this method, increases with the length of the analysis period and the results cannot be reproduced.

A second possibility consists of estimating the maximum demand based on statistical laws. If the demand complies with a specific distribution shape, then the probability of the demand not exceeding a certain value within the replenishment time can be determined. For the standard normal distribution, which the demands in Table 8.2 are based on, the following values apply (Table 8.3, [Schö-07*]).

Based on a demand value, the left half of the table provides the probability that this demand value will not be exceeded during the statistic period. The statistic period refers to the length of the period in which the individual demand values were compiled. The right half represents the reverse case. If an enterprise wishes to know for example which value the demand will fall below in 99% of the cases, it can read

**Table 8.3** Demand values and probability with standard normal distribution (according to Schönsleben)

| Demand Value [units] | Probability of DMD < DMD Value [%] |
|---|---|
| $DMD_m$ | 50.00 |
| $DMD_m + 0.5\ DMD_s$ | 69.15 |
| $DMD_m + 1.0\ DMD_s$ | 84.13 |
| $DMD_m + 1.5\ DMD_s$ | 93.32 |
| $DMD_m + 2.0\ DMD_s$ | 97.73 |
| $DMD_m + 2.5\ DMD_s$ | 99.38 |
| $DMD_m + 3.0\ DMD_s$ | 99.86 |
| $DMD_m + 4.0\ DMD_s$ | 99.997 |

| Probability of DMD < DMD Value [%] | Demand Value [units] |
|---|---|
| 50 | $DMD_m$ |
| 65 | $DMD_m + 0.385\ DMD_s$ |
| 80 | $DMD_m + 0.842\ DMD_s$ |
| 90 | $DMD_m + 1.282\ DMD_s$ |
| 95 | $DMD_m + 1.645\ DMD_s$ |
| 98 | $DMD_m + 2.054\ DMD_s$ |
| 99 | $DMD_m + 2.326\ DMD_s$ |
| 99.9 | $DMD_m + 3.090\ DMD_s$ |

$DMD_m$ : mean demand [units]
$DMD_s$ : standard deviation of the demands [units]

the corresponding calculation from the table. In this case, it applies to the sum of the means and 2.326 times the standard deviation.

In order to be able to derive information about the maximum demand during the replenishment time, the standard deviation of the demand during the replenishment period has to be calculated. According to Schönsleben, the following equation can be used to determine the standard deviation of the demand during the replenishment time from the standard deviation of the demand during the statistic period (here: 1 week) [Schö-07*]:

$$DMD_{s,TRP} = \sqrt{\frac{TRP}{P_{stat}}} \cdot DMD_{s,stat} \tag{8.8}$$

where

| | |
|---|---|
| $DMD_{s,TRP}$ | standard deviation of the demand during the replenishment time [units], |
| TRP | replenishment time [SCD or weeks], |
| $P_{stat}$ | statistic period [SCD or weeks], |
| $DMD_{s,stat}$ | standard deviation of the demand during the statistic period [units]. |

The equation demonstrates once again the significance of shorter replenishment times: The shorter the replenishment time, the lower the maximum demand during the replenishment time is and thus also the required safety-stock level.

In the example, the demand during the statistic period (here: 1 week) is based on a standard deviation of 30 units. Accordingly, the value for the standard deviation of the demand during the replenishment time is calculated as 43 units (Alternative A) and 60 units (Alternative B). Corresponding values for the maximum demand can be derived from Table 8.3.

Finally, both methods can of course only approximately estimate the maximum future demand. The maximum demand rate that is implemented for calculating the necessary safety-stock is thus an entrepreneurial decision. It expresses up to what demand rate there should be no decrease in the service level.

### 8.3.2 Replenishment Time

Similar to the delivery time, the replenishment time can be dimensioned according to Eq. 3.1 as:

$$TRP = TTP + TP + TDEB + TS + LDS + TAD \tag{8.9}$$

where

| | |
|---|---|
| TRP | replenishment time [SCD], |
| TTP | throughput time [SCD], |
| TP | procurement time [SCD], |
| TDEB | delivery time buffer [SCD], |
| TS | shipping time [SCD], |
| LDS | load shift [SCD], |
| TAD | administration time [SCD]. |

Some of the components of the replenishment time are inapplicable with the Order Point System. The process for dimensioning each of these components will be discussed in the following.

*Throughput time:* The planned throughput time of the supplied production units can be determined with a Logistic Positioning based on the routing sheet and Logistic Operating Curves of the workstations. The procedure for doing so is extensively described in Sect. 3.2.2.3.

*Procurement time:* The procurement time should be considered when materials need to be procured for a replenishment order, which the enterprise does not have stores of. This is, however, seldom the case.

*Delivery time buffer:* A delivery time buffer should not be taken into consideration with the Order Point System; rather the fluctuations in the delivery time should be buffered by the safety-stock.

*Shipping time:* If the Order Point System is applied within one of the enterprise's sites the shipping time corresponds to the time needed to place goods in storage. It can thus frequently be ignored. If the supplier and the stores are located on different sites, then it needs to be considered.

*Time for the load shift:* Usually time for shifting the load is not taken into consideration with the Order Point System. In order to do so, the load of the production has to be measured and the replenishment time—and thus the order points—constantly updated. A procedure such as this would increase the order point with a higher load thus triggering additional replenishment orders and further increasing the production's load. It would thus have a procyclical effect and amplify the load fluctuations of the production.

*Administration time:* The time between which the order point is attained and the order for replenishment is generated should be as short as possible. To do so, the stock levels have to be recorded without delays and the orders have to be directed to the supplier as quickly as possible. In numerous cases the administration time is short enough to be ignored.

### 8.3.3 Parameters Dependent on the Output Lot Size

The average lot size of a store output influences the efficiency of the Order Point System. Frequent withdrawals with smaller lot sizes are advantageous (Fig. 8.7a). With large store output lots the stock level falls, to some extent significantly, below the given stock order point at the order time point (Fig. 8.7b). During the replenishment time, the stock available for serving incoming demands is thus correspondingly less. Accordingly, the stock order point has to be increased in order to attain the desired stock level at the time of placing orders.

In the unfavourable case:

$$WIP_{\min,TO} = SOP - Q_{out,\max} + 1 \tag{8.10}$$

where

$WIP_{min,TO}$ minimal WIP at time of ordering [units],
SOP stock order point [units],
$Q_{out,max}$ store output quantity per output event (output lot-size) [units].

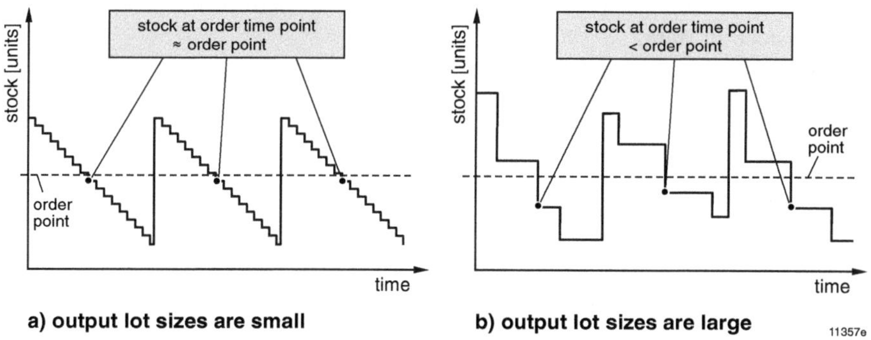


**Fig. 8.7** Impact of the lot size of the store outputs on the stock levels at the order point time

### 8.3.4 Order Lot Size

There are numerous methods for determining the order lot size for make-to-stock productions. The following discussion is limited to presenting a few aspects which should be taken into consideration when determining it. These comments supplement the discussion on determining lot sizes in Sect. 4.1.3.1. Afterwards, we will examine examples for three different methods of establishing the lot size.

**The Lot Size's Range of Influence**

*Setup Times and Setup Costs:* Lot sizes determine the number of setup operations on the production's workstations and thus the total setup time and costs that arise. The larger the lot sizes are, the fewer setup operations there are and the lower the setup costs are.

*Inventory Costs:* The larger the lots, the higher the stock levels in the warehouses and production. This argues in favor of keeping lot sizes as small as possible.

*Attainable WIP Levels and Throughput Times:* Lot sizes determine the work content and, thus, the course of the Logistic Operating Curves for the workstations' of a preceding production. They thus indirectly influence the economically attainable WIP levels and throughput times of the production. A good method for determining lot sizes, therefore, tries to keep the mean and variance of the work content as small as possible in order to target a favorable course of the Logistic Operating Curves.

*Lot Sizes of Adjacent Production Levels:* In multi-level production processes the lot sizes should be coordinated with one another in order to facilitate a closer interlocking of the production levels.

*Capacity Limits:* The smaller the lots are, the more setup times arise which reduces the productive capacities of the workstation. With limited capacities the production's output can thus be influenced by the lot size.

**Methods for Determining Lot Sizes** Most of the usual methods for calculating lot sizes consider at most part of these aspects. The most well-known lot sizing formula, the economic order quantity, minimizes the sum of the setup costs and costs of stock in store.

The lot size is calculated as:

$$LS_{EOQ} = \sqrt{\frac{200 \cdot DMD_{year} \cdot CSU}{CH \cdot CU}} \tag{8.11}$$

where

| | |
|---|---|
| $LS_{EOQ}$ | lot size for minimum costs according to the economic order quantity [units], |
| $DMD_{year}$ | yearly demand [units], |
| CSU | setup costs [€], |
| CH | holding costs [%], |
| CU | cost per unit [€]. |

The lot sizes calculated using this method tend to be too large. They do not take into consideration either the costs of the WIP in the production nor the impact on the attainable throughput times and thus on the course of the Logistic Operating Curves.

Nyhuis extends the standard model for calculating the lot size [Nyhu-91]: With *Throughput Oriented Lot Sizing* (TOLS) he also considers the capital tied-up during the operation and inter-operation times of the replenishment orders. In addition to the pure information about costs, the user provides the desired flow rate for the replenishment orders. The lot size is then calculated as [Nyhu-91]:

$$LS_{TOLS} = \sqrt{\frac{200 \cdot DMD \cdot CSU}{CH \cdot CP + \frac{(CP + CM) \cdot DMD \cdot RF \cdot CH}{60 \cdot TREF} \cdot \sum_{i=1}^{NOP} \frac{t_{p,i}}{CAP_i}}} \tag{8.12}$$

where

| | |
|---|---|
| $LS_{TOLS}$ | lot size with minimum costs calculated with Throughput Oriented Lot Sizing [units], |
| DMD | demand in reference period [units], |
| CSU | setup costs [€], |
| CH | holding costs for reference period [%], |
| CP | production costs/unit [€/unit], |
| CM | material costs/unit [€/unit], |
| RF | flow rate [−], |
| $T_{p,i}$ | processing time for operation *i* [min/units], |
| TREF | reference period [SCD], |
| $CAP_i$ | capacity of the workstation on which operation *i* is conducted [hrs/SCD], |
| NOP | number of operations [−]. |

The lot sizes calculated using this equation are generally smaller than those determined with the standard model. The greater the selected flow rate and work content are, the greater the reduction. As a result, the method harmonizes the work contents, thus creating a more favorable course for the workstation's Production Operating Curves.

The latter is the explicit goal of a method developed by Greiling which plans the lot sizes for all of the variants simultaneously [Grei-97]. The simultaneous planning requires more effort than the simplifying isolated lot size determining, but, it is more precise because it can map the interactions between the variants that are competing on a workstation.

When capacities are limited, small lot sizes can cause the capacity load to exceed the available capacities. The planned production plan can then not be realized. In this case it is possible to determine the maximum frequency of setups on a workstation and to distribute it among the variants. The procedure for doing so is explained for the Kanban System in Chap. 9.

### 8.3.5 Parameters for the Period Batch System

The parameters for the Period Batch System are the order interval and the maximum stock. The order interval provides the length of time between two order points and is measured in shop calendar days. It influences the order lot size and the maximum stock of a variant. Both increase along with the order interval. The order interval is usually set as a function of the delivery rhythm and is thus influenced by, for example, considerations regarding transportation logistics.

For an ideal storage process the resulting stock trend is shaped like a saw-tooth (Fig. 8.4). The dependence of the maximum stock on the order interval is expressed by the following equation:

$$SL_{\max} = (TRP + OI) \cdot RDMD_m + SSL \tag{8.13}$$

where

| | |
|---|---|
| $SL_{max}$ | maximum stock level [units], |
| TRP | replenishment time [SCD], |
| OI | order interval [SCD], |
| $RDMD_m$ | mean demand rate [units/SCD], |
| SSL | safety-stock level [units]. |

Consequently, the maximum stock level increases parallel to the replenishment time and order interval. The stock level reaches the maximum stock level, however, only when there are no store outputs during the replenishment time. If the store output corresponds to the mean demand rate, the stock level only reaches the planned maximum stock level.

$$SL_{\max,plan} = OI \cdot RDMD_m + SSL \tag{8.14}$$

where

| | |
|---|---|
| $SL_{max,plan}$ | maximum planned stock level [units], |
| OI | order interval [SCD], |
| $RDMD_m$ | mean demand rate [units/SCD], |
| SSL | safety-stock level [units]. |

The safety-stock serves to buffer delayed supplies, under-deliveries and fluctuations in the demand. Similar to with the Order Point System it can thus be set. However, unlike the Order Point System, the demand fluctuations are buffered here by a length of time equal to the sum of the order interval and replenishment time. This length of time is longer than the replenishment time. The required safety-stock is thus usually higher than with the Order Point System. The enterprise therefore requires a higher stock level in order to attain the same service level than with the Order Point System.

### 8.3.6 Dynamically Adjusting Method Parameters

Over time it is necessary to adjust the method parameters (especially the order point) to the changing influences; however, it is also a risky task. We will first turn to *the necessity of adjusting parameters* based on key factors which influence the order point. Following that we will consider the *risks involved with adjusting* them.

**Necessity of Adjusting Parameters** If an enterprise was to maintain the same parameters for extended periods it would not be able to react to changes in the various factors which influence the production. The enterprise would store too much stock when the factors become more favorable and would not be able to ensure the necessary service level in the inverse situation. Following is a brief discussion of how changes to key factors impact the order point:

*Variance of the Demand:* If the variance of the demand grows, the safety-stock should be increased in order to be able to ensure a consistently high service level. If the demand is more constant, the safety-stock and with that the order point can be reduced.

*Mean Demand:* If the mean demand climbs, the order point should be increased for two reasons: First, the demand during the replenishment time also climbs. Second the due date deviations also have a greater impact so that the safety-stock also needs to be increased. The first effect can, where applicable, be compensated for by reducing the replenishment time.

*Replenishment Time:* If the enterprise is successful at permanently shortening the replenishment time, it can reduce the order point for two reasons: On the one hand, the mean demand during the replenishment time falls and on the other hand, the maximum demand decreases as well. The latter reduces the required safety stock.

*Maximum Positive Due Date Deviation:* If the variance of the due date deviations is permanently reduced, the safety-stock can be decreased. Otherwise, the enterprise would not use the potential created by the improvement. If the variance worsens over the long term, increasing the safety-stock can ensure the delivery reliability from the customer's point of view.

**Risks Involved with Adjusting Parameters** Each change to the order point is basically connected to a risk as they temporarily decouple the orders from the trend of the customer demands. If the order points are increased, more orders are generated than would be required for serving customer demands. With reduced order points, the orders fall below customer demands. If the supplier interprets the changed order input as a result of changes in the customer demand, he makes a mistake in estimating the demand trend and, thus, is likely to make wrong decisions. Since the order point has a tendency to increase with higher demand and to decrease with a sinking demand, adjusting the safety-stock amplifies the demand fluctuations (see discussion on the bullwhip effect in Sect. 6.2).

Increasing the order points is especially problematic when the production is working at the capacity limits. In this case, it cannot manage any additional load. Increasing the order points is then counter-productive. Order points should thus generally only be increased when the production has sufficient capacities for dealing with the resulting greater load.

Changing the order point causes particularly pronounced load peaks when the changes are made in larger time intervals and simultaneously for all products. When it comes time to make adjustments, the Order Point System then generates replenishment orders for all products whose updated order point is higher than the current stock. The production can then only manage the resulting additional load when it can very quickly increase capacities. It is therefore recommendable, to adjust the order point of the products either in different time intervals or so frequently that the order points only change gradually.

## 8.4 Implementing the Order Point System Across the Supply Chain

The Order Point System is frequently applied in order to control production throughout the entire supply chain or network of enterprises. The purchaser then maintains a store in the receiving area with the Order Point System and triggers an order when the store reaches or falls below the order point. This order is conveyed to the supplier who either reproduces it or serves it from their stores.

Generally speaking, implementing the Order Point System across the supply chain leads to the same problems as when applying it within an enterprise. In particular, it generates replenishment orders without taking into consideration the capacity load of the suppliers. The differences result primarily from the frequently more difficult exchange of information between enterprises and—especially in buyers markets—from the stronger relationship between the customer and supplier:

1. *More difficult information exchange:* The enterprises' borders impede the exchange of information between the production and stock management. It is thus more difficult to recognize causes of problems resulting from the coordination of the production and stock management. Within an enterprise the production can protest if the stock management increases the safety-stock and by doing so generates numerous replenishment orders leading to a load peak in the production. In contrast, between two enterprises there is the danger of misconstruing the sudden increase in incoming orders as a success in sales and a positive development in the market and, possibly, even lead to investing in extending capacities.
2. *Higher priority of replenishment orders:* The supplier perceives replenishment orders as customer orders. Accordingly, they are often prioritized above the

enterprise's internal replenishment orders. This can positively impact the due date compliance of the external replenishment orders.

3. *Limited possibilities for reacting to demands during the replenishment time:* Within an enterprise, the production of replenishment orders can be accelerated (or delayed) if the demand during the replenishment time is higher (or lower) than planned. As a result the necessary safety-stock can be reduced. This is frequently not possible when implementing the Order Point System across a supply chain since the supplier does not have information about the trend of the customer's demand and fixed delivery times are usually agreed upon for the orders.
4. *Insufficient agreement between the order lot size and the production:* If the purchaser sets the order lot size, it usually does not take into consideration the supplier's production.
5. *Allocating stock to specific purchasers early:* If a number of purchasers withdraw (order) the same product, it is frequently practical to maintain only a very small store at the purchasers and instead establish a central store managed by the supplier. Otherwise, when each of the purchasers maintain their own stores, stock is always allocated to a specific purchaser and when there are supply bottlenecks, the purchasers cannot help each other out.
6. *Increased significance of transportation:* Transportation plays a more significant role in enterprise-external delivery relationships than in internal ones. On the one hand the transport time has to be considered when calculating the replenishment time. On the other hand, enterprises can generate orders so that a number of orders are delivered at the same time. This is especially efficient when a purchaser orders a number of products from one supplier. With a procedure such as this the advantage of lower transport stands in contrast to the disadvantages of a more strongly dispersed demand trend. Vollmann, Whybark and Berry provide an overview of extensions to the Order Point System which target generating replenishment orders based on transportation [Voll-91*]. A well known method suitable for coordinating the transportation is the described Period Batch System.

Generally, implementing the Order Point System across the supply chain is therefore linked to increased risks. These can, however, be limited when the purchaser and supplier closely cooperate and intensively communicate information with one another.

## 8.5 Application

The most outstanding aspect of the Order Point System is its simplicity. The method only uses information about stores for generating orders. Accordingly, it is particularly suitable for applying in situations where the demand can only be planned with a large amount of effort, if at all. In practice, this is frequently the case in make-to-stock productions manufacturing an extensive number of variants for a large number of customers. The mean demand rate for each variant is usually relatively low, whereas the variance is generally high. In most applications the value of the product is usually low. Under such conditions, the costs for improving the demand planning frequently exceeds the anticipated benefits.

There are countless applications of the Order Point System in the practice. The following section describes the example of a mechanical engineering firm which had difficulties targeting a high service level within a positive economical environment (Sect. 8.5.1). The section following that explains how the Order Point System can be implemented without PPC software (Sect. 8.5.2).

### 8.5.1 Case Study: Applying the Order Point System in a Mechanical Engineering Firm

The mechanical engineering plant is a mixed production. Part of the products are made-to-order, however, with regards to sales, the majority of products are made-to-stock. The enterprise controls these variants with the Order Point System. The market is positive and the enterprise has a large number of incoming orders. Nevertheless, the customers and management of the firm complain about the insufficient service level for many of the products. Moreover, the stock levels of the enterprise are too high and correspondingly tie-up a large amount of capital. An analysis uncovers the following causes for missing logistic targets:

*Large production backlogs due to insufficient capacities:* The demand for the products exceeded the available capacities. Consequently, a large backlog develops in the production and the orders enter into the store with considerable delays. Despite the large backlog the enterprise finds itself compelled to further take on custom orders.

*Strongly dispersed throughput times:* The throughput times of the orders are strongly dispersed. The orders thus enter into the store with very different delivery date deviations. One of the reasons for this is the fluctuating WIP in the production; another is that the operators in the production are changing the sequence the orders are processed in.

*Unsuitable methods for determining the safety-stock levels:* The enterprise sets the safety-stock levels for the orders as a multiple of the demand for one month. It thus fails to consider the variances in the demand rate as well as the different delivery date deviations and replenishment times of the orders. Since all three influential factors deviate to some degree quite strongly from product to product, the calculated safety-stock levels are not suitable for ensuring that the logistic targets are consistently attained. For a large percentage of the articles, the safety-stock is dimensioned excessively high. This applies especially to articles with high although steady demand. For another percentage of articles, the safety-stock is not sufficient enough to be able to ensure a high level of service. Combined, these two factors explain the frequently observed phenomenon in which despite having large stores, the enterprise is unable to attain a sufficient service level.

*Dynamically Adjusting the Safety-Stock:* The enterprise calculates the average demand of the article with the method known as 'exponential smoothing'. As a results and due to the growing demand the enterprise automatically increases the order points and safety-stock levels so that additional replenishment orders are generated. Since the production is already operating at the limits of its capacities this only serves to increase the backlogs and the WIP in the production. Increasing the safety-stock is therefore counter-productive.

*Determining the lot size:* The enterprise sets the lot size as a function of the month's demand. It thus fails to consider different setup and production costs as well as the resulting variance of the work contents.

*Determining the planned throughput times:* The planned throughput times of the orders are set at seemly high values. In addition to causing production backlogs, this also leads to long replenishment times.

*Lack of a backlog control:* On the one hand, the enterprise does not have the necessary room to manoeuvre as far as capacities are concerned in order to reduce backlogs. On the other hand, however, the firm also does not have any set procedure as to how to react to production backlogs.

*Large demand fluctuations:* For certain articles, the enterprise's marketing department amplifies the demand fluctuations by having sale campaigns and offering customers reduced prices for these articles. Especially when the production is already overloaded, deals such as this need to be critically assessed.

Measures suitable for increasing the frequency with which the enterprise attains its logistic targets can be derived by analyzing the causes. It is extremely important to *extend the capacities and/or flexibility of the production's capacities* in order to serve the increased customer demand and to reduce the production's backlogs. Attention needs to be paid when increasing capacities, however, so that the enterprise can still profitably operate when demands are lower.

Furthermore, the *method for determining the safety-stock* needs to be improved. In particular, the method should consider different demand fluctuations, replenishment times and delivery date deviations for the orders (see Sect. 8.3.1). The safety-stock should also not be increased as long as the production has not reduced the backlog.

*Calculating the lot size* should take into consideration the costs related to the lot sizes for the different variants and contribute as much as possible to harmonizing the work content. In order to do so, an enterprise could, for example, implement the Throughput Oriented Lot Sizing method (Sect. 8.3.4).

A *WIP control* would help the enterprise to not overload the production with orders and WIP. This frequently has a positive impact on the *sequencing compliance*, which incidentally, should also be expected from the operators more even without a WIP control. Especially when the delivery readiness is unsatisfactory and there is a backlog in the production, it is important to complete important orders punctually.

Finally, the *planned throughput times* should be reduced via a Logistic Positioning (for a more extensive discussion see Sect. 3.2.2.3). This has two outcomes: First, it will reduce the replenishment times and with that the order points (see Eq. 8.5). Through that, fewer replenishment orders will be generated for a transitional period and the production's load will sink for a period. As a consequence, the actual replenishment times will also be reduced. Second, shortening the planned throughput times also postpones the planned completion dates of the operations for the orders that have already been generated and are in production (with the exception of the last operation). As a result, the planned output is decreased and thus too the backlog. Especially when the bottleneck workstations are in the initial part of the material flow, reducing the planned throughput times offers the ("one time only") opportunity to reduce backlogs by shifting the planned schedule.

## 8.5.2 Implementing the Order Point System Without EDP

As a rule the Order Point System is implemented with PPC software. However, especially when only a very few variants are controlled it can be implemented quite well visually or with the help of cards (Fig. 8.8). Both increase the transparency for the production's workforce.

Figure 8.8a depicts the visual implementation of the Order Point System. The order point is specified for the store of each variant. If the stock meets or falls below the order point, the responsible operator reproduces a lot of that variant. In the example, this is the case for Variants 1 and 3.

Figure 8.8b shows how to implement the Order Point System using cards. A card marks the position of the order point in a tier of stock. When the stock reaches the order point, the card is transported to the generation area of the production, thus triggering the replenishment of the variant [Mond-98*]. The card contains

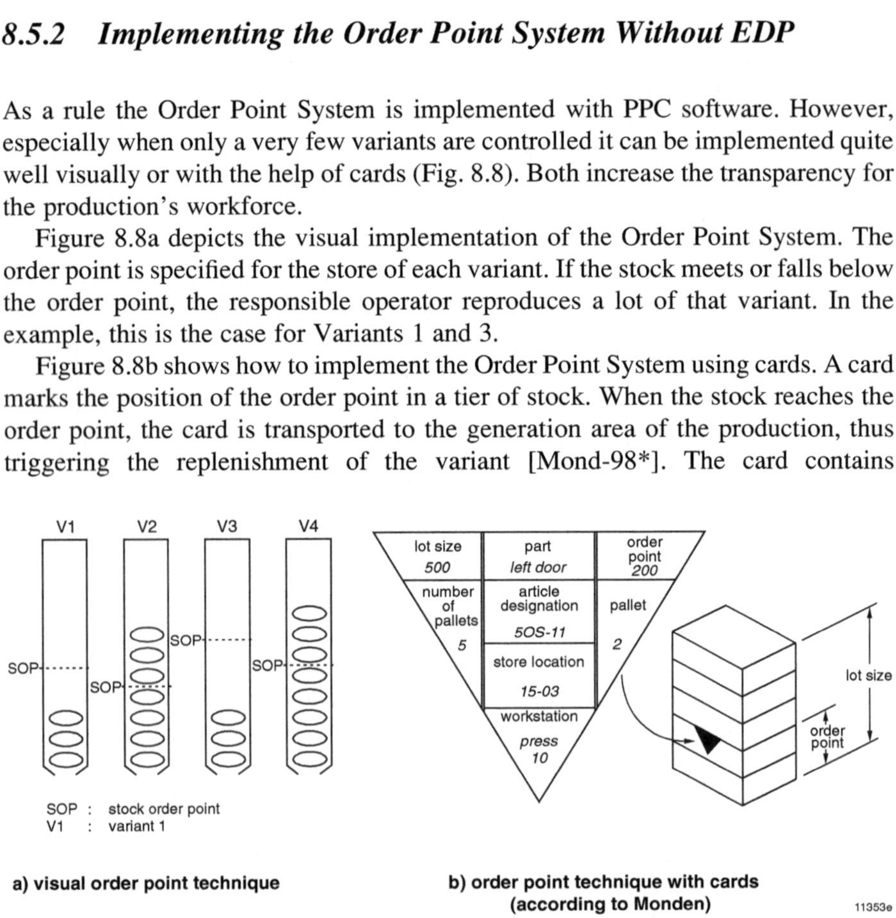


**Fig. 8.8** Possibility for implementing the order point system

information such as the lot size, the article designation and the production area. In this form, the Order Point System is similar to the Kanban System which can also be interpreted as a special form of the Order Point System. In the next chapter, we will take a closer look at the Kanban System.

## References


[Grei-97] Greiling, M.: Verbesserung der Produktionslogistik durch Losgrößenharmonisierung. Ein bedientheoretisch basierter Ansatz. Scientific Reports of the Institute for Material Handling and Logistics, University of Karlsruhe (TH), Karlsruhe (1997)

[Lutz-02] Lutz, S.: Kennliniengestütztes Lagermanagement. VDI Progress Reports, Series 13, No. 53, (2002)

[Mond-98*] Monden, Y.: Toyota Production System. An Integrated Approach to Just-In-Time. Engineering & Management Press, Norcross (1998)

[Nyhu-09*] Nyhuis, P., Wiendahl, H.-P.: Fundamentals of Production Logistics. Theory, Tools and Applications. Springer, Berlin (2009)

[Nyhu-91] Nyhuis, P.: Durchlauforientierte Losgrößenbestimmung. VDI Progress Reports, Series 2, No. 225, Düsseldorf (1991)

[Schö-07*] Schönsleben, P.: Integral Logistics Management. Operations and Supply Chain Management in Comprehensive Value-Added Networks, 3rd edn. Auerbach, Boca Raton (2007)

[Voll-91*] Vollmann, T.E., Berry, W.L., Whybark, D.C.: Manufacturing Planning and Control Systems. Irwin McGraw-Hill, Homewood/Boston (1991)

# Chapter 9
# Kanban

Kanban (*Kanban* is Japanese for card) is an important component of the Toyota Production System [Ohno-88*]. It delegates the responsibility for the stock to the production operators and establishes control loops for every variant between the production's workstations as well as the suppliers and purchasers.

Because Kanban is so well known and successful, business and production managers frequently have high expectations when introducing it. However, in many cases Kanban cannot attain the desired results without intensively restructuring the production and to some degree the products. The general conditions of the production are primary reason the Toyota Production System is so efficient. Essential elements include very short setup times, organization according to the one-piece-flow principle, controlled processes and very thoroughly planned productions.

There is an almost incalculable selection of publications about Kanban which describe the numerous variations of the method [Berk-92*, Golh-91*, Kim-85*, Kimu-81*, Masc-91*, Mora-91*, Sugi-77*, Uzso-90*]. Within German speaking areas, Wildemann has been one of the pioneers of this method [Wild-84]. In this next chapter we will examine the basic method as well as a number of variations (Sect. 9.1). As mentioned in the previous chapter, Kanban can be understood as a special case of the Order Point System. Accordingly, Sect. 9.2 is dedicated to comparing the methods. The procedural guidelines are then discussed and evaluated with regards to aspects of production logistics (Sect. 9.3) whereas, the fourth section introduces the pull principle that is typical of Kanban. The pull principle avoids allocating inventory too early to one of the consumers. Subsequently, the procedure for deriving the method's parameters based on the logistic objectives will be described (Sect. 9.5) followed by a discussion about applying a Kanban control loop across a supply chain (Sect. 9.6), and a description of a Kanban application (Sect. 9.7).

H. Lödding, *Handbook of Manufacturing Control*,

DOI 10.1007/978-3-642-24458-2_9, 

## 9.1 Method Description

*Preliminary Remarks:* Generally, a workstation is understood as a machine or group of machines arranged parallel to one another and which execute the same type of production process. In comparison, with Kanban the part flow is organized according to the one-piece-flow concept, thus, in our discussion of Kanban the term 'workstation' also includes manufacturing cells and production lines.

The basic idea of Kanban is that each workstation reproduces that which the subsequent (i.e. downstream) workstation has consumed (see Fig. 9.1 for a basic depiction of the method). In this way an over-production – which according to T. Ohno the founder of the Toyota Production System is one of the seven types of waste [Ohno-88*] – should be prevented while, nonetheless, attaining a high service level.

The flow of information is regulated using Kanbans. Every lot is allocated a Kanban at every point in time, so that the number of Kanbans limits the stock of materials.

In the following a Kanban system using one type of card (production Kanbans) will first be described (Sect. 9.1.1). It will then be extended into a Two-Card-Kanban with transport Kanbans (Sect. 9.1.2). Furthermore, Visual Kanban (Sect. 9.1.3), Bin Kanban (Sect. 9.1.4) and Minimal Blocking (Sect. 9.1.5) will be discussed.

### *9.1.1 One-Card-Kanban*

Every variant on a workstation is allocated a defined number of (production) Kanbans (Fig. 9.2). The Kanban moves back and forth between the workstation and the output store of the workstation. On the workstation it authorizes the reproduction of the corresponding variant, whereas in the store it identifies the variant. A Kanban contains at least the following information:

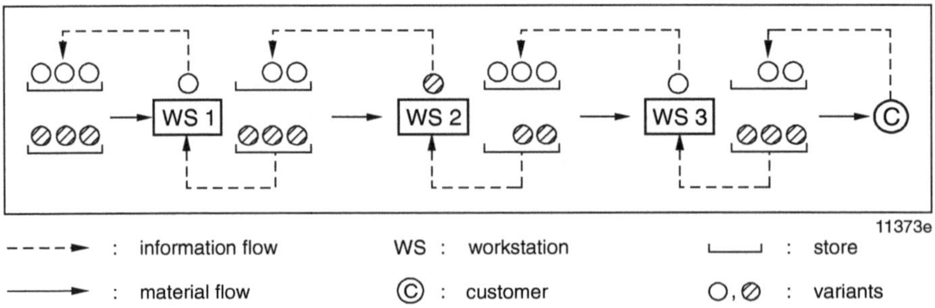


**Fig. 9.1** Principle of Kanban

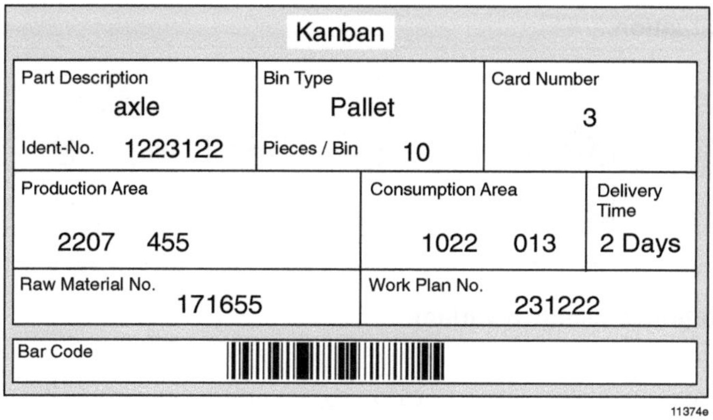


**Fig. 9.2** Composition of a production Kanban (Wiendahl)

- relevant workstation (production area),
- relevant variant (identification number),
- lot size and/or bin contents, and,
- running number of the Kanban.

The running number allows the number of circulating Kanbans to be controlled. Occasionally, the total number of Kanbans are also indicated. This has the advantage of a greater transparency, however, should the enterprise change the number of Kanbans, it also has the disadvantage of requiring the effort to change this information on all of them.

In the example depicted in Fig. 9.3 the basic information is supplemented with additional information such as the delivery time and the consumer area. The latter is only possible when there is only one area that consumes the part. In order to simplify the electronic inventory control, the Kanban also contains a bar code, which depicts the article number in bar form.

The following procedural rules describe how to control production using Kanban:

1. The operator of a workstation is only allowed to process a lot when there is a Kanban for the corresponding variant and the reproduction is thus authorized.
2. If there are Kanbans for a number of variants, the operator then selects the variant with the highest priority. The basic method assigns priority according to the sequence of the Kanbans input. Nonetheless, other sequencing guidelines may be applied with Kanban.
3. The operator verifies whether or not the corresponding materials are available in the receiving store. This should generally be the case. The operator then withdraws the material and the allocated Kanban. The Kanban is then conveyed to the preceding workstation, where it then authorizes the reproduction of the output materials. If the materials are *not* available (e.g., because the preceding workstation has been subject to an extended disruption), the operator selects the

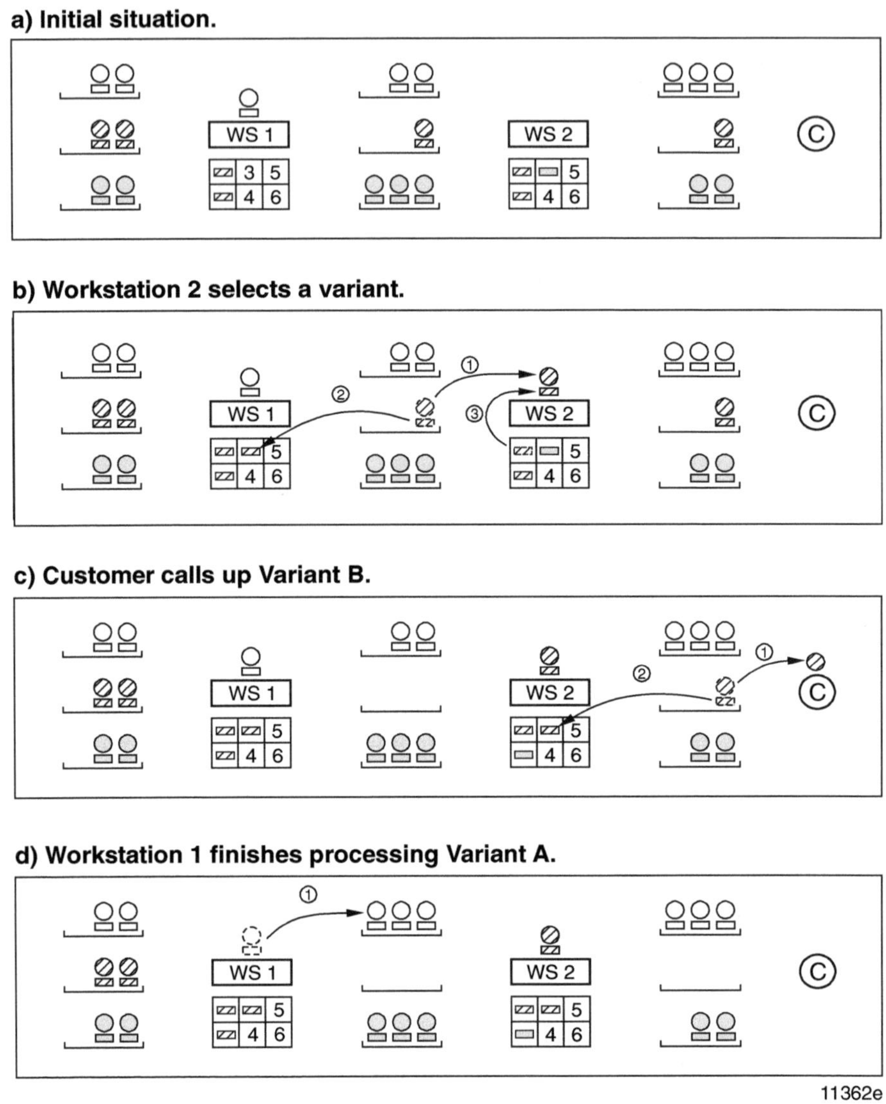


**Fig. 9.3** How Kanban functions

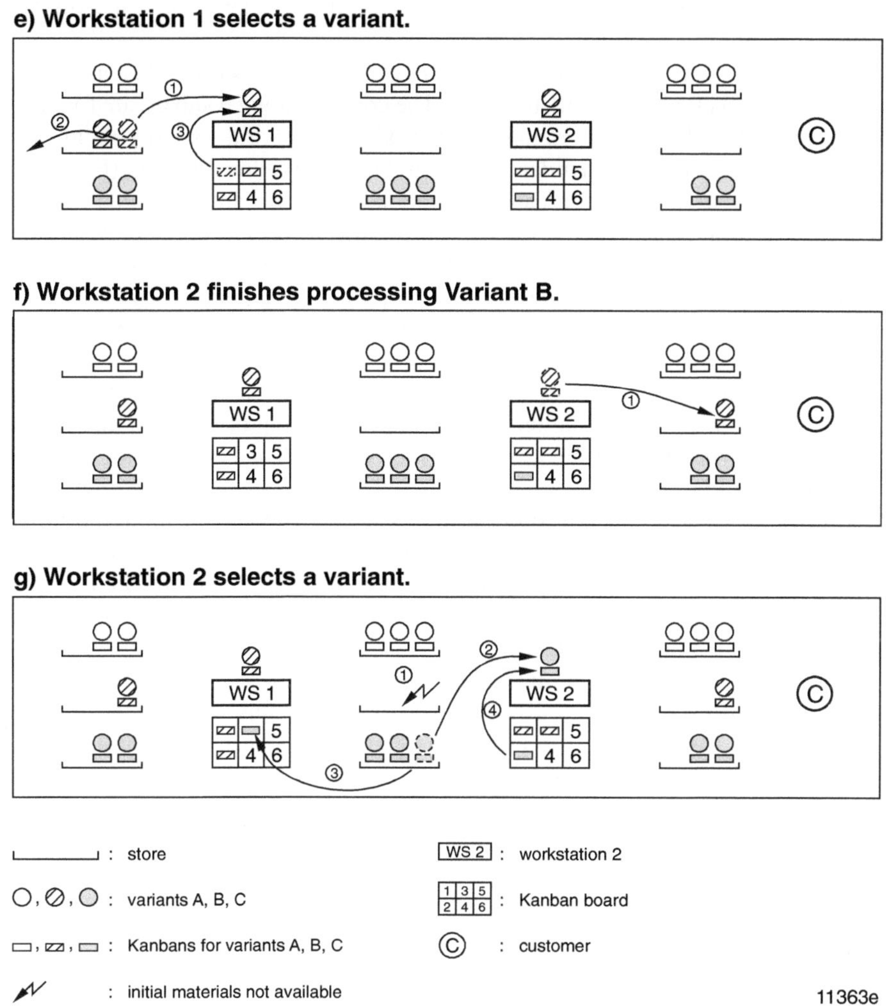


**Fig. 9.3** (continued)

next variant. If none of the materials are available for any of the variants for which a Kanban from one of the subsequent workstations has signaled a reproduction, then the workstation cannot be operated. It is blocked until either a Kanban from a variant arrives for which the materials are available or until the material for one of the already present Kanbans are delivered.

4. If the material as well as the Kanban is available for a variant, the operator starts producing the variant. After it is processed the operator conveys the Kanban and the relevant transport bin with the material to the store of outgoing goods and then verifies whether or not another variant can be processed (Steps 1–4 are then repeated).

I would now like to use an example involving two workstations and three variants (Fig. 9.3) to explain these procedural guidelines more thoroughly. For the sake of simplicity it is assumed that the production lots and transport lots correspond with one another.

In the initial state, the finished goods store for Variant A is full, whereas storage room for Variants B and C is still available. The corresponding Kanbans are found on Workstation 2 and are authorizing the reproduction of the variants there. The workstation just completed an order and is searching for a follow-up order (Fig. 9.3a).

The replenishment order for Variant B entered the workstation first and thus has the highest priority. The material for Variant B is available. The operator withdraws the material from the store (Step 1 in Fig. 9.3b) and directs the corresponding Kanban back to the preceding workstation (Step 2). The operator takes the Kanban from the Kanban board and secures it on a transport bin (Step 3). Finally Workstation 2 starts processing the order.

A little later a customer calls up Variant B (Step 1 in Fig. 9.3c). The corresponding Kanban then returns to Workstation 2 and authorizes the reproduction of that variant (Step 2). Next Workstation 1 finishes processing Variant A and conveys the lot to the store (Step 1 in Fig. 9.3d).

Afterwards, it selects the next order (Fig. 9.3e). Variant B has the highest priority. The operator withdraws the material from the received goods store (Step 1) and sends the corresponding Kanban back to the preceding workstation (Step 2). The Kanban from Workstation 1 is then secured to a transport bin (Step 3) and the processing of the order is begun.

Once Workstation 2 has completed the processing of Variant B, the operator conveys the material to the finished goods store and thus restores the on-time delivery for the variant (Step 1 in Fig. 9.3f). Variant B still has the highest priority. The outgoing material though is not available due to a disruption (Step 1 in Fig. 9.3g). Instead the operator reproduces Variant C. In order to do so he in turn withdraws the material from the store and sends the corresponding Kanban back to the preceding workstation. He brings the workstation's Kanban to the bin and begins processing the variant (Steps 2–4).

### *9.1.2 Two-Card-Kanban*

With Two-Card-Kanban there are usually WIP buffers for a variant both in the area where it is produced as well as where it is consumed. This is especially required when there is a considerable distance between the producer and the consumer. In order to control this sort of systems, two types of Kanbans are required:

The *transport Kanban* triggers the transport of materials from the producer's store of outgoing goods to the receiving goods store of the consumer. The *production Kanban* authorizes the reproduction of the variant. It thus corresponds to the Kanban in the One-Card-Kanban.

The transport Kanban contains at least the following information:

- place of storage,
- area requiring it, and,
- part number.

The way the Two-Card-Kanban functions is described here based on a simple example with only one variant (Fig. 9.4).

In the initial situation all of the stores are full (Fig. 9.4a). At a specific time a customer calls up a product from the finished goods store (Fig. 9.4b).

The Kanban bin is withdrawn from the finished goods store (Step 1) and the production Kanban is returned to Workstation 2, thus authorizing the reproduction of a further lot (Step 2). The operator on the workstation withdraws the materials from the store of received goods (Step 3). The allocated transport Kanban goes back to the outgoing goods store at the preceding workstation and authorizes the transport of another lot to the receiving goods store at Workstation 2 (Steps 4 und 5).

The production Kanban that has now become free in the store of outgoing goods at Workstation 1 authorizes the reproduction of a lot on Workstation 1 (Step 6). The corresponding operator withdraws the materials from the store of raw materials (Step 7). The transport Kanban that is freed returns to the (not depicted here) supplier (Step 8) and triggers the transport of another lot into the raw goods store.

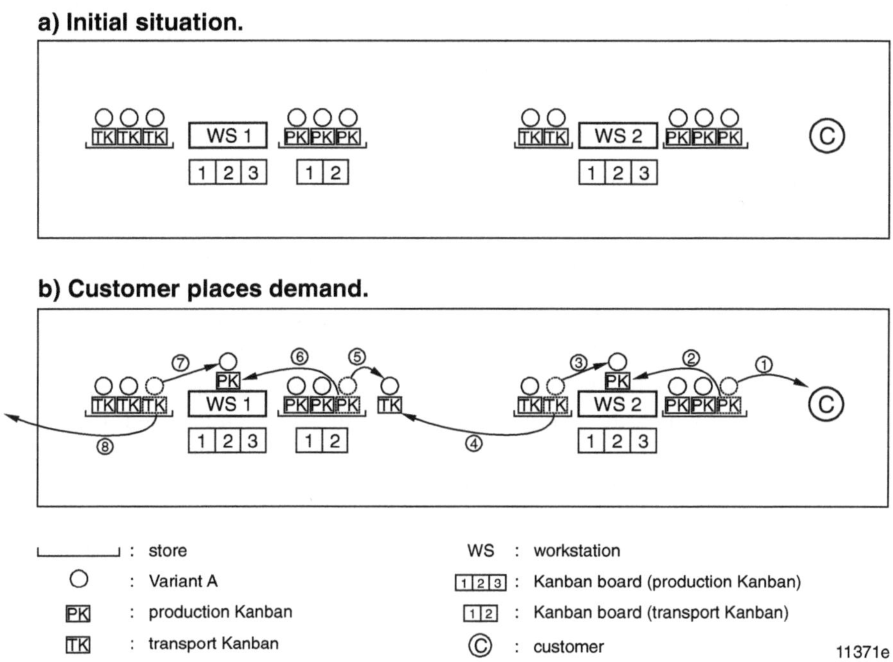


**Fig. 9.4** How the two card Kanban system functions

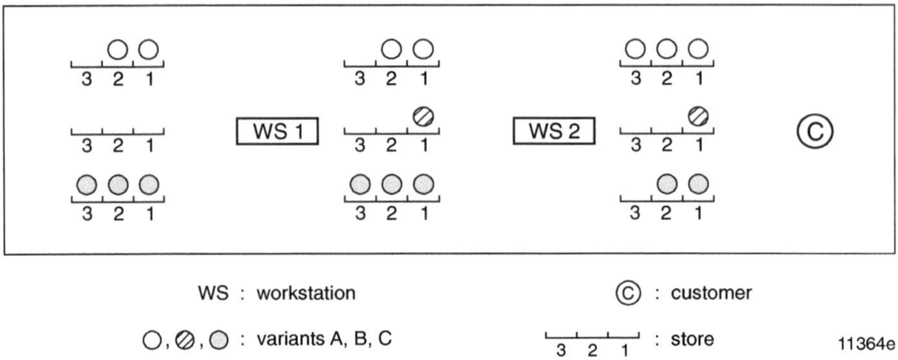


Fig. 9.5 How visual Kanban functions

## 9.1.3 Visual Kanban

Visual Kanban functions similarly to One-Card-Kanban with production Kanbans described in Sect. 9.1.1. However, the use of cards as a means of information is refrained from. The workstation and the store are found directly next to one another. The operator thus sees how many lots or bins are missing in comparison to the maximum stock level. In order to facilitate this, the storage places of the variants are frequently labeled.

With Visual Kanban, a workstation reproduces a variant in the case where

1. the stock level falls below the maximum level, and,
2. the materials are available.

If this applies to a number of variants, the workstation selects the variant with the highest priority. This is generally the variant with the lowest stock level or the lowest stock range. Nevertheless, other sequencing guidelines can also be implemented.

In the example from Fig. 9.5, Workstation 2 could produce Variants B and C, but not Variant A for which the maximum stock has already been attained. Using the described sequencing rule (lowest stock level) Workstation 2 reproduces Variant B. Workstation 1 is only able to reproduce Variant A since the maximum stock for Variant C is already attained and the materials for Variant B are not available.

When the conditions for implementing the Visual Kanban are fulfilled, it is very simple to implement and promotes a high level of transparency.

## 9.1.4 Bin Kanban

With Bin Kanban the transport bin simultaneously serves to convey information. Bins are allocated specifically to one variant and move back and forth between the workstations producing and consuming that variant. Accordingly, the number of transport bins limits the store of the variants. An empty transport bin authorizes the reproduction of the corresponding variant on the production work station. In order to better understand how Bin Kanban functions, let us consider an example in which two variants A and B are produced from the same raw materials (Fig. 9.6).

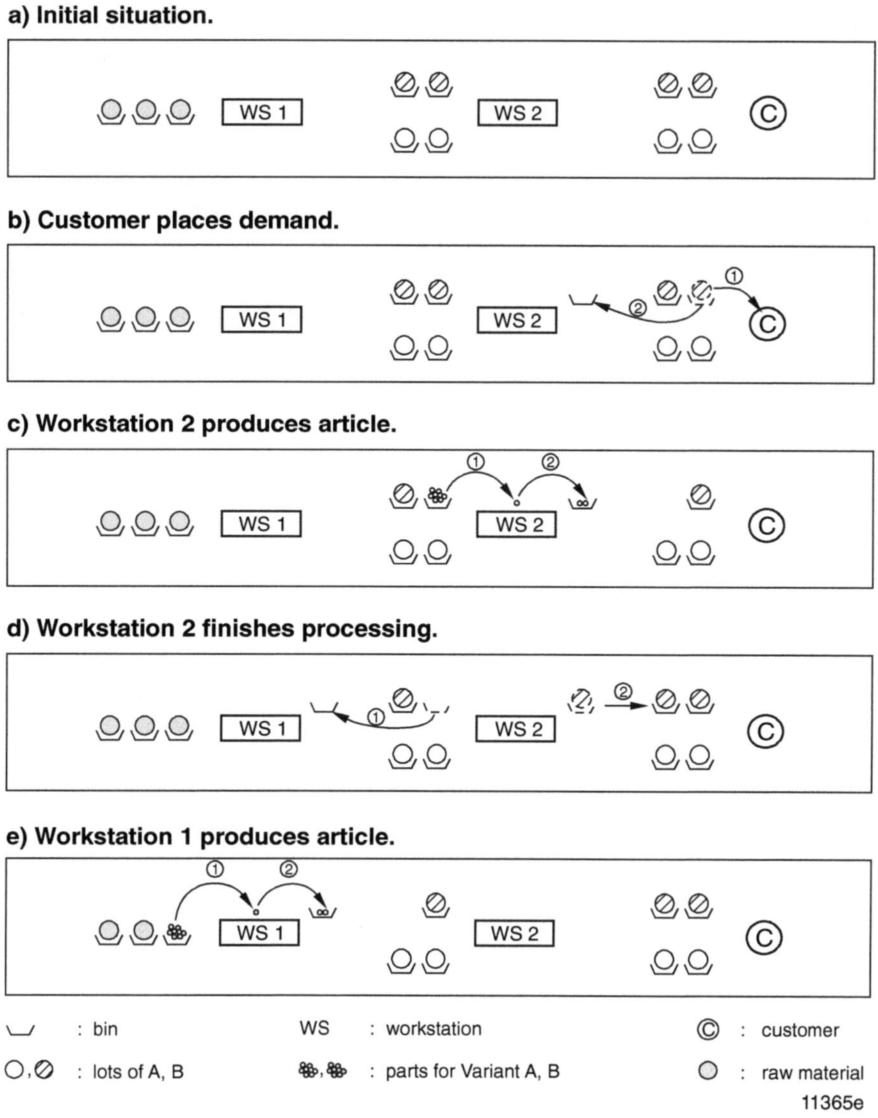


**Fig. 9.6** How Bin Kanban functions

In the initial situation, all of the bins are to be found in the receiving store of the consumer's workstation (Fig. 9.6a). If the customer demands a variant the contents of the bin is sent to the customer (Step 1 in Fig. 9.6b). The empty transport is returned to Workstation 2 and authorizes the reproduction of that variant there (Step 2). The operator on the workstation begins processing the order. She withdraws the initial material piece for piece from the transport bin in the receiving store, processes it and places it then in the bin in the workstation's outgoing store (Steps 1 and 2 in Fig. 9.6c).

Only after all its contents are processed does the operator direct the empty transport bin back to Workstation 1, where it then authorizes the reproduction of the output materials (Step 1 in Fig. 9.6d). The lot that has been completed goes to the finished goods store (Step 2) and Workstation 1 begins reproducing the variant (Step 1 and 2 in Fig. 9.6e).

In the described form the Bin Kanban differs from the usual variations of Kanban in two fundamental aspects. First, the material is stored at the consumer's workstation ('bring' principle). Second, the reproduction of the outgoing materials is delayed by the time required to process the variant on the consumer workstation. Both factors can reduce the method's efficiency (see Sect. 9.4).

### *9.1.5 Minimal Blocking*

Minimal Blocking is a variation of Kanban which has been researched in particular by So and Pinault [So-88*, So-90*], Mitra and Mitrani [Mitr-90*] as well as Bonvik, Couch and Gershwin [Bonv-97*]. The first two groups of authors showed that with a comparable WIP, Minimal Blocking is more efficient than conventional Kanban. In contrast, the latter group's findings did not support a clear advantage.

All of the authors examined a manufacturing line with only one product. In order to extend the application range of the method, a production with a number of variants is considered here.

Unlike the One-Card-Kanban, Minimal Blocking allocates each workstation – if not physically than at least in the method's logic – a receiving and an outgoing store (Fig. 9.7).

The number of the Kanbans limits the number of orders of a variant in the receiving store, in the processing and in the outgoing store of the workstation.

Minimal Blocking can be described with the following rules:

1. The operator of a workstation is allowed to process all of the orders in the receiving store of the workstation.
2. If there are orders for a number of variants, the operator selects the variant with the highest priority. This can, for example, be the variant with the lowest stock

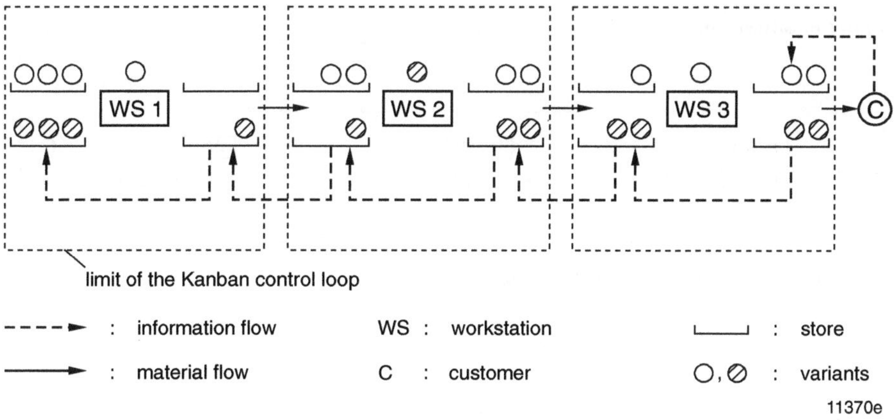


**Fig. 9.7** Principle of minimal blocking

range in the outgoing store and in the receiving store of the subsequent workstation.[1]

3. Each order for a variant is allocated a variant and workstation specific Kanban. This Kanban remains with the order in the receiving store, during the processing and in the outgoing store. Accordingly, the number of Kanbans for a variant on a workstation limits the number of orders for the variant.
4. If there is a Kanban free on the subsequent workstation, a lot of the same variant – when there is one available – is transported from the outgoing store of the observed workstation to the receiving store of the subsequent workstation. As a result, a Kanban is then freed on the observed workstation and the transported order is allocated a Kanban on the subsequent workstation.

Again, I would like to take a closer look at these procedural rules using an example (Fig. 9.8) in which a production with three workstations and two variants is considered. Each variant is allocated three Kanbans on each workstation.

In the initial situation (Fig. 9.8a) Workstation 3 is blocked. There is a Kanban for Variant A, but the material is not available. There is a lot of Variant B in the outgoing store of the preceding workstation, however, there is no Kanban that would authorize the production of it. Workstations 1 and 2 are processing an order for Variant A.

First, a customer places a demand for a lot of Variant B (Step 1 in Fig. 9.8b). The Kanban that is thus freed authorizes the transport of a lot of Variant B from the outgoing store on Workstation 2 to the receiving store on Workstation 3. Since the workstation is free and no other orders are available the workstation begins

[1] Sequencing is not discussed in the research, since one of the conditions is always a production line with only one variant.

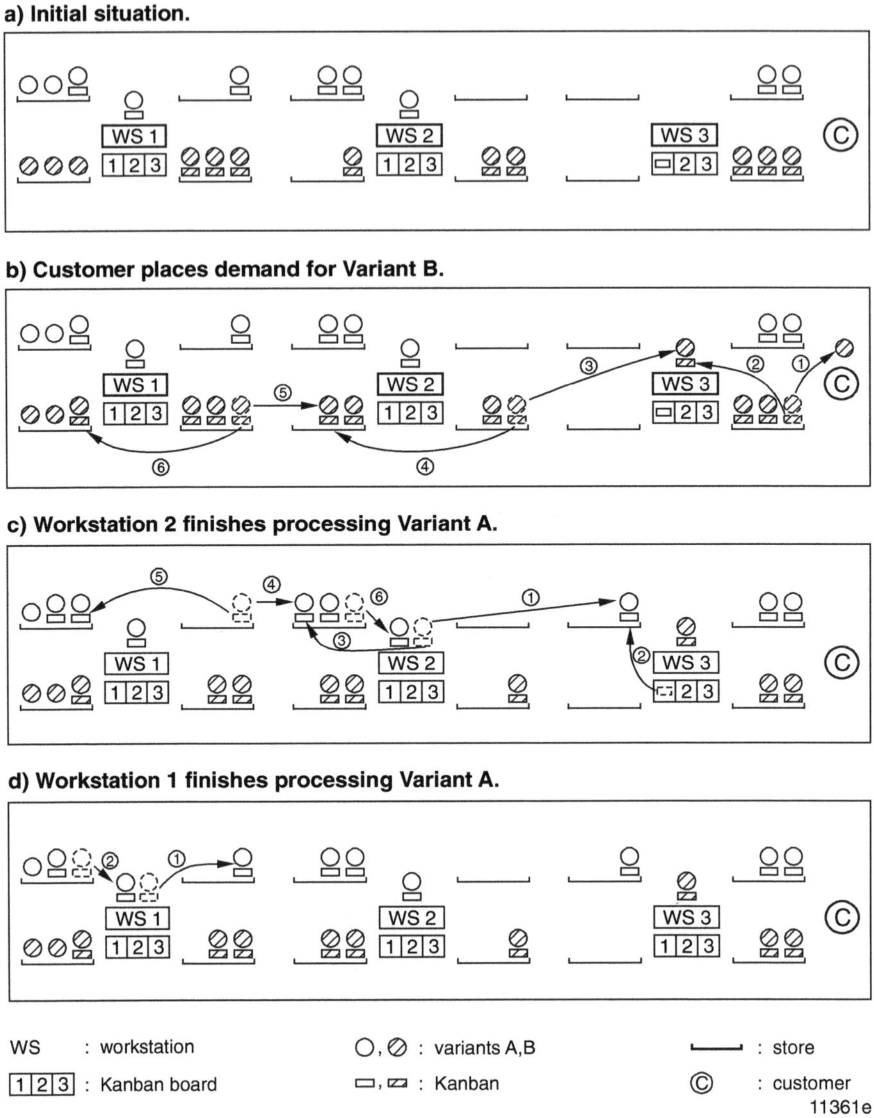


**Fig. 9.8** How the minimal blocking control functions

immediately to process the order. The Kanban and lot are therefore transported directly to the workstation (Steps 2 and 3).

In the outgoing store from Workstation 2 a Kanban is freed (Step 4). It triggers the transport of a Variant B lot from the outgoing store on Workstation 1 to the receiving store on Workstation 2 (Step 5).

As a result, a Kanban in the outgoing store from Workstation 1 is in turn freed (Step 6) and allocated to a lot in the raw materials store. The Kanban there indicates that Workstation 1 is allowed to process Variant B. The raw materials store is the only store in which there can be lots without Kanbans.

Next Workstation 2 completes processing an order for Variant A (Fig. 9.8c). Since a Kanban for Variant A is free on Workstation 3, the operator directs the lot to the receiving store on Workstation 3 (Step 1), where it is allocated the free Kanban (Step 2). The Kanban that becomes free on Workstation 2 (Step 3) triggers the transport of a lot from the outgoing store of the preceding Workstation 1 to the receiving store on Workstation 2 (Step 4). As a result a Kanban in the outgoing store from Workstation 1 is freed, which in turn is allocated to a lot in the raw materials store (Step 5). Workstation 2 can now produce an order for Variant A as well as one for Variant B. In this case, Variant A has the higher priority which is why the operator begins processing it (Step 6).

Finally, Workstation 1 completes processing Variant A (Fig. 9.8d). Since there is no available Kanban for this variant on Workstation 2, the lot and Kanban are conveyed to the outgoing store (Step 1). Workstation 1 in turn selects the variant with the higher priority (in this case Variant A) and begins processing it (Step 2).

When implementing Minimal Blocking on the shop floor, the outgoing store of the one workstation and the receiving store of the subsequent workstation would probably not be physically separated. With a common store, the transport between the stores does not apply; instead, only the Kanbans are moved.

The more extensive Kanban flow which is involved with Minimal Blocking, allows greater WIP fluctuations before the workstation, however, limits the total WIP on the production line. As a result the blockage of a workstation resulting from a disruption on a downstream workstation is delayed in comparison to conventional Kanban controls. The same applies when the subsequent workstation temporarily becomes a bottleneck due to dispersed processing times.

The maximum WIP of a variant before a workstation (in the receiving store as well as in the outgoing store of the preceding workstation) corresponds to the sum of the Kanbans for the variant on the observed workstation and preceding workstation:

$$WIPO_{max,i} = NKan_{i-1} + NKan_i \tag{9.1}$$

where

| | |
|---|---|
| $WIPO_{max,i}$ | maximum WIP level in number of orders on Workstation i [-], |
| $NKan_i$ | number of Kanbans on Workstation i [-]. |

Whether or not the logistic targets can actually be more consistently attained with Minimal Blocking than with conventional Kanban depends on the characteristics of the production. The more disruptions influence the production and the greater the variance of the work content between the workstations is, the more likely Minimal Blocking will actually be able to target a better performance. It is obvious that those representing the traditional just-in-time philosophy try to minimize these factors. Practitioners also point out that the procedural rules are comparatively more complicated and make implementing the method more difficult.

## 9.2 Comparison of Kanban and the Order Point System

As previously mentioned, Kanban can be understood as a special case of the Order Point System (Chap. 8). In order to do so the Order Point System should be parameterized as follows:

1. Fixed order quantity (order quantity = bin contents), and,
2. Order point = maximum stock – 1 (maximum stock corresponds to a multiple of the bin contents).

When this is the case, no fundamental difference between the methods can be assumed nor can the superiority of one over the other be postulated. Instead, the identity of the procedural rules which results from selecting the corresponding parameters causes us to realize that it is not Kanban itself, but rather the beneficial logistic conditions that are typical when implementing it which lead to a better attainment of logistic targets. The latter has also been supported by research using simulations [Rees-89*].

The necessary logistic boundary conditions include:

- minimal number of variants,
- small lot sizes and short setup times,
- organized as manufacturing cells (one piece flow),
- controlled processes,
- minimal consumption fluctuations, and,
- sufficient capacity reserves and flexibility.

In comparison, there are considerable differences in implementing Kanban and the Order Point System. Two of the fundamental differences are explained here:

*Delegation of the Responsibility for the WIP in the Production*: Kanban delegates the responsibility for the WIP to the operators in the production and frequently to the supervisor. This allows the suitability of the WIP level to be evaluated immediately on-site. In contrast, with the Order Point System, the order point and safety-stock are usually calculated automatically by planning software, thus making it more difficult to gradually decrease the WIP within the context of a continual optimization process. Secondly, there is the risk of amplifying fluctuation in the demand rate when the safety-stock is not suitably set. Thirdly, the employees who set the method parameters generally do not work in the production and are thus not directly confronted with the impact of their decisions.

*Visual Production Management:* Those who support Kanban emphasize the importance of greater transparency. This transparency is frequently not present when implementing the Order Point System due to the large number of variants and because orders are generated and parameters are set centrally. Last but not least, the material is stored in the production with Kanban, whereas it is frequently stored away from the production when implementing the Order Point System. As a result it is not possible for the production to directly access information about the stock situation.

## 9.3 Discussion of the Procedural Rules

We now come to our discussion about the procedural rules oriented on the guiding principles for designing manufacturing control systems (see Sect. 3.3.2).

1. *A manufacturing control method should allow the WIP of a production or workstation to be regulated at a defined level.*

Kanban limits the WIP on a workstation by the number of circulating Kanbans. Precisely controlling the (available) WIP on a workstation is nonetheless quite difficult. That is, there is no WIP that is not allocated a Kanban, however, the reverse situation does arise: The Kanban board contains Kanbans that are not allocated to any WIP. If the number of free Kanbans varies, then the WIP on the system also changes.

2. *A manufacturing control method should reduce the WIP fluctuations to a minimum.*

The WIP on a workstation is determined by the number of Kanbans that are allocated to a lot. If all of the Kanbans are allocated, the maximum WIP is then attained. If all of the Kanbans are on the Kanban board, then the WIP is zero (see Fig. 9.9).

As a result, it is not possible to precisely control the WIP on a workstation with the Kanban system. The WIP, therefore, can fluctuate considerably.

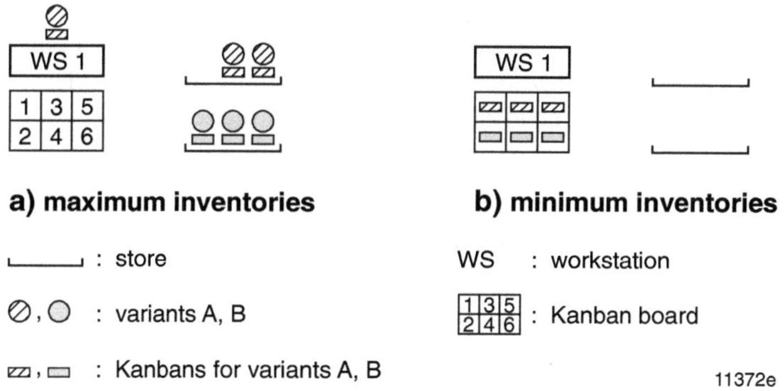


**Fig. 9.9** Limiting cases for the WIP with Kanban

3. *A manufacturing control method should cause as little blocked WIP as possible.*
With the Kanban control, the material in a workstation's outgoing store for which there is no production order on the subsequent workstation – in the form of a free Kanban – is blocked. The extent of the blocked WIP corresponds to the safety-stock when the demand is absolutely constant and the replenishment times and quantities are maintained exactly (see Sect. 9.5.2). The higher the number of variants produced on a workstation, the smaller the demand per variant and with that mathematically also the number of required Kanbans per variant. The number of Kanbans, however, cannot fall below a minimum number per variant (1 or 2 Kanbans), so that the total number of Kanbans increases with the number of variants. Accordingly, the amount of blocked WIP also increases with the number of variants. This is one reason why Kanban can only be economically implemented when there are only a few variants. Some practitioners say that there is a limit of six to eight variants.

4. *A manufacturing control method should contribute to closely aligning the load to the capacities.*
Load balancing accelerates or delays the release or throughput of orders with the goal of preventing breaks in the material flow (acceleration) or a build-up of WIP (delays). The Kanban system does not balance the load in any way that makes the release of orders dependent on the load of all the workstations they are routed through. This is – at least with the Toyota Production System – also not necessary because it requires the load planning to be coordinated with the capacities.

Kanban prevents the WIP, however, from building-up on the subsequent workstation in that it delays the processing of materials for which there is no Kanban available. This is an advantage in comparison to other order generation methods which implement long control loops for generating orders and therefore do not take into consideration the capacity restrictions on a workstation.

Kanban's load balance though is not as efficient as the load balance made possible by order release methods. As a variant specific method, Kanban controls different variants in separate control loops. If there are a number of variants produced on one workstation, it always makes decisions based only on a variant specific portion of the WIP. Accordingly, it requires a generally higher WIP in order to attain a specific utilization of the workstation than order release methods that make decisions about releasing orders based on the total WIP.

5. *A manufacturing control method should cause as few sequencing interchanges as possible.*
The Kanban system does not cause any sequencing interchanges that would negatively impact the service level.

6. *A manufacturing control method should be able to balance schedule deviations within the context of backlog control.*
It is the responsibility of the manufacturing control to prevent a threatening backlog and to reduce those backlogs which have developed. In order to do so capacities are temporarily increased. The procedural guidelines for Kanban itself fail to consider a

backlog control. The Toyota Production System, however, places great value on flexible capacities. Moreover, the work hours can also be temporarily increased [Ohno-88*].

The flexibility of the work hours serves to buffer fluctuations in the demand rate or the yield of the production. Frequently the resource capacities are designed so that the production usually operates in two shifts. It is thus possible to control the capacities when utilizing Kanban either with a Backlog Control (Chap. 27) or with the Inventory based Capacity Control (Sect. 28.4).

The flexibility of the capacities is considerably limited when an enterprise dimensions the capacities on the small side in order to minimize the investment costs. The production is then usually operated in three shifts and lacks the flexibility to react to demand fluctuations.

7. *A manufacturing control method should take into consideration the principle of bottlenecks.*

The rules for Kanban do not explicitly consider the throughput bottleneck of a production. Nonetheless, the production bottleneck provides the clock cycle for the production. If the preceding workstation has a greater capacity the Kanbans are frequently found in the outgoing store; moreover, it is frequently blocked. When the bottleneck system withdraws goods from the store, it removes the blockade and initiates the reproduction of materials. It is thus supplied with sufficient materials, without a particularly high WIP level being required.

Furthermore, the bottleneck workstation itself is only rarely blocked by subsequent workstations. If the bottleneck station is to be paid special attention when designing the parameters, the number of Kanbans on the preceding workstations and on the bottleneck station itself should be designed particularly high. The first should prevent breaks in the material flow and the latter should prevent blockades due to subsequent workstations.

The Toyota Production System attempts, nevertheless, to avoid bottlenecks when designing the processes and planning the capacities. In the ideal case all of the workstations work in the same rhythm and accordingly the significance of the bottleneck is reduced.

8. *A manufacturing control method should be as simple as possible.*

When the conditions for implementing Kanban are met, implementing Kanban is very simple. It does not even generally require a computer based information system. This is increasing the acceptance of the method by the production's workforce. The operators immediately see the necessity to reproduce a variant when the store is about to be emptied. A state of affairs visualized in this way, is fundamentally more urgent than a prompt from an electronic information system which for the operators is frequently less transparent.

9. *A manufacturing control method should communicate information about demands as quickly and clearly as possible.*

Generally, the same disadvantages apply to the Order Point System as to Kanban (see Sect. 8.2). However, a number of factors that arise when implementing Kanban tend to defuse these shortcomings: First, Kanban is frequently connected with smaller lot sizes. These are made possible on the one hand due to short setup times and on the other by separating the transportation and production lots. Second, the demand rate for a variant is frequently high with Kanban. Small (transportation) lots and a high demand rate cause demand information to be frequently conveyed. Third, when implementing Kanban in a multi-stage production the lot sizes at the different stages are coordinated with one another. As a result, additional distortion of the demand information – such as that which can surface due to the impact of MRP-nervousness – is avoided (see Chap. 7). Fourth, Kanban is frequently integrated into a higher level production plan, which is coordinated between the partners in the supply chain and conveys demand information over a number of stages. Due to this difference in the implementation, Kanban is also suitable for generating orders in supply chains.

## 9.4 Pull Principle Versus Bring Principle

With the bring principle, materials are brought to the subsequent workstation or workstations directly after they are completed. In contrast, with the pull principle the material remains at the workstation that generates it and is only brought to a consumer workstation when the workstation logs a demand for it. The pull principle has two advantages in comparison to the bring principle:

1. it delays the decision about the use of materials, and,
2. it increases the WIP transparency.

**Delaying the Decision About Usage** The difference between the pull principle and the bring principle is mainly significant when a number of workstations access the same production. In this case, the pull principle requires less WIP than the bring principle in order to ensure that the subsequent workstations are supplied with a specific variant. This is clarified by a simple example (Fig. 9.10).

If a variant is brought to a specific workstation immediately after being processed, then it is no longer available for other workstations. Accordingly, supply bottlenecks or breaks in the material flow can arise although the WIP is available within the entire system (Fig. 9.10a on Workstation 2). The bring principle thus shifts the generation of a variant unnecessarily forward.

If on the other hand the material remains on the workstation that generates it (Fig. 9.10b), all of the workstations requiring it can access it. Accordingly, the WIP can always be allocated to the workstation being threatened with a supply bottleneck. With the same WIP, the entire system can thus attain a higher service level with the pull principle than with the bring principle. If only one workstation accesses a product then the material can be transported to the subsequent workstation

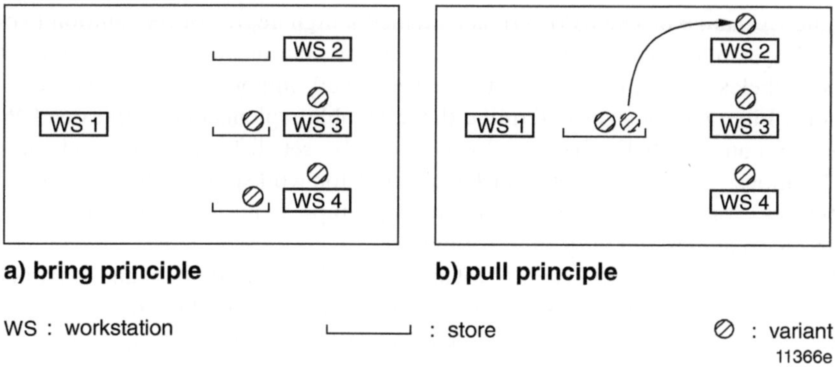


**Fig. 9.10** Difference between the bring principle and the pull principle

directly after it has been completed (however, please refer to the following remarks about WIP transparency).

**Increasing the WIP Transparency** With the bring principle, the material is brought to the subsequent workstation directly after it has been completed. If this is outside the line of sight, the operator does not see when a higher WIP has accumulated on the workstation or when there is a threat of a break in the material flow. With the pull principle, the material remains at the workstation where it is generated. The operator is therefore constantly informed about the WIP situation on the subsequent workstation.[2]

## 9.5 Setting the Method's Parameters

The three essential parameters for Kanban are the lot size, bin contents (Sect. 9.5.1) and the number of Kanbans (Sect. 9.5.2).

### *9.5.1 Lot Size and Bin Contents*

Methods for determining the lot size are only considered slightly important in traditional research on Kanban. Instead, the aim is to radically reduce the setup times so that extremely small lots are economical. Accordingly, the authors concentrate on describing methods for reducing setup times. The so-called Single

[2] One could argue that for this the WIP transparency is lacking at the consuming workstation. Since this cannot influence its material supply anyways, it is not as important as at the producing workstation.

Minute Exchange of Die (SMED) has attained a high degree of recognition [Shin-85*]. A systematic approach is recommended for example by Frühwald [Früh-90].

Nonetheless, determining the lot size is also very important for Kanban for two reasons: First, the lot sizes determine the attainable replenishment times and WIP levels. Second, with Kanban the lot size can be set differently for each of the workstations. Thus larger lots can be selected for workstations that are capacity bottlenecks or have particularly intensive setups than for workstations with reserved capacities and very short setup times.

In practice, it is typical to set the lot size as the daily requirement and to launch a run of each variant once a day [Ever-99]. This has the advantage of constant and easily planned replenishment times. The approach is, nevertheless, connected to obvious disadvantages:

1. When the setup times are very short, it does not exploit the potential of smaller lot sizes. WIP levels and replenishment times could be further reduced.
2. With long setup times this approach can cause high setup costs and where applicable capacity bottlenecks.
3. The approach does not differentiate between variants. Generally, it is practical to run variants with a higher demand more frequently than those with a low demand.

In the following a more pragmatic approach for determining the Kanban lot sizes will be recommended. It is divided into three steps:

1. Selecting the method for determining the lot size,
2. Calculating the uncoordinated lot size, and,
3. Coordinating the lot sizes by determining the bin contents.

**Selecting the Method for Determining the Lot Size** In order to select the appropriate method for determining the lot size for each workstation, the workstations are first differentiated according to setup time and capacity utilization: The two criteria open up a portfolio with four fields. Workstations with shorter or longer setup times and with lower or higher utilization of capacities are distinguished from one another (Fig. 9.11).

The Toyota Production System strives to position all of the workstations in the quadrant with the relative low utilization of capacities and short setup times. The portfolio serves three purposes. First, it identifies the workstations on which measures for reducing the setup times urgently need to be conducted (Quadrants I and II). Second, it indicates which workstations seem to require capacity increases (Quadrants II and IV). And third, it provides information about which method would be suitable for setting the lot size. In this regards, three cases can be distinguished:

***1. Workstations with Short Setup Times (Quadrants III and IV)***

If the setup time is short, the lot size should be set as small as possible (Quadrants III and IV, Workstations 1, 2 und 3). It then always corresponds to the content of a

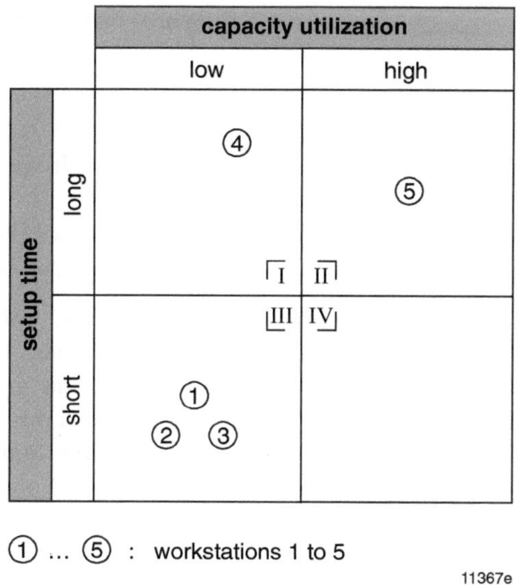


**Fig. 9.11** Capacity utilization and setup time portfolio

transport bin and is thus first set in the last step of the method described for determining the lot size.

2. ***Workstations with Long Setup Times and Low Capacity Utilization (Quadrant I)***

With longer setup times but sufficient capacities (Workstation 4) the lot sizes should be calculated according to an optimization approach. The method should take into consideration at least the costs related to setups and stock. The *setup costs* are generally dominated by personnel costs. Under no circumstances should depreciations of machinery be taken into consideration in these quadrants. The *inventory costs* are proportional to the interest rate, material costs and the mean turn-around-time on the workstation or in the store. One possible method is the Economic Order Quantity. This method has a tendency to calculate lot sizes that are too large (see Sect. 4.1.3.1), however, this can be compensated for by choosing a higher interest rate or a lot size below the calculated optimum. The latter alternative uses the minimal increase of the total cost function from the EOQ formula close to the optimum costs. After that a (moderate) decrease in the lot size increases the costs only to a small degree [Nyhu-91]. However, the area of the cost function in which the costs increase steeply should be shunned.

3. ***Workstations with Long Setup Times and High Utilization of Capacities (Quadrant II)***

If setup times are long and the utilization of capacities high (Workstation 5), the capacity limits can frequently be exceeded by frequent setups. As a consequence, the production program might not be able to be implemented. In this case the maximum setup frequency is calculated as:

$$SUF_{max} = \frac{CAP_{avail} - LD_{plan}}{t_{su}} \tag{9.2}$$

where

| | |
|---|---|
| $SUF_{max}$ | maximum setup frequency [1/SCD], |
| $t_{su}$ | setup time [hrs], |
| $CAP_{avail}$ | available capacities [hrs/SCD], |
| $LD_{plan}$ | planned load [hrs/SCD]. |

With this calculation, the capacity load only considers the individual processing times for the variants and not setup times. The setup frequency is subsequently to be distributed among the variants that are produced on one workstation (see [Burm-97]). The lot sizes for the variants then result from here. In most cases it is practical not to overestimate the capacities in order to be able to buffer disruptions as needed.

**Calculating Uncoordinated Lot Sizes** For each variant and each workstation the uncoordinated lot size is then determined. For the workstations entered in the capacity utilization and setup time portfolio (Fig. 9.11) the following lot sizes result for a Variant A[3] (Table 9.1):

**Determining Coordinated Lot Sizes** Each lot size results as a multiple of the contents of a transport bin. The transport bin should be dimensioned so that the resulting lot size corresponds as closely as possible to the uncoordinated lot size. Furthermore, criteria such as size and weight of the workpieces should be taken into consideration.

In the example, the following lot sizes are one possibility (Table 9.2):

The selected lot sizes are *production* lot sizes. The *transport* lot sizes always correspond to the bin contents. This allows an overlapped production between the workstations. A workstation is only allowed to start with the production when the number of production Kanbans corresponds to the lot size. In the example,

**Table 9.1** Uncoordinated lot sizes for Variant A on Workstations 1–5

| Workstation | 1 | 2 | 3 | 4 | 5 |
|---|---|---|---|---|---|
| Lot size | Bin contents | Bin contents | Bin contents | 75 | 133 |

**Table 9.2** Coordinated lot sizes for Variant A on Workstations 1–5

| Workstation | 1 | 2 | 3 | 4 | 5 |
|---|---|---|---|---|---|
| Lot size | 15 | 15 | 15 | 75 | 135 |

[3] The lot sizes were calculated for Workstation 4 with the EOQ lot-sizing formula and derived on Workstation 5 from the maximum setup frequency.

Workstations 1–3 therefore require one Kanban for Variant A, Workstation 4 – five Kanbans and Workstation 5 – nine Kanbans. The transport bins, however, are withdrawn individually from the store, so that the times at which Kanbans enter the preceding workstation are staggered, thus extensively avoiding a distortion of the demand information. The preceding workstation can thus already start with reproducing materials while the observed workstation still processes the order (overlapped production). Moreover, it can also reproduce other variants in between where necessary.

If the number of Kanbans for none of the variants corresponds with the planned lot size, the workstation (especially when it is a capacity bottleneck) can begin reproducing the variant for which the most production Kanbans are present. Generally, the missing production Kanbans will arrive while the workstation is still processing that variant. If this is not the case, then the workstation ends the processing before the planned lot size is attained.

## *9.5.2 Number of Kanbans*

The following section initially describes a basic method for setting the number of Kanbans per variant (Sect. 9.5.2.1, an alternative process for doing so is described by Bard and Golany [Bard-91*]). Sections 9.5.2.2 and 9.5.2.3 discuss how to calculate the replenishment time and safety-stock. These two parameters are the most significant when it comes to the number of required Kanbans.

### 9.5.2.1 Method

Stock serves two purposes when it comes to Kanban, it: protects workstations from a break in material flow and it ensures the on-time delivery. Let us take a look at these two functions more closely.

***1. Protects Workstations from a Break in Material Flow (WIP)***
If the WIP is not high enough to protect the workstation from breaks in the material flow, there is a loss of utilization on the workstation. The operators cannot be employed productively. Ultimately, this also endangers the availability of the variant that is produced on this workstation. The WIP required to utilize the workstation can be calculated with the assistance of Logistic Operating Curves and is variant neutral.

***2. Ensures the On-Time Delivery (stock of variant = safety-stock)***
The stock in the outgoing store of a workstation serves to ensure the availability of the variant. It buffers demand fluctuations and ensures the on-time delivery when the replenishment times fluctuate. This stock can be calculated with the assistance of Storage Operating Curves and is variant specific.

The number of Kanbans should be set by the supervisor or team leader of the production process that is controlled by the Kanbans in order to delegate the responsibility for improving the process and attainment of logistic targets [Mond-98*].[4] The aim is to gradually reduce the number of Kanbans to a minimum. According to this understanding, the number of Kanbans is a measure of how good the operators control the process.

At least for the purpose of planning a factory it is necessary to estimate the number of Kanbans in advance. Determining the number of Kanbans with the aid of a logistic model however is not simple. Among others, the following two aspects are problematic:

1. The number of circulating Kanbans regulate the sum of the WIP and stores. How the stock is distributed between these two, cannot be exactly controlled and is subject to fluctuations.
2. The stock available for ensuring the utilization is the stock for which there are both materials and Kanbans. Material that is available in the stores, but for which there is no Kanban on the Kanban board indicating a demand for it on the subsequent workstation, is blocked WIP (the subsequent workstation is not allowed to process the stock). It therefore does not ensure the utilization of the workstation.

The number of Kanbans can be calculated according to the following formula (see Monden [Mond-98*], Ohno [Ohno-88*] and Wildemann [Wild-84] for similar equations):

$$NKan = \underbrace{\frac{TRP}{BiC} \cdot RDMD_m}_{WIP} + \underbrace{\frac{SSL}{BiC}}_{Safety-Stock} \tag{9.3}$$

where

| | |
|---|---|
| NKan | number of Kanbans [-], |
| TRP | replenishment time [SCD], |
| BiC | bin contents [units], |
| $RDMD_m$ | mean demand rate [units/SCD], |
| SSL | safety-stock [units]. |

The first sum essentially corresponds to the Kanbans that form the WIP. These are either allocated to lots already being processed or waiting on the Kanban board to be processed. When the turn-around-times for information are significant some

[4] "The final authority to change the number of Kanbans is delegated to the supervisor of each process... If it is desired to inspire improved managerial ability, authority to determine the number of Kanbans must first be delegated." [Mond-98*]

of the Kanbans can also be waiting for transportation to the workstation. These should then be considered separately when calculating the number of Kanbans.

The second sum corresponds to the safety-stock. Given the lot sizes and bin contents, determining the number of Kanbans is limited to determining the planned replenishment time (Sect. 9.5.2.2) and safety-stock (Sect. 9.5.2.3).

#### 9.5.2.2 Calculating the Replenishment Time

The replenishment time is comprised of the following components:

$$TRP = TTP_{prod} + TTP_{inf} + TAD \tag{9.4}$$

where

| | |
|---|---|
| TRP | replenishment time [SCD], |
| $TTP_{prod}$ | production throughput time [SCD], |
| $TTP_{inf}$ | information throughput time (turn-around-time) [SCD], |
| TAD | administration time [SCD]. |

The information turn-around-time and administration time should be determined in the enterprise. Frequently, they are negligible in comparison to the production throughput time. Information turn-around-times can result for example, from transporting the Kanbans.

The mean throughput time of the production can be determined from a Logistic Positioning conducted with the support of Logistic Operating Curves. The corresponding procedure is extensively discussed in Sect. 3.2.2.3. An important input parameter is the ideal minimum WIP with which Kanban can be simplified according to the following formula:

$$WIPI_{min} = \frac{\sum_{i=1}^{n} (PRF_i \cdot WC_i^2)}{\sum_{i=1}^{n} (PRF_i \cdot WC_i)} \tag{9.5}$$

where

| | |
|---|---|
| $WIPI_{min}$ | ideal minimum WIP [hrs], |
| $PRF_i$ | production run frequency of Variant *i* during the planning period [-], |
| $WC_i$ | work content of Variant *i* [hrs]. |

The equation applies to a single workstation without a minimum inter-operation time (e.g., cooling or transport times). The process of calculating Logistic Operating Curves for manufacturing cells is explained in [Lödd-01b].

#### 9.5.2.3 Dimensioning the Safety-Stock Level

The level of the required safety-stock depends on the following factors (see Sect. 3.1.3):

1. fluctuations of the demand rate during the replenishment time,
2. maximum positive lateness, and,
3. maximum negative quantity deviation.

When these factors are known, the safety-stock required for a service level of 100% is calculated according to [Nyhu-09*]:

$$SSL_{SERL=100\%} = \sqrt{\left(DEL^{+}_{max} \cdot RDMD_{m}\right)^{2} + \left(Q^{-}_{max}\right)^{2} + \left(\left(RDMD_{max} - RDMD_{m}\right) \cdot TRP\right)^{2}} \tag{9.6}$$

where

| | |
|---|---|
| $SSL_{SERL=100\%}$ | safety-stock level [units], |
| $DEL^{+}_{max}$ | maximum positive delivery lateness (delay) [SCD], |
| $RDMD_{m}$ | mean demand rate [units/SCD], |
| $Q^{-}_{max}$ | maximum negative quantity deviation (under delivery) [units], |
| $RDMD_{max}$ | maximum demand rate [units/SCD], |
| TRP | replenishment time [SCD]. |

This equation requires the variables to be statistically independent of one another. If the replenishment time can be successfully coordinated with the demand rate, then a lower safety-stock level can be selected. A corresponding procedure will be discussed at the end of this section. First though, we turn to the factors which influence the safety-stock.

**Demand Rate Fluctuations** One of the fundamental aims of the just-in-time philosophy is to prevent fluctuations in the demand rate as much as possible. Toyota, therefore, plans the variants of the final assembly so that the variants alternate one after the other. The finite planning of the final assembly thus provides a constant demand rate for the variants. In the ideal case, no safety-stock is required in order to buffer the demand rate fluctuations. Furthermore, the final assembly plan offers a reliable foundation for planning the demand rate of components and for determining the number of Kanbans. Should an enterprise or supply chain abstain from controlling the demand in a stabilizing way such as this, the fluctuations in the demand rate will increase. Accordingly, a higher safety-stock level will be required.

On the shop floor, there are two methods available for setting the required safety-stock for demand fluctuations during the replenishment time. First, the demand rate fluctuations can be measured during the replenishment time. The safety-stock should correspond to at least the maximum positive deviation of the demand during the replenishment time from its mean thus compensating for the demand rate fluctuations:

$$SSL_{RDMDF} = (RDMD_{max} - RDMD_m) \cdot TRP \qquad (9.7)$$

where

| | |
|---|---|
| $SSL_{RDMDF}$ | safety-stock level for demand fluctuations [units], |
| $RDMD_{max}$ | maximum demand rate [units/SCD], |
| $RDMD_m$ | mean demand rate [units/SCD], |
| TRP | replenishment time [SCD]. |

Second, the maximum demand rate of a workstation is limited by the capacity restriction of the subsequent workstation. When there is only one subsequent workstation, an enterprise can refrain to some extent from using safety-stock to buffer demand fluctuations. This is the case, when the work content for all the variants on the subsequent workstation is longer than (or the same as) on the preceding workstation. In this case, the preceding workstation can punctually reproduce the orders when there are no disruptions and the information turn-around-times are short.

**Delivery Lateness and Quantity Deviations** An additional factor which generally causes uncertainty is the delivery lateness which results from fluctuations in the replenishment time. Possible causes for this include:

- If the total demand for the variants produced on a workstation (temporarily) increases, a backlog of orders can then develop. If the hours the workstation is operated are not temporarily adjusted to the total demand, the replenishment times increase and with that the delivery lateness (see the above comments).
- If machines experience longer malfunctions or supply bottlenecks with the base materials, replenishment times are extended.
- If the operator does not comply with the given processing sequence, fluctuations in the replenishment time result.

Enterprises should therefore try to reduce unplanned fluctuations in the replenishment time as much as possible through flexible capacities, secure procurement and production processes as well as closely complying with the planned sequence. Moreover, they can measure the variance of the replenishment times in order to gain a typical value for dimensioning the safety-stock.

The impact of *quantity deviations* can often be ignored.

**Coordinating the Replenishment Time with the Demand Rate** If it is possible to coordinate the replenishment time with the demand rate, then less safety-stock is required for ensuring a high service level. When there is an above average demand rate, the replenishment times should be reduced, whereas with a lower demand rate the replenishment times can be extended.

The question as to what extent it is possible to coordinate the replenishment times with the demand rate thus arises. Let us consider this with regards to two cases:

1. the demand rate increases for all of the variants (total demand increases), and,
2. the mix of the variant changes, but the total demand remains the same.

*Increasing Total Demand:* When the total demand is rising, the enterprise has to increase the capacities in order to prevent backlogs and ensure the service level. There are two possibilities for doing so: First, overtime can be authorized. Second, the enterprise can increase the output rate by implementing additional operators for example. If the enterprise maintains the WIP as a constant, both measures will decrease the replenishment time.

*Changing Mix of Variants while the Total Demand Remains Constant:* A change in the mix of variants while the total demand remains unchanged when the demand for a number of variants increase while the demand for other variants decreases by the same extent. With changes in demand that are effective over the long term, the number of Kanbans needs to be adjusted in order to permit a steady production. Short term fluctuations in the variant mix can however usually be counteracted alone through sequence interchanges. Unlike with the basic Kanban system, a workstation then produces the variants not in the order that the Kanbans arrive (this leads to a constant mean inter-operation time for each variant), but rather pulls the variant with the increased demand rate forward and holds back on the variant with the decreased demand rate. This shortens the replenishment time for the variant with the increased demand rate and lengthens the replenishment time for the variant whose demand rate remains lower than the planned value. This type of behavior can also be targeted in that a workstation always produces the variant which is the next to be threatened with a supply bottleneck. With short term fluctuations in the variant mix the number of Kanbans for the variants also does not have to be changed with this alternative. It is especially easy to attain this type of sequencing with Visual Kanban, since in this case, each of the operators are immediately informed about stock levels.

## 9.6 Implementing Kanbans Across the Supply Chain

Kanban is frequently implemented across supply chains and usually then serves to fulfill an enterprise's orders as efficiently as possible. Many enterprises refrain from physical Kanbans and instead call up orders electronically. Occasionally they also install a digital camera to communicate the stock levels on the internet. The supplier can then always see the current stock level similar to Visual Kanban. However, a warning also needs to be issued about exaggerated expectations in implementing Kanban: If the supplier requires high stock levels in order to fill the Kanban triggered call-up within the requested short replenishment time, then there is little gained. Accordingly, Monden – one of the leading advocates of the Toyota

Production System – points out that the supplier also needs to introduce a lean production system in order to ensure that Kanban is implemented efficiently. In this next section, the discussion is oriented on his presentation of integrating suppliers [Mond-98*].

**Presentation and Discussion of the Procedural Rules**

Kanban is generally integrated within a higher level production program that provides a (e.g., monthly) production quantity for each of the variants. This information allows the supplier to make the necessary arrangements for smoothly fulfilling the production program.

The actual Kanban retrieval could still deviate from the given production program due to changes in the end customer's demand, however, based on experience the deviation is usually minimal. If deviations were refrained from in order to comply exactly with the production program, it would result in much larger fluctuations in the demand rate between the production programs [Mond-98*].

The Kanban retrieval itself occurs according to the structure depicted in Fig. 9.12 (see [Mond-98*]):

1. The empty containers are brought together with the Kanbans to the supplier (Fig. 9.12a).
2. In the supplier's store, the driver hands over the empty transport bins and Kanbans, and, changes vehicles. The Kanbans authorize another production run of the variants on the last workstation of the supplier (Steps 1–3 in Fig. 9.12b). In the meantime, the operator on Workstation 1 at the manufacturer completes an order (Step 4). Another transport is loaded with an empty transport bin (Step 5).
3. The driver returns to the manufacturer's receiving store (Fig. 9.12c). Meanwhile, Workstation 1 at the manufacturing company starts processing an order of Variant A and triggers the corresponding movement of materials and Kanbans (Step 1–3). The supplier's last workstation completes an order (Step 4 and 5).
4. The driver arrives at the manufacturer, changes his vehicle again and returns the empty transport bins along with the Kanbans back to the supplier (Step 1 in Fig. 9.12d). The materials he delivered to the manufacturer are stored (Step 2). In the meantime, the supplier loads another transport with completed components (Steps 3 and 4). The last workstation at the suppliers selects an order for another production run (Steps 5–7) and the customer demands a lot of Variant A (Steps 8 and 9).

Figure 9.13 depicts the possible layout of a Supply Kanban [Mond-98*]. In addition to the article identification, the supply Kanban contains the exact delivery and store locations for both the supplier (store) as well as the purchaser (store and assembly). Furthermore, the exact delivery times and the delivery cycle is provided. The designator 1–6–2 for the delivery cycle means that the part is delivered six times daily and has to be delivered to the receiver two delivery time points after the Kanban has arrived at the suppliers (replenishment time).

Supply Kanbans can be directly utilized as production Kanbans for the last production stage at the suppliers. In this case, it is disadvantageous when a number

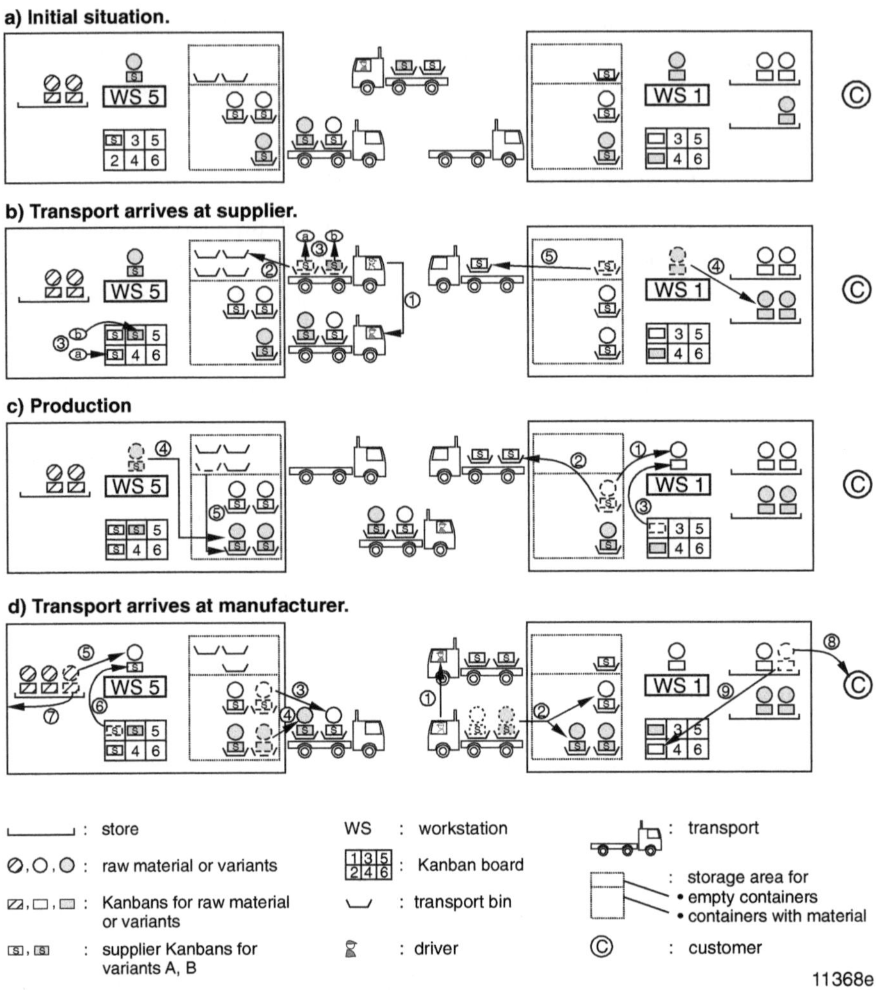


**Fig. 9.12** How supply Kanban functions

of Kanbans arrive at a workstation at the same time due to transportation. Conveying the production Kanbans electronically is thus believed to be advantageous. This shortens the information turn-around-times and supports a steady demand trend. A number of advocates of Kanban though prefer physical Kanban, because up until now it has been difficult to prevent operators working with electronic Kanban signals from re-ordering more frequently than Kanban guidelines allow.

Integrating suppliers via Kanban is particularly efficient when the suppliers are located close to the purchasers. On the one hand, transport times are shorter and the stock between the supplier and purchaser can be tightly dimensioned. On the other hand, it is then frequently possible to bundle the transportation of a number of

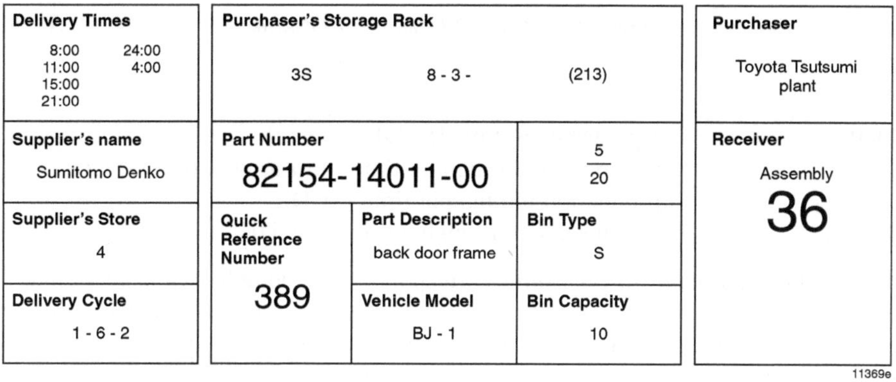


**Fig. 9.13** Sample supply Kanban (Monden)

suppliers (so-called milk runs). The suppliers can thus attain a high utilization of transportation resources despite a high transport frequency. This limits the transport costs and the traffic load through a just-in-time delivery.

From a logistics point of view, Kanban has two fundamental advantages: First, it provides a continuous, steady retrieval of supplies. Second, when interlocked with the production of the supplier, it limits the WIP. Both of these characteristics, counteract the amplification of demand fluctuations and with that the bullwhip effect. In this form, Kanban requires a close cooperation between the purchaser and its suppliers as well as very stable processes.

## Dimensioning the Supply Kanbans

When dimensioning the supplier Kanbans, a few particularities need to be considered. These result especially from the geographic distance between the supplier and purchaser along with the resulting transport times and costs.

### *(a) Lot Size*

In addition to the setup costs of the resources the transportation costs also need to be considered. These usually decrease as the lot sizes per transport unit increase. This has a tendency to lead to larger lot sizes. Enterprises could, however, frequently decrease the costs of smaller lot sizes by bundling transports (along the supply chain).

### *(b) Number of Kanbans*

As with an enterprise internal application, the number of Kanbans can be calculated according to Eq. 9.3. In addition to the demand rate, the fundamental parameters that influence this include the replenishment time and the safety-stock.

The *replenishment time* is comprised of the information throughput time (turn-around-time), the production throughput time and the transportation throughput time (see Eq. 3.1 for possible additional components).

$$TRP = TTP_{inf} + TTP_{prod} + TTP_{transp} \tag{9.8}$$

where

| | |
|---|---|
| TRP | replenishment time [SCD], |
| $TTP_{inf}$ | information throughput time (turn-around-time) [SCD], |
| $TTP_{prod}$ | production throughput time [SCD], |
| $TTP_{transp}$ | transportation throughput time [SCD]. |

The *information turn-around-time* corresponds to the length of time between the withdrawal of materials at the purchaser's and the point at which the supplier becomes aware of the demand. With electronic Kanban signals it can be ignored. If the information is conveyed with traditional (physical) Kanbans, it corresponds to at least the time needed to transport the Kanban from the purchaser to the supplier.

Values for the *production throughput time* can be derived with the aid of Logistic Operating Curves (see Sect. 3.2.2.3).

The *transportation throughput time* is comprised of three components: the time between the end of production and the start of transportation, the actual travel time from the supplier to the purchaser and the portion of time required to load and unload and/or store and withdraw the variant at the suppliers and purchaser. The first component is generally dependent on the frequency of transports. The travel time results from the distance and average speed. The last component frequently takes comparably little time.

The *safety-stock* is generally dimensioned larger with a supply chain Kanban system than with an enterprise internal Kanban system. The main reasons for this are the additional uncertainty that results from the transport of materials: Poor weather conditions, closed roads, traffic, breakdowns or accidents can increase transport throughput times considerably. Accordingly, a failed delivery should not cause a material related break in the production.

## 9.7 Application

Kanban is usually applied in productions with few variants. The mean demand rate of the controlled variants is high and there are minimal fluctuations in the demand rate. Generally there are also few customers. The demand rate can usually be more reliably planned than when applying the Order Point System, so that a smaller amount of safety-stock is sufficient. Accordingly, also highly valuable products can be controlled with Kanbans. Unlike a typical application of the Order Point System, Kanban is integrated into a detailed, higher level production program plan. This production program is implemented across multiple stages and in the ideal case across the entire plant or supply chain. In an environment such as this Kanban controls the interfaces between a consumer and a production division.

This method is applied in a great many enterprises, whereby the main emphasis is clearly in the automobile industry (see z. B. [Bodl-00]). Spencer and Larsen however, also describe its application in the heavy engineering industry [Spen-98*]. Reports of applications very early on can be found in [Wild-84].

## References


[Bard-91*] Bard, J.F., Golany, B.: Determining the number of Kanbans in a multiproduct, multistage production system. Int. J. Prod. Res. **29**(5), 881–895 (1991)

[Berk-92*] Berkley, B.J.: A review of the Kanban production control research literature. Prod. Oper. Manag. **1**(4), 393–411 (1992)

[Bodl-00] Bodlien, H.-G., Christmann-Jacoby, H.: Kanban-Prinzipien optimieren den Materialfluss bei DaimlerChrysler. PPS-Management **5**(3), 56–59 (2000)

[Bonv-97*] Bonvik, A.M., Couch, C.E., Gershwin, B.S.: A comparison of production-line control mechanisms. Int. J. Prod. Res. **35**(3), 789–804 (1997)

[Burm-97] Burmeister, M.: Auslegung der Verbrauchssteuerung bei vernetzter Produktion. VDI Progress Reports, Series 8, No. 658, Düsseldorf (1997)

[Ever-99] Eversheim, W. et al.: Ausprägungen der Produktionsplanung und -steuerung. In: Eversheim, W., Schuh, G. (eds.): Produktion und Management, vol. 4, pp. 14–60 - 14–83. Springer, Berlin (1999)

[Früh-90] Frühwald, C.: Analyse und Planung produktionstechnischer Rüstabläufe. VDI Progress Reports, (1990)

[Golh-91*] Golhar, D.Y., Lee Stamm, C.: The just-in-time philosophy. A literature review. Int. J. Prod. Res. **29**(4), 657–676 (1991)

[Kim-85*] Kim, T.-M.: Just-in-time manufacturing system: a periodic pull system. Int. J. Prod. Res. **23**(3), 553–562 (1985)

[Kimu-81*] Kimura, O., Terada, H.: Design and analysis of pull system. A method of multi-stage production control. Int. J. Prod. Res. **19**(3), 241–253 (1981)

[Lödd-01b] Lödding, H., Hernández, R., Wiendahl, H.-P.: Dimensionierung und Bewertung von Fertigungsinseln mit logistischen Kennlinien. Io Manage. **70**(1–2), 25–32 (2001)

[Masc-91*] Mascolo, M., et al.: A Unified Modeling of Kanban Systems Using Petri Nets. Int. J. Flex. Manuf. Syst. **3**(3/4), 275–307 (1991)

[Mitr-90*] Mitra, D., Mitrani, I.: Analysis of a Kanban discipline for cell coordination in production lines I. Manag. Sci. **36**(12), 1548–1566 (1990)

[Mond-98*] Monden, Y.: Toyota Production System. An Integrated Approach to Just-In-Time. Engineering & Management Press, Norcross (1998)

[Mora-91*] Moras, R.G., Jalali, M.R., Dudek, R.A.: A categorized survey of the JIT literature. Prod. Plan. Control **2**(4), 322–334 (1991)

[Nyhu-91] Nyhuis, P.: Durchlauforientierte Losgrößenbestimmung. VDI Progress Reports, Series 2, No. 225, Düsseldorf (1991)

[Nyhu-09*] Nyhuis, P., Wiendahl, H.-P.: Fundamentals of Production Logistics. Theory, Tools and Applications. Springer, Berlin (2009)

[Ohno-88*] Ohno, T.: Toyota Production System. Beyond Large-Scale Production. Productivity Press, Portland (1988)

[Rees-89*] Rees, L.P., Huang, P.Y., Taylor, B.W.: A comparative analysis of an MRP lot-for-lot system and a Kanban system for a multistage production operation. Int. J. Prod. Res. **27**(8), 1427–1443 (1989)

[Shin-85*] Shingo, S.: A Revolution in Manufacturing. The SMED System. Productivity Press, Portland (1985)

[So-88*] So, K.C., Pinault, S.C.: Allocating buffer storages in a pull system. Int. J. Prod. Res. **26**(12), 1959–1980 (1988)

[So-90*] So, K.C.: The impact of buffering strategies on the performance of production line systems. Int. J. Prod. Res. **28**(12), 2293–2307 (1990)

[Spen-98*] Spencer, M.S., Larsen, D.: Kanban implementation between a heavy manufacturing department and foundry suppliers. Prod. Plan. Control **9**(3), 311–316 (1998)

[Sugi-77*] Sugimori, Y., Kusunoki, K., Cho, F., Uchikawa, S.: Toyota production system and Kanban system – materialization of just-in-time and respect-for-human system. Int. J. Prod. Res. **15**(6), 553–564 (1977)

[Uzso-90*] Uzsoy, R., Martin-Vega, L.A.: Modelling kanban-based demand-pull systems: a survey and critique. Manuf. Rev. **3**(3), 155–160 (1990)

[Wild-84] Wildemann, H.: Flexible Werkstattsteuerung durch Integration von Kanban-Prinzipien. CW-Publikationen, Munich (1984)

# Chapter 10
# Capacity Oriented Materials Management

Capacity Oriented Materials Management (CORMA) was developed by Schönsleben especially for mixed productions in which both make-to-stock and make-to-order productions utilize the same set of resources (e.g., plants and/or equipment). It can, however, also be applied equally well in an exclusively make-to-stock production. CORMA primarily aims to constantly utilize the workstations of a production. To do so, make-to-stock orders that are routed through workstations with critical capacities are generated and released early for workstations that have available capacities.

Moreover, CORMA calculates the priorities of all orders by constantly updating the scheduling of the operations to facilitate sequencing at a workstation. This provides support for coordinating customer and stock orders in the production, while at the same time taking into consideration the current stock levels when scheduling. CORMA can therefore consider demand fluctuations during the replenishment time and accelerate the completion of critical orders.

CORMA is thus a combination of methods for generating and releasing orders as well as sequencing. An extensive discussion of this method by Schönsleben can be found in [Schö-95*, Schö-07*]. However, for the purposes of this book, the procedural rules are also described (Sect. 10.1) and discussed (Sect. 10.2) here. The setting of the parameters for the method – which is to some degree similar to the Order Point System – is outlined in Sect. 10.3. Section 10.4 discusses how the basic ideas of CORMA can also be applied across the supply chain (Sect. 10.4), whereas, Sect. 10.5 provides an overview of the implementing of CORMA in an enterprise in Switzerland.

## 10.1 Method Description

The method's planning process consists of four components, of which Schönsleben combines the first two.

H. Lödding, *Handbook of Manufacturing Control*,
DOI 10.1007/978-3-642-24458-2_10, © Springer-Verlag Berlin Heidelberg 2013

1. calculating the look-ahead-time (also known as 'anticipation time') for variants made-to-stock,
2. due date oriented order release with load balancing,
3. calculating the planned end date for released stock orders, and,
4. re-scheduling released orders with probability scheduling.

In the following, we will review each of the components of CORMA individually before discussing how they are applied together in an example.

*1. Calculating the Look-Ahead-Time for Variants Made-to-Stock*
CORMA calculates the look-ahead-time for every variant that is made-to-stock. This is the length of time before which the stock level is expected to fall below the order point (Fig. 10.1).
The look-ahead-time is calculated as follows [Schö-07*]:

$$T_{LA} = \frac{SL + \sum SO_{open} - SOP}{RDMD} \tag{10.1}$$

where

| | |
|---|---|
| $T_{LA}$ | look-ahead-time [SCD], |
| SL | stock level [units], |
| $SO_{open}$ | total quantity of open stock orders [units], |
| SOP | stock order point [units], |
| RDMD | demand rate [units/SCD]. |

Stock orders – in cases where there are available capacities – can when necessary be generated and released early. Unlike with the Order Point System used more commonly on the shop floor this allows the loads to be balanced.

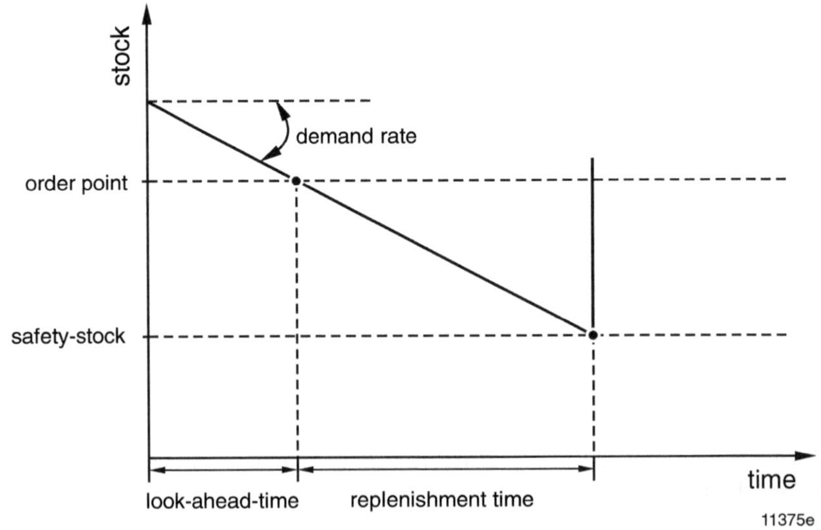


**Fig. 10.1** Calculating the look-ahead-time (Schönsleben)

*2. Due Date Oriented Order Release with Load Balancing*
CORMA initially releases all orders whose planned start date has been reached (Due Date based Order Release). In case workstations have available capacities which are typically well utilized, CORMA also releases additional orders that are usually processed on these workstations early. This results in a constant utilization of the workstations in the production which have critical capacities and requires the loads to be balanced. This load balancing, however, is not based on a defined algorithm. It can either be conducted by a production planner within the context of planning due dates and capacities or automated by a suitable release algorithm. Generally speaking, order release methods which involve balancing the loads specifically for each workstation are well suited (see Chaps. 20, 21 and 22).

*3. Calculating the Planned End Date for Released Stock Orders*
During every planning session, CORMA determines the expected planned end date for all released stock orders on which the stock level will probably fall below the safety-stock. This time point is calculated as follows (see [Schö-07*]):

$$EDO_{plan} = DPL_0 + \frac{SL - SSL}{RDMD} \tag{10.2}$$

where

| | |
|---|---|
| $EDO_{plan}$ | planned end date of order processing [SCD], |
| $DPL_0$ | planning date [SCD], |
| SL | stock level [units], |
| SSL | safety-stock level [units], |
| RDMD | demand rate [units/SCD]. |

By constantly adjusting the planned end date of the stock orders, CORMA takes into consideration the demand rate during the replenishment time. If the demand rate is higher than planned, the end date is moved up and delayed when the demand rate is lower than planned. This allows the replenishment time to be adjusted to the demand rate during the replenishment time and consequently to reduce the store of safety-stock.

In a variation of this method, the time point is calculated at which the stock is anticipated to have been absorbed. Accordingly, for the planned end date of the order:

$$EDO_{plan} = DPL_0 + \frac{SL}{RDMD} \tag{10.3}$$

where

| | |
|---|---|
| $EDO_{plan}$ | planned end date of order processing [SCD], |
| $DPL_0$ | planning date [SCD], |
| SL | stock level [units], |
| RDMD | demand rate [units/SCD]. |

*4. Re-Scheduling Released Orders with Probability Scheduling*
CORMA re-schedules the operations for all orders during the course of every planning. The so-called 'probability scheduling' that is employed, distributes the operations of an order yet to be completed between the planning date and the end date. In doing so, it takes into consideration the planned inter-operation times of the workstation. The calculated operation start and end dates serves as a measure for the priority of an order on the relevant workstation.

The probability scheduling multiplies all the planned inter-operation times with a standard stretch factor. This stretch factor is calculated as a ratio of the available and planned inter-operation times for the remaining operations of an order (see [Schö-07*] as well as Fig. 10.2):[1]

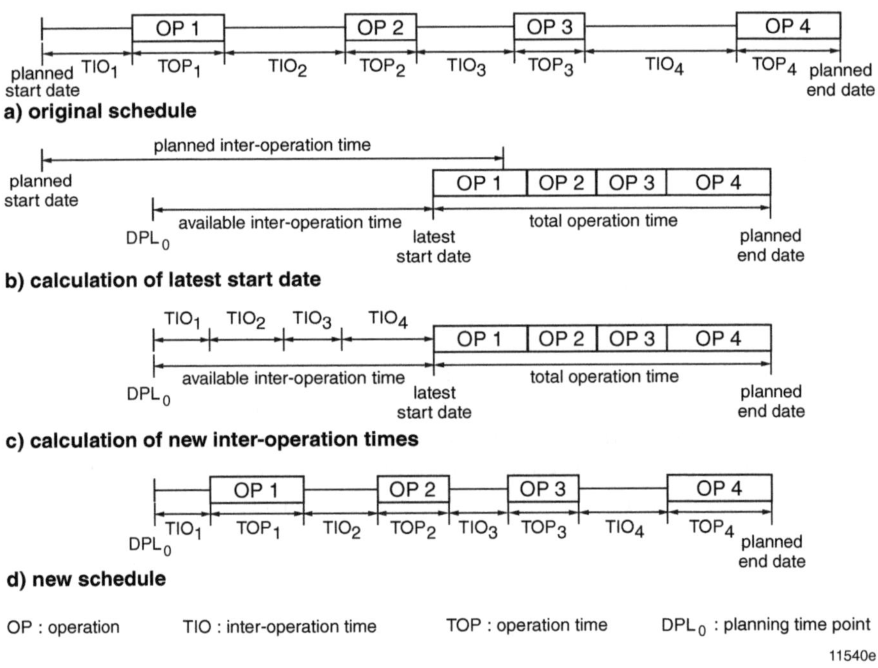


**Fig. 10.2** Principle of probability scheduling

[1] With multi-stage products the critical path can be shifted due to the re-scheduling. In this case, a number of iterations may be required in order to determine the stretch factor [Schö-07*].

$$SF = \frac{TIO_{avail}}{TIO_{plan}} = \frac{DLS - DPL_0}{\sum_{i=k}^{n} TIO_{plan,i}} = \frac{EDO_{plan} - \sum_{i=k}^{n} TOP_i - DPL_0}{\sum_{i=k}^{n} TIO_{plan,i}} \tag{10.4}$$

where

| | |
|---|---|
| SF | stretch factor [−] |
| $TIO_{avail}$ | available inter-operation time [SCD], |
| $TIO_{plan}$ | planned inter-operation time (remaining operations) [SCD], |
| DLS | latest start date of the order [SCD], |
| $DPL_0$ | planning date [SCD], |
| $TIO_{plan,i}$ | planned inter-operation time of the $i^{th}$ operation [SCD], |
| $EDO_{plan}$ | planned end date of order's processing [SCD], |
| $TOP_i$ | operation time of the $i^{th}$ operation [SCD], |
| k | index for the next, not yet completed operation [−], |
| n | number of operations [−]. |

The stretch factor is greater than 1 when there is more time for completing the remaining operations than planned. In this case the planned inter-operation times are extended. If, on the other hand, the completion of the order is delayed in comparison to the plan, the inter-operation times are shortened and the stretch factor is then smaller than one. When the stretch factor is zero the time until the planned due date corresponds with the sum of the inter-operation times of the remaining operations. A negative stretch factor requires negative inter-operation times. These can – to a limited degree – be obtained with overlapped production of the operations (see [Schö-07*] as well as Sect. 5.3).

Probability scheduling assigns delayed orders a higher priority than orders which are flowing through the production faster than planned. Furthermore, it allows comparable due dates for the operations to be allocated to both stock and customized orders. The planned end date for the customized order results from the promised planned delivery date. The procedure for calculating the planned end date for stock orders was described above as a third component of CORMA.

### Description of the Procedural Rules Based on an Example

Based on an example, the procedural rules for CORMA will be illustrated in the following. An enterprise produces Parts A, B and C make-to-stock, whereas additional parts are make-to-order based on specific customers' wishes. The production consists of five workstations, WS 1–WS 5.

Table 10.1 lists the current stock levels, the order points and the average demand rate for the parts. Based on this information the look-ahead-time can be calculated according to Eq. 10.1.

CORMA generates a stock order for each part. The stock orders are entered into the list of known, but not yet released orders (Table 10.2). In addition to the stock orders this list also contains the not yet released customer orders and is organized

**Table 10.1** Calculating the look-ahead-time

| Part | Stock + open orders [units] | Order point [units] | Demand rate [units/SCD] | Look ahead time [SCD] |
|---|---|---|---|---|
| A | 310 | 250 | 25 | 2.4 |
| B | 205 | 100 | 20 | 5.3 |
| C | 610 | 200 | 15 | 27.3 |

**Table 10.2** List of known but not released orders

| Run. Nr. [–] | Order number [–] | Type of order [–] | Part [–] | Planned start date [SCD] | Workstations routed through [–] |
|---|---|---|---|---|---|
| 1 | 4,911 | Make-to-order | XY | 100.0 | 1, 4, 5 |
| 2 | 8,712 | Make-to-order | XZ | 102.0 | 2, 3, 4 |
| 3 | 4,743 | Make-to-stock | A | 102.4 | 2, 4, 5 |
| 4 | 4,814 | Make-to-stock | B | 105.3 | 3, 4, 5 |
| 5 | 5,715 | Make-to-order | XX | 108.0 | 1, 3, 5 |
| 6 | 4,816 | Make-to-stock | C | 127.3 | 1, 2, 3,4 |

Planning date: $DPL_0$ = SCD 100

**Table 10.3** List of released stock orders

| Order number [–] | Part [–] | Stock level [units] | Safety-stock level [units] | Demand rate [units/SCD] | Planned due date [SCD] |
|---|---|---|---|---|---|
| 5,441 | C | 110 | 80 | 15 | 102.0 |

Planning date: $DPL_0$ = SCD 100

according to the sequencing of the planned start dates. The planned start date of the stock orders results from the sum of the planning date (here: SCD 100) and the look-ahead-time.

CORMA first releases all of the orders whose planned start date has been reached (here: Order 1). In the next step, it identifies the workstations that are generally well utilized but which have available capacities at the time of the planning. In the example this is the case for Workstation 2. Orders 2, 3 and 6 from the list of known but not released orders are routed through this workstation. Since Order 2 has the earliest planned start date, it has the highest priority and is released early. If this sufficiently utilizes Workstation 2, no further orders are released. Otherwise, Order 3 and where necessary Order 6 would also be released into production early.

In the next step, CORMA calculates the current planned end date for all released (and not yet completed) stock orders according to Eq. 10.2. In the example, only a stock order for Part C is circulating as WIP in the production (Table 10.3).

The stock for Part C is anticipated to reach the safety-stock level of 80 units in 2 days. Therefore, CORMA determines the planned end date to be shop calendar day 102 (see Eq. 10.2). This date is the initial point for the probability scheduling which CORMA conducts in the last planning step for the remaining operations of

**Table 10.4** Planning and feedback data from Order 1

| Order 5,441, part C | Planning date: | $DPL_0$ = SCD 100 | | | | | |
|---|---|---|---|---|---|---|---|
| | Planned end date: EDOP = SCD 102 | | | | | | |
| OP [–] | WS [–] | EDOP [SCD] | TOP [SCD] | $TIO_{plan}$ [SCD] | $TIO_{new}$ [SCD] | $SDO_{new}$ [SCD] | $EDOP_{new}$ [SCD] |
| 1 | 1 | 96 | 0.4 | 2.0 | | | |
| 2 | 2 | 99 | 0.2 | 1.5 | | | |
| 3 | 3 | | 0.5 | 1.0 | 0.4 | 100.4 | 100.9 |
| 4 | 4 | | 0.4 | 1.5 | 0.7 | 101.6 | 102.0 |

*OP* operation, *WS* workstation, *EDOP* end date of operations processing, *TOP* operation time, $TIO_{plan}$ planned inter-operation time, $TIO_{new}$ newly planned inter-operation time, $SDOP_{new}$. newly planned start date of operation processing, $EDOP_{new}$ newly planned end date of operation processing

all the released orders. This is explained using the example of the stock order for Part C (Table 10.4).

The first two operations of the order are already processed. Operations 3 and 4 need to be re-scheduled. In order to do so the stretch factors for changing the inter-operation time is first determined. According to Eq. 10.4 these are calculated:

$$SF = \frac{EDO_{plan} - \sum_{i=k}^{n} TOP_i - DPL_0}{\sum_{i=k}^{n} TIO_{plan,i}} = \frac{102.0 - (0.5 + 0.4) - 100.0}{1.0 + 1.5} = 0.44$$

where

| | |
|---|---|
| SF | stretch factor [–], |
| $EDO_{plan}$ | planned end date of order processing [SCD], |
| $TOP_i$ | operation time of the $i^{th}$ operation [SCD], |
| $DPL_0$ | planning date [SCD], |
| $TIO_{plan,i}$ | planned inter-operation time of the $i^{th}$ operation [SCD], |
| k | index for the next, not yet completed operation [–], |
| n | number of operations [–]. |

Consequently, the planned inter-operation times have to be reduced by 56%. With that the new dates that are listed in Table 10.4 for Operations 3 and 4 of the stock order for Part C are given. This calculation is conducted for all of the released but not yet completed orders and concludes the CORMA's planning process. As a result of this planning process newly calculated planned dates for the start of processing as well as the end of processing are available for all operations. These can be considered as priority indexes for the production operators. Shortly after the planning process, the WIP on Workstation 3 consists of the following orders (Table 10.5).

**Table 10.5** List of orders present as WIP on Workstation 3

| Run. nr [–] | Order [–] | Type of order [–] | Part [–] | $SDOP_{new}$ [SCD] | $EDOP_{new}$ [SCD] |
|---|---|---|---|---|---|
| 1 | 1,115 | Make-to-order | ZZ | 100.0 | 100.5 |
| 2 | 5,441 | Make-to-stock | C | 100.4 | 100.9 |
| 3 | 3,243 | Make-to-order | XV | 101.4 | 101.9 |
| 4 | 3,339 | Make-to-order | XZ | 101.5 | 101.8 |

$SDO_{new}$ new planned start date of operation processing, $EDOP_{new}$ new planned end date of operation processing

The operators on the workstation can use this list to set the sequencing in which they will process orders. In the most simple case, they produce orders in the sequence of the new planned processing starts. The procedural rules for CORMA, however, do not provide any set sequencing process.

## 10.2 Discussion of the Procedural Rules

CORMA is unusual in that it coordinates stock and customer orders both during the order release phase as well as during the orders' throughput. The corresponding components of CORMA – in particular the calculation of the look-ahead-time and the probable planned due date for the stock orders – are evaluated in the following based on the guidelines for designing manufacturing control methods (see Sect. 3.3.2).

1. *A manufacturing control method should allow the WIP of a production or workstation to be regulated at a defined level.*

A WIP control releases orders as soon as the WIP falls below the planned WIP. Consequently, orders can enter into production both before or after the planned release date. CORMA does not control the WIP directly. In particular, it does not hold back any orders in case the WIP is (very) high. Thus, if the production ends up with a backlog, the WIP builds up. This is also acknowledged by Schönsleben. He lists the ability of the enterprise to finance high WIP levels as one of the conditions for implementing CORMA [Schö-07*]. There is, however, no problem combining the CORMA with a WIP control.

2. *A manufacturing control method should reduce the WIP fluctuations to a minimum.*

Due to the lack of a WIP control CORMA cannot prevent WIP fluctuations. If it is supplemented with a WIP control, however, it provides very good conditions for reducing WIP fluctuations: It generates stock orders before the stock falls below the order point, allowing stock orders to be released early. This prevents the WIP from falling below the planned WIP if the order input is temporarily lower than the production's output.

*3. A manufacturing control method should cause as little blocked WIP as possible.*
CORMA does not cause any blocked WIP.

*4. A manufacturing control method should contribute to closely aligning the load to the capacities.*
Load balancing accelerates or delays the release or throughput of orders with the aim of preventing breaks in the material flow (accelerates) or the build-up of WIP (delays). CORMA emphasizes the first case, in that it releases those orders early which are routed through workstations with free capacities. The effectiveness of the method is dependent on the suitability of the applied algorithm or the production planner's skill. Since CORMA itself does not provide any algorithm it is possible to combine the system with other manufacturing control methods. For example, one could consider combining it with the Load Oriented Order Release approach or one of the other methods that support balancing the loads on each specific workstation (Chaps. 20, 21 and 22).

*5. A manufacturing control method should cause as few sequencing interchanges as possible.*
While orders are routed through the production, CORMA does not cause any sequencing interchanges. The operation start and end processing dates calculated with the probability scheduling are, however, practical criteria for determining the sequencing on the workstations.

In contrast, when releasing orders CORMA does cause sequencing interchanges in order to support a balanced load. As a consequence, urgent and non-urgent orders compete for the production's capacities. This cannot be avoided when loads are centrally balanced. By providing the planned processing times of the operations CORMA is however able to coordinate the orders in the production.

*6. A manufacturing control method should be able to compensate for schedule deviations within the context of backlog control.*
It is the responsibility of manufacturing control to prevent a backlog and to reduce any backlogs that do develop. To do so capacities can be temporarily increased. CORMA does not include a backlog control as a component. To ensure a high degree of schedule reliability CORMA should therefore be combined with a backlog control.

*7. A manufacturing control method should take into consideration the principle of bottlenecks.*
One of the basic ideas of CORMA is to protect the bottleneck workstation from breaks in the material flow. In order to do so it releases stock or customer orders early when a bottleneck workstation has available capacities. The efficiency and effectiveness of the method is dependent on the release algorithms or the production planners' skill.

*8. A manufacturing control method should be as simple as possible*
In comparison to the MRP II control or to the Order Point System, CORMA is relatively complex. In the course of each planning session it calculates the look-ahead-time of the variants made-to-stock and re-schedules all of the orders.

The effort invested in these calculations, however, allows stock orders to be incorporated when balancing the load or when controlling the WIP. Moreover, it contributes to coordinating stock and customer orders and to aligning the replenishment time of the stock orders with the demand rate during the replenishment time. Nonetheless, it is necessary to inform the production operator immediately about changed priorities. Since the progress of the work also has to be compiled precisely, CORMA requires operational data to be collected efficiently and in great detail.

The additional effort supports the logistic performance of the production. Moreover, CORMA can be integrated into a PPC system which in turn limits the effort required to implement it.

9. *Methods for generating orders should communicate information about demands as quickly and clearly as possible.*

Generally, the same statements made about the Order Point System apply to CORMA (see Sect. 8.2). Nonetheless, the system avoids the disadvantages of the Order Point System in that it generates the orders before reaching the order point and calculates an anticipated planned start date. With that the planned start date includes information about the demand between the two order points – information that is not available with the Order Point System. The production can thus better estimate which load it has to manage in the future.

## 10.3 Setting the Method's Parameters

Important parameters for CORMA include the *planned inter-operation time* of the workstation and the *order points* of the variants that are made-to-stock. Determining the planned inter-operation time within the context of a Logistic Positioning is extensively discussed in Sect. 3.2.2.3. How the order points can be set based on a model is the subject of Sect. 8.3. The safety-stock, however, can be dimensioned lower than with other manufacturing control methods that do not correlate the replenishment time with the demand rate during the replenishment time.

The *planning frequency* determines how often CORMA's planning process can be conducted. The higher the planning frequency, the more up to date the considerations of the stock orders and production progress of the orders are. Accordingly, shorter plan cycles are beneficial. This applies particularly if the planning runs can be highly automated and thus allow the effort involved to be limited. Without claiming to be universally valid, daily planning runs should be suitable for many enterprises.

Furthermore, it is possible to supplement CORMA with an advance release window. This indicates the maximum look-ahead-time with which an order is allowed to be released. This prevents the orders from being released far too early. Selecting a suitable advance release window is discussed in Chap. 21 for the Load Oriented Order Release method.

## 10.4 Implementing CORMA Across the Supply Chain

Similar to the Order Point System, CORMA can also be implemented across the supply chain. In order to do so it is necessary to communicate to the suppliers either the actual stock level or the look-ahead-time and the current planned end of processing dates for the variants. Doing so allows the suppliers to first, balance the load, and second, align the delivery time points with the demand trend during the replenishment time. In this way, both the supplier as well as the purchaser would profit from the exchange of information: The supplier would have the advantage of being able to steadily utilize the production, whereas, the purchaser could reduce the safety-stock for the variants and thus reduce tied-up capital.

Implementing CORMA across the supply chain requires mutual trust amongst the partners. Moreover, an efficient exchange of information between the enterprises needs to be ensured.

## 10.5 Application

CORMA reduces the deficiencies of the Order Point System in the production planning and uses the possibilities of controlling the production to increase the service level for the variants. CORMA thus improves the efficiency of the Order Point System, however, does not differ from it in the range of application (see Sect. 8.5).

In [Schö-07*], Schönsleben provides an account of applying this method at Trox Hesco Inc. in Switzerland. The enterprise estimates in particular, the time at which the store of a variant is anticipated to be exhausted. It then uses this information to determine the priority of the stock orders in the production.

## References

[Schö-95*] Schönsleben, P.: Corma: capacity oriented materials management. In: Proceedings of the APICS World Symposium in Auckland, Australasian Production and Inventory Control, pp. 160–164. (1995)

[Schö-07*] Schönsleben, P.: Integral Logistics Management. Operations and Supply Chain Management in Comprehensive Value-Added Networks, 3rd edn. Auerbach, Boca Raton (2007)

# Chapter 11
# Synchro MRP

Synchro MRP combines a centralized production planning and control with the decentralized Kanban System. The method was developed by the Japanese enterprise Yamaha with the intention of transferring the advantages of Kanban to a manufacturing environment characterized by frequently modified and changing products as well as by a large number of variants. Manufacturing environments such as these are generally more typical in many industries than that of the original Toyota production system. Synchro MRP became known primarily through publications by Hall [Hall-86*]. The method functions similarly to a Two-Card Kanban. A workstation, however, is only allowed to reproduce a variant if the planned output of that variant is not yet attained. The planned output per variant is given in the daily production program.

The following section describes the procedural rules for Synchro MRP. Subsequently, we will take a closer look at these based on the guidelines for designing manufacturing control methods (Sect. 11.2). How the parameters for Synchro MRP are set is the subject of Sect. 11.3 whereas its application across plants or supply chains is described in Sect. 11.4. Finally, Hall describes an application of this method at Yamaha (Sect. 11.5).

## 11.1 Method Description

Synchro MRP is based on two components:

- a daily production plan generated by a higher level PPC system, and
- a conventional Two-Card Kanban.

The method thus contains elements from both a central manufacturing control as well as a decentralized control loop within the production. Figure 11.1 depicts the central flow of information from the production program planning to the production's workstations and the decentralized flow of information between the workstations and the stores.

H. Lödding, *Handbook of Manufacturing Control*,
DOI 10.1007/978-3-642-24458-2_11, 

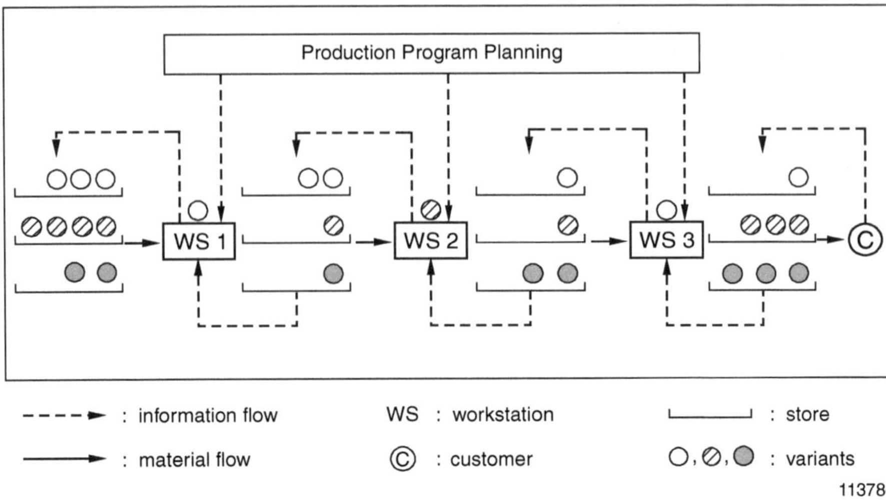


**Fig. 11.1** Principle of synchro MRP

A workstation is allowed to reproduce a variant if,

- the daily production program for that variant is not yet fulfilled, and,
- a production Kanban for that variant is present.

The daily production program and the Two-Card Kanban are therefore the essential components of Synchro MRP. A more extensive discussion of these follows.

**Daily Production Program** The enterprise initially plans a detailed production program for all of the products or models for a planning period. This production program is fixed and is divided as evenly as possible among the individual days in the planning period. The starting point is the planning of an assembly line, which serves here as an example (Fig. 11.2).

From the primary requirements of the assembly line, the material requirements are derived according to the components. In order to distribute the demand for components as evenly as possible, the individual models are continually produced in alternation to one another on the assembly line (thus in the example from Fig. 11.2 perhaps: A – B – A – C – A – B – A – D . . .).

An MRP algorithm can be utilized for calculating material requirements. However it is also possible to apply Cumulative Production Figures[1] (Chap. 13).

Conditions for the continuation of this example include that Model A and D each use Component C1 once and Models B and C each use Component C2 once (Table 11.1).

[1] Cumulative Production Figures are also known as progress numbers.

| | Day | | | | | | | | | | |
|---|---|---|---|---|---|---|---|---|---|---|---|
| Model | 1 | 2 | 3 | 4 | 5 | 6 | 7 | 8 | 9 | 10 | Sum |
| A | 210 | 240 | 210 | 210 | 210 | 240 | 210 | 240 | 210 | 210 | 2190 |
| B | 153 | 153 | 153 | 153 | 153 | 153 | 153 | 153 | 153 | 153 | 1530 |
| C | 60 | 60 | 60 | 90 | 60 | 60 | 60 | 60 | 60 | 90 | 660 |
| D | 30 | 0 | 30 | 0 | 30 | 0 | 30 | 0 | 30 | 0 | 150 |
| Sum | 453 | 453 | 453 | 453 | 453 | 453 | 453 | 453 | 453 | 453 | 4530 |

total demand for 10 day period : 4530 units
units per day for leveled load : 4530 / 10 = 453 units / day
cycle time required to determine line rate : 1.06 min / unit

$$\text{cycle time} = \frac{\text{480 assembly minutes per day}}{\text{453 units per day}} = \text{1.06 assembly minutes per unit}$$

11377e

**Fig. 11.2** Daily production program for an assembly line for a planning period of 10 SCD (according to Hall)

**Table 11.1** Allocation of components and models

| Model | Component |
|---|---|
| A | C1 |
| B | C2 |
| C | C2 |
| D | C1 |

The result of the manufacturing resource planning is a daily production program for the components. In order to simplify matters it is assumed in the example that the produced components are assembled on the same day. Thus, the daily demand for the components results directly from the sum of the daily requirements of the model in which they are incorporated (see Table 11.2).

**Two-Card Kanban** In addition to controlling production with a daily production program, Synchro MRP is based on a conventional Two-Card Kanban System.[2]

**Table 11.2** Daily production program for Components C1 and C2

| Components | Demand per day [units] | | | | | | | | | |
|---|---|---|---|---|---|---|---|---|---|---|
| | **1** | **2** | **3** | **4** | **5** | **6** | **7** | **8** | **9** | **10** |
| C1 | 240 | 240 | 240 | 210 | 240 | 240 | 240 | 240 | 240 | 210 |
| C2 | 213 | 213 | 213 | 243 | 213 | 213 | 213 | 213 | 213 | 243 |
| Total | 453 | 453 | 453 | 453 | 453 | 453 | 453 | 453 | 453 | 453 |

[2] Instead of a Two-Card Kanban other variations of the Kanban system can also be applied. For example Deleersnyder et al. describe a variation of the method with a One-Card Kanban system [Dele-92*].

The Two-Card Kanban System is extensively described in Sect. 9.1.2. The most important procedural rules are repeated here:

1. An operator at a workstation is only allowed to reproduce a variant if there is a production Kanban present and thus an indication of a demand for that variant on a subsequent workstation. With Synchro MRP it is also necessary that the daily production program for that variant is not yet fulfilled.
2. If a workstation is allowed to produce a number of variants, the operator selects the variant with the highest priority. In the basic method the variants are produced in the sequence which the production Kanbans arrive at the workstation.
3. The operator withdraws a lot of the variant that is to be reproduced from their receiving goods store and exchanges the transport Kanban for a production Kanban. The processing of the lot is then begun.
4. The transport Kanban that has become free is transported to the supplying workstation. There a lot of the variant is withdrawn from the outgoing goods store. The responsible operator exchanges the production Kanban for a transport Kanban and transports the lot to the following workstation. The freed production Kanban authorizes the reproduction of a further lot of this variant.

We will now take a closer look at these procedural rules based on an example which considers a production with two workstations and two variants (Fig. 11.3). The daily production program for Variant A is 100 units; for Variant B it is 25 units. The production lot size for both variants corresponds on both workstations to 25 units.

In the initial situation (Fig. 11.3a), Workstation 2 has already fulfilled the daily production program for Variant B and is thus not allowed to produce any more of that Variant. Workstation 1 and Workstation 2 are both completing the processing of a lot (Fig. 11.3b). Workstation 1 directs a lot of Variant B to the outgoing goods store (Step 1). The daily output of the variant thus increases to 25 units and reaches the target for the daily production program (Step 2). Although there is a production Kanban available for the variant and it has a higher priority than Variant A, the workstation is not allowed to start another lot of Variant B. Instead the workstation reproduces a lot of Variant A for which all of the necessary conditions for initiating the production are fulfilled (Step 3 and 4). Workstation 2 completes a lot of Variant A (Step 5), through which the daily output for the variant is increased to 75 units. Since there is no production Kanban authorizing the reproduction of Variant A the workstation is consequently blocked.

This also does not change when the customer places a demand for Variant B (Fig. 11.3c). Although a production Kanban is then freed and transferred to the Kanban board of the workstation (Step 2), the earliest the variant can be reproduced is the following day. This is because the daily production program for the variant is already fulfilled.

The blocked workstation is only released when a customer places a demand for Variant A (Step 1 in Fig. 11.3d). The freed production Kanban authorizes the

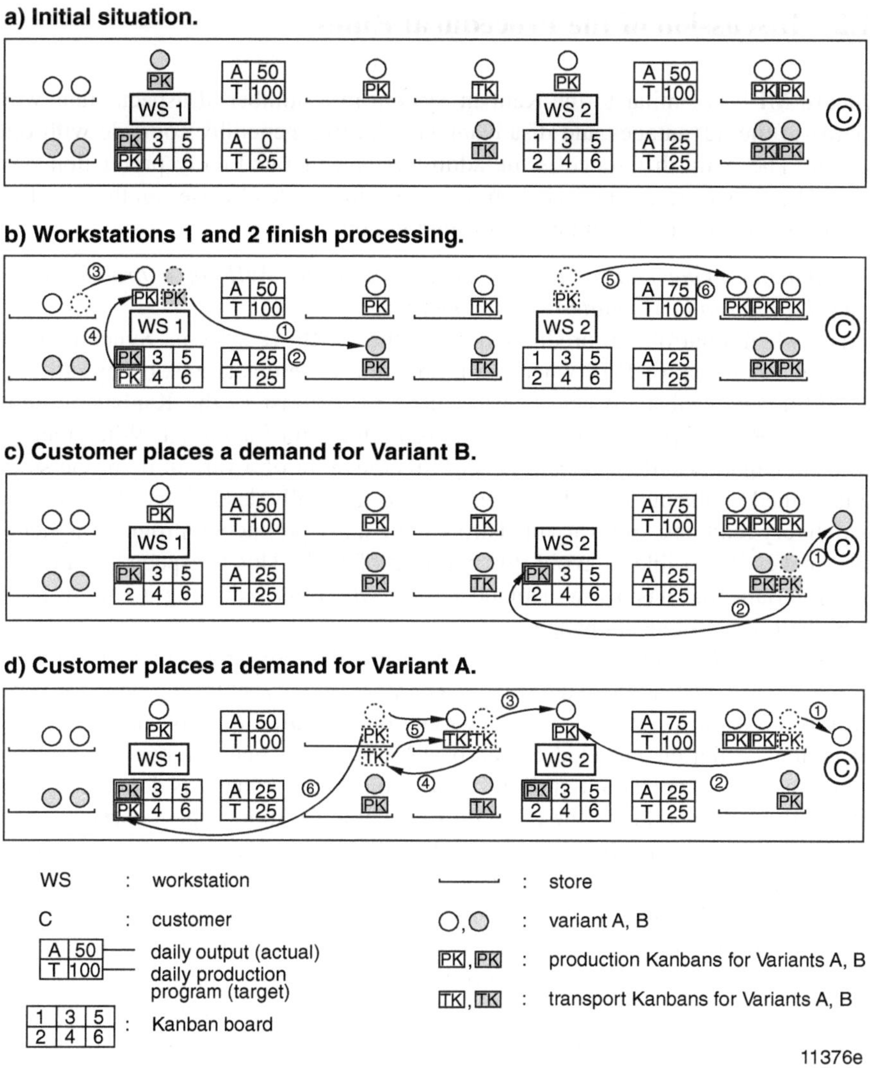


Fig. 11.3 How synchro MRP functions

reproduction of Variant A (Step 2). The operator withdraws the materials from the receiving goods store (Step 3) and sends the transport Kanban to the outgoing goods store of the supplier's Workstation 1 (Step 4). There the transport Kanban authorizes the transport of a lot of that variant from the outgoing goods store from Workstation 1 to the receiving goods store of Workstation 2 (Step 5). The freed production Kanban is then sent to the Kanban board for Workstation 1 (Step 6).

## 11.2 Discussion of the Procedural Rules

Synchro MRP is similar to the Kanban System in a number of aspects, thus with regards to the advantages and disadvantages the two generally coincide with one another. The main difference is the additionally generated daily production program. The following discussion about the procedural rules focuses on the resulting differences regarding the suitability of the method.

1. *A manufacturing control method should allow the WIP of a production or workstation to be regulated at a defined level.*

Unlike when controlling production with the MRP II system, Synchro MRP prevents a build-up of WIP when the production develops a backlog by utilizing the additional Kanban System. Furthermore in contrast to the Kanban System, Synchro MRP can prevent the automatic reproduction of a variant: When there is sufficient capacities, the Kanban System fills the store of a variant up to the stock limit defined by the number of Kanbans. This is generally desired in order to ensure the delivery readiness of a variant. Through the daily production program, Synchro MRP makes it possible to avoid this build-up of WIP. This is especially practical with a variant that has a low demand rate or when discontinuing a variant (e.g., when there is a technical change).

*Variants with lowered demand rates*: Variants that are rarely demanded are not produced daily, instead there are usually a number of days between production runs. A Kanban System would reproduce a variant directly after it is consumed and thus complete it earlier then needed. Synchro MRP, however, makes it possible to delay the production of the variant until the actual point in time that it is demanded. This reduces the WIP and keeps the capacities available for more urgent variants.

*Discontinuing a Variant*: If a variant is being discontinued, the manufacturing control has to prevent the variant from being reproduced by the self-regulating Kanban System beyond the planned discontinuation point. Synchro MRP offers a simple possibility for this: The variant is no longer taken up in the daily production program and accordingly no longer produced. To ensure a controlled discontinuation of a variant, the daily production program of the individual workstations need to be coordinated with one another.

2. *A manufacturing control method should reduce the WIP fluctuations to a minimum.*

The WIP cannot exceed a maximum value that results from the number of variants. The additional manufacturing control via the daily production program can prevent a WIP build-up, which – particularly for variants that are not required daily – results from the immediate reproduction of variants that occurs with the Kanban System.

3. *A manufacturing control method should cause as few WIP disruptions as possible.*

Synchro MRP causes more blocked WIP than the Kanban System. In addition to the lots for which there are no production Kanbans, this method for controlling production blocks the WIP for variants whose daily production program is already filled. Consequently, it prevents a build-up of WIP on the subsequent workstation. It also prevents, however, the production (or the production bottleneck) from exploiting a temporarily increased productivity to manufacture articles in advance.

4. *A manufacturing control method should contribute to closely aligning the load to the capacities.*

A chronological load balance accelerates or delays the release or throughput of orders with the aim of preventing breaks in the material flow (accelerates) or preventing the build-up of WIP (delays). Synchro MRP limits the WIP in the production and through that prevents the build-up of WIP. Otherwise the possibilities for a chronological load balance in the manufacturing control are few. Nonetheless, it is highly significant in the production program planning.

5. *A manufacturing control method should cause as few sequencing interchanges as possible.*

Synchro MRP can force sequencing interchanges when,

- no production Kanban authorizes the reproduction of a variant,
- the daily production program for a variant is filled.

In both cases a WIP build-up on the subsequent workstation should be prevented by the sequencing interchange. The sequencing interchanges, however, do not endanger the enterprise's delivery readiness.

6. *A manufacturing control method should be able to compensate for schedule deviations within the context of backlog control.*

It is the responsibility of manufacturing control to prevent a threatening backlog from developing and reducing a backlog once it has developed. In order to do so capacities can be temporarily increased. The procedural rules for Synchro MRP do not have a formalized backlog control in mind. However, the detailed daily production program suggests a very simple form of a backlog control: Work is continued until the daily production program is fulfilled. This requires on the one hand, flexible work hours and on the other hand, a realistic daily production program.

7. *A manufacturing control method should take into consideration the principle of bottlenecks.*

Through the Kanban System, Synchro MRP prevents the workstations before the bottleneck workstation from operating faster than the bottleneck capacities allow. It thus prevents an uncontrolled build-up of WIP in the production. Otherwise Synchro MRP does not take into consideration the bottleneck principle.

8. *A manufacturing control method should be as simple as possible.*

The increased effort involved in planning represents the general disadvantage of Synchro MRP. Hall repeatedly emphasizes the considerable effort that Yamaha undertakes in order to apply this method [Hall-86*]. In comparison to the Kanban

System the loss of the programs flexibility is especially obvious: The Kanban System can counteract (smaller) deviations from the planned production program easily through the Kanban control loop. The targets set by the daily production program annul this self-regulating mechanism of the Kanban System. In order to react to changes in the market demand, the production program itself has to be changed.

9. *Methods for generating orders should communicate information about demands as quickly and clearly as possible.*

Via the daily production program Synchro MRP targets a high level of transparency about the demand situation over a number of production levels. This information is, however, limited in that the daily production program is created for a complete planning period. Changes in the demand during this planning period cannot be communicated (or can only be conveyed by changing the production program – something which is not intended with this method).

## 11.3 Setting the Method's Parameters

The key parameters for Synchro MRP are the daily production program as well as the number of Kanbans per variant and workstation. How the number of Kanbans can be set is extensively described for the Kanban System in Sect. 9.4. For the daily production program, Hall's recommendation to distribute the load as evenly as possible is typical for just-in-time systems i.e. [Hall-86*].

## 11.4 Implementing Synchro MRP Across the Supply Chain

Yamaha uses transport Kanbans and a supply plan with detailed delivery quantities and delivery dates in order to integrate their suppliers [Hall-86*]. The suppliers deliver the materials directly to the workstation in the production. Among other functions, the transport Kanbans fulfill the following:

*Name of Location:* The transport Kanbans identify the workstation and the exact location of the delivery.

*Inventory Control:* The transport Kanbans are supplied with a barcode that identifies the part number. When it is delivered the barcode is read and the delivery is booked in the material management system. When the factories that are delivering and receiving are far away from one another, the output is also compiled by the supplier [Hall-86*].

## 11.5 Application

The same conditions for application apply to Synchro MRP as those that apply to the Kanban System (see Sect. 9.7). Through the more strict integration into a higher level production program plan though, a higher number of variants can be controlled.

Synchro MRP was developed by Yamaha and therefore on the shop floor. Hall writes an extensive report about its application and goes into details for example about the electronic inventory control as well as dealing with rejects [Hall-86*].

Deleersnyder brought together a group of authors who describe a variation of Synchro MRP in which a daily production program is provided only for specific workstations in the production. The remaining workstations in the production are controlled exclusively through Kanbans. The authors report about the application of this variation of the method in industrial enterprises [Dele-92*].

## References

[Dele-92*] Deleersnyder, J.-L., et al.: Integrating kanban type pull systems and MRP type push systems: insights from a Markovian model. IIE Trans. **24**(3), 43–56 (1992)

[Hall-86*] Hall, R.W.: Synchro MRP: combining kanban and MRP. The Yamaha PYMAC system. In: Driving the Productivity Machine: production Planning and Control in Japan, pp. 43–56. APICS, Falls Church (1986)

## Applications

## References

# Chapter 12
# Hybrid Kanban/CONWIP Control

The basic idea behind a hybrid Kanban/CONWIP control is to combine the decentralized WIP control from Kanban with the centralized WIP control from CONWIP. This method for controlling production was developed and evaluated in simulation studies by Bonvik, Couch and Gershwin [Bonv-97*]. In simulation studies with comparable WIP levels the method achieved a noticeably higher output rate than Kanban and a slightly higher output rate than CONWIP.

As is typical for one of the areas of research published on manufacturing control methods, the authors decided for reasons of simplicity to assume that only one variant was being produced. This limits the practicality of the studies, but simplifies the mathematic modeling and simulation of the manufacturing control method. Furthermore, it allows a number of general conclusions about the suitability and characteristics of manufacturing control methods. With the assumption that only one variant is produced, the difference between Kanban and CONWIP is reduced to the length of the control loop. It is then also not significant that Kanban utilizes variant specific WIP buffers and *generates* orders, while CONWIP implements variant neutral WIP buffers and *releases* orders.

In the following discussion, the procedural rules are transferred to a case that includes a number of variants. The aim here is to extend the range of applicability for the control method. To do so the control method is interpreted so that the variant specific buffer of Kanban remains intact.[1]

As with the previous methods, we will first take a look at the procedural rules for the hybrid Kanban/CONWIP control (Sect. 12.1), following that we will consider the procedural rules in view of the guidelines for designing manufacturing control methods (Sect. 12.2). Setting the method's parameters will be the focus of our discussion in Sect. 12.3 and is quite similar to Kanban. Finally, we will find that applying the Kanban/CONWIP control across the supply chain has an advantage

[1] This is a result of the author's explanation that the method can be understood as a variation of Kanban in that the finished goods store's Kanban is directed not to the last, but to the first workstation in the production.

H. Lödding, *Handbook of Manufacturing Control*,
DOI 10.1007/978-3-642-24458-2_12, © Springer-Verlag Berlin Heidelberg 2013

over Kanban in that the demand information is more quickly communicated to the suppliers (Sect. 12.4).

## 12.1 Method Description

Bonvik, Couch and Gershwin clarify that the method is a variation of Kanban in that the finished goods store's Kanban is forwarded directly to the first workstation in the production, authorizing the release of an additional order there [Bonv-97*]. Figure 12.1 depicts the centralized flow of information from the finished goods store to Workstation 1 and the decentralized flow of information between the workstations and the stores.

The Kanbans from the finished goods store are referred to in the following as finished goods Kanbans. The following procedural rules describe how a production is controlled using the Kanban /CONWIP:

1. If a variant is withdrawn from the finished goods store, the finished goods Kanban is directed to the first workstation that the variant runs through. The finished goods Kanban remains with the variant during its entire throughput in the production right up to the finished goods store.
2. The operator of a workstation is only allowed to process a lot when both the finished goods Kanban as well as the conventional Kanban signals the demand for a variant. The operator on the last workstation can generally process all the lots. The last workstation thus takes on a special status leading to a number of the method's unique characteristics which we will later discuss.
3. If the materials, finished goods Kanbans and conventional Kanbans are present for a number of variants, the operator selects the variant with the highest priority and begins processing it. If these conditions are not met for any of the variants, the workstation is blocked and the operator waits until the conditions are met for

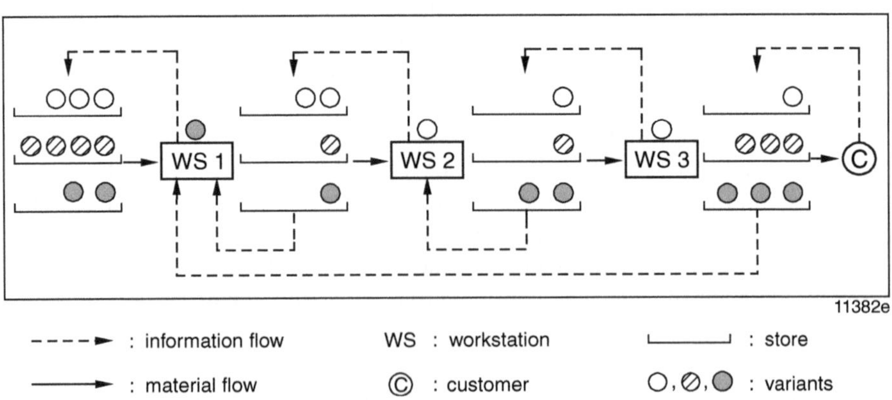


**Fig. 12.1** Principle of the Kanban/CONWIP control

one of the variants. As is typical of a sequencing rule for Kanban systems, the orders are processed in the sequence in which the Kanbans signal the demand for a variant.[2] At the first workstation, it has to be decided whether the finished goods Kanbans or the conventional Kanbans determine the sequence (in the following example, the second alternative is chosen). No conventional Kanbans are utilized on the last workstation. Thus, they can also not dictate the processing sequence. It is absolutely mandatory that the current demand be taken into consideration in the sequencing (see the discussion on the procedural rules for more information about this). A suitable sequencing rule for example would be to reproduce the variant with the lowest stock range in the finished goods store. If the finished goods stores for a number of variants are already exhausted, the variant with the largest backlog or the greatest demand should be produced first.

4. The operator withdraws the material from the outgoing goods store of the preceding workstation (or when it is the first workstation from the raw materials store). He directs the corresponding conventional Kanban to the preceding workstation where it triggers the reproduction of the outgoing materials.
5. The operator begins with the processing of the order. After it is processed the operator directs the lot and the corresponding Kanban to the outgoing goods store. The first workstation also adds a finished goods Kanban to the lot. The operator on the last workstation does not add any Kanban (the finished goods Kanban is already there and a conventional Kanban is not required on the last workstation). After the processing is completed the operators on all of the workstations verify whether or not they can begin processing another order (Steps 3–5 are then repeated).

Using an example, we can examine these rules for the Kanban/CONWIP control more closely. The focus of the example is a production system with three workstations WS 1 to WS 3 and three variants A, B and C (Fig. 12.2).

Each variant is allocated eight finished goods Kanban as well as three conventional Kanbans on Workstation 1 and 2. In the initial situation, Workstation 1 is blocked. There is a conventional Kanban present for Variant A, however, there is no corresponding finished goods Kanban. The situation for Variant B and C is the exact reverse: Finished goods Kanbans are present, but no conventional Kanbans are. The blocked workstation is freed when the customer calls-up Variant A from the finished goods store (Step 1 in Fig. 12.2b). The corresponding finished goods Kanban is directed to the first workstation in the production (Step 2). There, both the conventional Kanban for the variant that is taken from the Kanban board is available (Step 3) as well as the outgoing materials supplied by the raw materials store (Step 4). The raw materials themselves are controlled via a conventional Kanban system. The corresponding Kanban is forwarded to the supplier (Step 5).

---

[2] The authors of this method do not address the problem of sequencing since a production with only one variant is a condition of their discussion.

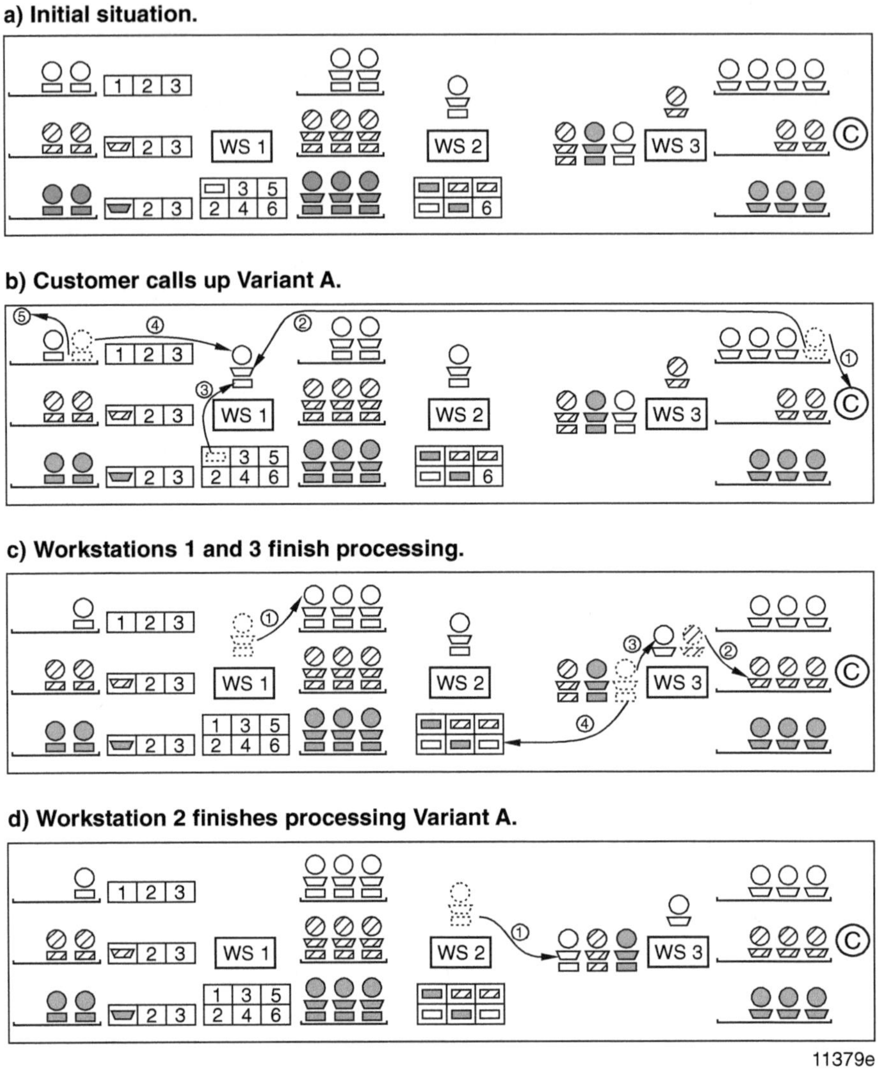


**Fig. 12.2** How the hybrid Kanban/CONWIP control functions

Next, Workstation 1 and 3 complete the processing of Variants A and B (Fig. 12.2c). The operator on Workstation 1 conveys the production lot including the Kanbans to the station's outgoing goods store (Step 1). Workstation 1 is initially unable to process any further orders because there are no conventional Kanbans present for any of the three variants. Furthermore, there is no finished goods Kanban for Variant A available. Workstation 3 directs the completed lot of Variant B with the finished goods Kanban to the finished goods store (Step 2).

**e) Workstations 2 and 1 select variants.**

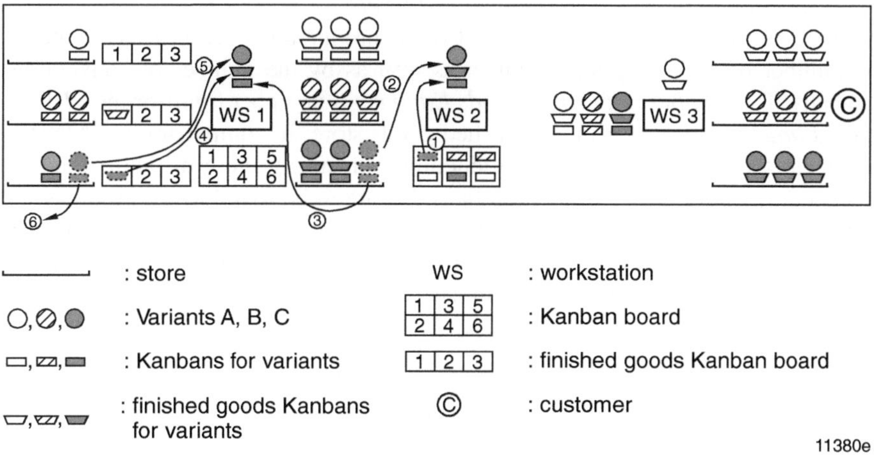


**Fig. 12.2** (continued)

Afterwards it selects the next order in the queue and begins processing it (Step 3). The conventional Kanban is directed to the preceding workstation (WS2; Step 4) indicating a demand for Variant A.

Once Workstation 2 has finished processing Variant A (Fig. 12.2d), the operator directs the material and Kanban to the queue for Workstation 3 (Step 1). Variant C now has the highest priority for Workstation 2.

The operator withdraws the corresponding conventional Kanban from the Kanban board (Fig. 12.2e, Step 1) and the material from the outgoing goods store at Workstation 1 (Step 2).

The operator then begins processing the order. The conventional Kanban of the outgoing materials is directed to Workstation 1 (Step 3). Once there it immediately triggers the processing of a new order. The finished goods Kanban and raw materials are made available (Steps 4 and 5) and the operator begins processing the order. Finally, the raw materials are re-ordered from the suppliers (Step 6).

## 12.2 Discussion of the Procedural Rules

One of the essential characteristics of the hybrid Kanban/CONWIP control is the hybrid WIP control, which regulates both the entire store of a variant as well as the WIP of the variant on the workstations. The following discussion about the procedural rules focuses especially on this aspect.

1. *A manufacturing control method should allow the WIP of a production or workstation to be regulated at a defined level.*

With the hybrid Kanban/CONWIP control the production's inventory is limited by the number of finished goods Kanban as well as by the number of conventional Kanbans. The *number of finished goods Kanbans* is an upper limit for the WIP of a variant *and* the stock of it in the finished goods store. In comparison to CONWIP, two factors make regulating the production's inventory more difficult: First, the percentage of the finished goods Kanbans that are to be found in the production or in the finished goods store cannot be exactly predicted. Second, the inventory increases with the number of variants produced. Thus, for the maximum total inventory in the production and the finished goods store:

$$STO_{\max,FGKan} = \sum_{k=1}^{NVar} NFGKan_k \tag{12.1}$$

where

| | |
|---|---|
| $STO_{max,FGKan}$ | maximum total inventory due to finished goods Kanbans (in number of Kanbans) [−], |
| $NFGKan_k$ | number of finished goods Kanbans for Variant *k* [−], |
| NVar | number of variants [−]. |

Moreover, the WIP is limited by the number of conventional Kanbans. The maximum WIP in the production is then attained when all of the Kanbans are allocated to a production lot (i.e., no Kanban is waiting for material or for a workstation to become available) and the last workstation in the production is processing an order (i.e. is utilized). Then:

$$STO_{\max,Kan} = \left( \sum_{k=1}^{NVar} \sum_{j=1}^{NWS-1} NKan_{k,j} \right) + Nw_{j=NWS} \tag{12.2}$$

where

| | |
|---|---|
| $STO_{max,Kan}$ | maximum total inventory due to conventional Kanbans (in number of Kanbans) [−], |
| $NKan_{k,j}$ | number of Kanbans for Variant *k* on Workstation *j* [−], |
| $Nw_{j=NWS}$ | number of work centers on the last workstation [−], |
| NVar | number of variants [−], |
| NWS | number of workstations [−]. |

Because the number of Kanbans on Kanban board can vary and are not (yet) allocated to an order, the production's WIP can also not be exactly controlled by conventional Kanbans.

Generally the production's WIP cannot exceed the lower of the two upper limits:

$$STO_{\max} = \min\left(STO_{\max,FGKan}; STO_{\max,Kan}\right) \tag{12.3}$$

where

| | |
|---|---|
| $STO_{max}$ | maximum total inventory of the production (in number of Kanbans) [−], |
| $STO_{max,FGKan}$ | maximum inventory due to finished goods Kanbans (in number of Kanbans) [−], |
| $STO_{max,Kan}$ | maximum inventory due to conventional Kanbans (in number of Kanbans) [−]. |

The hybrid WIP control opens two possibilities for delaying the release of an order:

(a) *There is no finished goods Kanban available:* In this case the stock level in the finished goods store is sufficiently high (or there are enough reproduction orders begun) so that no further orders need to be generated.
(b) *There is no conventional Kanban available:* In this case the production's WIP level is so high that releasing a further order would only increase the WIP without increasing the production's output.

The first case arises particularly in times when there is low demand, the second when there is high demand. If the demand is (clearly) lower than the production's capacities, the WIP in the production builds-up with conventional Kanban: Each workstation is then able to fill its outgoing goods store. The hybrid Kanban/CONWIP control in contrast reduces the WIP in the production when the demand is low and stores the majority of stock in the finished goods warehouse: Since the last workstation in the production is allowed to process all of the queued orders, the production can maintain the output at a higher level than the demand – as long as there is still available WIP. As soon as the production has run empty, the lower input then determines the output rate.

The additional central WIP control allows the hybrid Kanban/CONWIP control to slightly increase the number of conventional Kanbans in comparison to Kanban without the actual mean WIP increasing (which is prevented by the finished goods Kanbans). However, when the demand is greater the additional Kanbans help attain a higher output. As a result in simulations with a comparable mean WIP, a higher output was obtained than with Kanban alone.

Since the last workstation in the Kanban/CONWIP control is generally allowed to process all of the orders, it can lead to a larger finished goods store than the conventional Kanban control.

2. *A manufacturing control method should reduce the WIP fluctuations to a minimum.*

The production's WIP cannot exceed a maximum value resulting from the sum of finished goods Kanbans and conventional Kanbans. In the simulation trials conducted by Bonvik, Couch and Gershwin [Bonv-97*] the hybrid WIP control beneficially impacts the WIP fluctuations. The possible WIP fluctuations increase along with the number of Kanbans and with that also with the number of variants and safety-stock.

*3. A manufacturing control method should cause as little blocked WIP as possible.*
As with Kanban, the materials in the outgoing goods store of a workstation are blocked when there is no free corresponding Kanban on the subsequent workstations. Unlike Kanban, the WIP in the outgoing goods store of the second to last workstation cannot be blocked because the last workstation is generally allowed to process all orders. Furthermore, a portion of the orders are released later than with conventional Kanban via the centralized WIP control. On the whole, the extent of the blocked WIP is presumably a little less with the hybrid Kanban/ CONWIP control than with conventional Kanban.

*4. A manufacturing control method should contribute to closely aligning the load to the capacities.*
Load balancing accelerates or delays the release or throughput of orders with the aim of preventing breaks in the material flow (accelerates) or the build-up of WIP (delays). The hybrid Kanban/CONWIP control prevents the WIP from building up in the production and on the workstations. By contrast, no orders are expedited when a throughput bottleneck is threatened by a break in the material flow. In order to do so it would be necessary to generate the orders before the planned start date (for more information see Capacity Oriented Material Management, Chap. 11).

As a variant specific method, the hybrid Kanban/CONWIP control always regulates the stock of a specific variant. If a number of variants are produced, it makes decisions about releasing orders both on the workstations as well as for the entire production only on the basis of part of the stock. In order to attain a specific output rate, it requires a higher WIP than variant neutral methods that decide about releases based on the total stock.

*5. A manufacturing control method should cause as few sequencing interchanges as possible.*
The sequencing rules and compliance on the last workstation of the production are particularly significant for successfully applying the hybrid Kanban/CONWIP control. If the remaining workstations process the orders in the sequence in which the Kanbans arrive – as is typical with Kanban – then the last workstation provides the sequence for the entire production. This leads to considerable problems when the processing sequence on the last workstation is not oriented on the demand. If the orders on the last workstation are for example processed according to the first-in-first-out rule, a sequence will be repeated once it is established. The reason for this is that when an order is being processed on the last workstation, the conventional Kanban of the variants is always queued at the end of Kanban board of the second to last workstation. In the example from Fig. 12.3 the following sequence cycle thus results: A–B–B–A–B–A.

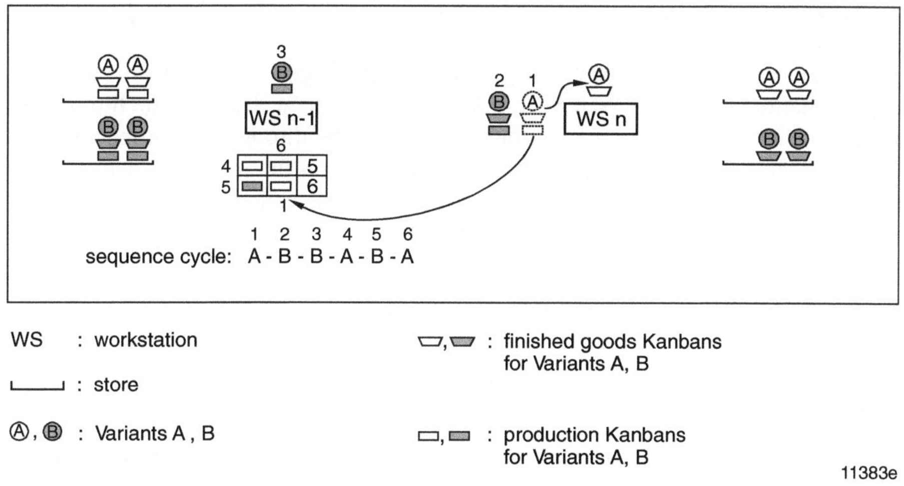


**Fig. 12.3** Sequence cycle with FIFO control on the last workstation

An established sequence is then first interrupted when a lack of material in the receiving goods store of the second to last workstation forces another sequence. This occurs at the latest when a lack of finished goods Kanban for a seldom demanded variant prevents the release of orders. Nonetheless, demand fluctuations for the individual variants can – when at all – only be reacted to with delays. For similar reasons sequencing rules such as the shortest or longest operating times or sequencing rules that optimize setup times also lead to insufficient results.

To be able to attain logistically satisfying results with the hybrid Kanban/CONWIP control the processing sequencing on the last workstation clearly needs to be oriented on the current demand. A suitable sequencing rule is for example, the stock range in the finished goods store. The last workstation then first reproduces the variant with the lowest stock range. If the stock for a number of variants is exhausted, the variant with the largest backlog or the greatest mean demand rate is reproduced.

The connection to the demand is also mandatory with conventional Kanban. However this is attained inherently by the system in that Kanban from the finished goods store is immediately directed to the last workstation in the production.

6. *A manufacturing control method should be able to balance schedule deviations within the context of backlog control.*

It is the responsibility of manufacturing control, to prevent a backlog and to reduce a backlog that has already developed. In order to do so capacities are temporarily increased. The procedural rules for the hybrid Kanban/CONWIP control do not include a backlog control. To be able to ensure on-time delivery even when there are schedule deviations, the method should thus be combined with a backlog control.

7. *A manufacturing control method should take into consideration the principle of bottlenecks.*

The discussion about the seventh guiding principle generally follows the same argumentation as Kanban (see Sect. 9.3). Given sufficient demand, the throughput bottleneck sets the production rhythm. To protect the throughput bottleneck from breaks in the material flow, the number of Kanbans on the preceding workstation can be set higher than usual. Under certain circumstances it is also practical to set the number of Kanbans on the workstation itself higher in order to prevent blocking the workstation.

8. *A manufacturing control method should be as simple as possible.*

The hybrid Kanban/CONWIP control is more complicated than conventional Kanban, however, this should not prevent it from being implemented in the industry.

9. *A manufacturing control method should communicate information about demands as quickly and clearly as possible.*

The hybrid Kanban/CONWIP control communicates information about the demand directly from the finished goods store to the first production stage. This seems at first to be advantageous, nonetheless, it is at least related to the problem of the last production stage not being informed about the demand. This should therefore be ensured through additional measures. The remaining production stages (that is the production stages between the first and last production stage) are not directly informed about the demand, but rather only indirectly via the Kanbans of each of the subsequent stages.

## 12.3 Setting the Method's Parameters

Bin contents, lot sizes and the number of Kanbans are essential parameters for the hybrid Kanban/CONWIP control. The procedure described in Sect. 9.5.1 for determining the bin content and lot size for Kanban can also be selected for the hybrid Kanban/CONWIP control. We will thus focus here on setting the number of Kanbans and in particular the number of finished goods Kanbans (Sect. 12.3.2). The section begins though with a brief summary of how to set the number of conventional Kanbans (Sect. 12.3.1).

### *12.3.1 Number of Conventional Kanbans*

Setting the number of conventional Kanbans is extensively discussed in Sect. 9.5.2, nevertheless a brief summary is provided here. Two basically different procedures for doing so can be distinguished. With the first procedure the responsible employee fixes the number of Kanbans at an initial value that is assumed to be sufficient. The number of Kanbans is then gradually decreased. In the ideal case one Kanban per variant is sufficient for ensuring the on-time delivery. The variant is then reproduced, while the withdrawn lot is being consumed. With the second procedure the number of Kanbans is determined based on a model in a Logistic Positioning. The input parameters include the safety-stock and the replenishment time. The safety-stock can be determined with the assistance of Storage Operating Curves and the replenishment time with the assistance of Logistic Operating Curves (see Sect. 9.5.2).

### *12.3.2 Number of Finished Goods Kanbans*

The general procedure for setting the number of finished goods Kanbans does not differ from the method for conventional Kanbans. The first alternative again consists of setting the number of finished goods Kanban at a value considered to be sufficient and then gradually reducing it within the context of a continual improvement process. The second alternative is the model based approach to setting the number of finished goods Kanbans.

**Gradually Reducing the Number of Finished Good Kanbans** The main difficulty with this procedure is determining the initial stock for the variants. There are two alternatives for doing this:

1. Each lot in the production and in the finished goods stores is allocated a finished goods Kanban when implementing the hybrid Kanban/CONWIP control. This procedure is very simple. However, problems can arise especially when the stock for a particular variant is quite low when it is being introduced (in the extreme case equal to zero).
2. In many cases planned values are saved in the PPC system for both the replenishment time of a variant as well as for the safety-stock. Where applicable, parameter or target values should also be agreed upon for one or both of these. The number of finished goods Kanban can then be derived according to Eq. 9.3 from the replenishment time, safety-stock, bin contents and mean demand rate.

**Setting the Number of Finished Goods Kanbans Based on a Model** In order to set the number of finished goods Kanbans based on a model, it is practical to differentiate between the circulating stock (i.e. the WIP) in the production and the safety-stock in the finished goods store. With Eq. 9.3:

$$NFGKan = \frac{TRP}{BiC} \cdot RDMD_m + \frac{SSL}{BiC} \tag{12.4}$$

where

| | |
|---|---|
| NFGKan | number of finished goods Kanbans [−], |
| TRP | replenishment time [SCD], |
| BiC | bin contents [units], |
| $\mathrm{RDMD_m}$ | mean demand rate [units/SCD], |
| SSL | safety stock level [units]. |

The *replenishment time* includes the length of time from when the finished goods Kanban is freed by the withdrawal of the allocated bin from the finished goods store up to the point in which a new lot arrives at the finished goods store. This duration can be comprised of the following components:

1. *Information Turn-Around-Time:* Information turn-around-times arise when a freed finished goods Kanban cannot be directly allocated to a new order due to organizational reasons. Information turn-around-times should be prevented or reduced as much as possible.
2. *Time Spent Waiting for Order Release:* If materials or conventional Kanbans are lacking for a variant, the corresponding finished goods Kanban cannot be directly allocated to an order. The time spent waiting for the materials can be ignored, if it only rarely results in material bottlenecks.
3. *Order Throughput Time up to but not Including the Last Workstation:* The WIP of a variant is limited in the production by the number of conventional Kanbans. When some of the Kanbans of a variant are not allocated to a production lot, the actual WIP is however lower than the number of conventional Kanbans. This is the case then when the workstations are well utilized and there are a number of Kanbans on the workstations' Kanban boards. The actual WIP of a variant at a workstation results on the one hand, from the WIP in the workstation's outgoing goods store and on the other hand, from the WIP that is at the workstation through the processing of an order. The WIP in the outgoing goods store can be estimated from the safety-stock of the variant at the workstation, whereas the WIP being processed can be estimated with Little's Law from the operation time and output rate. Accordingly, for the number of finished goods Kanbans that are bound during the throughput time:

$$NFGKan_{TTP} = \sum_{j=1}^{NWS-1} \left( \frac{SSL_j}{BiC} + \frac{TOP_j \cdot RDMD_m}{BiC} \right) \tag{12.5}$$

where

| | |
|---|---|
| $\mathrm{NFGKan_{TTP}}$ | number of finished goods Kanbans (order throughput time ) [−], |
| $\mathrm{SSL_j}$ | safety-stock on workstation *j* [units], |
| BiC | bin contents [units], |
| $\mathrm{TOP_j}$ | operation time on workstation *j* [SCD], |
| $\mathrm{RDMD_m}$ | mean demand rate [units/SCD]. |

The safety-stock level on the workstations is determined when calculating the conventional Kanbans. This value can be established when determining the number of finished goods Kanbans.

4. *Operation Time on the Last Workstation:* The WIP at the last workstation results directly from the application of Little's Law. Thus, for the number of tied finished goods Kanbans:

$$NFGKan_{TOP} = \frac{TOP_{j=NWS} \cdot RDMD_m}{BiC} \tag{12.6}$$

where

| | |
|---|---|
| $NFGKan_{TOP}$ | number of finished goods Kanbans (operation time) [−], |
| $TOP_{j=NWS}$ | operation time on workstation *j* [SCD], |
| $RDMD_m$ | mean demand rate [units/SCD], |
| BiC | bin contents [units]. |

If the first two components of the replenishment time are negligible, the number of finished goods Kanbans required for covering the replenishment time results directly from the sum of the Kanbans calculated according to Eq. 12.5 and 12.6. The *safety-stock in the finished goods store* then needs to be added to this value.

The *safety-stock* in the finished goods store can be set with the help of a Logistic Positioning based on the Storage Operating Curves Theory. The equations for calculating this are found in Sect. 3.1.3.

The following comments are thus limited to a key parameter of the computational equation – the replenishment time. The replenishment time is not as clearly defined with the hybrid Kanban/CONWIP control as it is with Kanban. Initially, it seems obvious to define the replenishment time as the length of time between the withdrawal of a lot from the finished goods store and the arrival of the order whose reproduction was triggered by it at the finished goods store. This would largely correspond to the throughput time of the related finished goods Kanban, as is used for calculating the number of finished goods Kanban that are tied into the production. The actual length of time that is spread between the withdrawal of a lot and the arrival of a new lot of the same variant is however usually shorter: Generally, the last workstation in the production can access existing stock of the variant in the outgoing goods store of the preceding workstation. There is then only one operation to be completed until the lot enters the finished goods store.

It thus seems obvious to set the planned replenishment time in a Logistic Positioning for the last workstation. To do so the planned WIP and planned inter-operation time that are to be attained on the workstation should be determined. The planned throughput time results from the sum of the operation time and the planned inter-operation time. The procedure for this is clarified in Sect. 3.2.2.3.

## 12.4 Implementation Across the Supply Chain

The hybrid Kanban/CONWIP control can be implemented as a variant-specific method within a manufacturing control system which spans a supply chain. It primarily serves to integrate suppliers. Three different variations of the application are shown in Fig. 12.4.

The variations differ from one another first with regards to whether the supplier and producer utilize finished goods Kanbans separately (variations *a* and *b*) or together (variation *c*). Second, the interfaces between the raw materials of the producer and the last workstation of the supplier can either be controlled via conventional Kanbans (variations *a* and *c*) or exclusively by a sequencing rule (variation *b*).

The focus is on the moment in which the customer withdraws a variant from the finished goods store. In *variation a*, the finished goods Kanban of the producer is directed to its raw materials store. When Workstation 1 processes the next order of that variant the raw materials are withdrawn. The supplier's corresponding finished goods Kanban is transported to the supplier's raw materials store and the conventional Kanban is sorted onto Kanban board of the supplier's last workstation. In variation *a*, the supplier's last workstation only processes an order when a conventional Kanban for the variant is present. As a result the supplier's last workstation can also be blocked. This is the case when material is available but there is no conventional Kanban which would authorize the production. This is advantageous

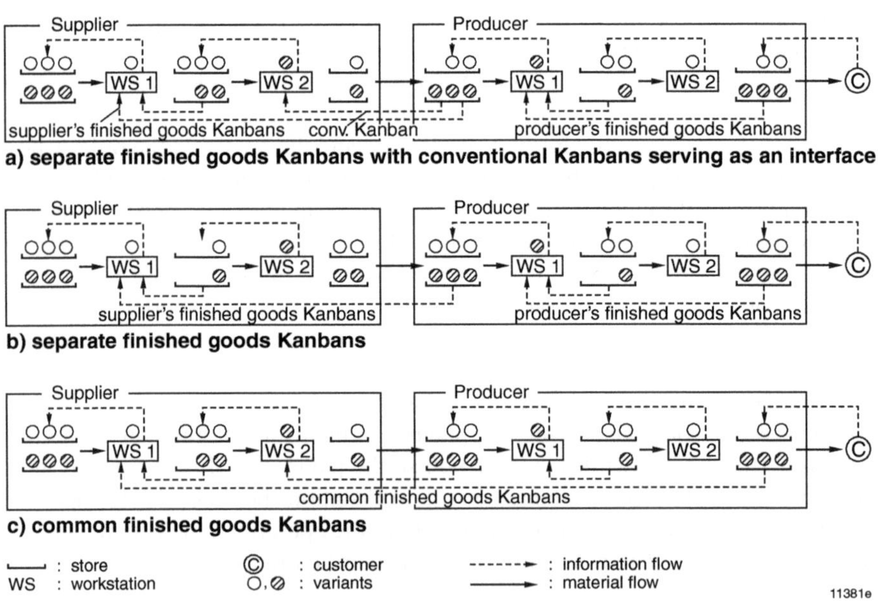


**Fig. 12.4** Implementing the hybrid Kanban/CONWIP control as part of a production control system across the supply chain

for the producer as a larger part of the stock is then kept at the supplier. Moreover, the Kanbans provide a practical processing sequence for the last workstation of the supplier.

In *variation b*, the conventional Kanban between the supplier and the producer is omitted. The last workstation of the supplier can thus process every order. There is a tendency for the supplier's stock to decrease and for the producer's store of raw materials to increase. One of the factors that play a role in the success of this variant is the sequencing rule on the supplier's last workstation (see also the discussion on the fifth guiding principle for designing manufacturing control methods).

In *variation c* the finished goods Kanban is transported directly from the producer's finished goods store to the supplier's raw materials store. The interface between the supplier's last workstation and the producer's raw goods store is controlled via conventional Kanbans. This variation is advantageous in that the customer's demand is conveyed directly to the supplier. With separate finished goods Kanbans this information is conveyed to some degree with delays for example, when materials or Kanbans are not available or the producer's first workstation processes orders with a higher priority first.

Controlling the interfaces between the supplier's store of outgoing goods and the producer's incoming goods store via transport Kanbans would be possible with all three variations. Moreover, with all three variations the supplier generally receives less distorted information about the end customer's demand more quickly than with a conventional Kanban system.

In the Toyota production system close communication between Toyota and their suppliers during the production program planning compensates for the information deficit that results from the conventional Kanban system [Woma-90*]. If the producer and supplier do not collaborate as closely, Kanban's delayed information could lead to problems.

# References

[Bonv-97*] Bonvik, A.M., Couch, C.E., Gershwin, B.S.: A comparison of production-line control mechanisms. Int. J. Prod. Res. **35**, 789–804 (1997)

[Woma-90*] Womack, J.P., Jones, D.T., Roos, D.: The Machine. That Changed the World. Harper Collins Publishers, New York (1990)

# Chapter 13
# Cumulative Production Figures

Cumulative Production Figures (abbreviated here as CPF) have been used to plan and control production in the automobile industry for decades. Developed by practitioners, Cumulative Production Figures (also known as 'progress numbers') synchronize the supplier's production with the automobile assembly. One of the first and foremost contributions in the field about this method was written by Heinemeyer at the University of Hannover's Institute of Production Systems and Logistics [Hein-88, Hein-94]. It is also discussed extensively in [Wien-97b]. Currently, the method is stimulated by the discussion around supply chain management.

Cumulative Production Figures are well suited for coordinating the production program planning in supply networks. They fulfill important – and currently vigorously debated – demands on PPC in supply chains such as the possibility of a central and coordinated sales plan. Moreover, they can contribute substantially to preventing or reducing the bullwhip effect.

Cumulative Production Figures are applied in repetitive manufacturing. However, especially in the area of manufacturing control, Cumulative Production Figures have not yet been algorithmitized. They do not provide any information about when an order should be generated or manufactured. Instead, this aspect is dependent on the user's specification or supplementation of the method. Thus on the one hand, controlling the production with Cumulative Production Figures is more flexible: They can be applied to coordinate very different types of production and can also be combined particularly well with other methods. On the other hand, it makes evaluations more difficult in that the degree to which the manufacturing control attains the logistic targets is fundamentally influenced by the method's specific design.

In this next section we will consider how a production is planned and controlled using Cumulative Production Figures (Sect. 13.1). For the latter, the method is supplemented with so-called 'control Cumulative Production Figures' (control CPF) which convert the concept of Cumulative Production Figures into an

H. Lödding, *Handbook of Manufacturing Control*, 
DOI 10.1007/978-3-642-24458-2_13, 

algorithmitized method for generating orders. Depending on how the control CPF are set, the method can represent different conventional control methods. They can thus be understood as a generic manufacturing control method. The procedural rules are discussed in Sect. 13.2 based on the guiding principles for designing manufacturing control methods, whereas, Sect. 13.3 is dedicated to setting the planned Cumulative Production Figures (planned CPF). How Cumulative Production Figures can be implemented across the supply chain is the subject of Sect. 13.4 and finally, in Sect. 13.5 their application is described.

## 13.1 Method Description

The description of the method is divided up into five sections:

- Definition of Cumulative Production Figures (Sect. 13.1.1),
- Program and Material Requirements Planning (Sect. 13.1.2),
- Planning Capacity Requirements (Sect. 13.1.3),
- Production Control (Sect. 13.1.4), and,
- Example (Sect. 13.1.5).

### *13.1.1 Definition of Cumulative Production Figures*

Cumulative Production Figures divide a production or supply chain into so-called 'control blocks' (Fig. 13.1).[1]

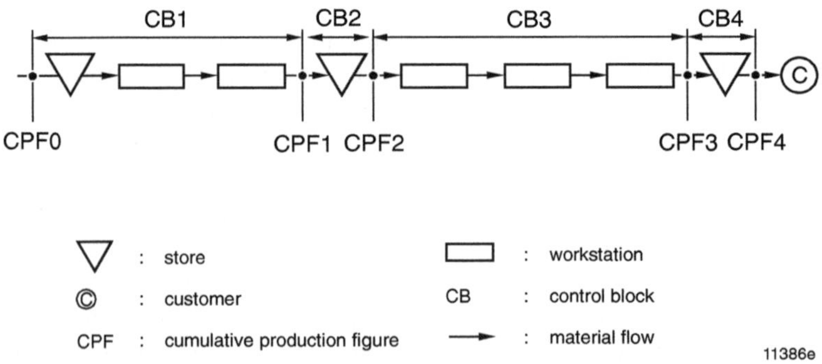


**Fig. 13.1** Definition of control blocks

[1] In the figure, both control blocks as well as Cumulative Production Figures are numbered in ascension in the direction of the material flow. This is beneficial in that the indices are identical thus facilitating the notation for the equations. In comparison, in research the Cumulative Production Figures are frequently numbered in ascension against the material flow.

The size of the control block can be freely selected and can for example be comprised of a single workstation, a store, a cost center, a production or an entire enterprise. The cumulated output for every variant is compiled in units (usually pieces) for every control block.

The cumulated output in pieces defines the (output) Cumulative Production Figure for the control block of a variant (Fig. 13.2a). The Cumulative Production Figure is thus a counter of a variant in a control block. At the same time, it corresponds to the input Cumulative Production Figure of the next control block.

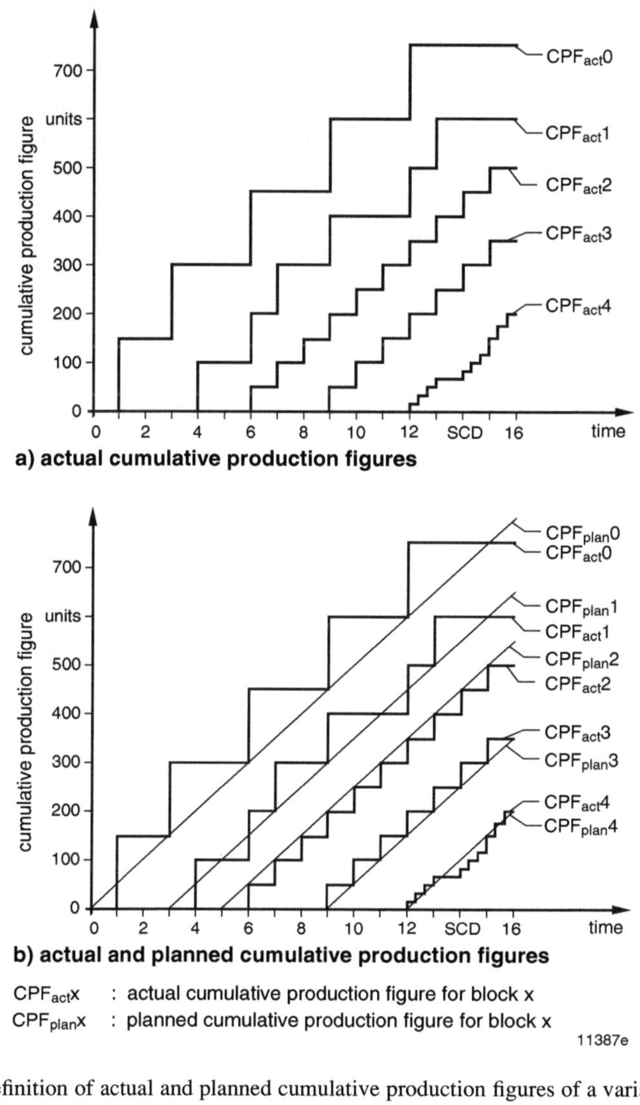


**Fig. 13.2** Definition of actual and planned cumulative production figures of a variant

The difference between the input and output Cumulative Production Figures corresponds at every point in time to the WIP of the variant in the control block. Thus,

$$WIPP_i = CPF_{i-1} - CPF_i = CPFIN_i - CPF_i \qquad (13.1)$$

where

| | |
|---|---|
| $WIPP_i$ | work-in-process (in pieces) for control block $i$ [units], |
| $CPF_i$ | Cumulative Production Figure for control block $i$ [units], |
| $CPFIN_i$ | input Cumulative Production Figure for control block $i$ [units]. |

Cumulative Production Figures thus allow the inventory of a control block to be calculated. Besides actual Cumulative Production Figures (actual CPF), planned Cumulative Production Figures which provide the planned output trend for a variant over time are also applied. These planned Cumulative Production Figures usually run continually (Fig. 13.2b); in order to distort demand information as minimally as possible no lot sizes are taken into consideration [Hein-88]. This has the additional advantage of allowing the lot sizes to be set locally and to vary both within a control block as well as between control blocks.

## *13.1.2 Production Program and Material Requirements Planning*

The starting point for planning the production program is planning the primary requirements for a reference control block which provides the basis for planning all of the other blocks. A reference control block can be e.g., the finished goods store or the final assembly. The planned Cumulative Production Figure for a variant of a reference control block is set as a function of the actual or predicted demand. In doing so an enterprise can – as in the section on planning the production program – generally follow a chasing strategy (where the seasonality of the demand is represented in the production program) or a leveling strategy where the capacities have a constant load. In each case the enterprise sets the planned output curve of the variants' primary requirements in the form of Cumulative Production Figures. These then form the basis for planning the production program in the preceding control blocks and for planning the material resources.

### Production Program in Preceding Control Blocks

Lead time offsets and WIP offsets are two standardized methods for deriving the planned CPF for dependent control blocks from the planned CPF curve of a reference control block.

*Lead Time Offsets:* Lead time offsets shift the planned Cumulative Production Figures curve for the reference control block by the planned throughput time (the so-called 'lead time') from the future back in the direction of the present in order to

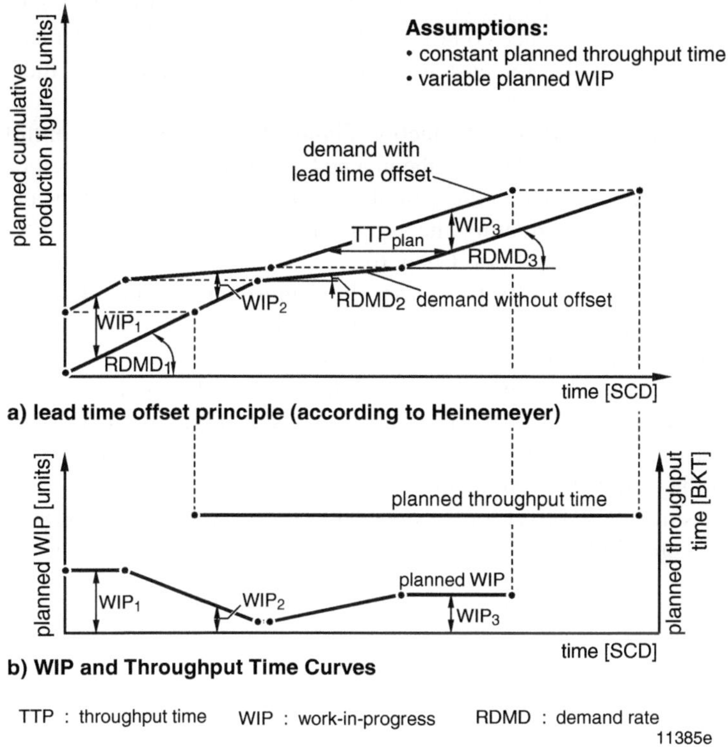


**Fig. 13.3** Lead time offset with cumulative production figures

determine the planned Cumulative Production Figures curve of the dependent control blocks (Fig. 13.3a).

$$CPF_{i,plan}\left(t - TTP_{plan}\right) = CPF_{\mathrm{ref,plan}}(t) \tag{13.2}$$

where

| | |
|---|---|
| $CPF_{i,plan}$ | planned Cumulative Production Figure for control block *i* [units], |
| t | time [SCD], |
| $TTP_{plan}$ | planned throughput time [SCD], |
| $CPF_{ref,plan}$ | planned Cumulative Production Figure for reference control block [units]. |

This leads to variable planned WIP levels in case the demand rate fluctuates (Fig. 13.3b). The planned stock corresponds at every point in time to the vertical distance between the curves of the Cumulative Production Figures.

In order to determine the planned CPF of the preceding control block *i*, the future planned CPF of the reference control block (and thus ultimately the demand for the end product) has to be forecast for a period that corresponds to the planned throughput time.

If the planned Cumulative Production Figures for the reference control blocks cannot be exactly predicted within the planned throughput time, the method causes a frequently sudden change in the Cumulative Production Figures of the dependent control blocks when correcting the prognosis. This in turn causes the demand fluctuations to be amplified (see Chap. 6).

*WIP Offset:* The WIP offset increases the planned CPF of the reference control block by the planned WIP in order to determine the planned CPF of the dependent control blocks (Fig. 13.4a).

The WIP offset thus assumes constant planned WIP levels. If the demand rate varies, then varying planned throughput times result (Fig. 13.4b). The throughput time corresponds at every point in time to the horizontal distance between the curves of the Cumulative Production Figures.

When using a WIP offset for the planned Cumulative Production Figures of the dependent control blocks:

$$CPF_{i,plan}(t) = CPF_{\mathrm{ref,plan}}(t) + WIPP_{plan} \tag{13.3}$$

where

| | |
|---|---|
| $CPF_{i,plan}$ | planned Cumulative Production Figure for control block *i* [units], |
| $CPF_{ref,plan}$ | planned Cumulative Production Figure for reference control block [units], |
| $WIPP_{plan}$ | planned work-in-process (in pieces) [units]. |

With that the WIP offset does not require the planned Cumulative Production Figure of the reference control block to be forecasted. However, the enterprise has to (be able to) adjust the throughput times of the variants in the control blocks when there are fluctuations in the demand rate. When discontinuing a variant it needs to be ensured that the Cumulative Production Figures of the dependent control blocks do not exceed the total planned production quantity, which should instead be set as a maximum value. Otherwise, more than required will be produced.

### Material Requirements Planning

By linking the Cumulative Production Figures of dependent item numbers, it is also possible to consider the relationships described by the bills of materials. To do so the CPF for the item numbers of the dependent demand are depicted as a function of the CPF for the item numbers of the primary requirements (or with multi-stage part lists also as a function of the higher level material requirements). In this way, changes in the primary requirements are automatically and immediately converted into changes in the material requirements. And, once again we can distinguish between a lead time offset and a WIP offset. For example, when employing a WIP

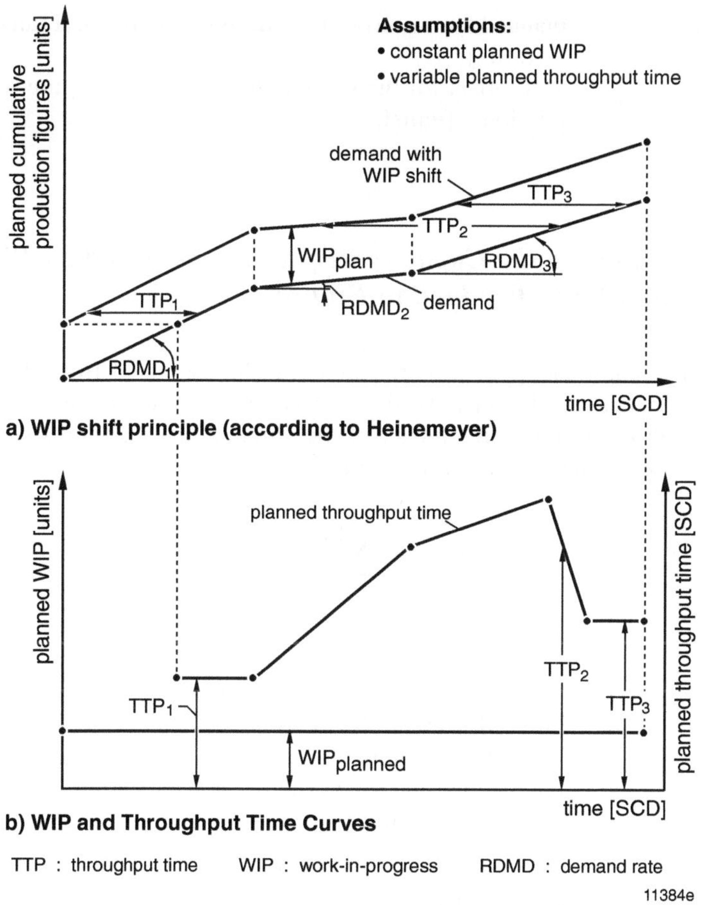


**Fig. 13.4** WIP offset with cumulative production figures

offset, the following equation can be used for the planned Cumulative Production Figures of a shaft that is built into two different transmissions and into each one twice:

$$CPF_{S,plan} = 2 \cdot CPF_{T1,plan} + WIPoff_{T1} + 2 \cdot CPF_{T2,plan} + WIPoff_{T2} \qquad (13.4)$$

where

| | |
|---|---|
| $CPF_{S,plan}$ | planned Cumulative Production Figure for shafts [units], |
| $CPF_{T1,plan}$ | planned Cumulative Production Figure for Transmission 1 [units], |
| $WIPoff_{T1}$ | WIP offset for the Cumulative Production Figure for Transmission 1 [units], |

$CPF_{T2,plan}$ planned Cumulative Production Figure for Transmission 2 [units],
$WIPoff_{T2}$ WIP offset for the Cumulative Production Figure for Transmission 2 [units].

### 13.1.3 Estimating the Capacity Requirements Based on Planned Cumulative Production Figures

The capacity requirements of a workstation can be estimated from the planned Cumulative Production Figures. To do so it is necessary to convert the planned Cumulative Production Figures for all of the variants that load a resource into a planned output for the resource via the processing times:

$$OUT_{plan,j} = \sum_{i=1}^{NVar} \frac{CPF_{i,j} \cdot t_{p,i,j}}{60} \tag{13.5}$$

where

$OUT_{plan,j}$ planned output for Workstation *j* [hrs],
$CPF_{i,j}$ planned Cumulative Production Figure for Variant *i* on Workstation *j* [units],
$t_{p,i,j}$ processing time for Variant *i* on Workstation *j* [min/units],
NVar number of variants [−].

An example of this procedure is depicted in Fig. 13.5. It should also be pointed out that the planned output calculated using this method does not contain setup times. Furthermore, it is necessary to provide a planned Cumulative Production Figure for each workstation.

### 13.1.4 Production Control with Cumulative Production Figures

There are diverse possibilities for applying Cumulative Production Figures in manufacturing control. Based on Heinemeyer [Hein-88] three functions will now be explained (Fig. 13.6):

*Determining the Range* (Fig. 13.6a)*:* Based on the Cumulative Production Figures diagram, it can be determined how much time it will take before the output falls below the planned Cumulative Production Figures when there is a loss of production.

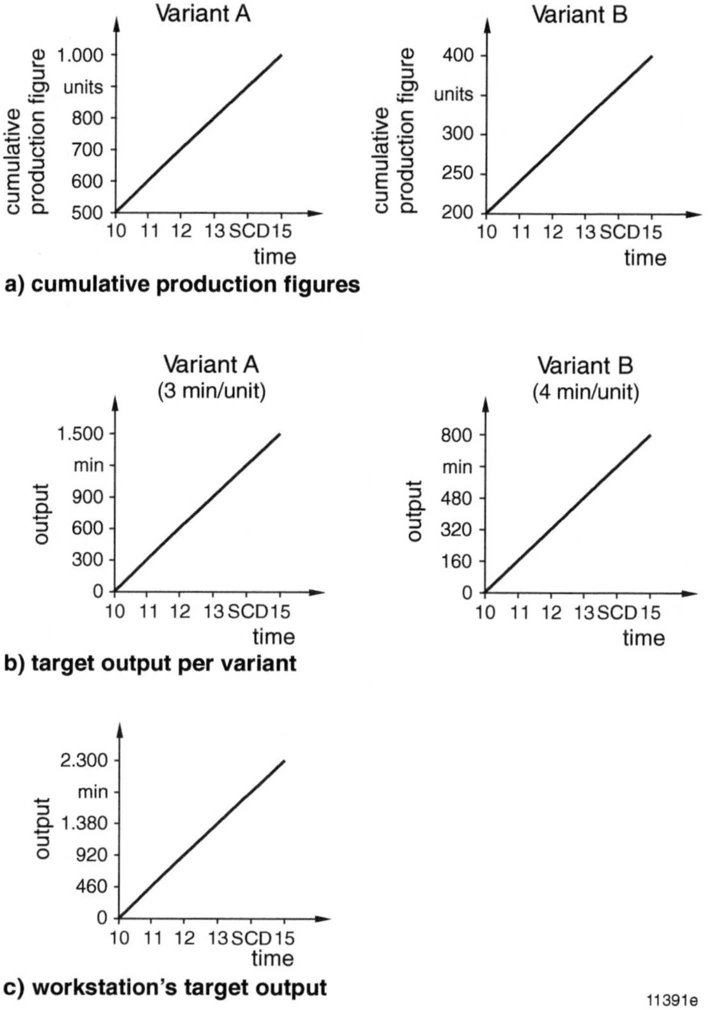


**Fig. 13.5** Deriving the planned output of a workstation based on planned cumulative production figures

*Program Control* (Fig. 13.6b): Whether or not a control block fulfils the planned progress for a variant can be determined directly. To do so, the deviation of the actual CPF from the planned CPF (the backlog) is calculated. This figure can be compiled for the entire supply chain and depicted in a diagram. It is thus also possible to identify the cause or causes of a supply bottleneck.

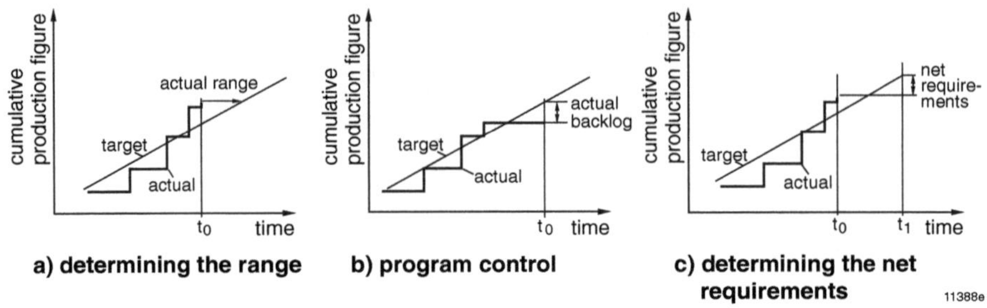


**Fig. 13.6** Applying Cumulative Production Figures in manufacturing control (Heinemeyer, cited according to Nyhuis)

*Determining Net Requirements* (Fig. 13.6c)*:* The employee responsible for a control block can determine the (anticipated) net requirements for each variant and each planning horizon. To do so, he compares the planned output for the (future) time $t_1$ to the actual output at the planned time $t_0$.

In traditional applications, Cumulative Production Figures serve as guidelines for manufacturing control, whereby the actual control decisions are however made locally. In the following, this method is modified so that it makes a portion of the decisions. In order to do so, so-called '*control Cumulative Production Figures*' (control CPF) are introduced. As soon as the control Cumulative Production Figure exceeds the actual Cumulative Production Figure of a variant, the reproduction of this variant in the control block is authorized. Inversely formulated, a control block is only allowed to produce a variant when the control CPF of a variant is greater than the actual CPF. Thus, the control CPF precedes the actual CPF by the replenishment time. With this convention, the Cumulative Production Figures are able to emulate the common methods for generating orders in make-to-stock productions. The replicated methods differ from one another in the following aspects:

1. *Definition of the Control Block:* The range and thus the limits of the control block have to be set.
2. *Setting the Control Cumulative Production Figure:* Next to defining the control block, setting the control Cumulative Production Figure is the most important characteristic of a method for generating orders based on CPF. Generally the control Cumulative Production Figure of the finished goods store corresponds to the cumulated demand, so that the demand is serviced from the store when there is stock. The control CPF from the production processes are usually linked with the customer demand or with the actual CPF of the subsequent control block. In the following, the process of setting the control Cumulative Production Figure will be depicted based on an example using Kanban and Base Stock.

*Kanban* sets the control CPF of a manufacturing control block as a function of the actual CPF of the subsequent store control block. Thus:

$$CPF_{i-1,control}(t) = CPF_{i,act}(t) + NKan \cdot BiC \tag{13.6}$$

where

| | |
|---|---|
| $CPF_{i-1,control}$ | control Cumulative Production Figure control block *i-1* [units], |
| $CPF_{i,act}$ | actual Cumulative Production Figure control block *i* [units], |
| NKan | number of Kanbans [−], |
| BiC | bin content [units]. |

*Base Stock* (which is extensively discussed in the next chapter) determines the control Cumulative Production Figure as a function of the customer demand:

$$CPF_{i,control}(t) = DC(t) + SB_i \tag{13.7}$$

where

| | |
|---|---|
| $CPF_{i,control}$ | control Cumulative Production Figure for the control block *i* [units], |
| DC | cumulative demand of the end customer [units], |
| $SB_i$ | base stock of the $i^{th}$ control block [units]. |

3. *Determining the Manufacturing Lot Size:* The manufacturing lot size is usually not provided by manufacturing control methods. Nevertheless, many methods make implicit assumptions about the manufacturing lot size: Kanban, for example, requires that the lot size is a multiple of the bin contents.
4. *Determining the Sequence:* The sequence, in which the variants are produced, is also generally not provided by manufacturing control methods.

### 13.1.5 Example

We can take a closer look at planning the production program and controlling the production using Cumulative Production Figures based on two examples. The first example will consider the production of product P which is delivered to customers from the finished goods store in two variants P1 and P2 (Fig. 13.7). The finished product is based on both components C1 (product variant P1) and C2 (product variant P2) and procurement parts PP, which will not be considered closer here. The components C1 and C2 are produced from the raw materials R1 and R2.

The production consists of manufacturing the parts using two workstations (WS 1 and WS 2) and assembling the parts produced in-house together with the procured

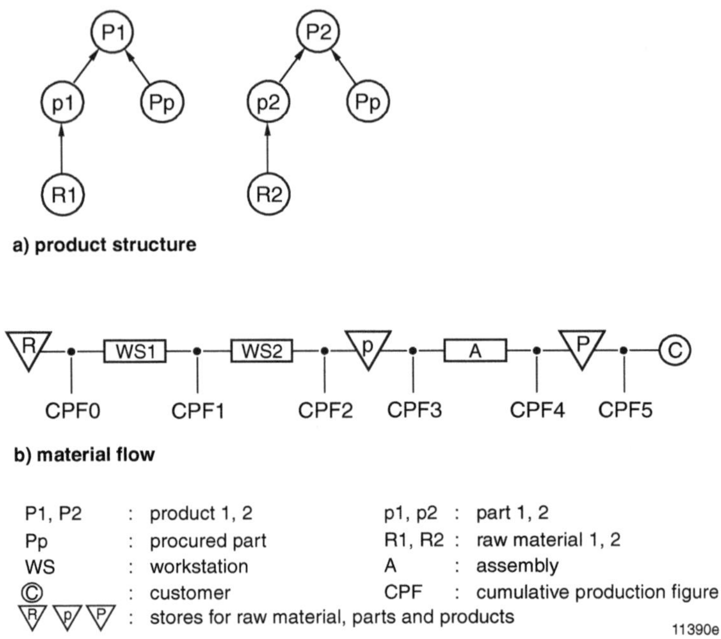


**Fig. 13.7** Product structure and material flow for the example

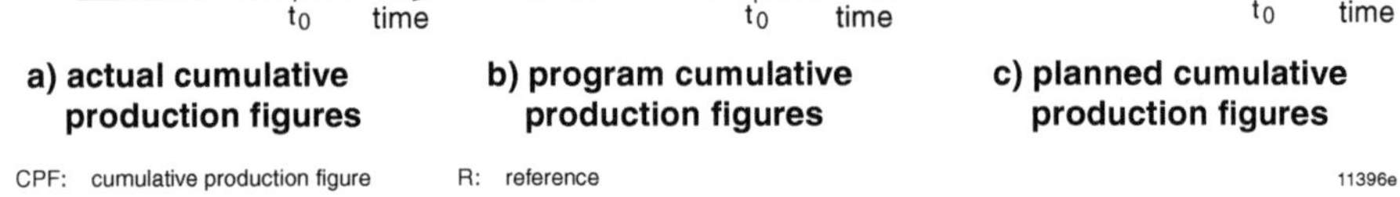


**Fig. 13.8** Representation of actual, program and planned cumulative production figures

parts (assembly A). A finished goods store buffers the assembly from demand fluctuations. In addition, there are both a raw materials store and a parts store.

The enterprise distinguishes three different types of Cumulative Production Figures: *Actual Cumulative Production Figures* represent the actual output of a variant (Fig. 13.8a), whereas, *program Cumulative Production Figures* represent the production program as it is established in the last planning round. The production program is created for 4 weeks at a time, but updated every 2 weeks. It represents the expected output and mainly serves as the basis for the production planning e.g., in order to set the planned capacities. Program Cumulative Production Figures are represented continually (Fig. 13.8b).

They are generated for the product based on a forecast. From these reference Cumulative Production Figures, the enterprise then calculates the program Cumulative Production Figures for the dependent control blocks using a WIP offset. Program Cumulative Production Figures are very limited with regards to their suitability for controlling production. Generally, the customer's requirements deviate from that which was forecast. In order to be able to quickly react to the deviations in the customer's demand, the enterprise thus also works with continually updated planned Cumulative Production Figures.

In the finished goods store these *planned Cumulative Production Figures* (Fig. 13.8c) correspond to the customer's actual cumulated demands. If – as in the example – all of the customer's demands can be punctually filled, then the planned CPF correspond to the actual CPF. Unlike the program Cumulative Production Figures, the planned Cumulative Production Figures in the example are thus not presented continually, but rather discretely. The rise corresponds to the quantity demanded by the customer. The planned CPF deviate further from the program CPF when the customer demand develops differently than planned. The planned Cumulative Production Figures of the preceding control block are established similarly to the program Cumulative Production Figures via a WIP offset and are continually updated. They provide information about the actual current customer demand. A customer demand immediately triggers an increase in the planned CPF in the finished goods store well as – via the WIP offset – on the production's workstations. The planned Cumulative Production Figures serve the enterprise as a basis for making decisions about sequencing and about temporarily adjusting capacities.

The enterprise measures the actual Cumulative Production Figures of the two product variants at six control points in the material flow (see Fig. 13.7):

- Raw materials store output (CPF 0)
- Workstation 1 output (CPF 1)
- Workstation 2 output (CPF 2)
- Components store output (CPF 3)
- Assembly output (CPF 4)
- Finished goods store output (CPF 5)

Program and planned Cumulative Production Figures are provided for the production (CPF 1 and CPF 2) as well as the assembly (CPF 4) and the finished goods store (CPF 5).

Figure 13.9 depicts the production program planning with Cumulative Production Figures for planning periods 3 and 4. The planning takes place at the end of Planning Period 2. In Planning Period 1 and 2 the demand for product P1 was somewhat less than originally planned, whereby the forecasts sales for product P2

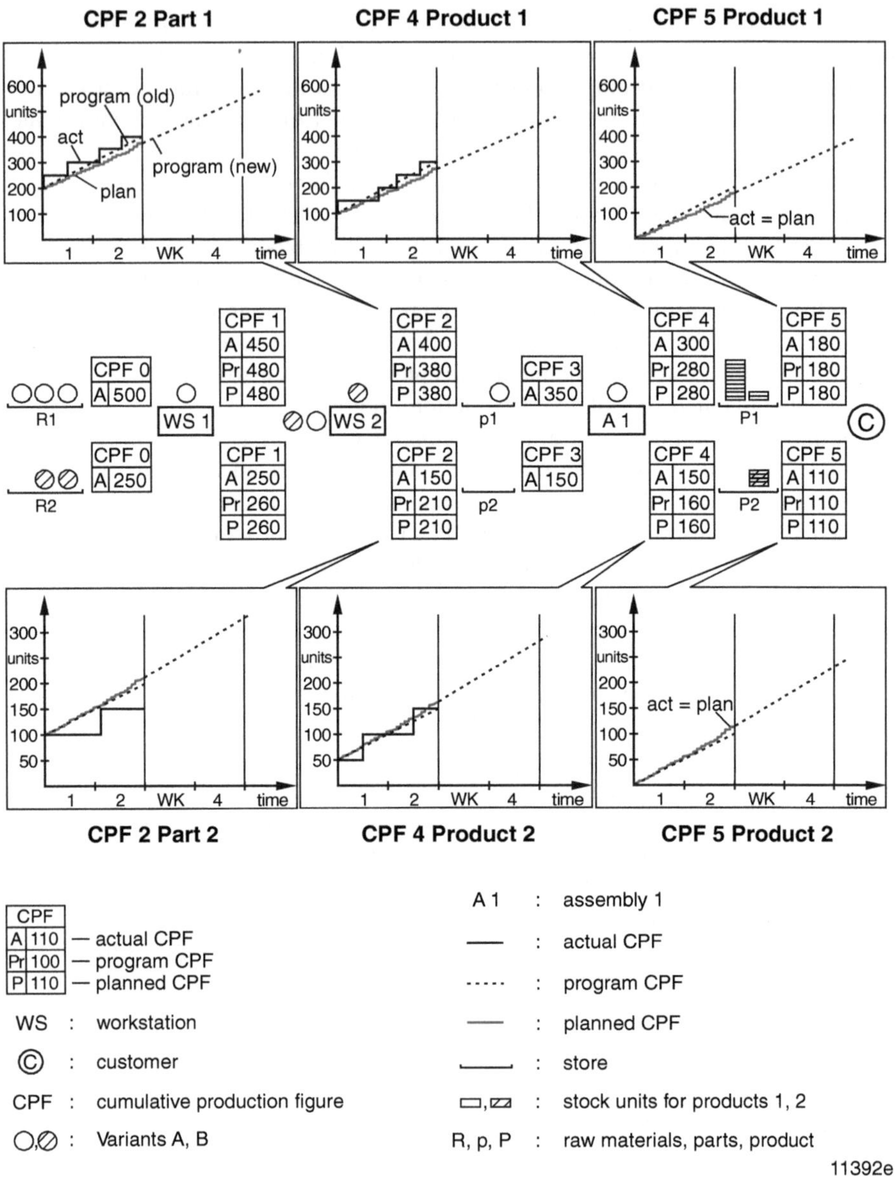


**Fig. 13.9** Production program planning with cumulative production figures

were in contrast exceeded. The deviations for product P1 and P2 can be read from the diagrams for CPF 5.

The new production program planning is based on the current *planned* Cumulative Production Figures. In other words, the forecast mistakes are corrected in each planning round, causing the program Cumulative Production Figures to generally be discontinuous (see the diagrams for CPF 5 for both products). Sales of 80 units/week are expected for product P1 and 60 units/week for product P2. As a result at the end of Period 4 the value for the program CPF 5 is 340 units for product P1 and 230 units for product P2.

The enterprise calculates the program of the preceding control blocks using a WIP offset. It increases the program CPF of the finished goods store for product P1 by 100 units (or 200 and 300 units) in order to calculate the program CPF for the assembly (and for Workstations 2 and 1). The corresponding WIP offset for product P2 is 50 units (assembly), 100 units (Workstation 2) and 150 units (Workstation 1).

The ensuing manufacturing control with the Cumulative Production Figures is clarified in Fig. 13.10.

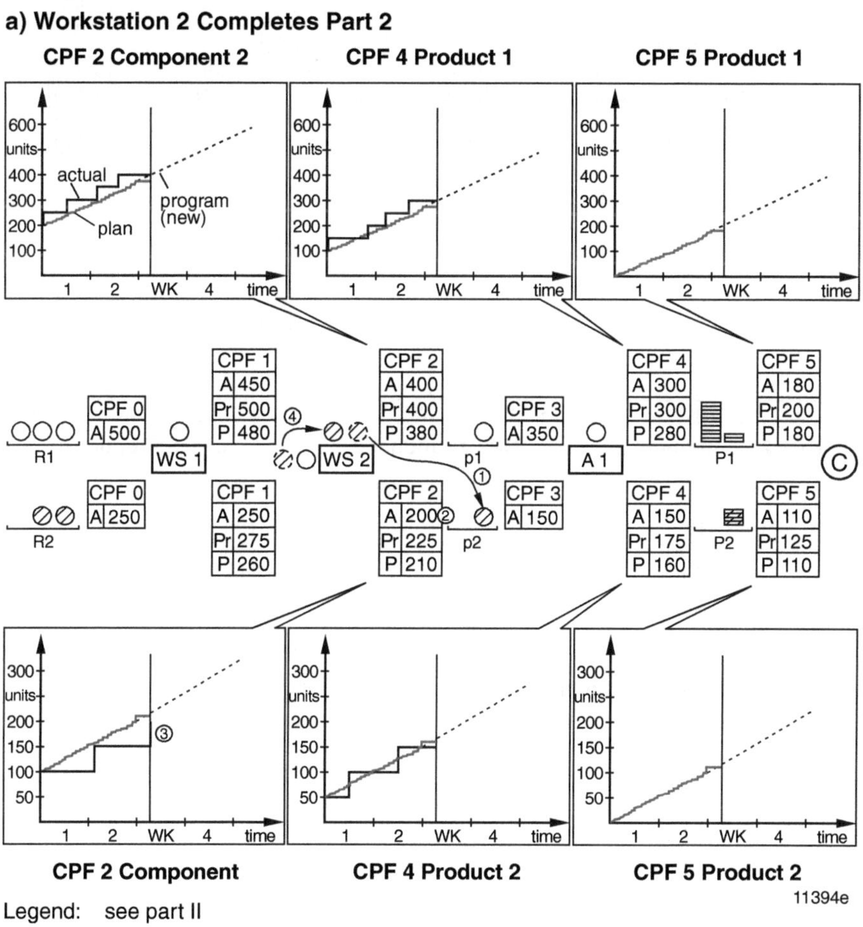


Fig. 13.10 Controlling production with cumulative production figures

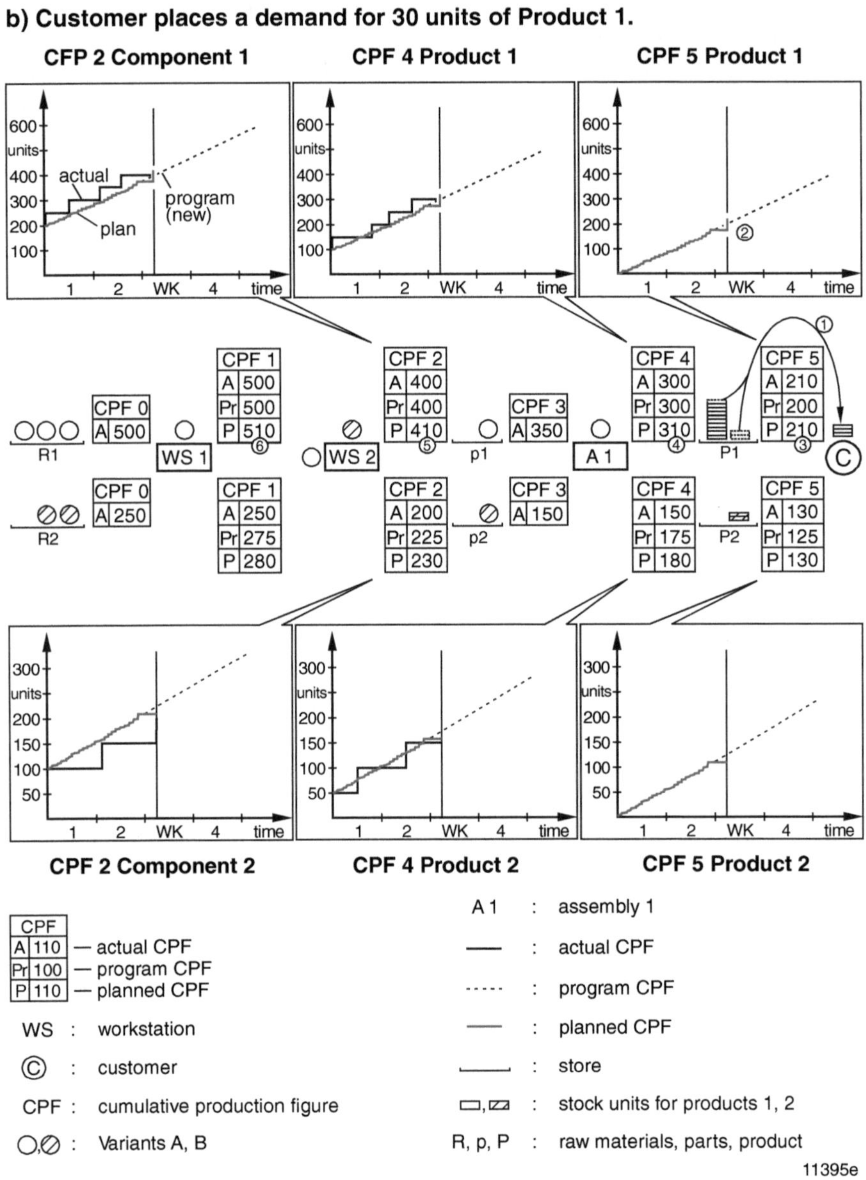


**Fig. 13.10** (continued)

Shortly after completing the new production program Workstation 2 finishes processing the order for component C2 (Step 1 in Fig. 13.10a). As a result the actual Cumulative Production Figure CPF2 increases on the workstation to 200 units (Step 2 and 3). The operator now considers which component she should produce next. Since the workstation has a backlog of 10 units for component C2 (planned Cumulative Production Figures CPF2 = 210 units), she produces another order for component C2 (Step 4).

A short time later, a customer places a demand for 30 units of product P1 (Step 1 in Fig. 13.10b). As a result the actual and planned Cumulative Production Figures for the finished goods store (Steps 2 and 3) as well as the planned Cumulative Production Figures for the assembly and the two manufacturing workstations each increase by 30 units (Steps 4–6).

The Cumulative Production Figures thus influence the enterprise's manufacturing control solely by the information they convey. However, they do not trigger any control decisions.

This changes fundamentally when *control* Cumulative Production Figures are applied. We will now take a look at a second example, which considers a production with three workstations and two variants, in order to explain how these operate. Actual and control Cumulative Production Figures are maintained for the workstations and the finished goods store (Fig. 13.11).

The enterprise provides the following rules for the manufacturing control:

1. The control Cumulative Production Figure for the finished goods store corresponds to the cumulated demand for the variants. This is a typical arrangement according to which the customer demands are immediately served from the store when stock is available.
2. The control Cumulative Production Figure of the last workstation is calculated as the sum of the control Cumulative Production Figure for the finished goods store and a WIP offset of 50 units set by the enterprise. The last workstation is thus oriented on the customer demand. The WIP offset regulates the stock in the finished goods store.
3. The control Cumulative Production Figures for Workstations 1 and 2 correspond to the actual Cumulative Production Figure of the last workstation and an additional WIP offset of 50 (Workstation 2) and 100 units (Workstation 1). This coupling with the output of the last workstation limits the WIP in the production.
4. The variants are manufactured in production lots of 25 units. This simplifies implementing production logistics with standard containers. In principal, however, Cumulative Production Figures are not dependent on given or constant lot sizes.
5. If a workstation or the assembly can reproduce a number of variants, it selects the variant with the actual Cumulative Production Figure that has the largest backlog in comparison to the control Cumulative Production Figure.

The control Cumulative Production Figures can also be set in many other ways. With a clearly defined bottleneck system for example, it is possible to utilize the Cumulative Production Figure for the bottleneck as a reference for the remaining

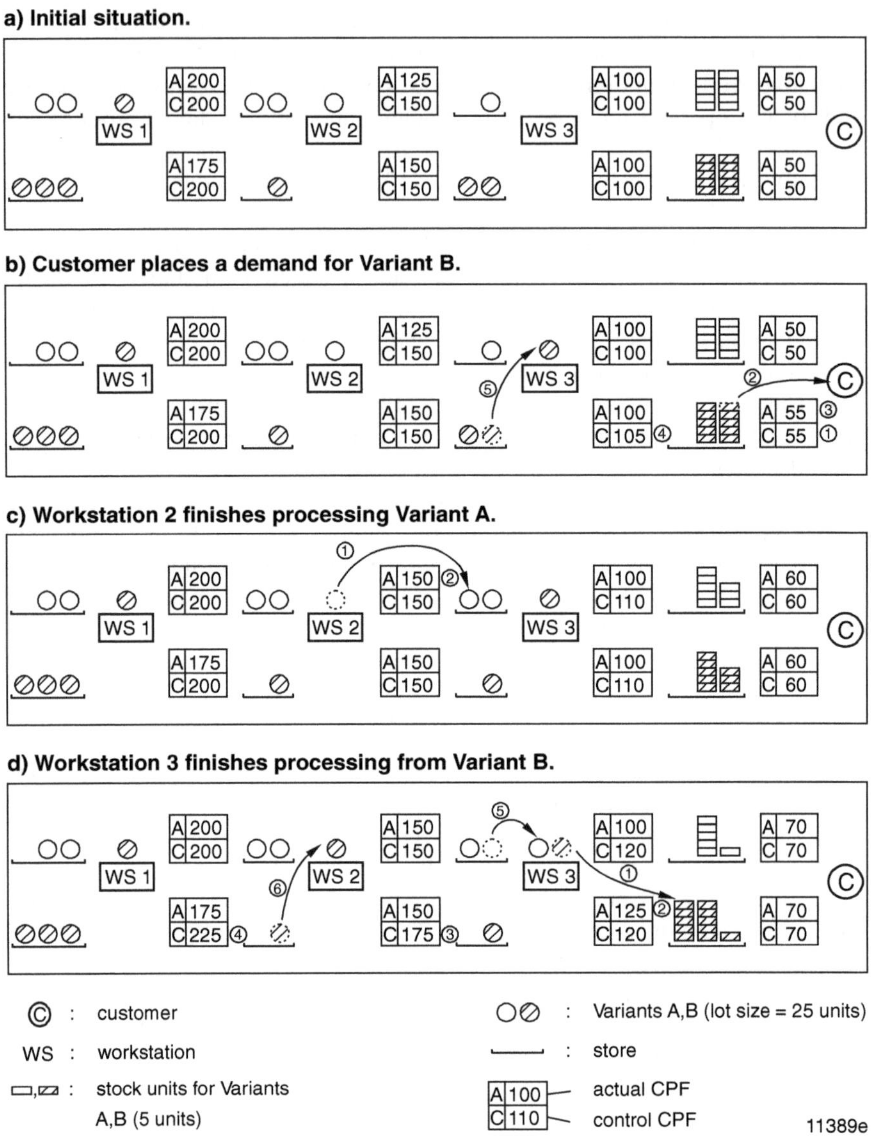


**Fig. 13.11** Controlling production with control cumulative production figures

workstations. The method for controlling production which results from the rules described here has the following advantages:

- It limits the WIP of the variants in the production.
- It regulates the store of the variants in the finished goods store.

- It forwards information about the demand comparatively quickly and accurately to the preceding control blocks (decoupling it only via the finished goods warehouse).

In the initial situation (Fig. 13.11a), the finished goods store for both variants is full. The control Cumulative Production Figure does not exceed the actual Cumulative Production Figure for any of the variants, thus Workstation 3 is blocked. Workstations 1 and 2 are each processing a lot with 25 units of Variant B (Workstation 1) and Variant A (Workstation 2).

If the customer places a demand for five units of Variant B (Fig. 13.11b), the control CPF for the finished goods store increases (Step 1) and the enterprise ships the corresponding quantity to the customer (Step 2). As a result the actual CPF for the variant climbs (Step 3). The increase in the control CPF of the finished goods store triggers a corresponding climb in the control CPF for Workstation 3 (Step 4). This therefore exceeds the actual Cumulative Production Figure and thus resolves the blockade on Workstation 3 which now reproduces Variant B (Step 5). Following that the customer places demands for further units of Variants A and B. The corresponding supply operations and resulting changes in the actual and control Cumulative Production Figures will however not be presented in detail here.

A little later Workstation 2 completes processing the lot of Variant A and directs that lot to its outgoing goods store (Step 1 in Fig. 13.11c). As a result the actual Cumulative Production Figure of the variant increases on Workstation 2 to 150 units (Step 2). Workstation 2 cannot produce any of the variants since the control CPF do not exceed the actual CPF.

Next, Workstation 3 completes processing Variant B (Step 1 in Fig. 13.11d). The actual Cumulative Production Figure for the variant thus increases to 125 units (Step 2). This immediately triggers an increase in the control Cumulative Production Figures for the variant on Workstations 1 and 2 (Steps 3 and 4). Workstation 3 reproduces Variant A (Step 5) and Workstation 2 can begin reproducing Variant B (Step 6) since the blockade on the workstation is lifted by the increase in the control Cumulative Production Figure.

## 13.2 Discussion of the Procedural Rules

One of the primary characteristics of Cumulative Production Figures are how they quickly and directly communicate demand information to the production levels.

1. *A manufacturing control method should allow the WIP of a production or workstation to be regulated at a defined level.*

Cumulative Production Figures can regulate the WIP of variants in the control blocks via a WIP offset. Especially when there are only a few variants, this also limits the WIP in a control block. However, the Cumulative Production Figures cannot regulate the WIP in manufacturing hours. This is only possible with the support provided by methods for releasing orders.

2. *A manufacturing control method should reduce the WIP fluctuations to a minimum.*

Cumulative Production Figures are limited in their suitability for preventing WIP fluctuations. Especially when there are a large number of variants and large control blocks, the WIP on the workstations in the production can also fluctuate more strongly.

3. *A manufacturing control method should cause as little blocked WIP as possible.*

Applying control Cumulative Production Figures to control the production, blocks the WIP in the incoming goods store of a control block whose actual CPF for the outgoing goods store is equal to or exceeds the control CPF. It thus resembles Kanban or the Order Point System with regards to the blocked WIP.

4. *A manufacturing control method should contribute to closely aligning the load to the capacities.*

Cumulative Production Figures generally do not actively contribute to balancing the load. The planned Cumulative Production Figures of the variants should however take into consideration the capacities on the workstations.

5. *A manufacturing control method should cause as few sequencing interchanges as possible.*

Cumulative Production Figures do not cause any sequencing interchanges that would interfere with the delivery reliability of the production.

6. *A manufacturing control method should be able to balance schedule deviations within the context of backlog control.*

Cumulative Production Figures visualize the backlog between the actual CPF and the planned CPF for the variants. These variant-specific backlogs can be converted into workstation-specific backlogs and thus into planned hours (for more information see Chap. 27). Cumulative Production Figures, thus, simplify measuring the backlog. Nonetheless, they do not make any decisions about adjusting capacities and should therefore be supplemented by a backlog control.

7. *A manufacturing control method should take into consideration the principle of bottlenecks.*

Cumulative Production Figures do not consider the bottleneck principle. Bottleneck workstations can, however, be considered separately when setting the planned Cumulative Production Figures (see the discussion on the Kanban control, Sect. 9.3).

8. *A manufacturing control method should be as simple as possible.*

One of the fundamental advantages of Cumulative Production Figures are their simplicity.

9. *A manufacturing control method should communicate information about demands as quickly and clearly as possible.*

Cumulative Production Figures communicate demand information quickly and – when a WIP offset is used – clearly. This is accomplished by refraining from lot size information with planned CPF and by simultaneously conveying the information to all of the production stages.

Cumulative Production Figures with a lead time offset (the goal here is to maintain constant planned throughput times) can amplify the demand fluctuations (see Chap. 6). Where possible, a WIP offset is therefore preferred. If for unavoidable reasons the throughput times are constant, then WIP fluctuations cannot be prevented when there are fluctuations in the demand. Accordingly, an amplification of the demand fluctuations can also not be prevented. In comparison to the WIP offset, the lead time offset then has the advantage of depicting reality better (when throughput times are constant).

## 13.3 Setting the Method's Parameters

Universally applicable and exact guidelines for setting the Cumulative Production Figures can hardly be made, instead, the procedure is dependent on the specifications of the method and in particular on the definition of the control blocks.

Generally, a planned WIP level has to be determined with a WIP offset and a planned throughput time with a lead time offset. Both parameters are linked to one another through Little's Law and are thus easily converted into one another.

Thus for the planned Cumulative Production Figures of a control block with a WIP offset:

$$CPF_{j,plan}(t) = CPF_{ref,plan}(t) + \sum_{i=j+1}^{R} WIPP_{i,plan} \tag{13.8}$$

where

| | |
|---|---|
| $CPF_{j,plan}$ | planned Cumulative Production Figures for control block $j$ [units], |
| $CPF_{ref,plan}$ | planned Cumulative Production Figures for the reference control block [units], |
| t | time point [SCD], |
| $WIPP_{i,plan}$ | planned work-in-process (in pieces) for control block $i$ [units], |
| R | index for the reference control block [−]. |

Accordingly for a lead time shift:

$$CPF_{j,plan}\left(t - \sum_{i=j+1}^{R} TTP_i\right) = CPF_{ref,plan}(t) \tag{13.9}$$

where

| | |
|---|---|
| $CPF_{j,plan}$ | planned Cumulative Production Figures for control block $j$ [units], |

| | |
|---|---|
| t | time point [SCD], |
| $TTP_i$ | throughput time for control block *i* [SCD], |
| $CPF_{ref,plan}$ | planned Cumulative Production Figures for the reference control block [units], |
| R | index for the reference control block [−]. |

The planned stock level and planned throughput time can be determined with the aid of a Logistic Positioning: For the storage processes this is conducted with the assistance of Storage Operating Curves (Sect. 3.1.3) and for the production processes based on Logistic Operating Curves (Sect. 3.2.2).

If a control block contains a number of (similar or different) processes, the planned throughput times or planned stock levels should be added in order to calculate the planned parameters of the control block. In order to clarify this, we will turn to an example. The control block of a variant is comprised of a store and three workstations WS 1, 2 and 3 (Fig. 13.12).

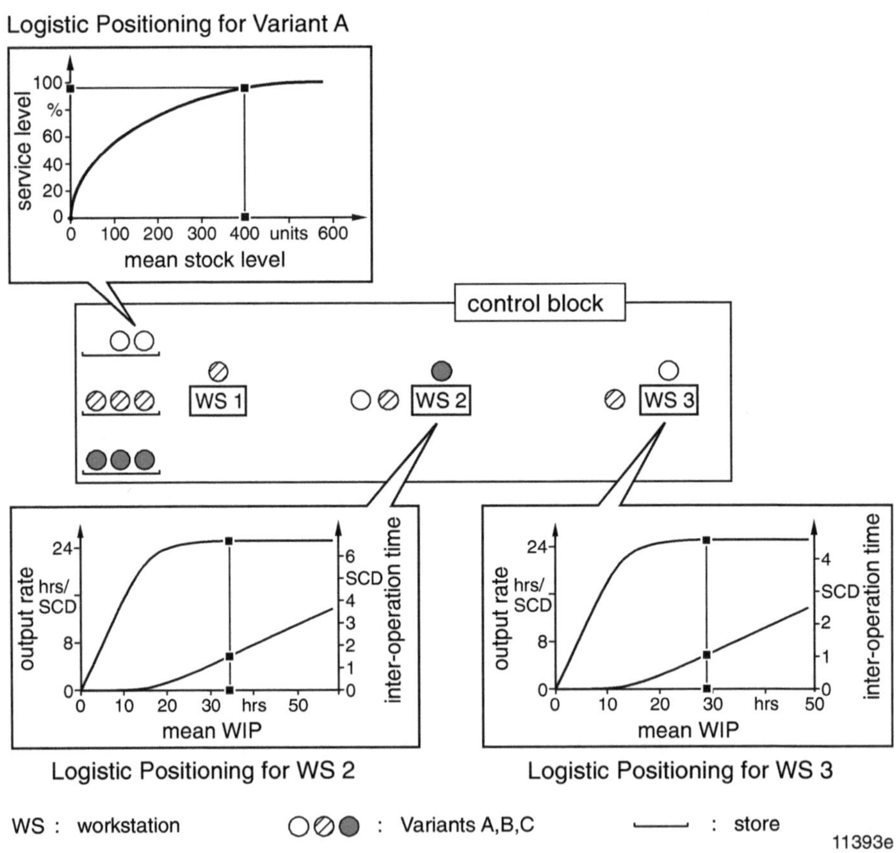


**Fig. 13.12** Setting the progress number for a control block

There is no (physical) queue at the first workstation to consider since it withdraws materials directly from the store. A Logistic Positioning needs to be conducted though for the store and for Workstations 2 and 3. In this example, the enterprise accepts a slight loss of utilization and a slight decrease in the service level. In calculating the planned Cumulative Production Figures, the planned stock level of the store and the planned inter-operation time for Workstations 2 and 3 are significant (Table 13.1).

**Table 13.1** Planned values for the control block

| Planning object | Planning parameter | Planned value |
|---|---|---|
| Store | Planned stock | 400 units |
| Workstation 2 | Planned inter-operation time | 1.5 SCD |
| Workstation 3 | Planned inter-operation time | 1.0 SCD |

The planned stock level of a Variant k on a Workstation j is calculated:

$$WIP_{plan,j,k} = \left(TOP_{j,k} + TIO_{j,plan}\right) \cdot RDMD_{m,plan,k} \tag{13.10}$$

where

$WIP_{plan,j,k}$ planned stock level for Variant $k$ on Workstation $j$ [units],
$TOP_{j,k}$ operation time for Variant $k$ on Workstation $j$ [SCD],
$TIO_{j,plan}$ planned inter-operation time Workstation $j$ [SCD],
$RDMD_{m,plan,k}$ mean planned demand rate for Variant $k$ [units/SCD].

With an operation time of 0.3 SCD and an expected demand rate of 100 units/SCD there is a planned WIP level for Workstation 2 of 180 units. In the same way, with an operation time of 0.2 SCD a planned WIP level of 120 units can be derived for Workstation 3. For Workstation 1, the planned inter-operation time does not apply (no queue) so that the planned WIP level results alone from the operation time and output rate. If the planned WIP levels of the objects being planned are determined, the planned WIP of the control block is calculated by simple addition (Table 13.2).

The control Cumulative Production Figures are set based on the planned Cumulative Production Figures. When calculating them the replenishment time from the order generation (when the actual Cumulative Production Figure is exceeded) up to the completion of the order needs to be considered. Once again this can occur via a lead time offset or a WIP offset. For the WIP offset:

**Table 13.2** Planned stock levels and planned throughput times in the control block (example)

| Planning object | Planned inventories [units] | Planned throughput time [SCD] |
|---|---|---|
| Store | 400 | 4.0 |
| Workstation 1 | 40 | 0.4 |
| Workstation 2 | 180 | 1.8 |
| Workstation 3 | 120 | 1.2 |
| Control block | 740 | 7.4 |

$$CPF_{i,control}(t) = CPF_{i,plan}(t) + WIPoff_{TREP} \tag{13.11}$$

where

| | |
|---|---|
| $CPF_{i,control}$ | control Cumulative Production Figure for control block *i* [units], |
| $CPF_{i,plan}$ | planned Cumulative Production Figure for control block *i* [units], |
| $WIPoff_{TREP}$ | WIP offset for the replenishment time [units]. |

With a lead time offset the control Cumulative Production Figure for a control block is calculated:

$$CPF_{i,control}(t - TREP_{i,plan}) = CPF_{i,plan}(t) \tag{13.12}$$

where

| | |
|---|---|
| $CPF_{i,control}$ | control Cumulative Production Figure for control block *i* [units], |
| t | time point [SCD], |
| $TREP_{i,plan}$ | planned replenishment time for control block *i* [SCD], |
| $CPF_{i,plan}$ | planned Cumulative Production Figure for control block *i* [units]. |

## 13.4 Implementation Across the Supply Chain

Cumulative Production Figures are well suited for coordinating the production across the supply chain. This possibility is primarily used in the automobile industry in order to integrate suppliers. Implementing it across the supply chain is not fundamentally different from applying it within an enterprise. Generally, however, transportation is given more weight and an information system that spans the supply chain needs to be installed. Two generally different levels of integration can be distinguished.

1. *Transmitting Planned Cumulative Production Figures:* If the data transfer between the enterprises is limited to transmitting planned Cumulative Production Figures (usually from the producer to the supplier), the application of Cumulative Production Figures is primarily aimed at coordinating the production program in the supply chain. The demands on the information system are then relatively minimal. The majority of current implementations of Cumulative Production Figures across supply chains are probably limited to this level of integration.
2. *Exchange of Planned and Actual Cumulative Production Figures:* If the enterprises exchange planned and actual Cumulative Production Figures (when possible in real-time) the Cumulative Production Figures can also be implemented

for controlling purposes and even for controlling production across the supply chain. A concept such as this places greater demands on the information system, however, it also offers a greater logistic potential.

## 13.5 Application

Cumulative Production Figures are suitable for coordinating the planning of the production program across a number of product stages. It thus seems obvious to apply them especially for managing the material flow across supply chains. They are particularly clear with a limited number of variants and a high mean demand rate. Similar to all of the other methods for generating orders, Cumulative Production Figures also benefit from minimal fluctuations in the demand rate. However, what is even more important is that it can limit the demand rate fluctuations of dependent demands when the method parameters are appropriately set. High value products can also be controlled with Cumulative Production Figures. Even more than with other methods, the setting of the parameters is crucial to the success of the application because the parameters determine the character of the resulting manufacturing control. On the one hand, this applies to the calculation of the dependent demands (WIP offset versus lead time offset). On the other hand, the degree to which decisions are to be automated has to be determined: By applying control Cumulative Production Figures this method can be used to generate orders. Otherwise the Cumulative Production Figures serve – as a more or less binding – orientation for decentralized decisions.

Whereas, Meyer and Schefenacker describe the experiences with EDP supported Cumulative Production Figures in the automobile industry [Meye-83], Daimler has extensive experience with the method [Hein-86].

## References

[Hein-86] Heinemeyer, W.: Fortschrittszahlen als Instrument zur Fertigungsplanung und -steuerung bei der Daimler-Benz AG, Presentation during the ACTIS seminar "Flexible Fertigungsorganisation durch DFÜ und Fortschrittszahlen"

[Hein-88] Heinemeyer, W.: Produktionsplanung und -steuerung mit Fortschrittszahlen für interdependente Fertigungs- und Montageprozesse. RKW-Handbuch der Logistik, 14th delivery XII, pp. 1–46. (1988)

[Hein-94] Heinemeyer, W.: Die Fortschrittszahlen als logistisches Konzept in der Automobilindustrie. In: Corsten, H. (ed.) Handbuch Produktionsmanagement. Strategie, Führung, Technologie, Schnittstellen, pp. 221–236. Gabler, Wiesbaden (1994)

[Meye-83] Meyer, B.E., Schefenacker, R.: Erfahrungen mit einem EDV-gestützten Fortschrittszahlensystem für Automobilzulieferer. ZWF **78**, 170–172 (1983)

[Wien-97b] Wiendahl, H.-P.: Fertigungsregelung. Logistische Beherrschung von Fertigungsabläufen auf Basis des Trichtermodells. Carl Hanser Verlag, Munich/Vienna (1997)

# Chapter 14
# Base Stock

The procedural rules of the Base Stock System were already published by Kimball in 1955 (see [Kimb-88*] for a reprint, as well as [Clar-60*, Grav-95*] as alternative sources) and is thus one of the oldest manufacturing control methods. The principle idea behind the method is to immediately convey each demand placed by an end customer to all of the workstations in a production. The corresponding demand signal authorizes the reproduction of the required variant. For each workstation a so-called 'base stock' is defined for each variant by which the cumulated production of the variant should exceed the cumulated customer demand. When the base stock is constant, this method prevents fluctuating demands from being amplified (bull-whip effect) and is thus suitable for coordinating supply chains. The possible WIP build-up that can occur as a result of a production backlog is, however, disadvantageous.

In the following, we will first review the procedural guidelines for the Base Stock System (Sect. 14.1) and then discuss these further based on the guiding principles for designing manufacturing control methods (Sect. 14.2). How the method's parameters – especially the base stock – can be set is the subject of Sect. 14.3. Implementing it across the supply chain and applying the Base Stock System are then discussed in Sects. 14.4 and 14.5.

## 14.1 Method Description

Figure 14.1 depicts the principle of the Base Stock System: Each demand placed by the customer is immediately relayed to the workstations in the production and authorizes the reproduction of the corresponding variant.

Variant-specific stores are maintained at all of the production's workstations. Similar to Kanban or the principle of Cumulative Production Figures, not all of the workstations in the production need to be included in the Base Stock System. Moreover, it is also possible to integrate a number of workstations into a control

H. Lödding, *Handbook of Manufacturing Control*,

DOI 10.1007/978-3-642-24458-2_14, 

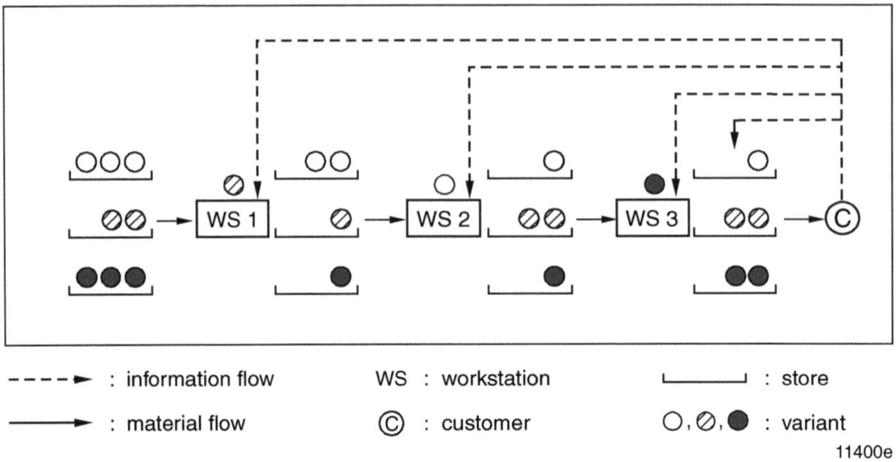


**Fig. 14.1** Principle of the base stock system

block (Fig. 13.1). This is especially practical when a number of workstations form a manufacturing cell.

The Base Stock System is easily explained with the aid of Cumulative Production Figures (Fig. 14.2 as well as Chap. 13).

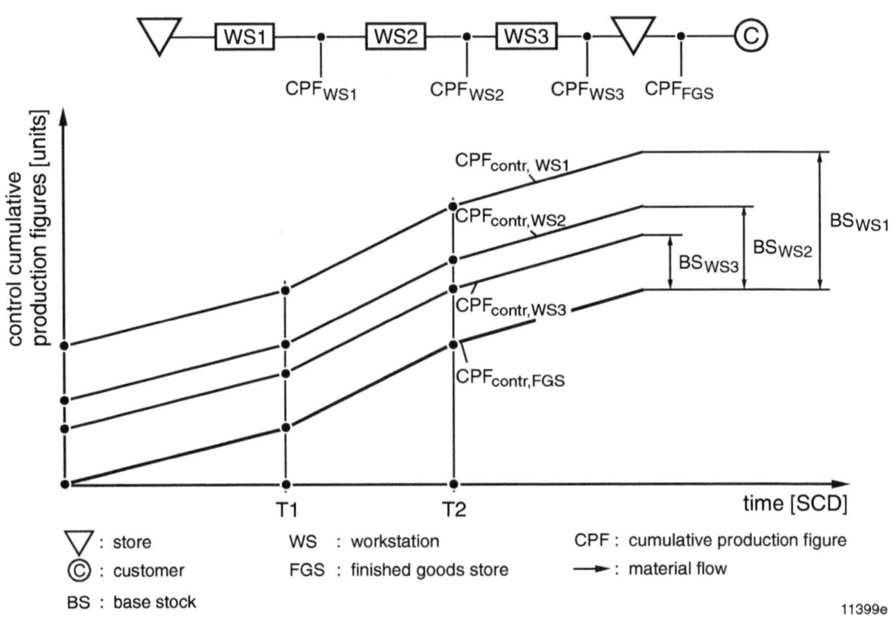


**Fig. 14.2** Definition of the control cumulative production figure with the base stock system

With the Base Stock System, the control Cumulative Production Figure of the finished goods store corresponds by definition to the cumulated demand of the customer or customers:

$$CPF_{control,FGS}(t) = \sum_{i=1}^{t} D(i) \tag{14.1}$$

where

| | |
|---|---|
| $CPF_{control,FGS}$ | control Cumulative Production Figure for the finished goods store [units], |
| $D(i)$ | demand on shop calendar day $i$ [units], |
| t | time of observation [SCD]. |

The control Cumulative Production Figures on the production's workstations result from the control Cumulative Production Figure of the finished goods store plus the base stock:

$$CPF_{control,j,k}(t) = CPF_{control,FGS,k}(t) + SB_{j,k} \tag{14.2}$$

where

| | |
|---|---|
| $CPF_{control,j,k}$ (t) | control Cumulative Production Figure for workstation $j$ for variant $k$ at time t [units], |
| $CPF_{control,FGS,k}(t)$ | control Cumulative Production Figure for the finished goods store for variant $k$ at time t [units], |
| $SB_{j,k}$ | base stock for variant $k$ on workstation $j$ [units]. |

The Base Stock System can be very easily described with this definition of the control Cumulative Production Figures.

1. The finished goods store fulfills a demand when stock is available.
2. A workstation is allowed to produce a variant when the actual Cumulative Production Figure is smaller than the control Cumulative Production Figure.
3. If a workstation is allowed to produce a number of variants it selects the variant with the highest priority. Various priority rules can be employed for making this decision (see Chap. 25).

An example which considers a production with two workstations and two variants will help to illustrate these procedural guidelines further (Fig. 14.3). The production lot size for both variants on both workstations is 50 units. The customers demand the variants in lots of 10 units. As soon as a production lot enters the store it is divided into 5 demand lots. For each workstation as well as the finished goods store, the control and actual Cumulative Production Figures of the variants are indicated on a board. The base stock on Workstation 1 is 150 units for both variants and 50 units on Workstation 2.

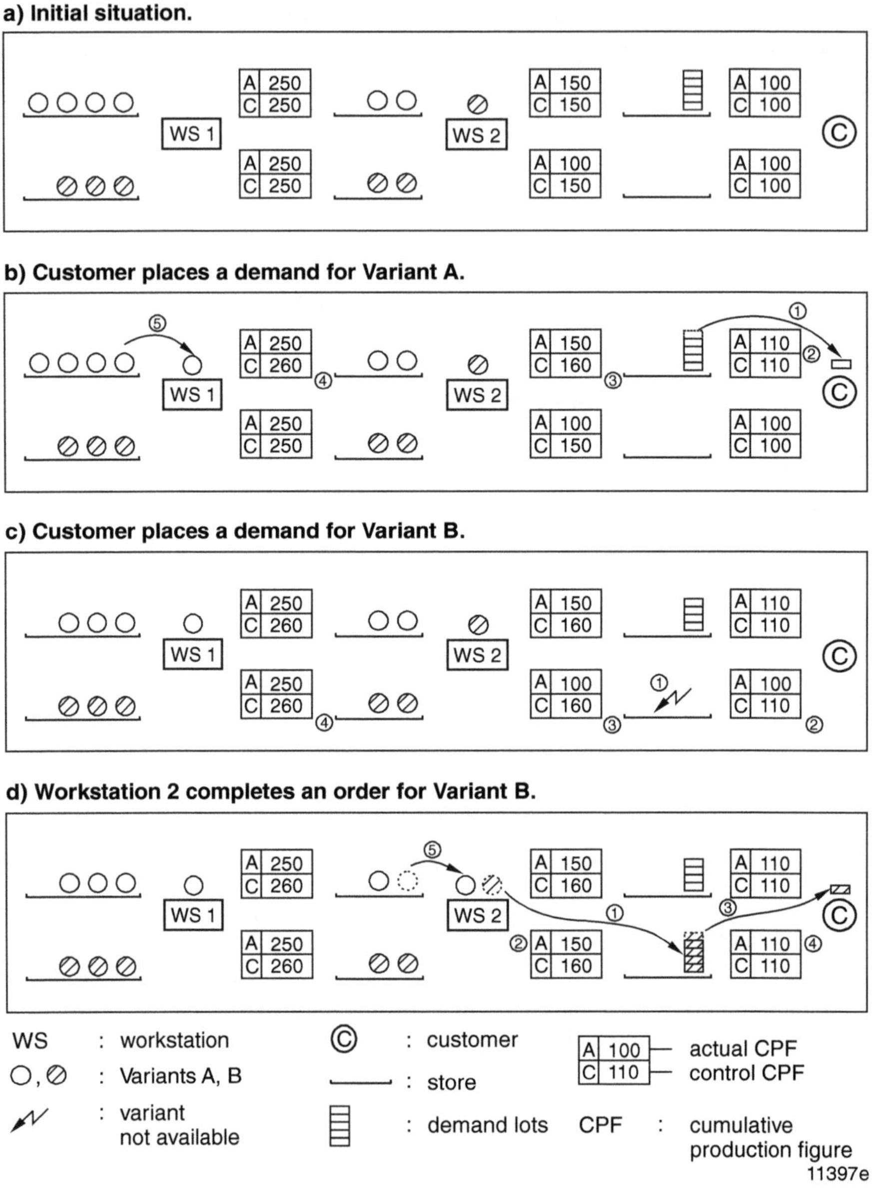


**Fig. 14.3** How the base stock system functions

In the initial situation (Fig. 14.3a) there are 50 units of Variant A in the finished goods store. There is no stock available for Variant B; a customer's demand could thus not currently be served. The finished goods store, however, does not have a backlog since the actual Cumulative Production Figure agrees with the control

Cumulative Production Figure. Workstation 2 is processing an order for Variant B, which upon completion should again restock the finished goods store. Workstation 1 is blocked: The control Cumulative Production Figure does not exceed the actual Cumulative Production Figure for either Variant A or Variant B. The workstation is thus not allowed to process an order.

If the customer places a demand for 10 units of Variant A (Fig. 14.3b), the variant is delivered to the customer (Step 1) and the actual Cumulative Production Figure as well as the control Cumulative Production Figure are increased (Step 2).

The demand also directly causes the control Cumulative Production Figures on Workstation 1 and 2 to be increased (Steps 3 and 4). As a result the actual Cumulative Production Figures on both workstations are exceeded and another production run of Variant A is authorized. Since Workstation 2 is already processing an order, only Workstation 1 begins producing Variant A (Step 5). It should be noted that the actual Cumulative Production Figure can exceed the control Cumulative Production Figure. This is the case for Workstation 1 when it completes the order before the customer has placed a demand for an additional 40 units of the variant.

Next, the customer places a demand for Variant B (Step 1 in Fig. 14.3c). The demand cannot be immediately satisfied since there is still no available stock of the variant. Nonetheless, the demand increases the control Cumulative Production Figures in the finished goods store and on the workstations (Steps 2–4). As a result the actual Cumulative Production Figure of the variant on Workstation 1 is exceeded and another production run is authorized. The workstation is at this time however already occupied so that the variant cannot be started.

When Workstation 2 completes the order for Variant B (Fig. 14.3d), the material enters into the finished goods store (Step 1) and the actual Cumulative Production Figure increases by the production lot size (Step 2). Since the finished goods store has a backlog the still open demand that was placed by the customer for Variant B is satisfied (Step 3) and the actual Cumulative Production Figure of the finished goods store is increased accordingly (Step 4). The actual Cumulative Production Figure from Workstation 2 for both variants is lower than the control Cumulative Production Figure. The workstation can thus produce both variants. It selects the variant with the highest priority[1] and begins processing it (Step 5).

## 14.2 Discussion of the Procedural Rules

The Base Stock System is characterized by three properties: the direct transmission of demand to all of the workstations in the production, the lack of WIP control and the variant-specific stock buffers. In the following, we will examine the procedural

[1] The Base Stock does not provide any set sequence. In this case the higher priority for Variant A was derived as the planned Cumulative Production Figure for Variant A exceeded the actual Cumulative Production Figure sooner than was the case for Variant B.

rules for the Base Stock System with regards to the guiding principles for designing manufacturing control methods.

*1. A manufacturing control method should allow the WIP of a production or workstation to be regulated at a defined level.*

The Base Stock System offers poor conditions for regulating the WIP in the production. On the one hand the production's WIP has a tendency to grow with the number of variants due to the variant-specific WIP buffers. If the production develops a backlog, it can lead to a build-up of WIP in the production. This is especially true when the first workstation has sufficient capacities for meeting the high customer demand. The correlations will be explained here based on an example in which the last workstation of a production does not have sufficient capacity reserves to manage an increased demand (Fig. 14.4).

From the onset of the increased demand (time $t_1$) up to time $t_2$ the higher demand can be served from the finished goods store. During the period, the WIP on Workstation 3 continually grows, while the stock in the finished goods store decreases. From time $t_2$ on the finished goods store develops a backlog and the customers' demands can only be served with delays. The backlog and WIP on Workstation 3 continue to grow.

*2. A manufacturing control method should reduce the WIP fluctuations to a minimum.*

Due to the lack of WIP control the Base Stock System cannot prevent WIP fluctuations. It is thus largely dependent on carefully planning and controlling capacities.

*3. A manufacturing control method should cause as little blocked WIP as possible.*

The Base Stock System blocks all of the WIP for which the control Cumulative Production Figure for the corresponding variant on the subsequent workstation is not greater than the actual Cumulative Production Figure. As with Kanban the amount of blocked WIP has a tendency to increase with the number of variants.

*4. A manufacturing control method should contribute to closely aligning the load to the capacities.*

The Base Stock System does not take the capacities into consideration. Accordingly, the method can also not contribute to balancing the loads. In order to successfully apply the Base Stock System both a thorough capacity planning and an effective backlog control are critical.

*5. A manufacturing control method should cause as few sequencing interchanges as possible.*

Due to the blocked WIP that occurs, the Base Stock System does cause sequencing interchanges. Nonetheless, when the base stock is carefully planned these do not impact the production's on-time delivery because they only block the WIP for variants for which there is already sufficient stock.

The Base Stock System should be combined with a sequencing rule that supports the production's on-time delivery. It should, for example, first produce the variant with the lowest stock range on the subsequent workstation.

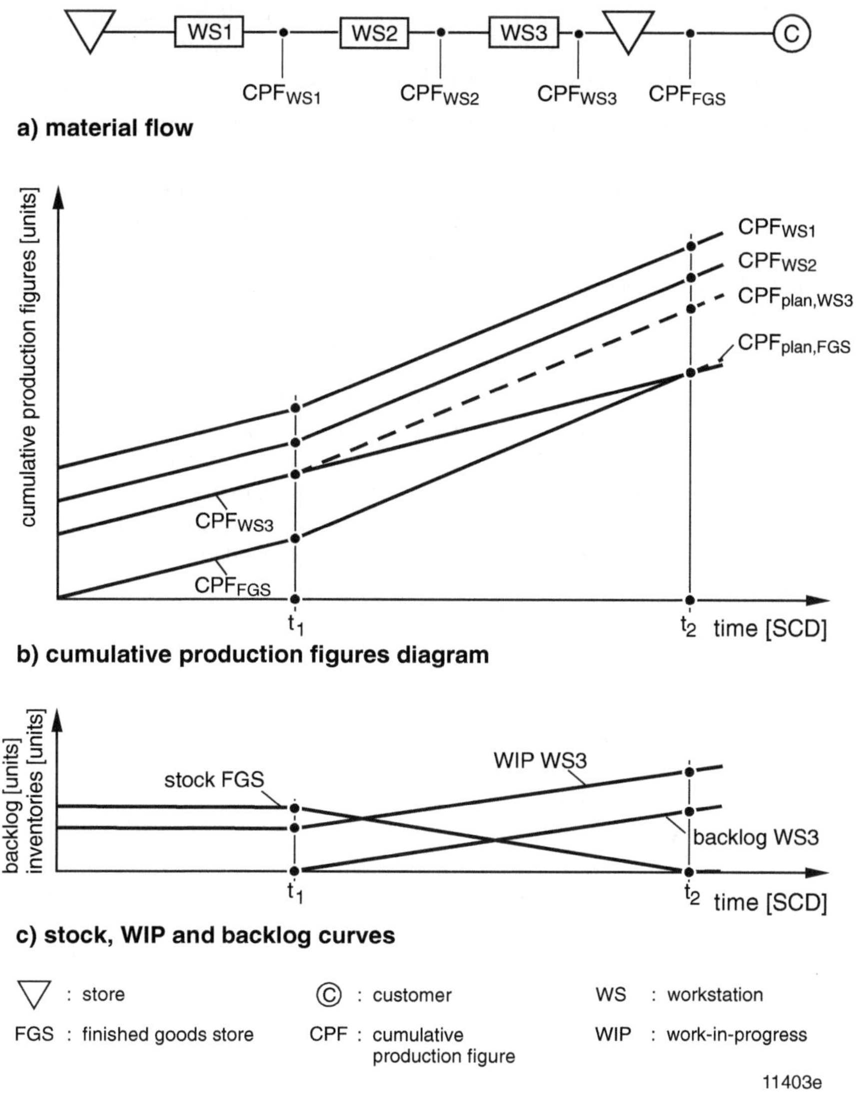


**Fig. 14.4** Cumulative production figures, as well as stock and backlog curves for a variant with insufficient capacities on workstation 3 (example)

6. *A manufacturing control method should be able to balance schedule deviations within the context of backlog control.*

It is the responsibility of manufacturing control to prevent a threatening backlog and to reduce those backlogs which have developed. The procedural rules for the

Base Stock System do not incorporate a backlog control. It is thus imperative to combine a backlog control with this method in order to ensure a high on-time delivery rate even when there are schedule deviations.

7. *A manufacturing control method should take into consideration the principle of bottlenecks.*

The Base Stock System does not take into consideration the bottleneck principle. The only possibility to orient the manufacturing control on an obvious bottleneck workstation is the setting of the base stock. In order to protect the workstation from blockades, the base stock for the bottleneck workstation may be set higher. Moreover, high base stock levels on preceding workstations should ensure that there is continually sufficient stock on the bottleneck workstation.

8. *A manufacturing control method should be as simple as possible.*

The procedural rules for the Base Stock System are easy to understand. It is, however, usually not possible to implement the method without an electronic information system. On the one hand this is because the demand information should be transmitted to all of the impacted workstations as close to simultaneously as possible. On the other hand, when the quantities demanded are small, demand information needs to be communicated very frequently.

9. *A manufacturing control method should communicate information about demands as quickly and clearly as possible.*

The Base Stock System communicates demand information quickly and without distorting it. It conveys demand information simultaneously to all production stages and conducts a WIP offset (see Chap. 13). It thus provides very good conditions for preventing the bullwhip effect.

The Base Stock System implicitly requires a variant-specific WIP control to be conducted and the possibility to shorten replenishment times if the demand increases. If this is not the case, the planned stock levels have to be increased when the demand grows and reduced when the demand sinks. Both then contribute to amplifying demand fluctuations, which however, are not caused by the Base Stock System.

## 14.3 Setting the Method's Parameters

The base stocks for the control blocks are the key parameters for the Base Stock System. These are determined from the planned stock levels for the production. There is thus a two stage procedure for determining the base stock:

1. setting the planned stock levels (Sect. 14.3.1), and,
2. setting the base stocks (Sect. 14.3.2).

### 14.3.1 Planned Stock Levels

The method for setting the planned stock levels is decisively determined by the processes that flow between two of the Base Stock System's control blocks. We will consider methods for three different types of control blocks (see Fig. 14.5): a single store, a store with a subsequent workstation and a store with a number of subsequent workstations.

**Store** If a control block is comprised solely of one store, the planned stock of a variant can be set in a Logistic Positioning with the aid of Storage Operating Curves. The corresponding procedure is described in Sect. 3.1.3.

**Store with One Workstation** If a control block consists of a store with one workstation, the stock of a variant is either in the store or being processed on the workstation. The planned stock in the store can again be determined in a Logistic Positioning with the aid of Storage Operating Curves. The mean planned WIPP on the workstation can be derived directly from Little's Law:

$$WIPP_{m,plan} = TOP \cdot RDMD_{m,plan} \tag{14.3}$$

where

| | |
|---|---|
| $WIPP_{m,plan}$ | mean planned work-in-process (in pieces) [units], |
| TOP | operation time [SCD], |
| $RDMD_{m,plan}$ | mean planned demand rate [units/SCD]. |

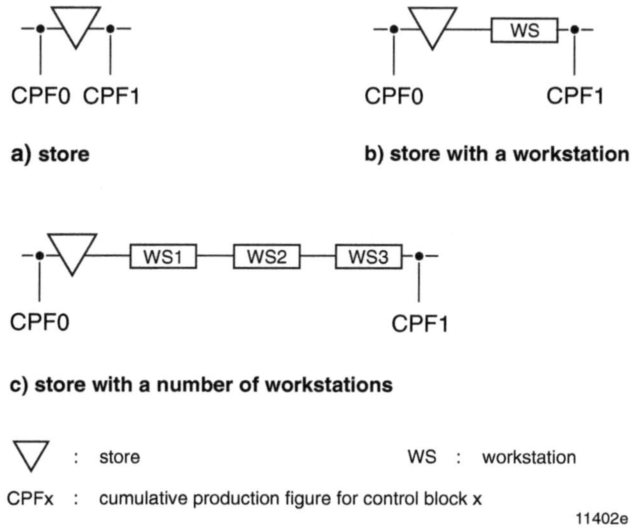


**Fig. 14.5** Possible control blocks for the base stock system (example)

**Store with a Number of Workstations** If the variants between two of the Base Stock System's control block are routed through a store and a number of workstations, the following procedure is possible: Once again, the planned stock in the store can be determined in a Logistic Positioning with the aid of Storage Operating Curves. The remaining stock is either being processed or queued before a workstation (according to definition, there can be no queue before the first workstation). Consequently, the mean WIP can be calculated according to Little's Law with the following formula:

$$WIPP_m = RDMD_m \cdot \left( TOP_1 + \sum_{j=2}^{NWS} TTP_{m,j} \right) \tag{14.4}$$

where

| | |
|---|---|
| $WIPP_m$ | mean work-in-process (in pieces) [units], |
| $RDMD_m$ | mean demand rate [units/SCD], |
| $TOP_1$ | operation time on the first workstation [SCD], |
| $TTP_{m,j}$ | mean throughput time on the $j^{th}$ workstation [SCD], |
| NWS | number of workstations [−]. |

The throughput time of the workstations can be determined in a Logistic Positioning with the aid of Logistic Operating Curves (generally described in Sect. 3.2.2) whereas the procedure for the Cumulative Production Figures control is explained in Chap. 14.

### *14.3.2 Base Stocks*

The base stocks for the controlled blocks can be derived from the planned stock levels. This is done against the material flow i.e., the base stock for the last control block is set and then the second to last control block etc.

With small lots and short replenishment times the base stock results directly from the planned stock. Thus,

$$SB_i = SB_{i+1} + WIPP_{plan,i+1} \tag{14.5}$$

where

| | |
|---|---|
| $SB_i$ | base stock for the *ith* control block [units], |
| $WIPP_{plan,i+1}$ | planned work-in-process (in pieces) of the *i + 1th* control block [units]. |

On the shop floor though, usually the *replenishment time* and where necessary the influence of a *difference in lot size* between the input and output of a control block needs to be considered.

**Influence of the Replenishment Time** With the Base Stock System the replenishment time is defined as the period of time between an order is triggered (by exceeding the actual Cumulative Production Figure of a variant) and its completion. Due to the replenishment time, the actual Cumulative Production Figures deviate from the control Cumulative Production Figures both in the input as well as the output of a control block (Fig. 14.6).

Thus, with Little's Law:

$$\Delta CPF_i = CPF_{control,i} - CPF_{plan,i} = TRP_{plan,i} \cdot RDMD_m \qquad (14.6)$$

where

| | |
|---|---|
| $\Delta CPF_i$ | deviation between the control and planned Cumulative Production Figure [units], |
| $CPF_{control,i}$ | control Cumulative Production Figure for control block *i* [units], |
| $CPF_{plan,i}$ | planned Cumulative Production Figure for control block *i* [units], |
| $TRP_{plan,i}$ | planned replenishment time for control block *i* [SCD], |
| $RDMD_m$ | mean demand rate [units/SCD]. |

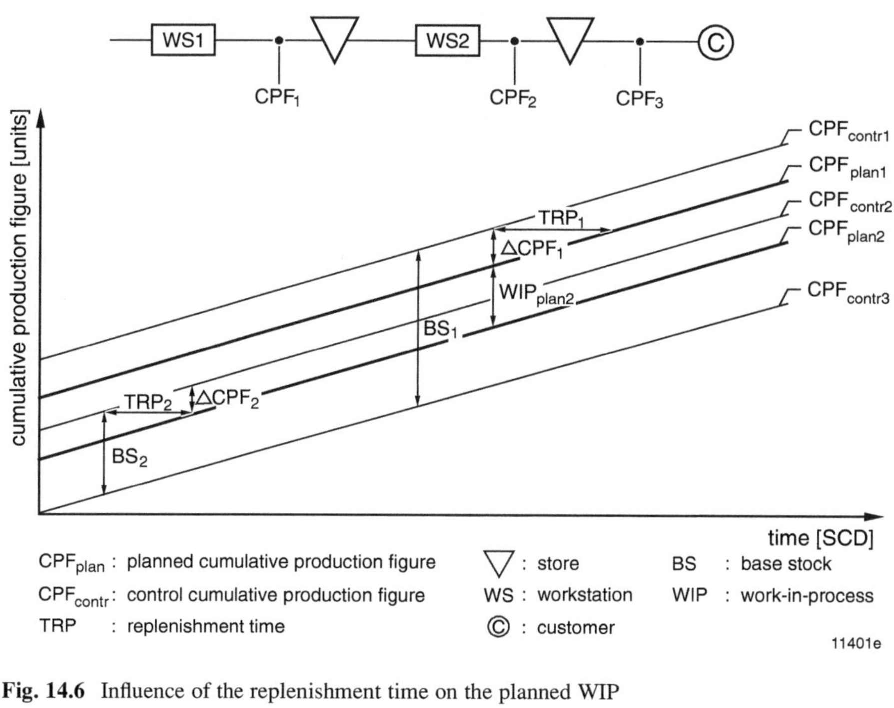


**Fig. 14.6** Influence of the replenishment time on the planned WIP

The deviation of the planned Cumulative Production Figure impacts the planned WIP:

$$WIPP_{plan,i+1} = SB_i - \Delta CPF_i - SB_{i+1} + \Delta CPF_{i+1} \tag{14.7}$$

where

| | |
|---|---|
| $WIPP_{plan,i+1}$ | planned work-in-process (in pieces) of the i + 1$^{th}$ control block [units], |
| $SB_i$ | base stock for the i$^{th}$ control block [units], |
| $\Delta CPF_i$ | deviation between the control and planned Cumulative Production Figure [units]. |

With a given planned WIP the following relation can be derived from Eqs. 14.6 and 14.7 for the base stock of a control block:

$$SB_i = WIPP_{plan,i+1} + SB_{i+1} + RDMD_m \cdot \left(TRP_{plan,i} - TRP_{plan,i+1}\right) \tag{14.8}$$

where

| | |
|---|---|
| $SB_i$ | base stock for the i$^{th}$ control block [units], |
| $WIP_{plan,i+1}$ | planned work-in-process (in pieces) of the i + 1$^{th}$ control block [units], |
| $RDMD_m$ | mean demand rate [units/SCD], |
| $TRP_{plan,i}$ | planned replenishment time for control block *i* [SCD]. |

According to definition, the base stock of the finished goods store is zero (Eq. 14.1). Based on the planned stock for the finished goods store all of the production's base stocks can be determined with the aid of Eq. 14.8. The planned replenishment times can be set in a Logistic Positioning with the aid of Logistic Operating Curves.

The last sum of Eq. 14.8 indicates that the planned WIP levels of the Base Stock System can also fluctuate with constant base stock when the demand rate changes. Here, two extreme cases should be considered:

1. *Constant Replenishment Times:* With increasing demand rates the planned WIP of a control block grows when its replenishment time is bigger than the preceding control block. In the inverse situation, the planned WIP decreases. When the replenishment times are equal in both control blocks the planned WIP remains unchanged. When considered across the entire production, the planned WIP decreases as the demand rate increases. The Base Stock System thus has a tendency to work anti-cyclical: It does not increase the planned WIP when there is a higher demand and thus does not lead to amplified demand fluctuations.
2. *Replenishment Times Change Contrary to Demand Fluctuations:* If an enterprise succeeds in shortening the replenishment times to the degree in which the demand rates increases, the planned WIP levels do not change. The Base Stock System thus behaves neutrally in comparison to the demand fluctuations.

**Influence of the Difference in Lot Sizes** When the lot sizes in the input and output of a control block differ from one another, this difference will also influence the planned WIP. Figure 14.7 depicts the correlations for an exemplary control block when the end customer's demand is continuous.

The planned Cumulative Production Figures are drawn in with or without consideration of the lot sizing. A lot-wise input to the control block increases the planned WIP, whereas a lot-wise output decreases the WIP. The WIP deviation corresponds to half of the input or output lot size. Thus,

$$SB_i = WIPP_{plan,i+1} + SB_{i+1} + RDMD_m \cdot \left(TRP_{plan,i} - TRP_{plan,i+1}\right) + \frac{1}{2} \cdot (LS_{i+1} - LS_i) \tag{14.9}$$

where

| | |
|---|---|
| $SB_i$ | base stock for the $i^{th}$ control block [units], |
| $WIPP_{plan,i+1}$ | (continuous) planned work-in-process (in pieces) for the $i + 1^{th}$ control block [units], |
| $RDMD_m$ | mean demand rate [units/SCD], |
| $TRP_{plan,i}$ | planned replenishment time for the $i^{th}$ control block [SCD], |
| $LS_i$ | lot size for the $i^{th}$ control block [units]. |

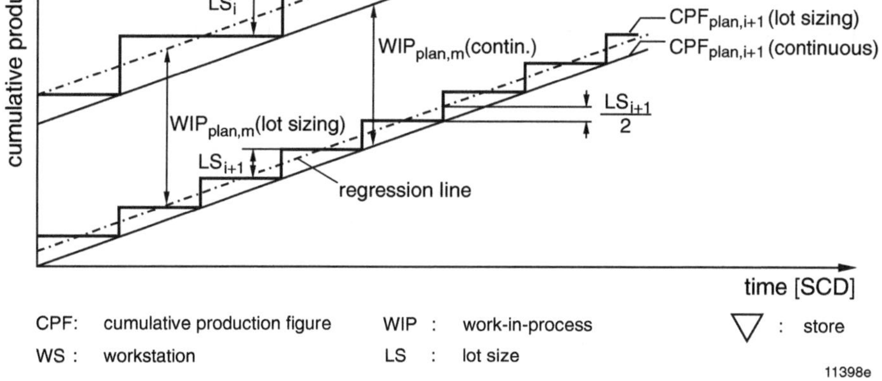


**Fig. 14.7** Influence of the lot sizing on the planned WIP of the control block

## 14.4 Implementation Across the Supply Chain

The Base Stock System is well suited for implementing across the supply chain. The transmission of information takes on a key role here, which could explain why up to now the Base Stock System has been applied rather infrequently: An enterprise has to forward the demand information not only to one contact, but rather to a number of them at the same time. This poses greater demands on the information system than most of the usual manufacturing control methods. However, with today's technology, this problem is solvable. Implementing the Base Stock System across the supply chain seems to be particularly well suited when there is a large supply volume between enterprises.

## 14.5 Application

The Base Stock System is a special case of the Cumulative Production Figures control. It is particularly well suited for generating orders over multiple stages. Here, it distinguishes itself from other controls with its quick and undistorted communication of demand information. It is also especially well suited when there are sufficient capacities available and the replenishment times can be reduced with increasing demand, or, when there are only minimal demand fluctuations (with regards to both aspects, see the discussion of the procedural rules in Sect. 14.2).

## References

[Clar-60*] Clark, A.J., Scarf, H.: Optimal policies for a multi-echelon inventory problem. Manag. Sci. **6**, 475–490 (1960)

[Grav-95*] Graves, R.J., Konopka, J.M., Milne, R.J.: Literature review of material flow control mechanisms. Prod. Plan. Control **6**(5), 395–403 (1995)

[Kimb-88*] Kimball, J.E.: General principles of inventory control. Int. J. Manuf. Oper. Manag. **1**(1), 119–130 (1988)

# Chapter 15
# Production Authorization Cards

PAC (Production Authorization Cards) is a generic method for controlling production. Similar to Cumulative Production Figures (Chap. 13) depending on how the parameters are set PAC can represent different methods for generating orders. The method was developed and described by Buzacott and Shantikumar [Buza-92*] and is especially interesting for software developers and researchers. In comparison, practitioners tend to orient themselves directly on the methods which PAC can represent.

The discussion here is limited to a description of the method. The logistic suitability and the properties of the method are dependent on how the very flexibly designed parameters are set. For this reason, unlike with other methods, both the discussion about the procedural rules as well as the usual description of setting the method's parameters will be omitted here. Nonetheless, depending on how the method is designed the remarks which correspond to the related control systems (e.g., Kanban or Base Stock System) can be transferred.

## 15.1 Method Description

A simplified form of the method is presented here. Buzacott and Shantikumar consciously designed the method to be more open, in order to develop a wide range of applicability. Thus, it is possible for example, to select different production and transport lot sizes or to take defects into consideration. The focus here will be on discussing and explaining the basic principle of PAC. In order to do so the following simplifications among others are made:

- The production consists of workstations comprising of one machine (Buzacott and Shantikumar assume production cells that can contain a number of parallel or sequentially arranged machines.)

H. Lödding, *Handbook of Manufacturing Control*,

DOI 10.1007/978-3-642-24458-2_15, 

- The lot sizes are provided and are the same for all workstations.
- Only PA cards, and what Buzacott and Shantikumar refer to as order tags and requisition tags will be considered. The other components mentioned by Buzacott and Shantikumar i.e., process tags, cancellation notes, surplus tags and defect order tags will be omitted.

With these simplifications PAC is based on only three elements:

*PA Cards:* PA cards are each allocated to a specific variant and a specific workstation. They authorize the reproduction of a product on the allocated workstation. The number of PA cards thus limits the number of open production orders on a workstation. Unlike with Kanban, after the order is completed, the PA cards do not remain with the lot in the outgoing goods store and are instead stored separately. Thus the triggering of an order can be separated from the material requisition.

*Order Tags:* Order tags provide information about the demand for a specific variant and are either generated by the end customer or from a workstation. Immediately after they are generated they are brought to the relevant supply store. If there is a PA card there for the corresponding variant, the order tag is disposed of and the PA card is immediately posted on the workstation's PA card board. Whereupon the workstation immediately produces both an order and a requisition tag for the variant. The order tag is brought directly to the preceding store, whereas the requisition tag follows it either immediately or after a delay (see below). If there is no PA card present, the order tags remains in the store until a PA card is freed.

*Requisition Tag:* The requisition tag is allocated to a specific variant and authorizes the withdrawal of a variant from a store and its provision to the subsequent workstation (or in the finished goods store the delivery to the customer). Requisition tags are generated by the subsequent workstation upon reception of a PA card (or for the finished goods store, directly from the customer). After it is generated the requisition tag does not have to immediately enter the store. Delaying it can prevent the materials from being supplied before a certain time or from being shipped too early to a customer. Once the requisition tag enters the store, the corresponding materials will be supplied immediately if they are present and the requisition tag is then disposed of. It the material is not available the requisition tag remains in the store until the corresponding material arrives and can be supplied.

The properties of PAC can be influenced by three parameters:

1. number of PA cards per variant and workstation,
2. the so-called 'base stock' of each variant and workstation, and,
3. the delay (or lag time) of the material requisition per variant and workstation (or end customer).

The base stock is most easily explained with a mental experiment: Imagine that at a specific point in time, there is no order for another production run at a workstation. The base stock then corresponds to the stock in the workstation's outgoing goods store.

Next, we will take a closer look at the procedural rules for PAC based on two examples. The first example emulates a make-to-stock production, the second a simple make-to-order production. Each production consists of three workstations, a raw materials store, two semi-finished goods stores and a finished goods store.

### Example 1: Make-to-Stock Production

Two variants are produced (Fig. 15.1). There is one PA card for each variant and each workstation. The base stock is three lots per variant and workstation. The delay (or lag time) for the material requisition is zero.

This method for setting the parameters follows a recommendation from Buzacott and Shantikumar [Buza-92*] and presents the so-called 'local control system', which can be understood as a Kanban variation. Each workstation is allocated a PA card board. Here, the PA cards for waiting orders are collected. A similar board, separated according to variants is found in the workstation's outgoing goods store. There the PA cards for which there are no order tags are collected. Furthermore, in the outgoing goods store, there is a requisitions board and an order tag board. The first is where the requisition tags for the demands of the subsequent workstation or customer which have not yet been delivered are collected. It is always empty when there is a lot of the variant in store (otherwise the requisition tag would authorize the delivery of the lot).

On the order tag board, the order tags for which there are no PA cards present are collected. As a result of the procedural rules, the PAC board and order tags board cannot be filled at the same time, since the order tags would immediately trigger the transport of the PA card to the relevant workstation and in turn trigger the disposal of the order tag.

In the initial situation (Fig. 15.1a) the finished goods store for Variant 1 is empty and the one for Variant 2 is completely filled. All of the workstations are processing an order. There is also a queued order at both Workstation 1 and 2.

When the customer places a demand for Variant 2 (Fig. 15.1b), he delivers a requisition tag as well as an order tag to the finished goods store (Steps A and B). Following that the material is delivered to the customer (Step C). The order tag generated by the customer triggers the transport of the PA card for Variant 2 to Workstation 3 (Step D). The workstation generates a requisition tag and an order tag for Variant 2 (Steps E and F) and forwards these to the incoming goods store. The requisition tag triggers the delivery of the material (Step G); the order tag is pegged on the order tag board of the incoming goods store since there is no PA card available in the store.

When Workstation 2 completes Variant 2 (Fig. 15.1c), the lot is transported to the outgoing goods store (Step A) and the PA card is initially placed on the store's PA card board (Step B). Since an order tag for the variant is available, the PA card however, is forwarded immediately to the workstation's PA card board (Step C) and the order tag is disposed of (Step D). After receiving the PA card the workstation generates a requisition tag and an order tag for Variant 2 and delivers these to the incoming goods store (Steps E and F). The requisition tag triggers the delivery of the materials (Step G); the order tag is placed on the incoming goods store's order tag board since there is no PA card in the store. Workstation 2 then begins

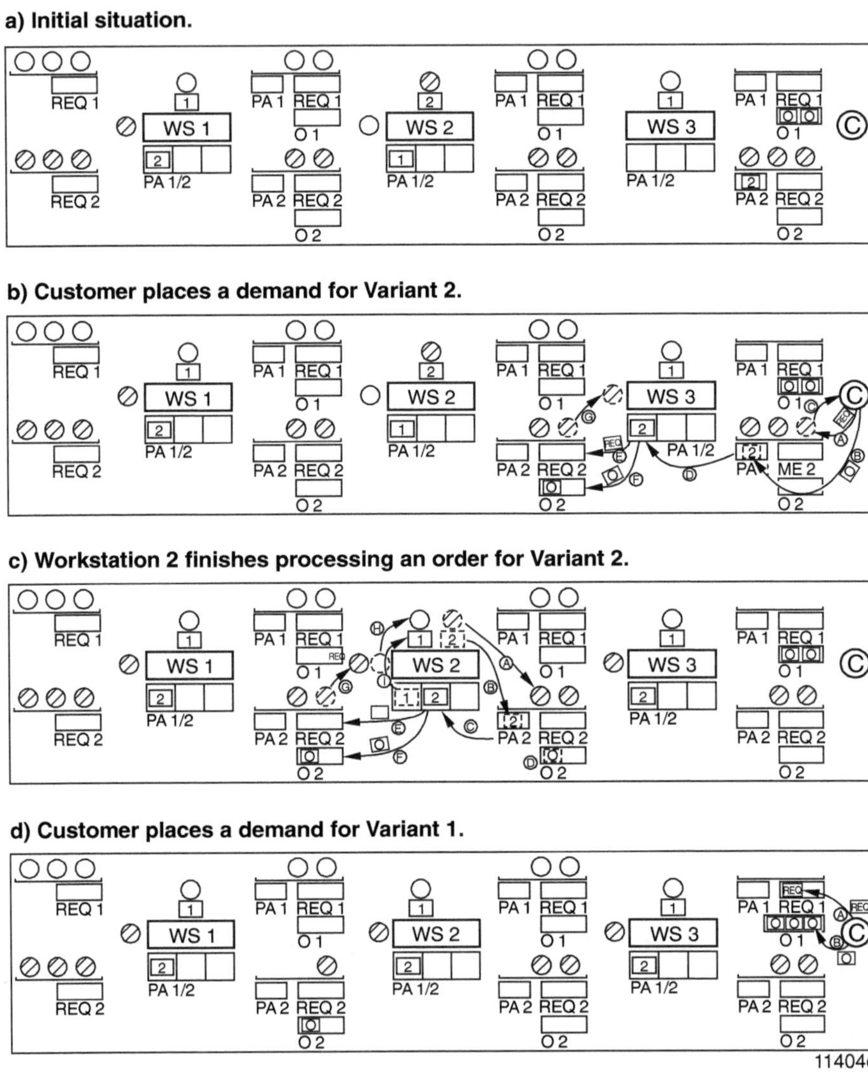


Fig. 15.1 How PAC functions, Example 1

**e) Workstation 3 finishes processing an order of Variant 1.**

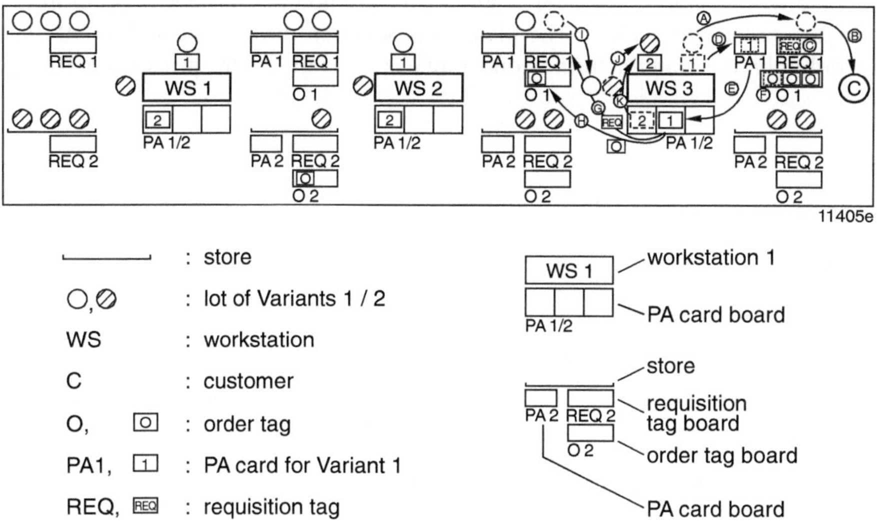


Fig. 15.1 (continued)

processing an order for Variant 1 (Step H). The corresponding PA card is directly allocated to the order while it is being processed (Step I).

Next, the customer places a demand for Variant 1, in doing so he generates an order tag and a requisition tag (Steps A and B in Fig. 15.1d). Since there is no material in the finished goods store, the requisition tag is placed on the corresponding board, thus visualizing the delivery backlog. The order tag is placed on the order tag board.

Workstation 3 completes the order of Variant 1 (Fig. 15.1e). The lot is initially transported to the finished goods store (Step A) and then delivered directly to the customer (Step B) because there is a requisition tag. The tag is then disposed of after the delivery is made (Step C). The PA card for Variant 1 initially goes on the PA card board in the finished goods store (Step D). However, there are order tags here, thus the PA card is returned to Workstation 3 (Step E) and an order tag is disposed of (Step F). After receiving the PA card, Workstation 3 generates a requisition tag and an order tag and forwards these onto the incoming goods store (Steps G and H). There the requisition tag triggers the delivery of the materials for Variant 1 (Step I). The order tag is placed on the order tag board since there is no PA card present. Workstation 3 begins processing the order for Variant 2 (Step J). The corresponding PA card is directly allocated to the order while it is being processed (Step K).

**Example 2: Make-to-Order Production** The same production as in the first example will be considered here. The variants are however produced according to customer orders (Fig. 15.2). The planned production throughput time is 10 days. Accordingly, the manufacturing control releases the orders for processing 10 days

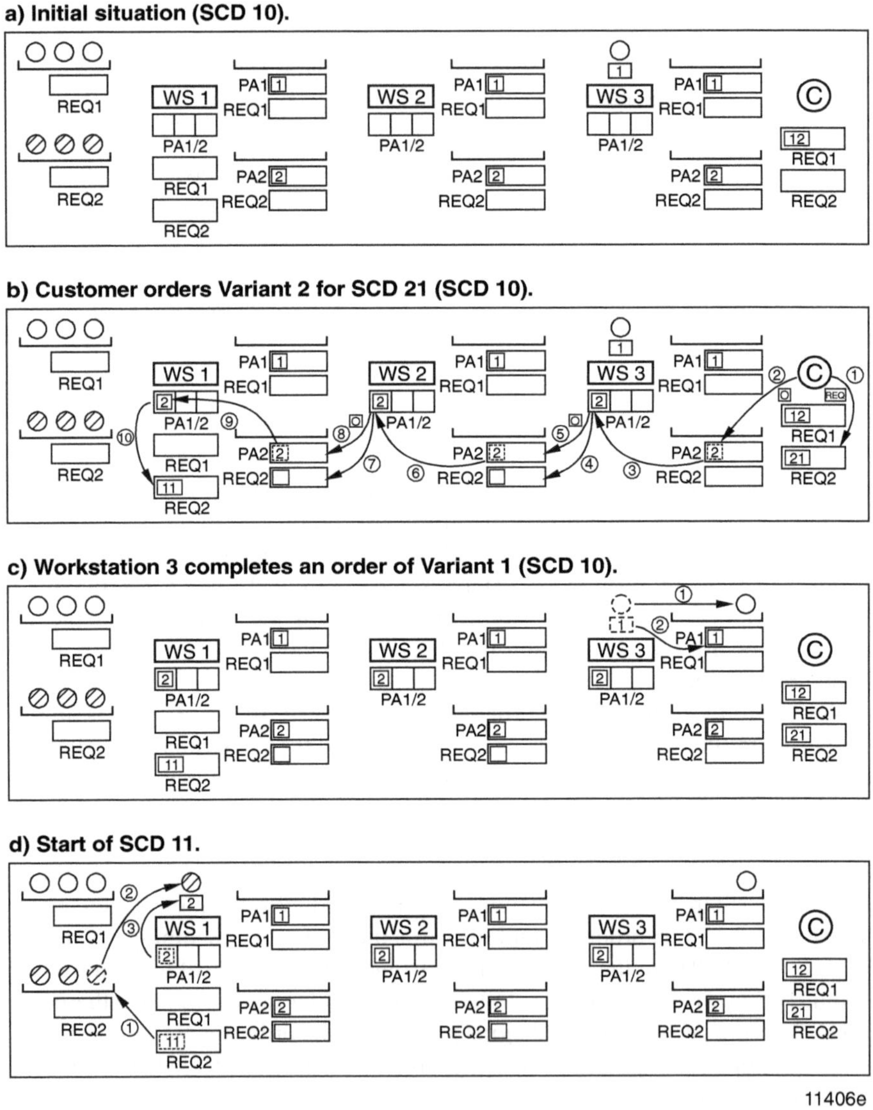


**Fig. 15.2** How PAC functions, Example 2

**e) Start of SCD 12.**

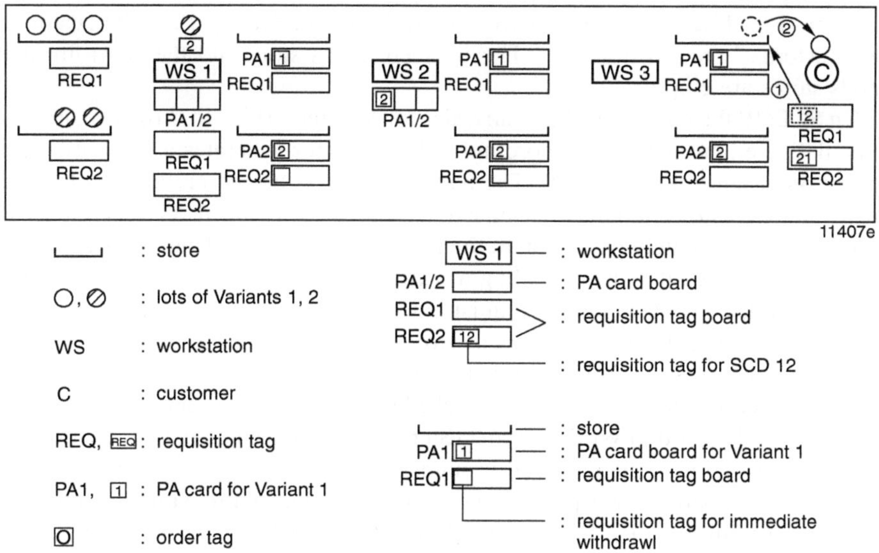


**Fig. 15.2** (continued)

before the planned delivery date. Orders completed early are not to be delivered early from the finished goods store to the customer.

A manufacturing control such as this can be represented by the following parameters:

- number of PA cards = unlimited for all workstations
- base stock = 0 for all workstations
- lag time of requisitions in the raw materials store = planned delivery date – planning date – planned delivery time
- lag time of requisitions in the finished goods store = planned delivery date – planning date
- lag time of requisitions in the semi-finished goods stores = 0

Workstation 1 and the customer are allocated boards on which the requisition tags whose lag times have not yet expired are collected.

On each of the requisition tags the shop calendar day on which the lag time expires is entered and the tag is passed on to the allocated store. The board for the order tags in the store can be omitted. Due to how the parameters are set, there are an unlimited number of PA cards, so the order tags never have to wait for a PA card to become available. There is always a PA card on the board as a representation of the theoretically unlimited number of cards. Otherwise the presentation of the example is the same as the first.

For the sake of clarity there is only one order in the production in the initial situation (Fig. 15.2a). It is to be delivered on shop calendar day 12 to the customer (see customer's requisition tag). The finished goods store and the semi-finished goods stores are empty.

On SCD 10 the customer places an order for Variant 2 (Fig. 15.2 b). The order is to be delivered on SCD 21. Accordingly, the customer generates a requisition tag and an order tag (Steps 1 and 2). The requisition tag is placed on the requisite tag board and only enters into the finished goods store on the planned delivery date. The order tag is immediately passed on to the finished goods store. There, it triggers the transport of a PA card to the preceding workstation (Step 3). Upon receiving the PA card the workstation generates a requisition tag and an order tag (Steps 4 and 5). Both directly enter the incoming goods store (lag time of requisitions in the semi-finished goods stores = 0). The requisition tag is placed on the corresponding board since there is no material in the semi-finished goods store; the order tag triggers the transport of a PA card to Workstation 2 (Step 6). Here, once again, a requisition tag and an order tag are generated and brought to the incoming goods store (Steps 7 and 8). The requisition tag is placed on the corresponding board, while the order tag triggers the transport of a PA card to Workstation 1 (Step 9). This immediately generates a requisition tag (Step 10) which remains on the workstation until SCD 11 (delivery date – planned throughput time = 21–10 = 11).

On the same day, Workstation 3 completes the order for Variant 1 (Fig. 15.2c). The order enters the finished goods store and the PA card returns to its board (Steps 1 and 2). Since there is no requisition tag the order cannot yet be delivered to the customer. The planned delivery date has not yet been attained.

At the start of SCD 11, Workstation 1 sends the requisition tag for Variant 2 to the raw materials store (Step 1 in Fig. 15.2d). The raw materials store in turn delivers the raw materials and Workstation 1 begins processing the order (Step 2). The PA card is allocated to the order for the processing duration (Step 3).

On SCD 12 the delivery date for the order of Variant 2 is attained. The customer gives the requisition tag to the finished goods store (Step 1 in Fig. 15.2e), which then immediately delivers the order to the customer (Step 2).

The mechanism for generating orders initially seems to require a great deal of effort. Nonetheless, it has an important advantage for the programming of PPC software: The same algorithm can be utilized for both make-to-order productions as well as make-to-stock productions.

With this review of PAC our discussion about methods for generating orders is concluded. In the next part of this book we will turn our attention to methods for releasing orders.

## Reference

[Buza-92*] Buzacott, J.A., Shantikumar, J.G.: A general approach for coordinating production in multiple-cell manufacturing systems. Prod. Oper. Manag. **1**(1), 34–52 (1992)

# Part C
# Methods for Releasing Orders

# Chapter 16
# Fundamentals of Order Release

The order release determines the point in time after which a production is allowed to process an order. Usually, it also immediately triggers the provision of the required materials. The material is thus specifically allocated to an order and is generally no longer available for other orders. The order release influences the WIP and the WIP dependent utilization of the production as well as the throughput time of the orders.

In the following chapter we will consider the three characteristics with which order releases can be classified:

- The *criteria* of the order release defines the conditions according to which decisions about the release of orders are made (Sect. 16.1).
- The order release's *degree of detail* determines if the order will be released as a whole or if a decision about the release for each individual operation will be made (Sect. 16.2).
- The order release's *trigger logic* determines the logic for when decisions about the order release will be initiated (Sect. 16.3). Decisions about releasing new orders can either be made at regular reoccurring time points (periodic order release) or after a defined event (event oriented order release).

Our discussion will conclude with a systematic overview of the order release methods discussed here (Sect. 16.4).

## 16.1 Criteria

Four criteria can be distinguished according to which orders are released for production.

1. *No Criteria (Immediate Release of Orders):* Orders are released immediately after they are generated (Sect. 16.1.1).

H. Lödding, *Handbook of Manufacturing Control*,
DOI 10.1007/978-3-642-24458-2_16, © Springer-Verlag Berlin Heidelberg 2013

2. *Planned Start Date:* The method for releasing orders based on due dates, releases an order as soon as the planned start date is reached (Sect. 16.1.2).[1]
3. *WIP in a Production or at a Work Station:* Order release methods oriented on regulating the WIP, release orders when the WIP in the production or at a workstation falls below a planned value (Sect. 16.1.3).
4. *Load of the Workstations:* Releasing orders with a workstation specific load balance takes into consideration the current or expected load of the workstations. This is measured via the WIP at a workstation between the time an order is released and its completion on the workstation. The order release with a load balance is thus a special case of the WIP oriented order release (Sect. 16.1.4).

An immediate order release is typical for order generating manufacturing control methods such as those extensively discussed in Part B of this book. In comparison the order release methods described in Part C use either the planned start date or the WIP as release criterion.

### *16.1.1 Immediate Order Release*

An immediate order release, releases a job directly after it is generated. As a result, this type of order release cannot specifically influence the WIP, throughput time and utilization of the production. It is solely dependent on the random order input and processing in the production.

In a make-to-order production this is connected to obvious disadvantages. The order release cannot differentiate between urgent and non-urgent orders. Accordingly, both classes of orders compete for the productions (scarce) resources. Since orders that are not urgent will also be released into production the WIP level is unnecessarily high. Long and frequently widely dispersed throughput times emerge and the ability to plan orders is reduced. Thus, in make-to-order productions orders are rarely released immediately after being generated. This method is common in make-to-stock productions though. Here the production is frequently allowed to process orders immediately after they are generated.

### *16.1.2 Due Date Based Order Release*

When releasing orders according to the due date, the planned input determines the actual input (Fig. 16.1a). It releases an order when the planned start date is reached.

[1] Various authors suggest order release methods that calculate the planned start date – to some degree based on the production's WIP. These methods will not be described here, since determining the planned due dates is considered the responsibility of production planning. Readers who are interested however can find a list of references for these methods in [Berg-97*].

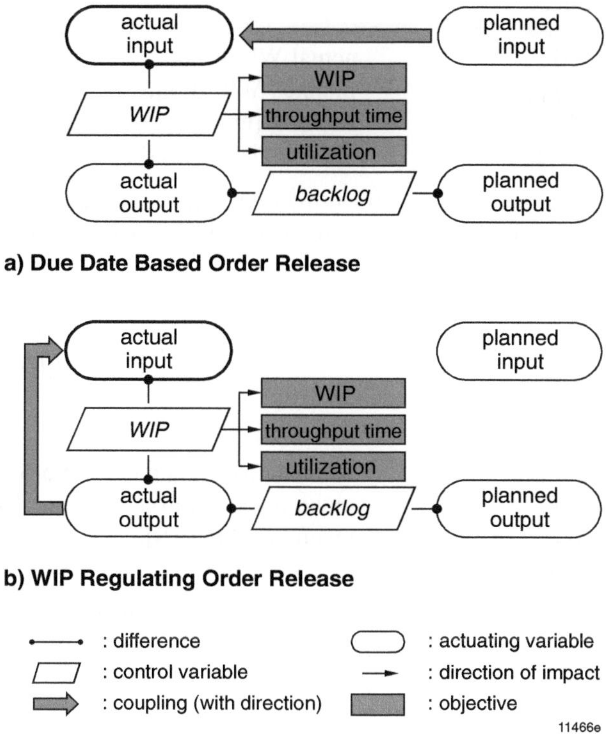


**Fig. 16.1** Comparison of Due Date Based Order Release and WIP Regulated Order Release

Due Date Based Order Release is the standard method in most PPC systems and with the MRP II system. This explains why it is so common. It aims to implement the production program exactly. It is thus particularly suitable when both the input and output of a production can be precisely planned and implemented.

In comparison, if the plan cannot be implemented due to disruptions in the input (e.g., materials arrive late) or output (e.g., machine malfunctions which cannot be compensated for by a backlog control), releasing orders according to the due date is less suitable. This is mainly because the WIP cannot be regulated by the Due Date Based Order Release. As soon as the actual output deviates from the planned output or the order input is disrupted, the WIP and throughput times, as well as to a lesser degree the utilization, deviate from the planned values.

When the disruptions in the order input are insignificant, then for the WIP deviations [Lödd-01]:

$$WIP_{act} = WIP_{plan} + BL \tag{16.1}$$

or

$$\Delta WIP = BL \tag{16.2}$$

Where

| | |
|---|---|
| $WIP_{act}$ | actual WIP [hrs], |
| $WIP_{plan}$ | planned WIP [hrs], |
| BL | backlog [hrs], |
| $\Delta WIP$ | WIP deviation [hrs]. |

If the impact on the utilization can be ignored, and the sequencing and scheduling of inter-operations times are independent of the work content, it follows that [Lödd-01]:

$$TTP_{m,act} = TTP_{m,plan} + \frac{BL_m}{ROUT_m} \tag{16.3}$$

Where

| | |
|---|---|
| $TTP_{m,act}$ | mean actual throughput time[SCD], |
| $TTP_{m,plan}$ | mean planned throughput time[SCD], |
| $BL_m$ | mean backlog [hrs], |
| $ROUT_m$ | mean output rate [hrs/SCD]. |

If the production develops a backlog, it increases its WIP and the throughput times of the orders become longer. The higher WIP allows additional sequencing interchanges in the production and the ability to plan the throughput times thus decreases.

The chances of a negative backlog (the actual output exceeds the planned output) developing are nearly impossible when orders are released according to due date. This would provide enterprises with the advantageous opportunity to exploit phases where the productivity is unexpectedly good [Hopp-96*]. In order to do so though, at least a part of the orders has to be released early, thus making a Due Date Based Order Release pointless.

A special form of a Due Date Based Order Release discussed particularly in theoretical publications is the so-called 'clockwork control'. It distributes the release times for the orders evenly over the release period. The basic idea is to thus attain a steady order input which positively influences the trend of the Logistic Operating Curves. The clockwork control is easy to simulate and therefore often serves as a reference for the logistic suitability of newly developed control methods (see for example [Glas-88*]).

We will take a closer look at controlling production by releasing orders according to due date in Chap. 17.

### *16.1.3 WIP Regulating Order Release*

With a WIP Regulating Order Release the actual output determines the actual input (Fig. 16.1b). The American, Oliver Wight, is often seen as the spiritual father of using the release of orders to regulate the WIP. Early on he promoted coupling input of a production with its output ("The input to a shop must be equal or less than the output" [Wigh-70*]) and discussed the advantages of this type of strategy. Despite extensive suggestions and a few tools for implementing an order release that would regulate the WIP Wight, however, never formulated an algorithmized control method. The CONWIP control (Chap. 18), comes very close to implementing his suggestions, nonetheless – with the exception of Due Date Based Order Release – the remaining order release methods described in the third part of this book also follow the logic provided by Wight.

A WIP Regulating Order Release authorizes orders (or an operation) when the planned WIP in the production (or at a workstation) falls below a planned value. In the following this planned value is referred to as the WIP limit. The release can occur before, after or at the planned release date. A WIP Regulating Order Release thus does not lead to an exact implementation of the planned input. Frequently, an advance release window is defined, within which orders are allowed to be released early. In this case a WIP Regulating Order Release is combined with a Due Date Based Order Release. In order to release an order then, two conditions have to be met:

1. actual WIP $<$ WIP limit
2. actual date $\geq$ planned start date – advance release window

There are at least eight advantages to WIP regulating order releases.

***1. Robust to Errors in Dimensioning the Planned WIP***

A WIP control is robust to errors made when setting the planned WIP. If the planned WIP is set too high or too low it will only impact the output of the production or workstation minimally in the typical operating range of the Logistic Operating Curve. This is due to the declining trend of the output in comparison to the WIP (Fig. 16.2a). In comparison, the impact on the throughput time is approximately proportional to the WIP deviation.

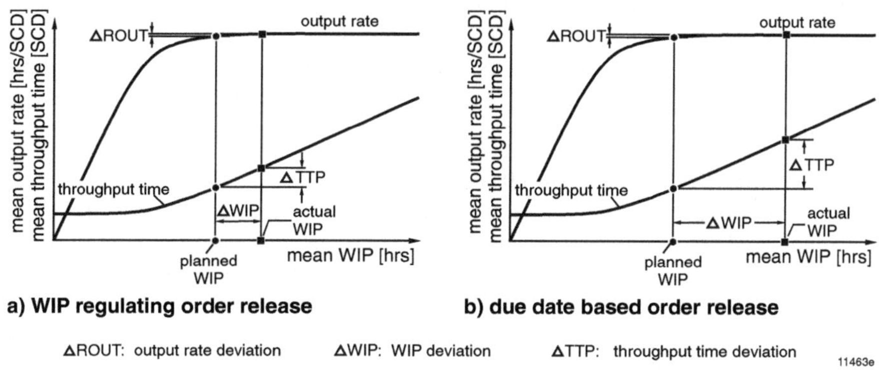


**Fig. 16.2** Impact of planning errors with WIP regulating order release and due date based order release

In comparison an error in estimating the capacities or load has a disproportionate impact on the WIP and throughput times when orders are released according to the due date (Fig. 16.2b).

The greater robustness of a WIP Regulating Order Release is emphasized especially by Hopp and Spearman [Hopp-96*].

***2. Robust to Plan Deviations in the Order Input***

When orders are released according to due date and the material for an order cannot be supplied punctually or an order is completely cancelled it is necessary to reschedule. The input into the production (or workstation) falls behind the planned input and there is possibly the threat of a loss of utilization. With a manual rescheduling an order is pulled forward in order to be able to utilize the capacities.

In comparison, a WIP control reacts automatically. If the WIP falls below the planned level, it releases the next available order. If the planned order is not available, it automatically releases the following order early. This is possible when the conditions for releasing it are met e.g., the required materials are available. A loss of utilization is then prevented. This is applicable at least when the order released early is processed on the same workstations as the order that was set aside.

***3. Predictability of the Throughput Times***

If a production's (or workstation's) WIP is maintained at a constant level, the throughput times fluctuate only minimally. This is applicable, nonetheless, only when sequencing interchanges can be prevented. Controlling the WIP therefore simplifies scheduling.

***4. Protects the Production from Extreme Turbulence***

A WIP Regulating Order Release limits the WIP in a production. It thus protects the production from fluctuations which can arise due to a variable demand or different preferred delivery times. Thus a WIP control can convert preferred delivery times into homogenous throughput times. This can be easily accomplished when the preferred delivery time is longer than the mean throughput time of the production. If there are isolated incidences where shorter delivery times are attained the turbulence created by these impacts the production. Homogenous throughput times are then no longer sufficient. Accordingly, sequencing interchanges cannot be avoided. H.-H. Wiendahl discusses this correlation extensively [Wien-02].

***5. Protects the Production from Overload***

If the production's or workstation's load exceeds the capacities, the WIP accumulates. The throughput times become longer and are frequently, more strongly dispersed; their predictability decreases. A WIP Regulating Order Release constantly maintains the WIP in the production and holds back orders when the production is overloaded. Enterprises can thus frequently take into consideration changes preferred by the customers *before* an order is released. Moreover, the order release can be used to prioritize the orders up to just before they are released in order to use the capacities for important orders.

***6. Protects the Production Bottleneck from Overload***

One aspect of overloading a production is exceeding the capacities of the production bottleneck. When orders are released according to due date, workstations with available capacities can temporarily produce more than the bottleneck can manage. As a result, the WIP on the bottleneck workstation strongly accumulates. With a WIP regulating order release, the output rate of the production equals the input.[2] Since the bottleneck determines the output rate of the production the input is automatically aligned with the production bottleneck.

***7. Protects the Production from Underload***

If the planned load of a production or workstation falls below the capacities, the WIP is reduced and the utilization decreases. A WIP Regulating Order Release prevents this by releasing orders early. It thus smoothes fluctuations in the planned load and can exploit phases in which the production can manage more than planned. This advantage of the WIP regulated order release is emphasized by Hopp and Spearman [Hopp-96*]. They show that a WIP regulated production attains a specific output rate at a lower WIP level than one with a Due Date Based Order Release.

***8. Less Tied-Up Capital***

Especially in comparison to methods that release orders immediately after they are generated, controlling the WIP leads to less capital being tied-up in current assets (see also Chap. 2).

Therefore, a WIP Regulating Order Release is generally connected to a number of important advantages. Nonetheless, not all of its properties are positive. Three possible disadvantages of controlling the WIP are:

- The WIP control has to be coordinated with the scheduling of the orders and thus gives it less room to maneuver. If as a result of scheduling the orders the planned WIP exceeds the WIP limit of the WIP control, then the scheduling cannot be implemented. The actual processes then inevitably deviate from the plan.
- If orders are released immediately after they are generated, a higher utilization can be attained than with a WIP control due to the increased WIP level.
- If necessary, the WIP control releases orders early in order to ensure the utilization of the production. Consequently, these orders are frequently completed early increasing the capital tied into the finished goods stores.

With the exception of the Due Date Based Order Release, all of the methods for releasing orders in the following discussion use WIP as criterion for the release. A number of these methods also lead to a (workstation specific) load balance (see the following section).

[2] In the long-run, this is also the case with a due date based order release (see the first basic law of production in [Nyhu-09*]). Over the short term though, the input and output rates can be set (to a large degree) independently from one another.

### 16.1.4 Order Release with a Workstation Specific Load Balance

The methods described here for combining a workstation specific load balance with the release of orders is a special form of a WIP regulating order release. It measures the WIP for each of the workstations in a production. In doing so it considers both the *direct WIP* at the workstation as well as the *indirect WIP* at the preceding workstations which will be a part of the workstation's load in the future. Thus, the total WIP of a workstation includes all of the orders from the point of the order release up to their completion on the workstation. There are a number of alternatives for this type of order release:

1. An order is released when the total WIP at all workstations falls below the WIP limit.
2. An order is released when the total WIP at a minimum of one workstation falls below the WIP limit (in the following this special form will only be considered on the periphery).
3. The orders are released so that the total WIP on all workstations deviates as little as possible from the target WIP.

Releasing orders with a load balance can be combined with the due date based method for releasing orders. In this case, an order is only allowed to be released when the planned start of processing is within a specific advance release window.

Unlike a simple WIP control, a workstation specific load balance inevitably leads to sequencing interchanges when releasing orders. It can thus endanger the production's schedule compliance. In order to minimize the effects of the sequencing interchanges, the planned completion dates should be taken into consideration when deciding about sequencing.[3]

The effectiveness of a load balance is dependent on five factors:

***1. Quality and Degree of Detail of the Due Date and Capacity Planning***
When the due date and capacity planning are properly conducted there are sufficient capacities available (with the exception of unexpected disruptions). Balancing the load later is then superfluous. When finite planning is applied for orders and operations, balancing the load based on heuristic rules can even have a negative impact. Accordingly, a load balance is especially practical when there is no due date and capacity planning or when it is not exact – as is the case for many enterprises. Especially in a make-to-stock production in which the orders are frequently generated automatically, the due date and capacity planning is almost inevitably based on estimations and tends to be more approximate.

---

[3] Under circumstances this can reduce the effectiveness of balancing the load. In many cases however the schedule reliability is the more important objective.

***2. Significance of Disruptions***
Disruptions can annul the validity of even the most thorough due date and capacity planning. The order release is the latest point in time at which disruptions can be considered before the processing starts. Releasing orders with a load balance is therefore particularly practical in a manufacturing environment characterized by disruptions.

***3. Flexibility of Capacities and Fluctuations in the Planned Load***
If the capacities are highly flexible, fluctuations in the load can be managed by adjusting the capacities. Accordingly, a chronological load balance is no longer necessary. Consequently, it is especially practical to release orders with a load balance when the capacities are not so flexible in comparison to the fluctuations of the planned load.

***4. Complexity of the Material Flow***
The basic idea behind load balancing is to hold back orders which flow through overloaded workstations and to release orders that flow through underloaded workstations earlier. This requires a certain degree of complexity in the material flow. If all of the orders flow through the same workstations, the load can only be balanced via the work content, if at all. Releasing the orders with a load balance is thus usually ineffective.

***5. Number of Orders at Disposal***
The effectiveness of balancing the load is dependent on the quantities and composition of the orders that are available for the load balancing. The *orders at disposal* include all of the known but not yet released orders within the advance release window. The more orders at disposal there are and the more heterogeneous their composition (material flow complexity), the more balancing the load can increase the production's or workstation's output. The number of orders at disposal is in turn influenced by four factors:

- *Order Throughput Time:* With a given delivery time the shorter the production's throughput time is the more orders at disposal there are.
- *Load Shift:* The greater the load shift, the more orders at disposal there are.
- *Backlog in the Production's Input:* In addition to orders that are planned to be released shortly, the orders at disposal include all of those whose planned processing start has already passed but which have not yet been released. The greater the backlog in the order input is, the larger this portion is. When the WIP is controlled in the production, the input backlog corresponds approximately to the backlog in the production's output.
- *Advance Release Window:* The number of orders at disposal increases along with the size of the advance release window. When the advance release window is unlimited, the orders at disposal include all of the known but not yet released orders.

## 16.2 Degree of Detail

The decision to release orders can either apply to the entire order (low degree of detail) or for one of the individual operations (high degree of detail). Differentiating between the varying degrees of detail is significant especially for the WIP regulating order release. In the following, a WIP Regulating Order Release with a low degree of detail will be referred to as a centralized WIP control and one with a high degree of detail as a decentralized WIP control (see also the discussion in [Wien-02]). Accordingly, the following definitions apply (Fig. 16.3).

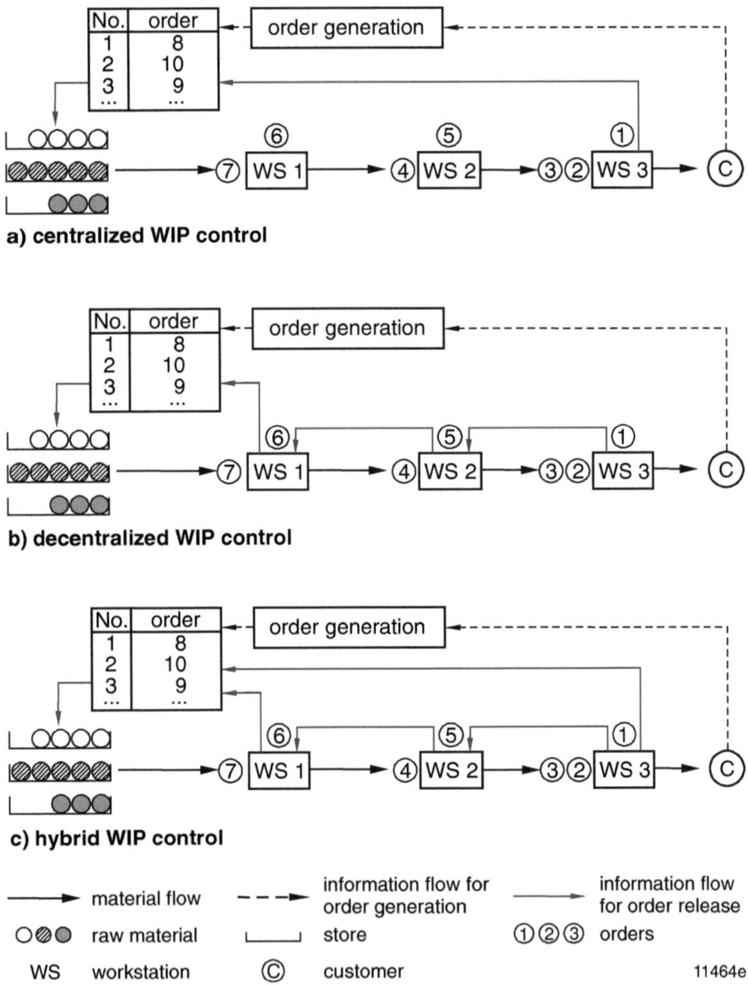


Fig. 16.3 Flow of information in centralized, decentralized and hybrid WIP controls

A *centralized WIP* control regulates the WIP in the entire production (Fig. 16.3a). It releases an order when the WIP falls below the planned level. This frequently occurs after the last workstation has completed an order and triggers the release of a further order. CONWIP (Chap. 18) is a common example of a centralized WIP control.

A *decentralized WIP control* regulates the WIP at the workstations (Fig. 16.3b). Accordingly, the order release is authorized for an operation and not for orders on the whole. POLCA and the Decentralized WIP Oriented Manufacturing Control are two examples of this.

A *hybrid WIP control* releases both the order as well as the operation (Fig. 16.3c). Whereas an order is released when the total WIP of the production falls below the planned WIP, an operation is authorized when the WIP at a workstation falls below the planned WIP. The only difference between the hybrid WIP control and the decentralized WIP control is in regards to the first operation. When necessary the hybrid order release will also hold back an order when the WIP on the first workstation falls below the planned WIP. This is the case when the WIP of the entire production exceeds its planned WIP. Similarly, in comparison to the centralized WIP control, a hybrid WIP control also holds back an order when the WIP in the entire production falls below the planned WIP but the WIP on the first workstation exceeds the planned level. The hybrid WIP control is thus more restrictive than both the centralized and decentralized WIP controls.

Decentralized WIP control has the following five advantages:

1. *Integrating Workforce:* With a decentralized WIP control the production's workforce can regulate WIP levels and throughput times on the workstations themselves. This is a good condition for continuous improvement processes.
2. *Differentiating Targets:* WIP can be regulated at an individually appropriate level on each of the workstations. It is therefore possible to set the planned WIP at the throughput bottleneck higher than at workstations utilized less.
3. *Identification of Bottlenecks:* The blockings in a decentralized WIP control provide information about the production's dynamic bottleneck. This information can be utilized to control backlogs and can be visualized in so-called "blocking diagrams" [Lödd-01].
4. *Balancing the Load:* As the orders flow through the production a decentralized WIP control tends to balance the load between the workstations (see Sect. 24.2 as well as [Lödd-01]). This balancing of the load causes the workstations to be more consistently utilized.
5. *Predictability of Throughput Times:* When the material flow is complex, throughput times tend to remain more predictable and therefore more plannable than with a centralized WIP control (see the discussion on advantages and disadvantages of centralized WIP control as well as [Lödd-01]).

Decentralized WIP control is also related to the following four disadvantages:

1. *Number of Parameters:* The number of parameters is considerably greater than with centralized WIP control and corresponds to at least the number of

workstations. In comparison, centralized WIP control can – in the ideal case – have only one parameter.
2. *Blocked WIP:* Decentralized WIP control inherently causes blocked WIP in the production. This can lead to a loss of utilization on the affected workstations.
3. *Deadlocked Workstations:* When the material flow is complex there is a danger that workstations will be deadlocked. In situations where the procedural rules do not exclude such deadlocks, the field of application for this control method is limited.
4. *Sequencing Interchanges:* The blocking of orders can cause sequencing interchanges. This makes throughput times more difficult to plan (see e.g., [Lödd-01]).
5. *Freedom for Scheduling:* Decentralized WIP control reduces the freedom for scheduling the orders more than centralized WIP control. Finite scheduling can only be implemented when the planned WIP on every workstation remains below the planned WIP level. With centralized WIP control this applies only to the WIP limit for the entire production.

The advantages and disadvantages of centralized WIP control unfold as a mirror image of the argument for or against decentralized WIP control: A disadvantage of the decentralized WIP control is an advantage of centralized WIP control and vice versa. The most important arguments are summarized below:

1. *Simplicity:* Centralized WIP control is simpler to implement than its decentralized counterpart. First, only a few employees are directly involved in centralized WIP control. Second, it has considerably fewer parameters and third, deadlocked workstations are generally excluded.
2. *Sequencing Interchanges, Load Balancing and Blocking WIP:* Centralized WIP control causes neither blocked WIP nor sequencing interchanges. This reduces the variance of throughput times, but does not lead to a temporary load balance. A load balance can however be targeted by combining it with a suitable sequencing rule (see Chap. 25).
3. *Integrating the Workforce:* With centralized WIP control, production workers are insufficiently integrated into regulating the WIP and throughput times.
4. *Predictability of Throughput Times:* When there is a complex material flow, order throughput times have a tendency to fluctuate more strongly than with decentralized WIP control. In order to be able to predict throughput times, it would be necessary to forecast WIP levels on individual workstations in the production. This is however not possible with centralized WIP control [Lödd-01].

## 16.3 Trigger Logic

As indicated in [Berg-97*] for example, decisions regarding order release can either be made at specific points in time (periodic order release) or according to defined events (even oriented order release). In the following section, we will discuss both these possibilities.

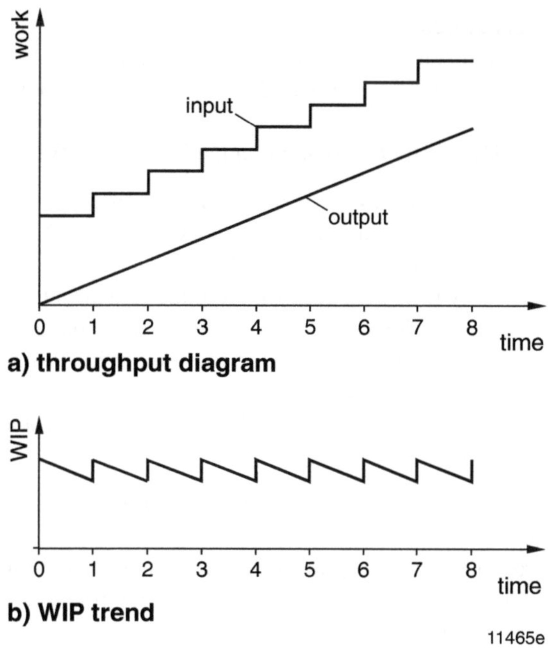


**Fig. 16.4** WIP fluctuations with periodic order release

**Periodic Order Release** With periodic order releases decisions about releasing new orders are made at predetermined time points. Between two release points, no orders are released. Typical release points are the start of a work day or a new shift. Periodic order releases inevitably lead to orders being released intermittently and thus to fluctuations in the WIP (Fig. 16.4). This applies especially when the work content of the orders is on average considerably less than the work content of an order release, otherwise, the variance of the work contents superposes the effect.

The longer the release period is, the stronger WIP fluctuations are. Furthermore, periodic order releases can also increase the delivery time of orders – on average by half of the period length: After being generated an order can usually not be immediately released, but rather is released, at the earliest, at the next release point. Finally, periodic order releases complicate the setting of WIP limits. The output is obviously dependent on the capacity during the release period. If this varies, then the WIP limit also has to be adjusted (for an example in which the number of working days in the release period varies due to holidays see [Brei-02b*]). This is particularly difficult when capacities can be temporarily adjusted after orders are released.

One advantage of a periodic order release is that it generally requires less effort than an event oriented order release.

**Event Oriented Order Release** Event oriented order releases make decisions about releasing orders after a certain event has occurred. These events generally include:

- An order being generated,
- Reaching the planned start date (or advance release window),
- The WIP falling below its designated limit once an order or operation is completed, or
- One of the method's parameters being changed (e.g., the WIP limit).

Event oriented order releases generally release orders into the production steadily. Thus, in comparison to a periodic order release, WIP fluctuations can be prevented. At least in principle, orders can be released immediately after being generated, so that delivery times are not extended due to waiting for the next release point. For event oriented order releases however, events have to be detected as quickly as possible. Thus the effort required to implement it is usually greater than a periodic order release [Berg-97*].

Occasionally it is difficult to differentiate between periodic and event oriented order releases:

- When release periods are very short, a periodic order release approximates an event oriented order release.
- When orders are released according to schedule it is typical to coordinate the release points with planned start dates. Orders are then released at the same point in time as with an event oriented order release.

Moreover, the trigger logic is not a characteristic of the method: Generally, the methods described here for releasing orders can release orders either periodically or based on events. Nevertheless, the methods are frequently associated with the trigger logic used by the developers when describing the procedure (see for example [Berg-97*]).

## 16.4 Classifying Order Release Methods

In order to facilitate the overview of the different methods for releasing orders Fig. 16.5 depicts a classification of the order release methods. Part A of the diagram outlines three classes of methods: immediate order releases, due date based order releases and WIP regulating order releases.

In the following, we will not be discussing the immediate release of orders further. As already mentioned it is only suitable in special cases. Instead, the next chapter will focus on Due Date Based Order Release. This method is implemented in numerous enterprises and thus is very significant for the industry.

The large majority of methods for releasing orders can be classified as WIP regulating order releases. These can then be further classified as to whether or not they support balancing the loads on individual workstations. Moreover, they can be differentiated as either centralized or decentralized approaches (Fig. 16.5b).

We will start by describing centralized methods beginning with methods that do not balance the loads on the individual workstations (CONWIP and Bottleneck

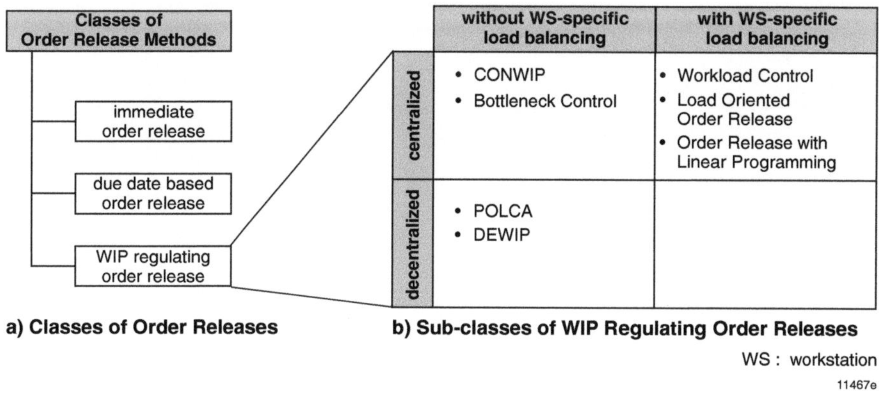


**Fig. 16.5** Classification of order release methods

Control in Chaps. 18 and 19). Centralized methods with workstation specific load balancing (Workload Control, Load Oriented Order Release and Order Release with Linear Programming in Chaps. 20 to 22) will follow that. Finally, we will end Part C of this book by taking a look at Decentralized methods for releasing orders (POLCA and Decentralized WIP Oriented Manufacturing Control in Chaps. 23 and 24).

## References

[Berg-97*] Bergamaschi, D., et al.: Order review and release strategies in a job shop environment. A review and a classification. Int. J. Prod. Res. **35**(2), 399–420 (1997)

[Brei-02b*] Breithaupt, J.-W., Land, M., Nyhuis, P.: The workload control concept: theory and practical extensions of load oriented order release. Prod. Plan. Control **12**(7), 625–638 (2002)

[Glas-88*] Glassey, C.R., Resende, M.G.C.: Closed-loop job release control for VLSI circuit manufacturing. IEEE Trans. Semicond. Manuf. **1**(1), 36–46 (1988)

[Hopp-96*] Hopp, W.J., Spearman, M.L.: Factory Physics. Irwin, Chicago (1996)

[Lödd-01] Lödding, H.: Dezentrale Bestandsorientierte Fertigungsregelung. VDI Progress Reports, Series 2, No. 587, Düsseldorf (2001)

[Nyhu-09*] Nyhuis, P., Wiendahl, H.-P.: Fundamentals of Production Logistics. Theory, Tools and Applications. Springer, Berlin (2009)

[Wien-02] Wiendahl, H.-H.: Situative Konfiguration des Auftragsmanagements im turbulenten Umfeld. Jost-Jetter Verlag, Heimsheim (2002)

[Wigh-70*] Wight, O.: Input/Output control. A real handle on lead time. Prod. Invent. Manag. **11**, 9–31 (1970)

# Chapter 17
# Due Date Based Order Release

Due Date Based Order Release is the standard approach in the large majority of PPC systems and MRP II, accordingly, the method is very common. Currently, a clear majority of all make-to-order productions still conduct a Due Date Based Order Release that is admittedly frequently overridden by manual interventions.

This chapter is divided into five sections. Section 17.1 discusses controlling production with a Due Date Based Order Release. The procedural rules for this method are then considered based on the guidelines for designing manufacturing control methods (Sect. 17.2). In doing so, a number of deficits related to controlling production with Due Date Based Order Release will be revealed. The chapter then continues with a discussion of the parameters involved in this method (Sect. 17.3) and concludes with a generalized case study of applying Due Date Based Order Release in the industrial practice (Sect. 17.4). The example will be used to point out typical mistakes made when applying it as well as to describe how the production should be controlled in order to also attain ambitious logistic targets.

## 17.1 Method Description

One of the prerequisites of Due Date Based Order Release is that a higher level production planning first generates the orders and determines the planned start dates. In many cases the orders are generated with an MRP II system. Generally speaking though, any method for generating orders can be combined with Due Date Based Order Release. A Due Date Based Order Release can be described by a single procedural rule:

*A Due Date Based Order Release, releases a production order once the order's planned start date has been reached or exceeded and the required materials are available.*

H. Lödding, *Handbook of Manufacturing Control*,
DOI 10.1007/978-3-642-24458-2_17, © Springer-Verlag Berlin Heidelberg 2013

**Table 17.1** Release list

| No. [–] | Order number [–] | Planned start date [SCD] | Item number [–] |
|---|---|---|---|
| 1 | 4,012 | 100 | 43 |
| 2 | 4,125 | 100 | 40 |
| 3 | 3,984 | 101 | 18 |
| 4 | 5,642 | 102 | 42 |
| 5 | 8,221 | 102 | 40 |

The most important control instrument is thus a list of the orders to be released; for practical sakes the list is organized according to planned start dates (Table 17.1).

Figure 17.1 depicts the flow of information for a Due Date Based Order Release. No information from the production is taken into consideration when releasing the orders.

In order to clarify how the production is controlled, let us consider an example with a production consisting of three workstations. The manufacturing control releases orders at the start of every day.

In the initial situation there are five orders in the production (Fig. 17.2a). The planned start date for Orders 11 and 12 has been reached. The PPC system verifies the availability of the raw materials and triggers their provision (Fig. 17.2b).

The workstations process the orders that are in the production up until the next release point; Orders 1 and 2 are completed. With the majority of other orders, at least one operation is completed (Fig. 17.2c).

In the meantime production planning generated an additional order, Order 10. It is released together with Orders 3 and 5 the following day (Fig. 17.2d).

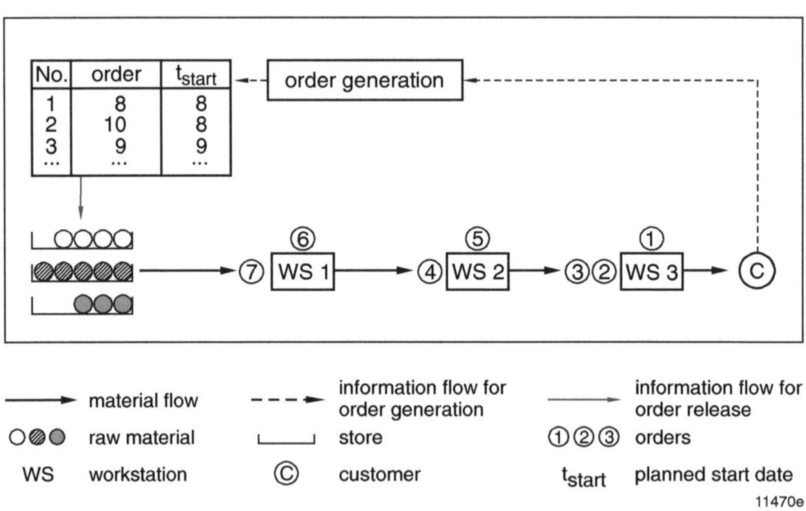


**Fig. 17.1** Principle of controlling production with due date based order release

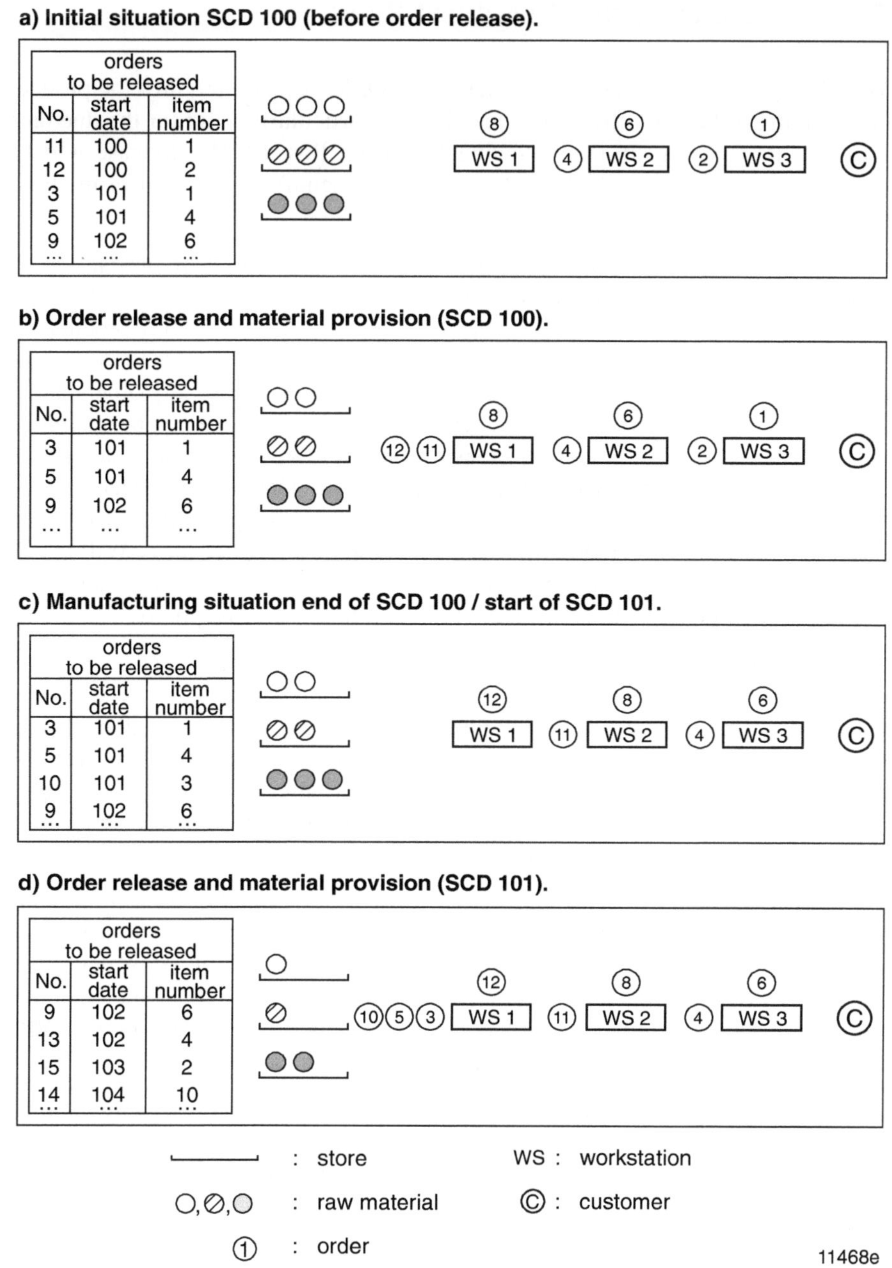


**Fig. 17.2** How manufacturing control with due date based order release functions

## 17.2 Discussion of the Procedural Rules

Controlling production with a Due Date Based Order Release requires the production planning to be carefully conducted. This is particularly valid for the due date and capacity planning. In the following Due Date Based Order Release will be discussed based on the guidelines for designing manufacturing control methods (see Sect. 3.3.2).

*1. A manufacturing control method should allow the WIP of a production or workstation to be regulated at a defined level.*

Due Date Based Order Release is unable to regulate the WIP in a production. Should a backlog develop, the WIP increases. If the production is able to exceed the planned output, Due Date Based Order Release does not release any orders early. Thus phases where there is above average productivity can only be exploited minimally since the production threatens to run idle. Even when the materials for some of the production orders are not available, Due Date Based Order Release does not release orders early to ensure the utilization of the production. Generally speaking, Due Date Based Order Release thus requires a higher mean WIP than a WIP regulating order release method to attain a comparable output rate. This correlation is discussed extensively by Hopp and Spearman [Hopp-96*].

*2. A manufacturing control method should reduce WIP fluctuations to a minimum.*

For the same reasons, controlling the production with a Due Date Based Order Release also does not provide the conditions for avoiding WIP fluctuations. Nonetheless, carefully planning the due dates and capacities in order to prevent both overloading and underloading the production contributes to a steady WIP trend. The due date and capacity planning is thus vital when controlling the production with Due Date Based Order Release.

*3. A manufacturing control method should cause as little blocked WIP as possible.*

While the order is flowing through the production Due Date Based Order Release does not cause any blocked WIP. Nonetheless, it does hold back all of the orders which have not yet reached their planned start date. Compared to an approach that immediately releases orders (see Chap. 16) this contributes to reducing the WIP. This applies in particular to multi-stage products.

*4. A manufacturing control method should contribute to closely aligning the load to the capacities.*

Load balancing accelerates or delays the release or throughput of orders with the aim of preventing breaks in the material flow (accelerates) or a build-up of WIP (delays). A Due Date Based Order Release limits the possibilities of balancing the load to the production planning phase. It is the task of the due date and capacity planning to schedule orders so that not only load peaks are avoided but also situations where workstations with critical capacities are not temporarily underloaded. Many enterprises however are not thorough enough when planning the due dates and capacities. Some basic reasons for this are:

- The PPC software does not offer sufficient support for automatically aligning load and capacities. This increases the manual planning effort.
- Outdated PPC systems calculate the capacity's load based on the planned throughput of the orders without considering the actual progress in the production. If the production develops a backlog, these PPC systems allocate the capacities that are still required for delayed orders to new orders. As a result production capacities are overloaded.
- A large number of enterprises promise unrealistic delivery dates. It is then frequently impossible to schedule orders so that all production orders can be completed on-time. Realistically scheduling orders would mean that some of them are planned so that promised delivery dates are infringed upon. Many of the enterprises shrink away from such decisions. Nevertheless, it is especially in these types of situations that transparency is required and orders need to be prioritized before being released into production. Otherwise, the production inevitably develops a backlog and it is then frequently just matter of chance which orders are finished on-time. Promising realistic delivery dates is at the very least just as important as realistically scheduling orders.
- Periodic production planning cannot immediately take into consideration schedule deviations that occur between planning sessions. Thus with long planning cycles, periodic production planning can in principle only have a delayed reaction to schedule deviations.

5. *A manufacturing control method should cause as few sequencing interchanges as possible.*

Due Date Based Order Release does not cause any sequencing interchanges.

6. *A manufacturing control method should be able to balance schedule deviations within the context of backlog control.*

It is the responsibility of manufacturing control to prevent a backlog and to reduce a backlog that has already developed. In order to do so capacities are temporarily increased. A Due Date Based Order Release does not involve any backlog control. It is thus practical to supplement this manufacturing control with one.

7. *A manufacturing control method should take into consideration the bottleneck principle.*

With a Due Date Based Order Release the bottleneck principle can only be taken into consideration when planning due dates and capacities. One possibility here would be to schedule a comparatively high WIP level for the bottleneck workstation in the production and to pay particular attention to it when planning due dates and capacities.

8. *A manufacturing control method should be as simple as possible.*

Controlling production with a Due Date Based Order Release is – with only one procedural rule – very simple. This, in addition to the large variety of PPC software that includes a Due Date Based Order Release is probably one of the fundamental reasons for the method being so common.

## 17.3 Setting the Method's Parameters

The planned start dates are the only manufacturing control parameters for a Due Date Based Order Release. These are usually set within the context of the due date and capacity planning (see Sect. 4.1.2.2). Generally, two types of scheduling can be distinguished: One with limited capacities and one with unlimited capacities. The Flow Rate Oriented Scheduling described in Sect. 3.2.2.3 belongs to the latter. For a discussion of scheduling with limited capacities I would suggest turning to other professional literature, for example [Pine-08*].

## 17.4 Application

A Due Date Based Order Release is especially practical for enterprises that are very thorough during the production planning and which determine coordinated, realistic, planned start dates. Moreover, the planned schedule should be able to be implemented. This requires not only reliable suppliers who can ensure that the materials are available at the planned start date, but also that the enterprise is able to either completely avoid disruptions in the processing or to quickly and reliably compensate for them through a backlog control.

Instead of a explaining a single example of releasing orders according to due dates, the following fictive case study involves elements from a number of industrial projects. The aim here is to illustrate why logistic targets are so frequently missed when this method is implemented on the shop floor. To some degree the case study is thus a worst case scenario. Even when there are of course examples of successfully applying a Due Date Based Order Release in the industry, basic elements from this case study can surely be found in many enterprises. The example is divided into three parts: the description of the enterprise and the manufacturing control (Sect. 17.4.1), the analysis (Sect. 17.4.2) and the catalogue of measures for improving the attainment of logistic targets (Sect. 17.4.3).

### *17.4.1 Manufacturing Control at Circuit Boards Ltd*

Circuit Boards Ltd. produces circuit boards for customers on a make-to-order basis. Their customers include enterprises in the automotive, mechanical engineering and electronic industries. The customers equip the circuit boards with electronic components and integrate them into their products. Since the electronic components are generally in stock, the circuit boards often determine the delivery time for the entire printed circuit board. The delivery time is thus an essential mark of quality with which the circuit board manufacturer can compete for customer orders. This is

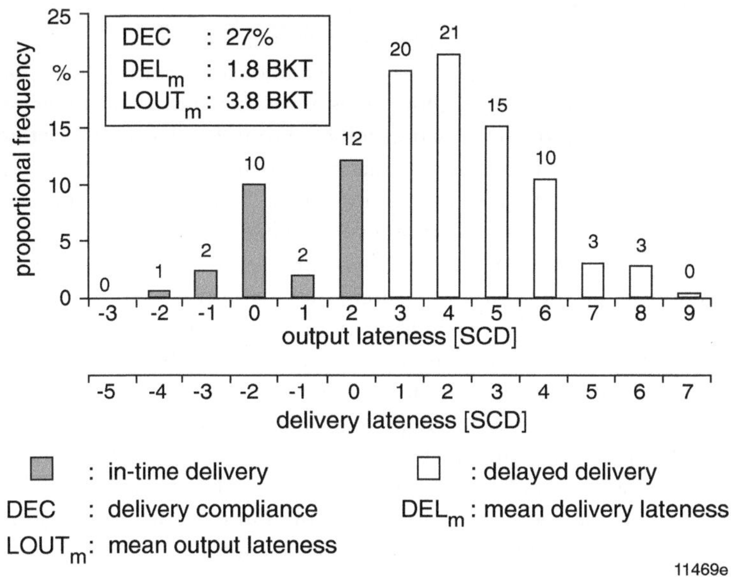


**Fig. 17.3** Distribution of Circuit Boards Ltd's delivery lateness

expressed in the considerable rush charges that circuit board manufacturers charge their customers for particularly short delivery times (see Fig. 2.3).

Corresponding to industry standards Circuit Boards Ltd. offers their customers a standard delivery time of 15 SCD. The second mark of quality is the delivery compliance, that is, the percentage of deliveries which are supplied before or on the promised delivery date (see Fig. 17.3 as well as the definition of delivery compliance in Sect. 2.1.2).

Poor delivery compliance is a fundamental concern for Circuit Boards Ltd. They only delivery a little more than a quarter of their orders in-time. The deliveries are delayed on average by two shop calendar days and in comparison to their internal schedule almost 4 days late (delivery time buffer = 2 SCD). Furthermore, the delivery lateness is highly variable: Some of the orders are completed earlier than required, whereas others are delayed significantly more than 1 week.

From their customers' perspective the poor delivery compliance weakens Circuit Boards Ltd. negotiation position. They have to make concessions in their prices and in order to compensate for delayed deliveries they have to commit to delivery dates for new orders that can only be kept by strongly prioritizing the orders. Moreover, they cannot charge the usual price premiums for such 'compensatory orders'. Despite a high utilization of resources (Circuit Boards Ltd. operates key workstations with three shifts) – they make only a minimal profit. The owners therefore regularly and with increasing emphasis threaten to sell the enterprise.

The sales and distribution area is responsible for the contact with customers. The manager attends to important customers and orders personally. If the sales department gains an order, he arranges the price and delivery time with the customer as well as the basic specifications of the order. This information is given to Circuit

Boards Ltd's order center. The customer conveys the exact product specifications to Circuit Boards Ltd. by means of electronic data transmission.

The order center schedules incoming orders and the order throughput times result from the sum of the operation throughput times. Operation throughput times are a fixed value for each of the production's workstations. From the planned order throughput and work content, the order center calculates the planned load for those capacities considered to be critical. If an overload is indicated, the order center informs sales.

Once the planned start date of an order is reached, the order center releases the order if the raw materials for the circuit boards are there. Due to the extensive standardization of the raw materials, this is almost always the case, even when temporary bottlenecks might arise during cyclical upswings.

Part of the orders are released by the order center even before the planned start date has arrived. This applies especially to orders that seem particularly important. The employees in the order center hope that the orders will be completed on-time due to the extra throughput time that is thus ensured. Moreover, orders are released early when workstations indicate that they have available capacities.

Once released the production orders flow through the workstations. The sequence in which the orders are processed on the workstations in the production is dependent on the individual operators and the workstations. Some of the typical sequence variations are:

*Setup Optimizing:* Some workstations have to be reset when the circuit board format is changed. Circuit boards with the same format can, however, be processed without converting the workstation. In order to avoid setups, operators usually try to process all of the orders for the one format before setting it for a new format.

*Longest Operating Time:* Many operators find setting up machines to be a tiresome task. They thus prefer to process large orders so that they (temporarily) have fewer setup operations. Furthermore, they can (temporarily) produce a larger number of circuit boards than when processing smaller orders. If an employee operates a number of machines, selecting larger orders also allows them to avoid idle machinery.

*Random or Non-Foreseeable Processing Sequence:* Some operators select the orders more or less randomly from the available WIP. In doing so, criteria such as the accessibility of an order or the compatibility of the processing duration for an order with the personnel's working hours play a role to some extent. The resulting processing sequence cannot be predicted.

*First-in-First-out:* A number of operators consider it important to process orders in the sequence that they arrive at the workstation. These operators ensure that incoming production orders are constantly put at the end of the queue.

*Urgency:* Many operators strive to process orders according to their urgency. Here, the orders' dispatch notes which include the planned end dates for the individual operations provide support. On some workstations, monitors also provide information about the urgency of queued orders.

*Priority Levels:* Circuit Boards Ltd. assigns different priority levels to production orders (e.g., normal orders, rush orders, management orders, etc.). These priority levels are often heeded even by those operators who no longer concern themselves with the planned end date of operations that have usually already been exceeded.

*Mixed Criteria:* The majority of operators do not process orders strictly according to one of the mentioned criteria, but rather combine them with one another, weighting the various criteria differently. Generally, the resulting sequencing cannot be or is not easily formalized.

When important customers complain about delayed deliveries, management will order special shifts from time to time on weekends in order to reduce the production backlog. Furthermore, the enterprise out-sources orders to another production enterprise which provides drilling services for the through-hole plating. Independent of the load and backlog, the enterprise also assigns operations that they cannot complete themselves due to technical reasons to their competitors. This, however, rarely occurs.

Once the orders are completed they are delivered to the customer. Local express services, who are happy to have their order books well filled, are usually engaged for delayed orders. The additional costs for an express delivery are usually considerable.

The enterprise evaluates feedback from the operational data acquisition system more or less regularly. An employee in the order center calculates the delivery reliability and production throughput times from the feedback data. The measured throughput times are then taken into consideration when setting planned throughput times.

### *17.4.2 Case Study Analysis*

In the following, the production planning and in particular the manufacturing control are examined for weaknesses. The analysis includes eight areas:

***1. Long Term Capacity Planning***

Despite the poor delivery reliability which acts to deter customers, Circuit Boards Ltd. can acquire enough orders to utilize production capacities. Nonetheless, they lack the flexible capacities required to be able to manage the fluctuations in the order input. Moreover, the enterprise does not fully utilize the available load flexibility. When there are a large number of incoming orders it neither refuses orders, nor extends the delivery times. The production thus develops a backlog and can only deliver the production orders late.

***2. Coordinated Sales and Production***

The failure to link the order acceptance to production capacities is obvious: Sales can then agree to delivery dates without considering the capacity load. Moreover, despite the order center's evaluation, sales is still insufficiently informed about the capacity load. In particular, there is no formalized procedure with which they can determine realistic delivery dates and the order center has no possibility of influencing the production's load.

***3. Scheduling Orders***
The planned throughput times of orders are not systematically determined: Every operation on a workstation receives the same planned throughput time. This ignores the difference in the work contents. Even more significant though is that the workstation throughput times are determined unsystematically. As is typical for the market, delivery times are one of the performance benchmarks as well as the measured throughput times. With the latter, there is the danger of gradually increasing the planned throughput times when the production develops a backlog (see also the so-called vicious cycle of PPC, [Wien-97b]).

***4. Operative Capacity Planning***
The future load of the capacities is estimated based only on the planned throughput of the orders. In many cases as soon as the actual throughput of the orders deviates from the planned, the future load that was calculated no longer coincides with the reality.

On the one hand, loads of orders not yet completed at the planning time are not considered, even though the planned end date has been reached. The required capacities are therefore underestimated. On the other hand, the capacity planning includes the load of orders completed early whose planned end date is not yet reached at the time of the planning. This leads to the required capacities being overestimated.

Generally then, the operative capacity planning is not able to realistically estimate the future capacity load. Even when the employees manage to identify in time that capacities will be overloaded in the future, the enterprise at best half-heartedly extracts consequences: It is not enough to just inform sales about the capacities being overloaded – this information also needs to be connected to clear instructions on how to proceed.

***5. Releasing Orders***
Circuit Boards Ltd. releases orders that are especially important early in order to ensure that they are completed in-time. As a result, the WIP increases and urgent orders compete with non-urgent orders for the limited production capacities. If the production's output is then generally not greater than planned, completing the orders early inevitably causes other orders to be delayed (see Sect. 3.2.1.3). When the production's workstations have available capacities, assessing the release of orders early is more complex. If there is a throughput bottleneck, then the output rate of the entire production can be increased. However, if workstations which have critical capacities do not register any free capacities, then releasing orders early only increases the WIP.

***6. Sequencing***
The sequencing interchanges in the production contribute to Circuit Boards Ltd's poor delivery reliability. Generally, the enterprise fails to provide clear instructions as to the sequence in which orders should be processed on a workstation. Moreover, they do not sufficiently implement and monitor the existing production plan.

***7. Backlog Control***
The possibilities for Circuit Boards Ltd. to reduce a backlog by increasing capacities are limited. Throughput bottlenecks in the production are already operating with three shifts. There are thus only two possibilities for increasing the capacity. First, by implementing weekend shifts and second, by out-sourcing operations with critical capacities to external enterprises. The enterprise waits too long before using the possibility of weekend shifts. It arranges them only after the production has developed a backlog and customers have complained about late deliveries. The second possibility of outsourcing operations with critical capacities is used by the enterprise only with the mechanical processing of the circuit boards i.e., when drilling the through-hole plating.

***8. Logistic Production Monitoring***
A systematic logistic production monitoring does not exist. By evaluating the throughput times and schedule reliability, Circuit Boards Ltd. concentrates only on attaining logistic targets. Although this is necessary, it fails to consider causes for insufficiently attaining them. In particular, it lacks an analysis of the production backlog and the sequencing discipline. Moreover, the logistic objectives are evaluated with considerable delays.

## *17.4.3 Deriving Measures for Improvement*

In the following we will examine measures suitable for increasing Circuit Boards Ltd.'s attainment of logistic targets. These are outlined according to the same eight areas used to analyze the causes of their failure to meet targets.

***1. Long Term Capacity Planning***
Circuit Boards Ltd. should consider expanding their capacity. In addition to the fact that their current capacities are already limiting the enterprises growth, sufficient capacities are a basic condition for promising the customer short delivery times. Furthermore, the enterprise has additional potential for shortening delivery times and increasing their delivery reliability (see below), thus, there is a good chance of gaining further orders for the additional capacities and for boosting sales revenues.

***2. Coordinating Sales and Production***
Circuit Boards Ltd. should implement a formalized system for determining delivery times. This system should consider the capacity load in particular. There are primarily two possibilities for implementing such a system:

(a) The employees in sales gain access to the enterprise's production planning – where possible via an electronic connection – and can independently log in orders. With this type of solution, the sales representatives take on part of the tasks of the order center. They can thus develop a sense of the actual load situation in the production and determine realistic delivery dates.

(b) The order center determines the delivery date of the orders (where necessary in agreement with sales) and thus maintains control over the production's load.

***3. Scheduling Orders***
Circuit Boards Ltd. should determine the planned throughput times based on a Logistic Positioning. A suitable method for doing so is Flow Rate Oriented Scheduling. The goal here should be to reduce order throughput times clearly below delivery times typical for the market. Should they succeed in doing so, the enterprise would gain a considerable logistic advantage over their competition.

***4. Operative Capacity Planning***
Circuit Boards Ltd. should consider the deviations of the actual throughput from the planned throughput when calculating the capacity load. In order to limit the effort required for this they can use the actual backlog at the time of the planning as a measure for the additionally anticipated capacity load. Depicting the planned and actual output of a workstation in a Throughput Diagram has proven to be helpful for such aggregated capacity planning (see Fig. 3.10).

***5. Releasing Orders***
Circuit Boards Ltd. should also release particularly important orders only at the planned start date. Over the mid-term the enterprise should create the conditions for being able to complete all production orders on-time. Special treatments for particularly important orders are then unnecessary. In the short-term, the firm can flag such orders specially. They thus obtain a higher priority as soon as the planned completion date of an order is reached. Before the planned completion date however, there is no special rules for the flagged orders. In this way the early completion of orders that displace other more urgent orders is prevented.

To better utilize the production, Circuit Boards Ltd. should only release orders early when a break in the material flow threatens a throughput bottleneck. As long as there is a backlog in the production, these cases should be rare.

***6. Sequencing***
Circuit Boards Ltd. should set clear guidelines for each of the workstations regarding the sequencing and communicate how important complying with these are for the enterprise's success. Furthermore, the enterprise should create conditions which allow operators to maintain the sequencing rules. This includes clearly flagging the priority of orders, where available via monitors at the work places. Finally, Circuit Boards Ltd. should control the compliance with the sequencing tasks and create suitable incentives for complying.

***7. Backlog Control***
Circuit Boards Ltd. should introduce a formalized method for controlling backlogs. In the simplest case, the enterprise could automatically authorize weekend shifts, as soon as the backlog in the production or on the throughput bottleneck workstation exceeds a defined limit.

If the enterprise had greater capacity reserves they could modulate the backlog control more precisely in order to compensate for load fluctuations and plan deviations. In this case, overtime is frequently sufficient for reducing a backlog. Controlling the backlog is discussed extensively in Chap. 27.

***8. Logistic Production Monitoring***
Circuit Boards Ltd. should expand their logistic production monitoring to include backlogs, sequencing discipline and WIP. Moreover, the enterprise should try to compile and evaluate the logistic figures more quickly. In the ideal case, calculating the most important logistic figures can be extensively automated. Independent of the daily controls, a more thorough report about attaining the logistic targets can be composed in greater intervals e.g., in the form of an audit [Trac-97]. This should present the mid to long term development as well as explain and discuss suitable measures for improving the attainment of logistic targets.

## References

[Hopp-96*] Hopp, W.J., Spearman, M.L.: Factory Physics. Irwin, Chicago (1996)
[Pine-08*] Pinedo, M.: Scheduling. Theory, Algorithms and Systems. Springer, Berlin (2008)
[Trac-97] Tracht, T.: Auditierung der Produktionsplanung und -steuerung. VDI Progress Reports, Series 2, No. 430, Düsseldorf (1997)
[Wien-97b] Wiendahl, H.-P.: Fertigungsregelung. Logistische Beherrschung von Fertigungsabläufen auf Basis des Trichtermodells. Carl Hanser Verlag, Munich/Vienna (1997)

# Chapter 18
# Constant Work in Process

Since the 1990s, CONWIP has increasingly gained attention in the area of manufacturing control methods. Spearman and Hopp who coined the catchy name (CONWIP = *Con*stant *W*ork *i*n *P*rocess) have played a considerable role in spreading the method and developing an understanding of its advantages [Hopp-96*, Spea-89*, Spea-90*, Spea-92*]. For further information on this release method, though with somewhat different language see e.g., [Bake-84*, Raga-88*, Wein-88*] and the bibliographies provided there. CONWIP is generally associated with controlling production lines. It can, however, also be applied with more complex material flows (see also [Lödd-01]). The basic idea behind the method is to maintain the WIP in a production or production line at a constant level.

This chapter is divided into the following sections:

- Section 18.1 explains the procedural rules based on an example.
- Section 18.2 discusses the procedural rules based on the guidelines for designing manufacturing control methods.
- Section 18.3 describes how to set the parameters for the method, especially the number of CONWIP cards.
- Section 18.4 discusses the rather unusual application of CONWIP across the supply chain.
- Section 18.5 provides a brief report about experiences applying CONWIP.

## 18.1 Method Description

The procedural rules for CONWIP are quite simple: It releases an order for a production line as soon as the WIP of the production line falls below a WIP limit (see Fig. 18.1 for a depiction of this principle).

The order with the highest priority is then selected from the release list. This release list contains all known orders that have not yet been released and whose planned start date is within a defined advance release window [Hopp-96*].

H. Lödding, *Handbook of Manufacturing Control*,
DOI 10.1007/978-3-642-24458-2_18, © Springer-Verlag Berlin Heidelberg 2013

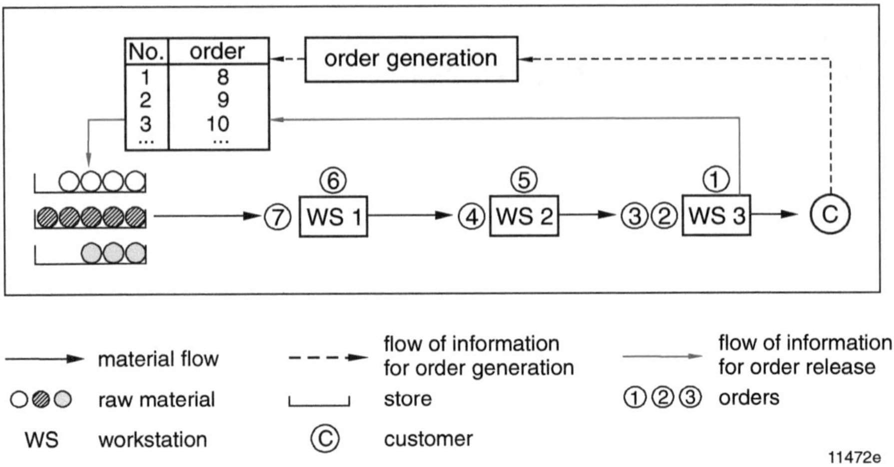


**Fig. 18.1** Principle of the CONWIP control

The condition for being included in the release list according to [Spea-90*] is:

$$TPL_0 + ARW \geq DPS \tag{18.1}$$

where

| | |
|---|---|
| $TPL_0$ | planning time point [SCD], |
| ARW | advance release window [SCD], |
| DPS | planned start date of order [SCD]. |

With a limited number of variants and a high frequency of repetitions the release list can often be presented in a simplified form. It then only contains the sequence in which the individual variants should be processed (Fig. 18.2a).

The WIP in the production is measured either in number of parts, number of orders or in planned hours [Hopp-96*]. In the last case the method is also referred to as the Maximum Load Limit (see e.g., [Meln-91*]). The planned times can be calculated using the following alternatives:

1. The planned time of the order corresponds to the sum of the planned times of its operations.
2. The planned time of the order corresponds to the sum of the planned times of the not yet completed operations [Meln-89*].
3. The planned time of the order corresponds to the planned time of the bottleneck workstation.

The following discussion will be based on the number of orders as the unit of measurement.

CONWIP, which is used to control the WIP, can be implemented with the aid of CONWIP cards (Fig. 18.2b, see also [Hopp-96*]). An order is not allowed to be

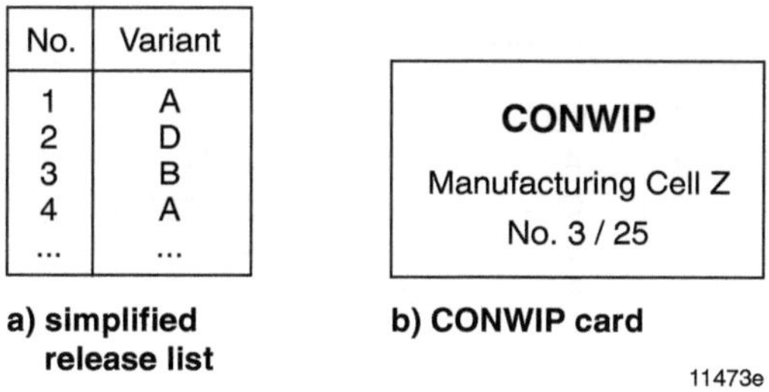


**Fig. 18.2** Aids for the CONWIP control

released into the production without an accompanying CONWIP card. The card remains with the order throughout its entire processing on the production line. After the order is completed, the CONWIP card is freed and authorizes the release of another order. The CONWIP cards thus limit WIP in the production line. The WIP corresponds to the number of CONWIP cards, when the following conditions are met:

1. The release list always contains at least one order when a CONWIP card is freed.
2. The orders are released immediately after a CONWIP card becomes available (information throughput time = 0).

CONWIP can control both orders triggered by a customer order as well as make-to-stock orders. Similar to with CORMA (see Chap. 10), if the make-to-stock orders are generated with the Order Point System, the orders can be generated already before the stock order point is reached. The planned start date for the make-to-stock orders is then:

$$DPS = TPL_0 + \frac{SL + \sum SO_{open} - SOP}{RDMD_m} \qquad (18.2)$$

where

| | |
|---|---|
| DPS | planned start date of order [SCD], |
| $TPL_0$ | planning time point [SCD], |
| SL | stock level [units], |
| $SO_{open}$ | open stock orders (already released orders) [units], |
| SOP | stock order point [units], |
| $RDMD_m$ | mean demand rate [units/SCD]. |

Based on an example (Fig. 18.3), let us now take a closer look at the procedural rules for CONWIP. The production line being controlled has four workstations WS 1 to WS 4 and five variants – labeled here from A to E. The number of CONWIP cards is limited to eight. In the initial situation all eight of the CONWIP cards (and correspondingly eight orders) are in production. The release list indicates the sequencing of the variants for the next five releases (Fig. 18.3a).

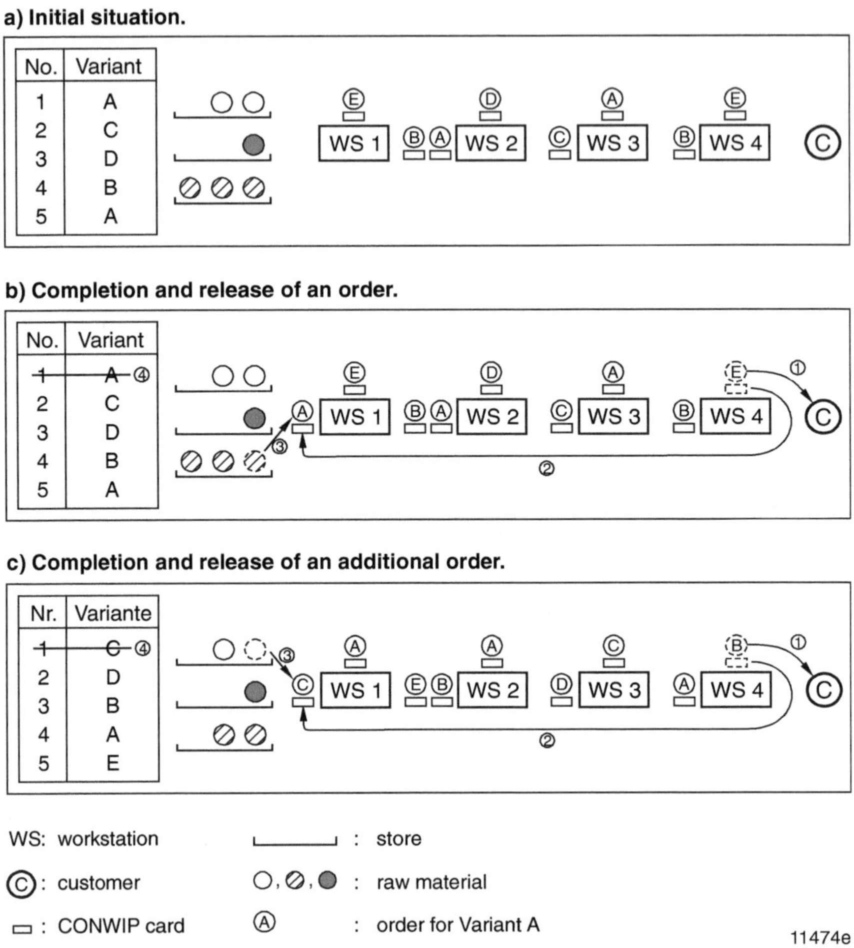


**Fig. 18.3** How CONWIP functions (example)

With the completion of the order for Variant E and Workstation 4 (Step 1 in Fig. 18.3b) a CONWIP card is freed (Step 2) authorizing the release of an order for Variant A as well as the provision of materials (Step 3). CONWIP thus does not necessarily release the same variant for production as was completed. The released order is removed from the release list (Step 4).

By the time the next order is completed (Step 1 in Fig. 18.3c) Workstations 1, 2 and 3 have each finished an operation and begun processing a new order. With the completion of the order a CONWIP card is freed (Step 2), so that CONWIP releases an order for Variant C into the production (Step 3). The released order is in turn removed from the release list (Step 4).

## 18.2 Discussion of the Procedural Rules

One of the characteristics of CONWIP is the WIP control. It thus eliminates one of the fundamental weaknesses of releasing orders immediately or releasing orders according to due date. Next, I would like to examine the procedural rules for CONWIP based on the guidelines for designing manufacturing control methods (see Sect. 3.3.2).

*1. A manufacturing control method should allow the WIP of a production or workstation to be regulated at a defined level.*

CONWIP allows WIP in a production line to be regulated at a defined level. The WIP corresponds to the number of CONWIP cards. Due to the constant WIP the throughput times of the orders in the production can be predicted well and thus easily planned. CONWIP, however, is not able to regulate the WIP on a workstation in the production at a defined level. In a production line though, this is also not necessary for the predictability of throughput times. The WIP levels and throughput times on the workstations are negatively correlated: Since WIP in the entire production line is constant, above average WIP levels and throughput times on one workstation are inevitably linked to low values on other workstations. WIP and throughput time fluctuations on individual workstations thus do not have an impact on the entire production line (see [Hopp-96*]).

In a production with a complex material flow, the orders flow through different workstations. WIP and throughput time fluctuations on the workstations therefore do not inevitably compensate for one another. Accordingly, the variance of the throughput times increases [Lödd-01].

*2. A manufacturing control method should reduce WIP fluctuations to a minimum.*

The CONWIP control's influence on WIP fluctuations generally results from the discussion of the first guideline: The control method minimizes WIP fluctuations for the entire production line, however, WIP levels on the individual workstations can vary.

*3. A manufacturing control method should cause as little blocked WIP as possible.*

CONWIP does not cause any blocked WIP in the throughput of orders in the production.

*4. A manufacturing control method should contribute to closely aligning the load to the capacities.*

Load balancing accelerates or delays the release or throughput of orders with the aim of preventing breaks in the material flow (accelerates) or a build-up of WIP (delays). CONWIP supports this type of load balance for the production line as a whole. It releases orders early if the WIP would otherwise fall below the planned WIP and holds back orders whose planned start date has been reached when releasing them would mean exceeding the planned WIP of the production line.

CONWIP, however, does not consider the load situation on the individual workstations. That is, it does not specifically release those orders which would load workstations threatened by a break in the material flow. Balancing the load on specific workstations in a production line can only be realized by taking into consideration the different work contents. With complex material flows, differences in the order throughput times can also be exploited (see the methods described in Chaps. 20, 21 and 22).

It is possible though to extend CONWIP to balance the loads on specific workstations. In order to do so the orders in the release list have to be sequenced so that a load balance is supported.

5. *A manufacturing control method should cause as few sequencing interchanges as possible.*

CONWIP does not cause any sequencing interchanges either when releasing the orders or when the orders are flowing through the production. In order to reduce the variance of order throughput times, Hopp and Spearman expressively recommend processing orders in the sequence of the orders' input to the production line (so called 'First-In-First-Out' rule [Hopp-96*]).

6. *A manufacturing control method should be able to balance schedule deviations within the context of backlog control.*

It is the responsibility of manufacturing control, to prevent a backlog and to reduce a backlog that has already developed. In order to do so capacities are temporarily increased. CONWIP does not include any component for regulating backlogs and should thus be combined with one in order to ensure a high schedule reliability even when there are plan deviations.

7. *A manufacturing control method should take into consideration the bottleneck principle.*

The procedural rules for CONWIP do not take into consideration the bottleneck principle. Hopp and Spearman argue that it is also not necessary to since the orders will automatically build-up at the current throughput bottleneck. The resulting WIP protects the throughput bottleneck from a loss of utilization [Hopp-96*]. However, when there is a clearly defined throughput bottleneck, it seems obvious that WIP should only be controlled up to the bottleneck. A number of simulation trials show that with a comparable WIP level such a variation of the production rules attains a higher output rate than CONWIP [Enns-02*, Gill-02*, Wein-88*]; Roderick, Philipps und Hogg [Rode-92*] however, obtained contradictive results. The resulting method is described in professional literature under various names [Grav-95*]. In this book it will be described as the Bottleneck Control in the next chapter.

8. *A manufacturing control method should be as simple as possible.*

CONWIP is very simple. The procedural rules can be easily communicated and a minimal number of parameters are involved: Only the advance release window and the number of CONWIP cards need to be set. Both of these facilitate implementing it on the shop floor.

## 18.3 Setting the Method's Parameters

The most important parameter for CONWIP is the number of CONWIP cards that control the WIP in the production line and thus also the order throughput times. In addition, the length of the advance release window has to be set. It determines the maximum length of time within which an order is allowed to be released early for production. The following two sections each describe different methods for determining the number of CONWIP cards. Section 18.3.1 explains how the number of CONWIP cards in use in the production can be determined. Section 18.3.2 introduces a model based method centered on a Logistic Positioning for determining the number of CONWIP cards. Finally, in Sect. 18.3.3 the factors which influence the length of the advance release window are discussed.

### *18.3.1 Number of CONWIP Cards (Operational Method)*

This method for setting the number of CONWIP cards is recommended by Hopp and Spearman [Hopp-96*]. It is oriented on a philosophy made known and proven by Kanban in which the WIP is gradually reduced in order to reveal and resolve any process disruptions (see Sect. 2.2.1). This procedure is depicted in Fig. 18.4.

In the first step, an enterprise sets an initial WIP level that is known to be sufficient (or assumed to be) and provides the corresponding number of CONWIP cards. Over time the enterprise lowers the number of CONWIP cards and measures

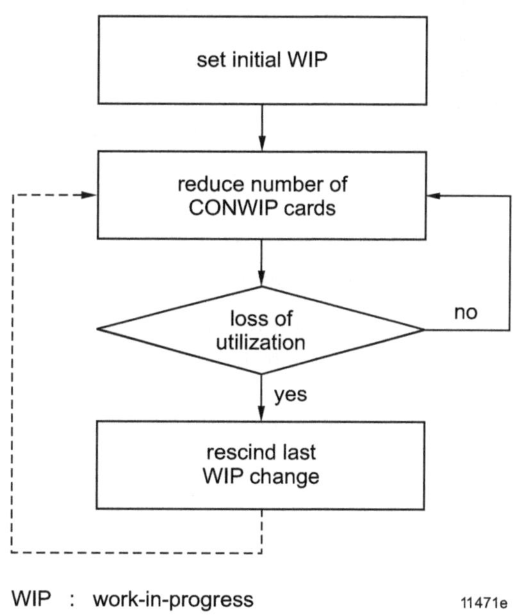


**Fig. 18.4** Gradually reducing the number of CONWIP cards

whether or not there is a loss of utilization or a threat of one. One measure of this is the length of queues before the workstations with critical capacities [Hopp-96*]. If there is no loss of utilization the enterprise can further reduce the number of CONWIP cards, otherwise, it increases the number of CONWIP cards again to the previous value. The enterprise should try at least over the medium term to further reduce the WIP (dashed line Fig. 18.4). In order to do so, measures which allow the production line to operate with a low WIP level should be introduced.

An enterprise can set the initial WIP at the number of orders found in the production when the method is introduced. With this procedure, each order in the production line then receives a CONWIP card. A new order is then only allowed to be released when an order is completed and a CONWIP card is freed [Hopp-96*].

A second procedure consists of deriving the number of CONWIP cards according to Eq. 18.3 from the applicable planned throughput times of the production line. When the production has a backlog, the resulting number of CONWIP cards is thus lower than the above recommendation if the orders up to now were released according to due dates.

It is very important to align the parameters of the higher level planning methods with the number of CONWIP cards. This applies in particular for the planned throughput times which are used by almost all general order generating methods for setting the method parameters. Only when the planned throughput times are coupled with the number of CONWIP cards can an enterprise exhaust the advantages of a WIP reduction in the production.

The planned throughput times of a production law can generally be determined according to Little's Law:

$$TTP_{m,plan} = \frac{NoCONWIP}{ROUT_{m,plan}} \tag{18.3}$$

where

| | |
|---|---|
| $TTP_{m,plan}$ | mean planned throughput time [SCD], |
| NoCONWIP | number of CONWIP cards [−], |
| $ROUT_{m,plan}$ | mean planned output rate [−/SCD]. |

The method described here for setting the number of CONWIP cards at an initial level and then gradually reducing them has three important advantages:

- It allows the philosophy of continual improvement to be transferred to manufacturing control.
- It is oriented on the real production line, thus compared to a model based method, no modeling errors or imprecision can arise.
- It is very simple.

However, there are also disadvantages to this method. The slope of the Utilization Operating Curve in the transitional zone is already quite low thus reducing the WIP leads to only a minimal loss of utilization. On the one hand, this causes a measurement problem: A relatively long evaluation period can then be necessary in order to precisely determine the utilization. On the other hand, the target conflict requires a Logistic Positioning. This is usually not consciously done with a gradual reduction of the number of CONWIP cards though. Rather it is dependent on the level of security needed by the operator responsible for setting the number of CONWIP cards. An operator who shies away from risks prefers a higher WIP level and therefore higher order throughput times than an operator who enjoys taking risks. In both cases, the WIP levels, utilization and throughput times do not inevitably agree with the strategic targets. Thus, it is generally practical to supplement the gradual reduction of the number of CONWIP cards with a model based procedure.

### *18.3.2 Number of CONWIP Cards (Model Based Procedure)*

The general idea of the model based method is to set the number of CONWIP cards within the context of a Logistic Positioning oriented on the Logistic Operating Curves. Four steps are required for doing so (see Fig. 18.5):

1. *Logistic Operating Curves are calculated for each workstation.*
   The calculation of the Logistic Operating Curves is extensively discussed in Sect. 3.2.2.3. It should be noted that one of the prerequisites here is that the orders are transported lot-wise between the workstations (see Sect. 5.3.1). When there is a one-piece-flow a Logistic Operating Curve has to be calculated for the entire production line [Lödd-01b].

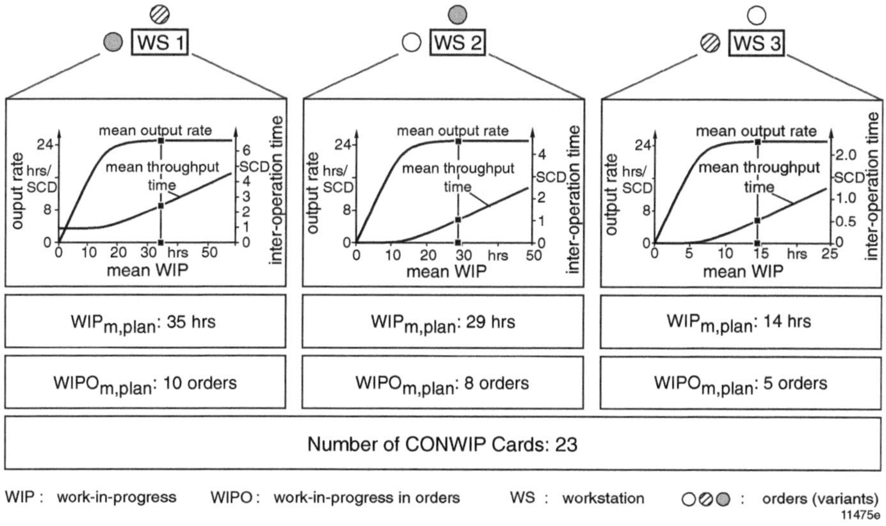


**Fig. 18.5** Deriving the number of CONWIP cards (Example)

2. *The planned WIP (in planned hours) is determined for each workstation in a Logistic Positioning.*
3. *The planned WIP for each workstation is converted into 'number of orders' (units).*
   The conversion is conducted using the following rule [Nyhu-09*]:

$$WIPO_m = \frac{WIP_m}{WC_m} - \frac{U_m \cdot WC_v^2}{100} \tag{18.4}$$

   where

| | |
|---|---|
| $\mathrm{WIPO_m}$ | mean WIP in number of orders [−], |
| $\mathrm{WIP_m}$ | mean WIP (in planned hours) [hrs], |
| $\mathrm{WC_m}$ | mean work content [hrs], |
| $\mathrm{WC_v}$ | coefficient of variance for the work content [−], |
| $\mathrm{U_m}$ | mean WIP dependent utilization [%]. |

4. *The planned WIP levels of the workstations are added together and where applicable rounded to a whole number.*
   The resulting throughput time is then derived according to Eq. 18.3. The number of CONWIP cards can also be very elegantly determined by a Logistic Positioning with the aid of a Manufacturing System Operating Curve for the entire production area being observed. A Manufacturing System Operating Curve depicts the output rate of a production area over the mean WIP. The basis for calculating the Manufacturing System Operating Curve is presented by Schneider [Schn-04, Schn-09*]. The advantage here is that the Logistic Positioning can then be conducted with the aid of only one Logistic Operating Curve.

### *18.3.3 Advance Release Window*

The advance release window determines how far in advance an order can be released for production before the actual planned start date has been attained. CONWIP releases an order early under two conditions:

1. *Problems with Supplying Materials:* If an order cannot be released because the materials (or resources) are not available, CONWIP pulls the next order forward. As a result it not only ensures the utilization of the production line but also that the input of the materials for the order held back does not delay the release of (also urgent) orders: The remaining orders are released early up until the materials arrive. A gap is thus created for the order that was held back. CONWIP leads to an automatic re-planning when the order input deviates from the schedule. Generally a shorter advance release window would sufficiently compensate for problems in supplying materials.

2. *The Actual Output Exceeds the Planned Output:* A negative backlog develops. CONWIP uses the available capacities to process available orders early. This can be preferable when the demand is high and any additional production can be sold on the market. In this case a greater value for the advance release window can be practical. In comparison, if the demand and thus the utilization of the production line are low, the disadvantage of orders being completed early and the resulting larger stores of finished goods predominate.

Generally, the advantage of an improved utilization of the capacities has to be carefully considered in comparison to the possible disadvantages of completing orders early. The latter can reduce the delivery reliability if the completion date is outside of the defined due date tolerance. The advance release window and the definition of the due date tolerance should thus be coordinated.

In the industrial practice, the discussion about setting the advance release window is unfortunately often only of academic value: When the production generally has a backlog, controlling the WIP releases the orders later than planned and the advance release window is not taken advantage of.

## 18.4 Implementation Across the Supply Chain

CONWIP is generally not implemented across a company or supply chain. This would require both a very close supply relationship between the enterprises as well as coordinated production lot sizes. An example of this would be the production of components which are begun in one enterprise and completed in another. Constellations such as these tend to be rare in the industry. When the conditions for the application are given however, CONWIP can also be beneficially implemented across the supply chain.

## 18.5 Application

CONWIP has – at least theoretically – a fairly broad range of applicability. Possible reasons for *not* applying CONWIP include:

- When the order generation is independent of the workstation capacities and the material flow is complex, releasing orders with a workstation specific load balance attains a higher output rate with a comparable WIP level than CONWIP.
- When the throughput bottleneck is clearly defined the Bottleneck Control attains a higher output rate with a comparable WIP level.
- When the utilization rate is low it is not necessary to implement CONWIP. Moreover, it may lead to releasing and completing orders early and thus increasing WIP levels.

- When finite scheduling is used for orders and capacities are limited it is at least theoretically possible that the WIP is temporarily greater than the WIP limit. CONWIP is then not able to implement the production plan.

## References


[Bake-84*] Baker, K.R.: The effect of input control in a simple scheduling model. J. Oper. Manag. **4**(2), 99–112 (1984)

[Enns-02*] Enns, S.T., Costa, M.P.: The effectiveness of input control based on aggregate versus bottleneck work loads. Prod. Plan. Control **12**(7), 614–624 (2002)

[Gill-02*] Gilland, W.G.: A simulation study comparing performance of CONWIP and bottleneck-based release rules. Prod. Plan. Control **13**(2), 211–219 (2002)

[Grav-95*] Graves, R.J., Konopka, J.M., Milne, R.J.: Literature review of material flow control mechanisms. Prod. Plan. Control **6**(5), 395–403 (1995)

[Hopp-96*] Hopp, W.J., Spearman, M.L.: Factory Physics. Irwin, Chicago (1996)

[Lödd-01] Lödding, H.: Dezentrale Bestandsorientierte Fertigungsregelung. VDI Progress Reports, Series 2, No. 587, Düsseldorf (2001)

[Lödd-01b] Lödding, H., Hernández, R., Wiendahl, H.-P.: Dimensionierung und Bewertung von Fertigungsinseln mit logistischen Kennlinien. Io Manage. **70**(1–2), 25–32 (2001)

[Meln-89*] Melnyk, S.A., Ragatz, G.L.: Order review/release: research issues and perspectives. Int. J. Prod. Res. **27**(7), 1081–1096 (1989)

[Meln-91*] Melnyk, S., Ragatz, G., Fredendall, L.: Load smoothing by the planning and order review/release systems. A simulation experiment. J. Oper. Manag. **10**(4), 512–523 (1991)

[Nyhu-09*] Nyhuis, P., Wiendahl, H.-P.: Fundamentals of Production Logistics. Theory, Tools and Applications. Springer, Berlin (2009)

[Raga-88*] Ragatz, G.L., Mabert, V.A.: An evaluation of order release mechanisms in a job-shop environment. Decis. Sci. **19**, 167–189 (1988)

[Rode-92*] Roderick, L.M., Phillips, D.T., Hogg, G.L.: A comparison of order release strategies in production control systems. Int. J. Prod. Res. **30**(3), 611–626 (1992)

[Schn-09*] Schneider, M.: Manufacturing System Operating Curves. In: Nyhuis, P., Wiendahl, H.-P.: Fundamentals of Production Logistics, Springer, Berlin, 2002, pp. 112–117

[Schn-04] Schneider, M.: Logistische Fertigungsbereichskennlinien. Ein Wirkmodell zur Beschreibung des logistischen Leistungsverhaltens von Fertigungsbereichen. Dissertation, University of Hanover, (2004)

[Spea-89*] Spearman, M.L., Hopp, W.J., Woodruff, D.L.: A hierarchical control architecture for constant work-in-process (CONWIP) production systems. J. Mfg. Oper. Mgt **2** (3), 147–171 (1989)

[Spea-90*] Spearman, M.L., Woodruff, D.L., Hopp, W.J.: CONWIP: a pull alternative to kanban. Int. J. Prod. Res. **28**(5), 879–894 (1990)

[Spea-92*] Spearman, M.L., Zazanis, M.A.: Push and pull productions systems: issues and comparisons. Oper. Res. **40**(3), 521–532 (1992)

[Wein-88*] Wein, L.M.: Scheduling semiconductor wafer fabrication. IEEE Trans. Semicond. Manuf. **1**(3), 115–130 (1988)

# Chapter 19
# Bottleneck Control

Bottleneck Control is the simplest way to convert the bottleneck principle into a manufacturing control method: Every time a bottleneck workstation completes an order, a Bottleneck Control releases an order for production. Bottleneck Control sub-divides the production into a WIP controlled section up until and including the bottleneck workstation and a not WIP controlled section after the bottleneck workstation. In the following we will discuss this based on the example of a production line.

In the discussion and descriptions of Bottleneck Control found in professional publications very different names are used, some of which are also related to other methods for planning and controlling production. The procedural rules for it are similar to CONWIP. Nonetheless, unlike CONWIP, Bottleneck Control only regulates the WIP up to the bottleneck of the production line. Bottleneck Control is especially practical when there is a clearly definable bottleneck workstation in the production line. Whereas [Enns-02*, Gill-02*, Wein-88*] found that in this case it attains a higher output rate than CONWIP with a comparable WIP, the simulations by Roderick, Philipps and Hogg found CONWIP fared better [Rode-92*].

The following section presents the procedural rules for Bottleneck Control (Sect. 19.1). In Sect. 19.2 we will then take a look at these in view of attaining the logistic targets, after which we will consider setting the parameters for this method in Sect. 19.3. Due to the extensive similarities with CONWIP, readers are referred to the corresponding explanations for CONWIP. The chapter concludes with a discussion about applying Bottleneck Control in order to control production across the supply chain (Sect. 19.4) as well as with a few very brief comments about experiences in applying it (Sect. 19.5).

A number of authors refer to the similarity of Bottleneck Control to the drum, buffer and rope concept that orients the program planning and scheduling of orders on the bottleneck workstations in the production. This philosophy, primarily developed by Goldratt, is described in detail in [Gold-84*, Schr-90*, Spen-95*]. The most important characteristics will be discussed in an excursus found in Sect. 19.6.

H. Lödding, *Handbook of Manufacturing Control*, 347
DOI 10.1007/978-3-642-24458-2_19, © Springer-Verlag Berlin Heidelberg 2013

## 19.1 Method Description

Following, three versions of Bottleneck Control will be described:

- the basic method (Sect. 19.1.1),
- bottleneck control with multiple bottlenecks (Sect. 19.1.2) and
- starvation avoidance (Sect. 19.1.3).

### *19.1.1 Basic Method*

Bottleneck Control releases an order as soon as the WIP in the production line up to and including the bottleneck workstation falls below the planned value (Fig. 19.1).

Similar to CONWIP, it selects the most urgent order from the list of orders to be released (see Table 17.1 as well as Fig. 18.2a). Furthermore, an advance release window can be defined within which orders are allowed to be released early (see Sect. 18.1).

The WIP is measured in either ‘number of parts’, ‘number of orders’ or ‘planned hours’ on the bottleneck workstation of the production line. Wein utilizes the latter and refers to the resulting method as ‘workload regulating’ [Wein-88*]. The following remarks however are based on the condition that WIP is measured in ‘number of orders’.

Bottleneck Control can be implemented with the aid of bottleneck cards without which no order is allowed to be released. The bottleneck card accompanies an order from its release up to the end of its processing on the bottleneck workstation. At this point the bottleneck card is freed and authorizes the release of a further order. The bottleneck cards thus limit the WIP in the production line up to the bottleneck workstation.

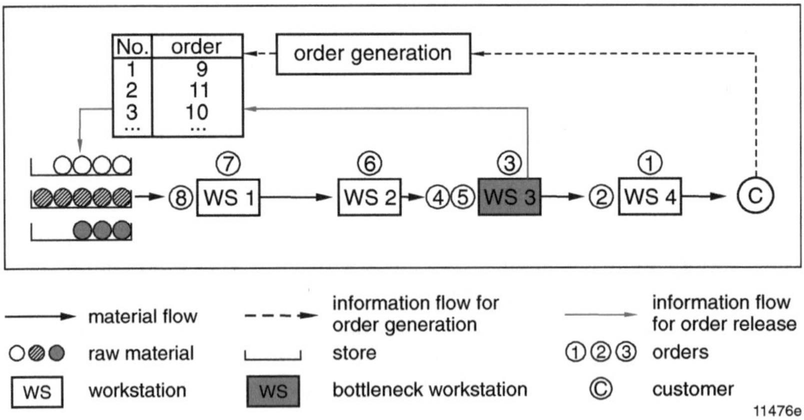


**Fig. 19.1** Principle of Bottleneck Control

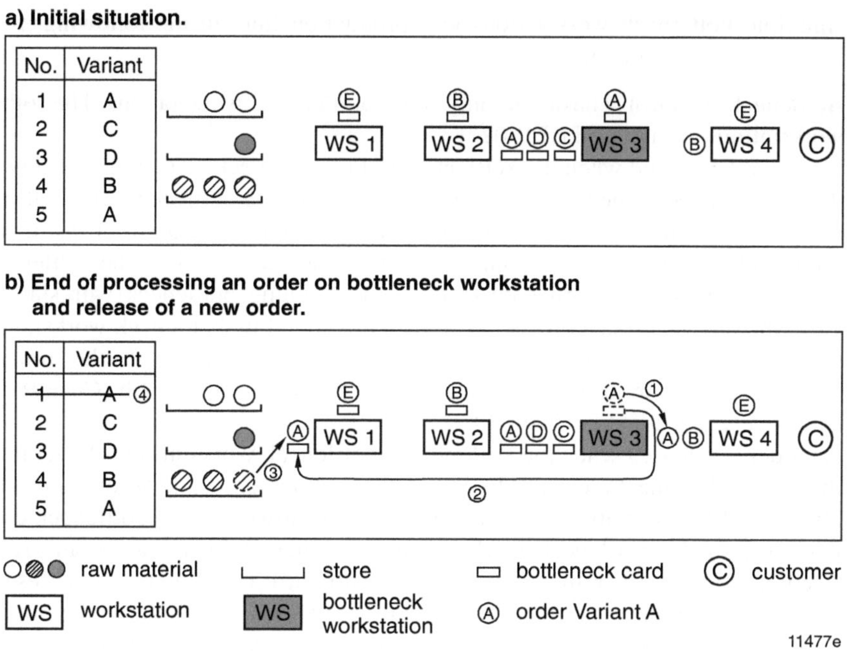


**Fig. 19.2** How bottleneck control functions (example)

In order to better understand Bottleneck Control let us consider the procedural rules by looking at an example (Fig. 19.2).

In the example, a production line with four workstations WS 1 to WS 4 is managed using a Bottleneck Control system. Workstation 3 is the production line‘s throughput bottleneck. In the initial situation, there are eight orders in the production line (Fig. 19.2a). Six of them are either on or before the bottleneck workstation and thus have a bottleneck card.

Once the bottleneck workstation has completed processing the order for Variant A (Step 1 in Fig. 19.2b), the order enters the queue for Workstation 4. The freed bottleneck card triggers the release of the next order from the release list (Steps 2 und 3). Once released the order is removed from the release list (Step 4).

### *19.1.2 Bottleneck Control with Multiple Bottlenecks*

There have been a number of methods developed and analyzed that extend the bottleneck control. These especially take into consideration situations where there are multiple bottleneck workstations in a production line. In the following, four variations will be outlined:

1. Bottleneck Control considers only one bottleneck workstation. The other workstations are ignored [Gill-02*].
2. An order is released when the WIP on all of the bottleneck workstations is lower than the planned value [Gill-02*]: In this case all of the bottleneck workstations are allocated different bottleneck cards. The bottleneck cards from *all* of the bottleneck workstations are required for the release of an order. The bottleneck cards remain with the order up to the corresponding bottleneck workstation. Once the order has been processed on the corresponding bottleneck workstation the bottleneck cards are freed and can authorize the release of further orders. This method is similar to a recommendation from Chang and Yih [Chan-94*], extensively discussed in Chap. 20.
3. An order is released when the WIP on any bottleneck workstation is lower than the planned value [Wein-88*]: All of the bottleneck workstations are also allocated different bottleneck cards in this case. However, *one* bottleneck card from any of the bottleneck workstations is sufficient for releasing an order. The bottleneck cards remain with the order up until it is done being processed on the corresponding bottleneck workstation. Once it is completed the bottleneck card is freed. If possible, the card is then allocated to one of the orders already released that does not have a bottleneck card for this workstation and which is either at or before the bottleneck workstation. Orders such as these have already been released due to bottleneck cards from other bottleneck workstations. If all of the released orders before this bottleneck workstation have a bottleneck card, then this bottleneck card releases a new order from the release list.
4. The WIP before all of the bottleneck workstations is added together and an order is released when the sum of this WIP falls below a defined planned value [Wein-88*]: In this case, the same bottleneck card is used for all bottleneck workstations. An order is only allowed to be released if the number of available bottleneck cards corresponds to at least the number of bottleneck workstations. As the order flows through the production line, a bottleneck card is freed each time an operation is completed on a bottleneck workstation.

In a simulation study with two bottleneck workstations Gilland compared Variations 1 and 2 as well as CONWIP. Variation 2 performed better than CONWIP and Variation 1 in all of the simulation trials. CONWIP attained better results with a high utilization rate than Variation 1, and poorer results than Variation 1 when there was a low utilization rate [Gill-02*]. In a simulation study by Wein, Variations 3 and 4 performed better than CONWIP [Wein-88*]. Van Enns and Prongué Costa deduce from simulations that the clearer the bottleneck is defined and the more complex the material flow, the greater the advantages of a Bottleneck Control [Enns-02*].

### 19.1.3 Starvation Avoidance

Starvation avoidance was developed by Glassey and Resende for the production of integrated circuits (ICs) [Glas-88*]. This variation of Bottleneck Control measures the WIP in planned hours on the bottleneck workstation. Starvation avoidance is generally suitable for all productions with a clearly defined bottleneck workstation and the following characteristics:

- The bottleneck workstation consists of a number of the same types of machines that are subject to disruptions.
- The orders flow through the bottleneck workstation a number of times.

Both of these characteristics are taken into consideration by special procedural rules.

***(a) Disruptions on the Bottleneck Workstations***

If one or more of the workstation's machines are subject to disruptions, the capacities of the bottleneck workstation are reduced. Correspondingly, a lower WIP is sufficient for ensuring the workstation is supplied. Glassey and Resende thus add a virtual WIP portion to the measured WIP. This corresponds at every point in time to the product of the mean time for repairs and the number of disrupted machines on the bottleneck workstation [Glas-88*]:

$$WIP_{virt} = WIP + NDeMa \cdot TTR_{m,bn} \quad (19.1)$$

where

| | |
|---|---|
| $WIP_{virt}$ | virtual WIP [hours], |
| WIP | WIP [hours], |
| NDeMa | number of defect machines at the bottleneck workstation [−], |
| $TTR_{m,bn}$ | mean time to repair at the bottleneck workstation [hours]. |

When one or more of the bottleneck workstation's machines are disrupted the release of an order is triggered later than when their are no disruptions. Correspondingly, a lower WIP in the production ensues.

***(b) Orders Flow through the Bottleneck Workstation Multiple Times***

Glassey and Resende consider orders that flow through a bottleneck workstation multiple times by calculating the virtual WIP as follows:

- An order is taken into consideration in the WIP on a workstation with a maximum of one operation (with the work content of the next operation on the bottleneck workstation).
- An order will not be considered in the (virtual) WIP if the anticipated length of time that the order requires before it enters the bottleneck workstation, is greater than the expected time it takes for a newly released order to enter the bottleneck workstation. As a measure of the length of time, Glassey and Resende use the

sum of the work content for the operations up to the workstation. That means, when the sum of the work content of the operations before the bottleneck workstation for an order exceeds a limit, the order is not taken into consideration. The limit corresponds to the sum of the work content for an average order from its release up to the operation on the bottleneck workstation.

In simulations, starvation avoidance performed better than the basic bottleneck control, CONWIP and when orders were released in set intervals (so-called 'Clockwork Control') [Glas-88*]. At a comparable WIP level, it attains a higher output rate. Yet another method for releasing orders is recommended by Lingayat, Mittenthal and O'Keefe and is based on the same general philosophy of starvation avoidance [Ling-95*]. It too attains better results than CONWIP in experiments.

## 19.2 Discussion of the Procedural Rules

Two properties characterize the bottleneck control: First, that it controls the WIP in a part of the production line and second, that it is oriented on the production's throughput bottleneck. In this next section, we will once again take a closer look at the procedural rules for Bottleneck Control in relation to the guidelines for designing manufacturing control methods (see Sect. 3.3.2). For our discussion here we will focus on the basic version.

*1. A manufacturing control method should allow the WIP of a production or workstation to be regulated at a defined level.*

Bottleneck Control divides the production line into an area up to and including the bottleneck workstation in which the WIP is controlled and an area following the bottleneck station in which the WIP is not controlled. It regulates the WIP at a defined level up to the bottleneck workstation. Generally, the WIP corresponds to the number of bottleneck cards.[1] Due to the constant WIP the order throughput times can be predicted well up to the production's bottleneck. When the WIP levels and throughput times for the production line following the bottleneck workstation do not vary greatly, this also applies for the total order throughput time. Bottleneck Control thus supports a high delivery reliability. Similar to CONWIP, Bottleneck Control is not able to regulate the WIP at a defined level on the workstations in the production line (see the discussion on the first guideline in Sect. 18.2).

*2. A manufacturing control method should reduce WIP fluctuations to a minimum.*

Bottleneck Control reduces WIP fluctuations to a minimum in the production line in the area up to the bottleneck workstation in which the WIP is controlled. The extent

[1] This applies under two conditions: (1) There is always one order available that can be released for production. (2) Immediately after the completion of an order on the bottleneck workstation, the bottleneck cards are available for releasing a new order.

of the WIP fluctuation is essentially influenced by the area in which the WIP is not controlled. The more frequently workstations in this area become temporary production bottlenecks, the greater the WIP fluctuations are.

3. *A manufacturing control method should cause as little blocked WIP as possible.*
Bottleneck Control does not cause any blocked WIP during the throughput of orders in a production.

4. *A manufacturing control method should contribute to closely aligning the load to the capacities.*
Aligning the load to the capacities accelerates or delays the release or throughput of orders with the aim of preventing breaks in the material flow (accelerates) or a build-up of WIP (delays). Bottleneck Control supports this for the portion of the production line up to and especially for the bottleneck workstation(s). It releases orders early in order to prevent the WIP from falling below the planned value. Conversely, it holds back urgent orders if releasing them would cause the WIP to exceed the planned value.

5. *A manufacturing control method should cause as few sequencing interchanges as possible.*
Bottleneck Control does not cause any sequencing interchanges.

6. *A manufacturing control method should be able to balance schedule deviations within the context of backlog control.*
It is the responsibility of manufacturing control, to prevent a backlog and to reduce a backlog that has already developed. In order to do so capacities are temporarily increased. Bottleneck Control does not include a component for regulating backlogs and should thus be combined with a backlog control in order to be able to ensure a high schedule reliability even when there are plan deviations.

7. *A manufacturing control method should take into consideration the bottleneck principle.*
The basic idea of Bottleneck Control is the orientation on the production bottleneck. I would thus like to extensively discuss this here and compare it to CONWIP which refrains from this bottleneck orientation.

In various simulation studies, Bottleneck Control attains a higher output rate than CONWIP when the bottleneck workstation is clearly defined and the WIP levels are comparable [Gill-02*, Wein-88*]. A workstation is clearly a bottleneck workstation when its mean utilization is notably higher than all of the remaining workstations. Even when there is a clear (static) bottleneck workstation in a production line, other workstations can be temporary (dynamic) bottlenecks due to the variance of the work content and due to disruptions.

The possible advantage of Bottleneck Control's output rate compared to CONWIP is mainly determined by two criteria.

(a) *Position of the Bottleneck Workstation:* The further upstream in the material flow the bottleneck workstation is, the greater the output rate advantage of

Bottleneck Control is. If the bottleneck workstation is the last workstation in a production line, then Bottleneck Control and CONWIP are identical. Accordingly, they are also similar with regards to attaining the logistic targets.

(b) *Utilization of the Non-Bottleneck Workstations:* If the non-bottleneck workstations of a production line are also highly utilized then the output rate advantage of Bottleneck Control disappears and CONWIP may even attain a higher output rate [Gill-02*].

The possible output rate advantages of a Bottleneck Control can be explained by the particular significance of the bottleneck workstation for the output of the entire production line. The loss of utilization on a bottleneck workstation inevitably leads to a loss of output for the entire production line. In order to avoid losses of utilization, sufficient WIP has to be maintained on the bottleneck workstation. With the bottleneck control, every output from the WIP on the bottleneck workstation (caused by the completion of an order on the bottleneck workstation) *immediately* triggers the release of an order in the production line (see Fig. 19.3a).

Therefore, the time between the completion of an order on a bottleneck workstation and the arrival of a new order is short. It corresponds to the sum of the throughput times for the operations before the bottleneck workstation. Moreover, the variance of this time is limited by controlling WIP up to the bottleneck.

In comparison, CONWIP releases a new order only when the entire order is completed (Fig. 19.3b). The length of time from the completion of an order on the bottleneck workstation to the arrival of the newly released order is increased by the information turnaround time. This corresponds to the throughput time of the operations after the bottleneck workstation.

The more upstream the bottleneck workstation is located in the material flow, the greater the information turnaround time is. Moreover, it varies with WIP levels on workstations following the throughput bottleneck. The variance of the WIP in this section of the production line is especially high when one of the workstations is temporarily a bottleneck for the production line or is subject to disruptions.

In both cases, WIP builds-up and the time between the end of an order being processed on the bottleneck workstation and the release of a new order is extended. Thus with CONWIP a disruption or a temporary bottleneck can also cause breaks in the material flow on the bottleneck workstation when the workstation subject to the disruption is positioned behind the bottleneck workstation in the material flow. This is excluded with Bottleneck Control. When the utilization of non-bottleneck workstations is high though, the disadvantages of Bottleneck Control can dominate. The limitation to only one bottleneck[2] and the lack of a WIP control in the production line following the bottleneck workstation are disadvantageous.

[2] A number of variations of the method are recommended for extending the bottleneck control to multiple bottlenecks; please refer to the comments in Sect. 19.1.2 regarding this.

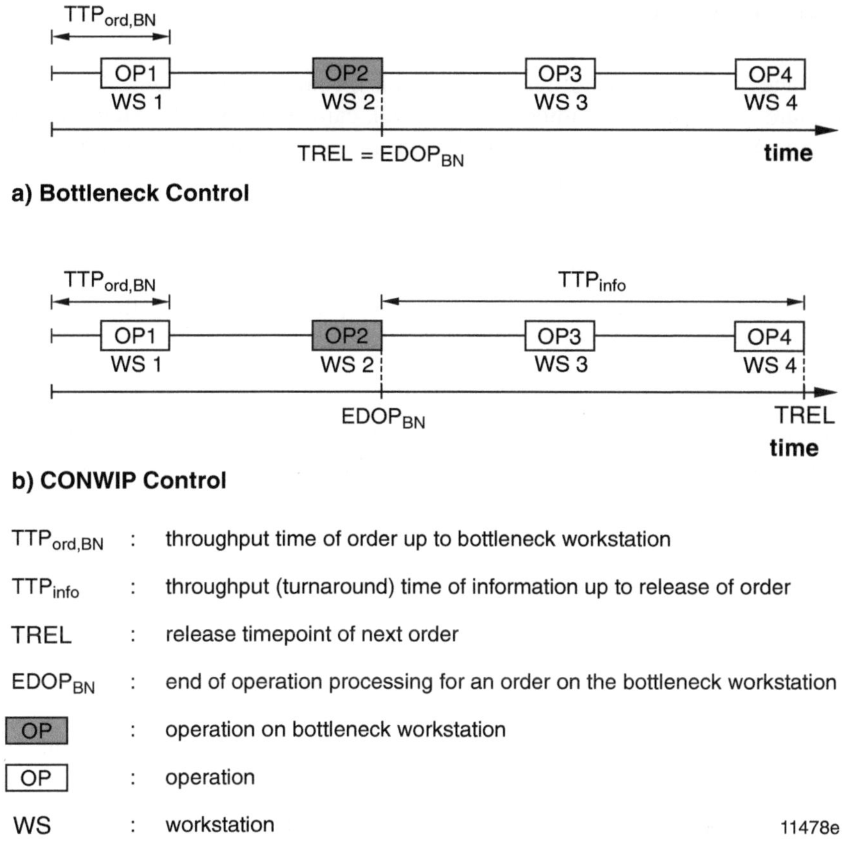


**Fig. 19.3** Comparison of the release times of Bottleneck Control and CONWIP

*8. A manufacturing control method should be as simple as possible.*
Bottleneck Control is very simple. The procedural rules can be easily communicated and the method has few parameters: Only the advance release window and the number of bottleneck cards need to be set. Both of these support implementing the Bottleneck Control on the shop floor.

## 19.3 Setting the Method's Parameters

In the following we will discuss how to set the method's parameters for both the basic method (Sect. 19.3.1) as well as for the starvation avoidance variation of the method (Sect. 19.3.2).

### 19.3.1 Basic Method

One of the parameters of Bottleneck Control is the number of bottleneck cards. The procedure for setting the number of bottleneck cards corresponds extensively with that of CONWIP. Unlike CONWIP though, the workstations after the bottleneck workstation remain unconsidered.

Due to the extensive similarities to CONWIP the procedure will only be explained very briefly here and can be supplemented with the corresponding section in the chapter on CONWIP (Sect. 18.3). In turn, two different procedures can be identified for setting the parameters: one takes place on the shop floor while the other is model based. On the shop floor the number of bottleneck cards is set at an initial value and then gradually reduced (see Fig. 18.4 as well as Sect. 18.3.1). The starting point in the model based procedure is the Logistic Positioning of the workstations with the aid of Logistic Operating Curves. Unlike CONWIP, workstations following the bottleneck workstation are not taken into consideration. Accordingly, the number of bottleneck cards is calculated:

$$NBNC = \sum_{i=1}^{BN} WIPO_{m,plan,i} \tag{19.2}$$

where

| | |
|---|---|
| NBNC | number of bottleneck cards [−], |
| $WIPO_{m,Plan,i}$ | mean planned WIP (in number of orders) on workstation $i$ [−], |
| BN | position of bottleneck [−]. |

It is reasonable to select a higher (relative) WIP for the positioning of the bottleneck workstation than for the remaining workstations in the production line. This WIP buffer should protect the bottleneck workstation from breaks in the material flow.

A significant difference from CONWIP also arises for the scheduling of the orders. With Bottleneck Control the order throughput time has two components: The throughput time for the part of the production up to and including the bottleneck workstation that is WIP controlled and the throughput time of the part following the bottleneck workstation, where the WIP is not controlled. In the part where the WIP is controlled, the planned throughput time is calculated according to Little's Law from the ratio of the planned WIP (which is equal to the number of bottleneck cards) and the planned output rate. For the remaining workstations the planned throughput time is set by means of a Logistic Positioning (see Sect. 3.2.2.3).

Accordingly, for the planned order throughput time:

$$TTP_{order,plan} = \frac{NBNC}{ROUTO_{m,plan}} + \sum_{i=BN+1}^{NWS} TTP_{plan,i} \tag{19.3}$$

where

| | |
|---|---|
| $TTP_{order, plan}$ | planned order throughput time [SCD], |
| NBNC | number of bottleneck cards [−], |
| $ROUTO_{m,plan}$ | mean planned output rate (in number of orders) [−/SCD], |
| $TTP_{plan,i}$ | planned throughput time of the $i^{th}$ workstation [SCD], |
| BN | position of the bottleneck workstation [−], |
| NWS | number of workstations [−]. |

## 19.3.2 Starvation Avoidance

In Glassey and Resende's variation of the method a new order is released for production when the virtual WIP falls below a WIP limit (see Sect. 19.1.3). The developers of the method suggest setting the WIP limit in analogy to the order point in the Order Point System. In order to so do they calculate the range of the virtual WIP and compare it with the expected throughput time from the point an order is released up to the time it first arrives at the bottleneck workstation. The virtual range results from the quotient of the virtual WIP and the output rate of the bottleneck workstation [Glas-88*].

$$R_{virt} = \frac{WIP_{virt}}{ROUT_{BN}} \tag{19.4}$$

where

| | |
|---|---|
| $R_{virt}$ | virtual range [SCD], |
| $WIP_{virt}$ | virtual WIP [hours], |
| $ROUT_{BN}$ | output rate of the bottleneck workstation [hours/SCD]. |

The virtual range is thus a measure of the length of time for which the bottleneck workstation is supplied with work.

The planned throughput time from the release of an order up until it first arrives at the bottleneck workstation corresponds to the sum of the planned operation throughput times before the bottleneck [Glas-88*].

$$TTP_{plan,1-BN} = \sum_{i=1}^{BN-1} TTP_{plan,i} \tag{19.5}$$

where

| | |
|---|---|
| $TTP_{plan,1\text{-}BN}$ | planned throughput time from the release of an order up until it first arrives at the bottleneck workstation [SCD], |
| $TTP_{plan,i}$ | planned throughput time of the $i^{th}$ operation [SCD], |
| BN | position of the first operation on the bottleneck workstation [−]. |

A new order should be released at the latest when the range of the WIP on the bottleneck workstation falls below this throughput time. In order to compensate for fluctuations in the throughput time or output rate of the bottleneck workstation, a throughput time buffer should also be considered. As a result for the WIP limit [Glas-88*]:

$$WIPL = (TTP_{plan,1-BN} + TTP_{buf}) \cdot ROUT_{BN} \tag{19.6}$$

where

| | |
|---|---|
| WIPL | WIP limit [hours], |
| $TTP_{plan,1\text{-}BN}$ | planned throughput time from the release of the orders up until it first arrives at the bottleneck workstation [SCD], |
| $TTP_{buf}$ | throughput time buffer [SCD], |
| $ROUT_{BN}$ | output rate of the bottleneck workstation [hours/SCD]. |

This procedure can be transferred to the basic Bottleneck Control as well. The planned throughput times of the operations before the bottleneck workstation can be determined with the help of a Logistic Positioning. The throughput time buffer should be set as a function of the throughput time variance.

It is especially important with this procedure to precisely and when possible continuously measuring the WIP. The range of the WIP is actually smaller than the calculated value particularly when an order has already been partially processed but is still counted towards the WIP with its complete work content. Consequently, it is possible that the danger of a break in the material flow is underestimated.

## 19.4 Implementation Across the Supply Chain

Bottleneck Control is only rarely implemented across the supply chain (for more on this see also the corresponding remarks for CONWIP). The basic idea of orienting the production planning and control on the throughput bottleneck can however be transferred to supply chains. The enterprise that represents the throughput bottleneck of the supply chain then needs to be identified and taken especially into consideration in the planning and controlling.

## 19.5 Application

The conditions for applying Bottleneck Control are similar to those for CONWIP (see Sect. 18.5). Unlike CONWIP though, Bottleneck Control requires a clearly defined production bottleneck and attains a higher output rate with a comparable WIP level. Generally, the question is raised as to how many enterprises fulfill the condition of a clearly defined bottleneck workstation. Many enterprises consider the different utilization rates when allocating operators to the workstations. Existing differences in the utilization of the material resource capacities are thus leveled by the allocation of workforce capacities.

Gilland reports that Bottleneck Control is applied by Intel [Gill-02*], however, fails to provide information about the success of the application.

## 19.6 Exkursus: Theory of Constraints

The Theory of Constraints (TOC) was predominantly developed by Goldratt who describes the basic principles of it in his novel "The Goal" [Gold-84*]. The most well known, but not the only area where TOC is applied is in planning and controlling production. Here, due to the persuasiveness of its underlying philosophy and successful marketing of its advocates, it has found a number of followers. Although TOC is not a method for controlling production, I will provide a descriptive outline of the aspect of TOC relevant to PPC. The discussion here is oriented on a paper by Spencer and Cox [Spen-95*].

The basic idea of TOC is the assumption that there is a bottleneck or "constraint"[3] in every production which limits the throughput of the production and therefore should be exploited as much as possible. In comparison, non-bottleneck workstations do not determine the throughput and are thus not critical to success.[4] A goal of TOC is thus identifying the production bottleneck and orienting the entire production planning and control on the bottleneck.

The main component of TOC is a five stage process for focusing on the constraint [Gold-84*]:

1. identify the constraint,
2. fully exploit the constraint's capacity,
3. subordinate all other processes to the constraint,
4. increase the capacities of the constraint in order to increase the throughput,
5. continually improve the system by avoiding inertia in management.

This method can be transferred to the production program planning and to scheduling. Spencer and Cox derive the following seven steps for doing so [Spen-95*]:

[3] In order to align the language of TOC with that used throughout this book and thus facilitate the discussion, we will be using the terms bottleneck and constraint interchangeably in the following discussion.

[4] Windt expands the bottleneck concept to the remaining logistic objectives and demonstrates that different workstations can form the bottleneck depending on the objective [Wind-01].

***1. Determine the constraint by analyzing the capacities.***
Based on the demand (forecast), the part list and the work plans the required capacities for the resources are roughly calculated (see Chap. 4): The production bottleneck (or constraint) is the workstation with the highest calculated utilization of capacities.

***2. Determine which components are routed through the constraint.***
This information can be derived from the work plans. The production program planning is first limited to these components.

***3. Use the profit margin per constraint minute to determine priorities.***
The profit margin and the processing time of the product on the bottleneck (as the sum of the processing times for the incoming components) are calculated for the end product. The product with the highest quotient has the highest priority (see Table 19.1). End products whose components are not routed through the bottleneck are not considered.

***4. Use the priorities along with Gantt charts to set the production program (master production schedule).***
In this step, a master production schedule, which is accurate 'to the minute', is calculated for the bottleneck. In doing so, the components for the end products with the highest priority are given preference when planning. The production schedule for the end products that form the bottleneck's load (through the incoming components) results from this planning. An example of this master production schedule is shown in Table 19.2.

**Table 19.1** Calculating the priority of the end product

| Rank [–] | Product [–] | Profit margin [€] | Processing time on bottleneck [min] | Profit margin per constraint minute [€/min] |
|---|---|---|---|---|
| 1 | 4 | 75 | 5 | 15 |
| 2 | 1 | 72 | 6 | 12 |
| 3 | 3 | 40 | 4 | 10 |
| 4 | 2 | 40 | 10 | 4 |

**Table 19.2** Master production schedule for the bottleneck (based on Spencer and Cox)

| No. | Component | End product | Processing time | Start | | End | |
|---|---|---|---|---|---|---|---|
| | | | | Day | Time | Day | Time |
| [–] | [–] | [–] | [min] | [SCD] | [hh:mm] | [SCD] | [hh:mm] |
| 1 | C4 | 4 | 175 | 2 | 00:00 | 2 | 02:55 |
| 2 | C1 | 1 | 105 | 2 | 02:55 | 2 | 04:40 |
| 3 | C4 | 4 | 435 | 2 | 04:40 | 2 | 11:55 |
| 4 | C3 | 3 | 435 | 2 | 11:55 | 2 | 19:10 |
| 5 | C1 | 1 | 243 | 2 | 19:10 | 2 | 23:13 |
| 6 | C2 | 2 | 305 | 2 | 23:13 | 3 | 04:18 |

5. ***Schedule all end products not containing components routed across the constraint evenly in the master production schedule.***

Such products are referred to as 'free goods' [Spen-95*]. These products play a subordinate role in the production planning and control and are continually produced in the required quantities.

6. ***Develop a material release plan using backwards scheduling from the constraint and include a constraint buffer.***

The planned start date of the production orders for the components result from the material release plan. They are determined using a simple backwards scheduling. In order to do so the planned throughput time from the release of the order up to its arrival at the workstation are deducted from the scheduled processing starts on the bottleneck workstation. This time is referred to as the constraint (or bottleneck) buffer.

Table 19.3 depicts the release times for the raw materials calculated from the master production schedule.

The bottleneck buffer is consciously generous, so that the master production schedule can be upheld even when there are schedule deviations. Spencer and Cox recommend starting with a value equal to five times the sum of the operation times on the preceding workstations [Spen-95*].[5]

7. ***Developing the shipping schedule by forward scheduling from the constraint and including a shipping buffer.***

This results in the planned delivery date for the end product. The planned throughput times for the remaining operations are added to the scheduled end of a components processing on the bottleneck. This is referred to as the shipping buffer.

**Table 19.3** Calculations for the release times (constraint buffer = 9 h) (based on Spencer and Cox)

| No. | Component | Raw material | Start | | Release | |
|---|---|---|---|---|---|---|
| | | | Day | Time | Day | Time |
| [–] | [–] | [–] | [SCD] | [hh:mm] | [SCD] | [hh:mm] |
| 1 | C4 | R4 | 2 | 00:00 | 1 | 15:00 |
| 2 | C1 | R1 | 2 | 02:55 | 1 | 17:55 |
| 3 | C4 | R4 | 2 | 04:40 | 1 | 19:40 |
| 4 | C3 | R3 | 2 | 11:55 | 2 | 02:55 |
| 5 | C1 | R1 | 2 | 19:10 | 2 | 10:10 |
| 6 | C2 | R2 | 2 | 23:13 | 2 | 14:13 |

[5] Schragenheim and Ronen recommend three times the average throughput time from the order release up to the arrival on the bottleneck [Schr-90*]. The suggestion from Spencer and Cox seems more appropriate as a starting value, however, since it prevents the vicious cycle of manufacturing control in that extending the planned throughput times causes extended actual throughput times.

**Table 19.4** Calculations for the planned delivery date (shipping buffer 12 h) (based on Spencer and Cox)

| No. | Component | End product | End | | Delivery date | |
|---|---|---|---|---|---|---|
| | | | Day | Time | Day | Time |
| [–] | [–] | [–] | [SCD] | [hh:mm] | [SCD] | [hh:mm] |
| 1 | C4 | 4 | 2 | 02:55 | 2 | 14:55 |
| 2 | C1 | 1 | 2 | 04:40 | 2 | 16:40 |
| 3 | C4 | 4 | 2 | 11:55 | 2 | 23:55 |
| 4 | C3 | 3 | 2 | 19:10 | 3 | 07:10 |
| 5 | C1 | 1 | 2 | 23:13 | 3 | 11:13 |
| 6 | C2 | 2 | 3 | 04:18 | 3 | 16:18 |

Table 19.4 depicts the calculations for the delivery date from the bottleneck's master production schedule.

The described planning algorithm can be divided into a production program planning and scheduling which are aligned with one another via the finite bottleneck scheduling. The scheduling portion is frequently referred to as the drum-buffer-rope approach. The production program planning is similar to the well established operations research algorithm for determining the production plan that is optimal for the profit margin (see [Müll-73]). In comparison to the described method, these have the added advantage of being applicable also when there are multiple bottlenecks. More extensive examples of the calculations for TOC method can be found in [Schr-90*, Spen-95*].

When there is a clearly defined bottleneck, TOC method has the following advantages:

- it limits the precision scheduling to the bottleneck and thus reduces the complexity of the planning problem.
- it calculates a practical production program and with that delivery dates that are at least generally realistic.

Nonetheless, just as Crabtree emphasized in a satirical critique [Crab-02*], it does seriously have to be questioned whether or not the profit margin is suitable as the sole criteria for determining the production schedule. Only three critical points are mentioned here: First, the danger of giving preference to rarely demanded variants instead of variants that frequently have lower profit margins but are in high demand increases especially when costs are calculated traditionally. The enterprise would then weaken specifically those products which enjoy a good market and additionally contribute to the highest profit margin sum. Second, a number of enterprises have overcapacities and thus could generally produce the entire marketable production schedule. And third, exclusively focusing on (short term) profit objectives neglects other aspects which enterprises frequently need to consider in planning the production program e.g., the interests of important customers.

# References


[Chan-94*] Chang, T.M., Yih, Y.: Generic Kanban systems for dynamic environments. Int. J. Prod. Res. **32**(4), 889–902 (1994)

[Crab-02*] Crabtree, D.: Capacity Bottlenecks: illusions and false gods. Control Mag. **28**(6), 24–26 (2002)

[Enns-02*] Enns, S.T., Costa, M.P.: The effectiveness of input control based on aggregate versus bottleneck work loads. Prod. Plan. Control **12**(7), 614–624 (2002)

[Gold-84*] Goldratt, E.M., Cox, J.: The Goal. A Process of Ongoing Improvement. Gower, Aldershot (1984)

[Glas-88*] Glassey, C.R., Resende, M.G.C.: Closed-loop job release control for VLSI circuit manufacturing. IEEE Trans. Semicond. Manuf. **1**(1), 36–46 (1988)

[Gill-02*] Gilland, W.G.: A simulation study comparing performance of CONWIP and bottleneck-based release rules. Prod. Plan. Control **13**(2), 211–219 (2002)

[Ling-95*] Lingayat, S., Mittenthal, J., O'Keefe, R.M.: An order release mechanism for a flexible flow system. Int. J. Prod. Res. **33**(5), 1241–1256 (1995)

[Müll-73] Müller-Merbach, H.: Operations Research. Methoden und Modelle der Optimalplanung. Vahlen, Munich (1973)

[Rode-92*] Roderick, L.M., Phillips, D.T., Hogg, G.L.: A comparison of order release strategies in production control systems. Int. J. Prod. Res. **30**(3), 611–626 (1992)

[Schr-90*] Schragenheim, E., Ronen, B.: Drum-Buffer-Rope shop floor control. Prod. Invent. Manag. J. **31**(3), 18–22 (1990)

[Spen-95*] Spencer, M.S., Cox, J.F.: Master Production Scheduling Development in a theory of constraints environment. Prod. Invent. Manag. J. **36**(1), 80–14 (1995)

[Wein-88*] Wein, L.M.: Scheduling semiconductor wafer fabrication. IEEE Trans. Semicond. Manuf. **1**(3), 115–130 (1988)

[Wind-01] Windt, K.: Engpassorientierte Fremdvergabe in Produktionsnetzen. VDI Progress Reports, Series 2, No. 579, Düsseldorf (2001)

# Chapter 20
# Workload Control

Workload Control is a simple technique for controlling production which balances the load of the workstations within a manufacturing system. This technique was published by Jendralski [Jend-78] at the Leibniz University of Hannover's Institute of Production Systems and Logistics as well as other (groups of) authors independently of one another and with slightly different procedural rules (Bertrand and Wortmann [Bert-81*], Hendry, Kingsman, Tatsiopoulos [Hend-91*, King-89*, Tats-93*] as well as Chang and Yih [Chan-94*]).[1] Up until now, a standard term for this method has not yet been established, however, based on Bertrand and Wortmann's terminology we will be referring to it here as 'Workload Control'.

The basic idea of Workload Control is to hold back orders which would be routed through workstations that are already overloaded. The workstation's WIP is the key criterion here. In addition to WIP on the workstation, the work contents of orders that have already been released and that will pass through the workstation are taken into consideration.

Following, we will briefly consider the procedural rules for Workload Control and the differences between the variations (Sect. 20.1). The procedural rules will then be discussed based on the guidelines for designing manufacturing control methods (Sect. 20.2). Whereas the workstations' WIP limits are the main parameters for the method, the process for setting these is explained in Sect. 20.3. The reason for why implementing this method across the supply chain would be unusual is briefly addressed in Sect. 20.4. Finally, Sect. 20.5 presents the conditions for applying this method and provides an account of an industrial application.

[1] In both Bertrand and Wortmann's work as well as that of Hendry and Kingsman, the release of orders forms only a section of a more comprehensive concept of production planning and control. Chang and Yih's work first conducts an event oriented order release.

H. Lödding, *Handbook of Manufacturing Control*,
DOI 10.1007/978-3-642-24458-2_20, 

## 20.1 Method Description

Our discussion of the procedural rules will begin with the basic method before moving on to a number of variations in Sect. 20.1.2.

### *20.1.1 Basic Method*

Figure 20.1 illustrates the principle of Workload Control: The orders are generated and scheduled by a higher level PPC system. The method regulates the release of orders with the current WIP on the workstations that an order is routed through. The order is only actually released when all of the workstations involved authorize its release.

Controlling the production with this method is thus based on three elements:

- a release list,
- the WIP accounts of the workstations, and
- the WIP limits of the workstation.

The *release list* contains known but not yet released orders which satisfy specific conditions that can be set by the user. For example it could be a condition that the planned start date has to fall within a set advance release time. It is generally practical to sort the orders in the list according to the planned started date.

A *WIP account* is kept for each workstation (Table 20.1). The work content of an operation is credited to this WIP account when the order is released for production

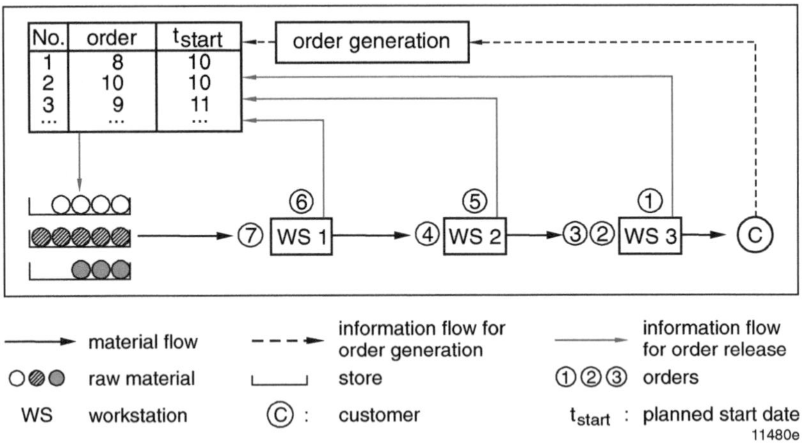


Fig. 20.1 Principle of workload control

**Table 20.1** WIP account for a workstation (Example)

| Order number [–] | Work content [hours] |
|---|---|
| 4,500 | 10 |
| 2,720 | 8 |
| 2,955 | 4 |
| 5,005 | 10 |
| Sum | 32 |

and debited once the operation on the workstation is completed. When the order is released it is thus simultaneously credited in the WIP accounts of all the workstations through which it will pass.

The WIP account of the workstation thus contains both WIP on the workstation itself (direct WIP) and so-called 'indirect WIP'. Indirect WIP refers to WIP that is still on preceding workstations and that will load the workstation in the future. The total WIP in the account is an important criterion for the release of further orders.[2]

In addition, each workstation is allocated a *WIP limit*. The WIP limit plays an important role in the release of orders: If WIP in the WIP account exceeds the WIP limit, the release of all orders routed through the impacted workstation is blocked.[3]

As noted in Chap. 16, two types of order release can be differentiated based on the trigger logic: *event oriented order release* and *periodic order release*.

**Event Oriented Order Release** Event oriented order release is applied by Chang and Yih as the standard for Workload Control. A decision about releasing an order is made when one of two events occur:

1. When an *order is entered* into the release list Workload Control verifies whether or not the WIP limit on any of the workstations the order will be routed through is exceeded. If this is the case, the method blocks the release of the order. Otherwise the order is released and is credited to the workstations' WIP accounts.
2. When an *operation is completed* on a workstation the order is debited from the WIP account. If as a result, the WIP falls below the WIP limit Workload Control verifies whether or not the blocking of orders in the release list can be lifted. If this is the case, the method first releases the most urgent of these orders and credits it in the WIP accounts of the impacted workstations. Subsequently, the method verifies the second most urgent order etc. Obviously, it is not necessary

[2] In Chang and Yih's variation of the method, WIP is not measured in planned hours but rather in the number of orders. Tatsiopoulos as well as Hendry and Kingsman divide the WIP account's total WIP by the planned capacity and use the resulting WIP range as criterion for releasing orders.

[3] In Chang and Yih's variation, the blocking is triggered when the WIP limit is reached. Tatsiopoulos as well as Hendry and Kingsman define a corresponding limit for the WIP range. This change in the method is not relevant to the technique used for releasing orders, however, it plays an important role when setting the capacities (see Sect. 20.2).

to verify the release of other orders once the WIP limit of the observed workstation is exceeded.

**Periodic Order Release** Orders are released at specific time points e.g. at the start of a work day or a shift. Beginning with the most urgent order, Workload Control verifies the release for each order one after the other. Released orders are credited to the workstations' WIP accounts. Blocked orders remain in the release list. Periodic order release is the standard used in variations of this method developed by Jendralski, Bertrand and Wortmann as well as from Tatsiopoulos, Kingsman and Hendry.

Let us turn our attention now to an example that we can use to expand upon the procedural guidelines of Workload Control. In doing so, the following values are chosen for the method's specifications:

1. the WIP is measured in planned hours.
2. the order release is event oriented.

A production with three workstations WS 1 to WS 3 forms the basis of our considerations (Fig. 20.2). The WIP account for each of the workstations is depicted with the WIP limits.

The figure also includes the release list. For each and every order it contains the planned start date and the workstations that orders are routed through. These workstations are marked by an 'x', thus Order 7 is part of the load for WS 1 and 3. In the initial situation (Fig. 20.2a), all of the orders are routed through at least one blocked workstation (WS 2 or WS 3). The Workload Control thus holds the orders back in order to not further increase the production's WIP.

When Workstation 2 completes Order 2 (Step A in Fig. 20.2b), the order is debited from the WIP account (Step B) causing the WIP to fall below the WIP limit (Step C). The workstation then begins processing Order 4 (Step D).

At the same time Workload Control verifies whether or not the blocking of the orders to be released can be lifted since the WIP limit on WS 2 is no longer exceeded (Fig. 20.2c). This applies to Orders 8 and 9. Due to the earlier planned start date, the method releases Order 8 first and removes it from the release list (Steps E and F). When the order is released the WIP accounts of the impacted workstations (WS 1 and 2) are credited with the respective work contents (Steps G to J). Consequently, the WIP limit of WS 1 is exceeded; even though the WIP limit on WS 2 is not yet attained Order 9 can thus not be released.

Next WS 3 completes Order 1 (Step A in Fig. 20.2d). The order is debited from the WIP account (Step B), whereby the WIP falls below the limit (Step C).

Since all of the orders in the list of those to be released are routed through WS 1 – on which the WIP limit is exceeded – no orders can be released. Workstation 3 begins processing Order 3 (Step D). The blocking of the orders is lifted when WS 1 finishes processing Order 6 (Step E) and the order is debited from the WIP account (Step F and G).

At this point in time, WIP levels on all of the workstations have fallen below the WIP limits. The method thus releases the most urgent order (Order 7) and credits

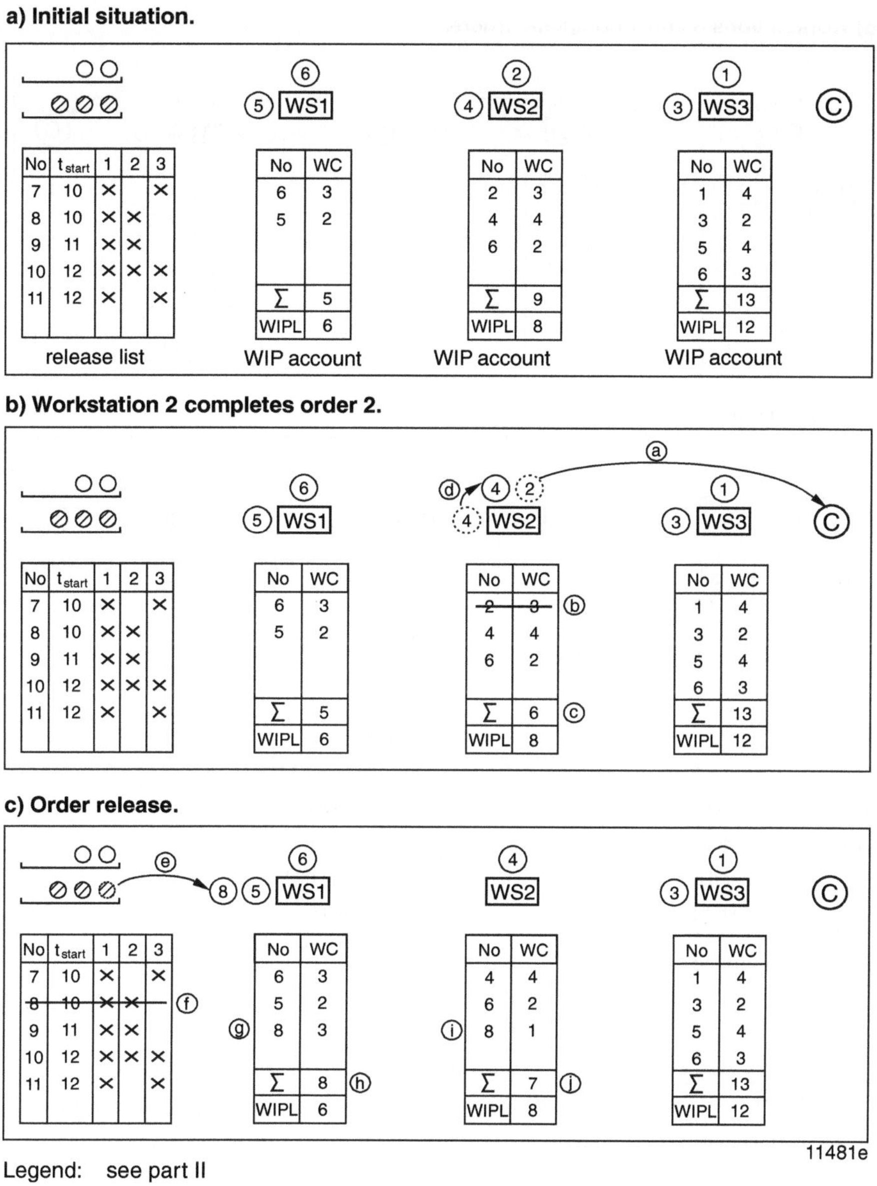


Fig. 20.2 How workload control functions

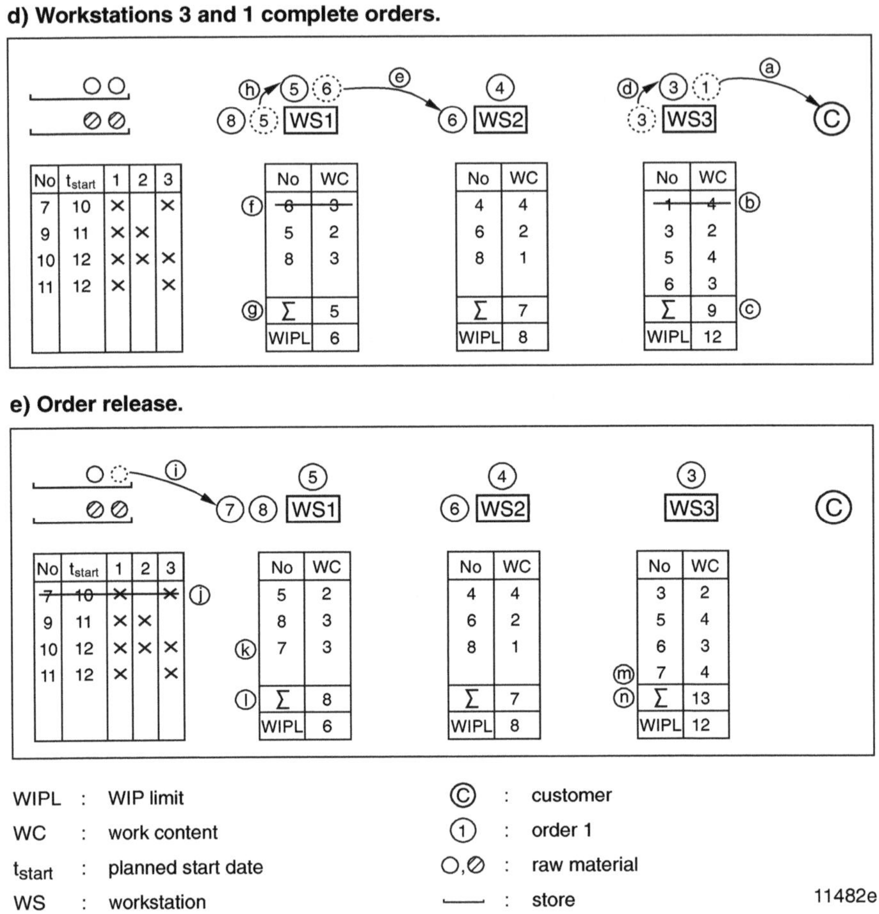


**Fig. 20.2** (continued)

the WIP accounts for WS 1 and 3 with the corresponding work contents (Steps I to N in Fig. 20.2e). As a result WIP limits on both workstations are exceeded and no further orders can be released.

## 20.1.2 Method Variations

We will turn our attention now to four variations of this method:

- Workload Control with modified pay points,
- Workload Control with modified pay points and modified booked work contents,
- Workload Control with cards,
- Workload Control with different release conditions.

**Workload Control with Modified Pay Points Workload** As a variation Tatsiopoulos suggests debiting the orders from the WIP account of the workstation only after the last operation is completed [Tats-93*]. This simplifies the implementation of the method since the completion of the operations no longer has to be compiled. Instead only feedback concerning an order is required.

Orders that have already been completed on a workstation are then also maintained in the stations' WIP accounts. The delayed completion of these orders thus also delays the release of new orders for this workstation. This in particular negatively impacts the method's suitability in cases where the bottleneck workstations are located in the upper part of the material flow and the throughput times in the lower part of the material flow strongly vary (see [Land-96*]).

**Workload Control with Modified Pay Points workload and Modified Booked Work Contents** The suggestions from Land and Gaalman [Land-96b*] are based on Tatsiopoulos: In order to simplify the feedback, this variation of the method is also limited to feedback for the entire order. In addition, the orders are booked in the accounts with a value that decreases with the position of the operation in the order's production schedule: In order to do so the work content is divided by the operation's position in the work schedule:

$$WC_{acct} = \frac{WC}{PosOP} \tag{20.1}$$

where

| | |
|---|---|
| $WC_{acct}$ | work content booked in WIP account [hrs], |
| WC | work content [hrs], |
| PosOP | operation's position in the production schedule [−]. |

An example will serve to explain this further: An order is routed through three workstations (see Table 20.2). When the order is released the work content is credited to the first workstation in its entirety, the second workstation is credited with half of the work content and the third workstation with a third of the work content.

**Workload Control with Kanbans** Chang and Yih implement Workload Control with the aid of workstation specific Kanbans [Chan-94*]. In order to do so, they allocate each workstation in a production a specific number of Kanbans which

**Table 20.2** Calculation of the booked work content

| Operation<br>[−] | Workstation<br>[−] | Operation's position<br>[−] | Work content<br>[hours] | Booked work content<br>[hours] |
|---|---|---|---|---|
| 10 | 1 | 1 | 10 | 10 |
| 20 | 3 | 2 | 4 | 2 |
| 30 | 2 | 3 | 6 | 2 |

corresponds to the workstation's WIP limit. An order can only be released when it has a Kanban from each and every workstation that it will pass through. The Kanbans remain with the order from the point it is released up until its completion on the workstation that the Kanban is allocated to. In other words, when an operation is completed on a workstation, a Kanban is freed and can therefore authorize the release of a further order.

The Kanbans can help defuse a problem typical to Workload Control. The method tends to delay orders that have a larger number of operations. The release of an order during the verification process is dependent only on whether the WIP limit is exceeded (order is blocked) on at least one of the workstations it is routed through or not (order is released). If only one WIP limit is exceeded the order's release is denied. Where possible Workload Control then pulls forward orders that are not routed through the overloaded workstation. By releasing these orders the WIP limits on other workstations which the deferred order would be routed through are frequently exceeded. The deferred order can thus often not be released when the WIP level on the workstation that originally caused it to be deferred falls below the WIP limit. Orders routed through a number of different workstations that have critical capacities tend to have a disadvantage when Workload Control is implemented without Kanbans i.e., there is thus a greater risk that the release of these orders will be delayed.[4]

The use of Kanbans can prevent this problem. The Kanbans are then always allocated to the most urgent orders. This is further illustrated by the example in Fig. 20.3 which depicts a release list with an integrated Kanban board.

Kanbans authorize the release of orders for WS 1 and 3. The Kanbans are allocated to Order 1, which cannot be released because there is no Kanban for WS 2. Although Order 2 which is not routed through WS 2 could be released with the available Kanbans, these Kanbans remain allocated to Order 1. In this way it is

| No. | $t_{start}$ | WS | | | Kanban | | |
|---|---|---|---|---|---|---|---|
| | | 1 | 2 | 3 | 1 | 2 | 3 |
| 1 | 10 | × | × | × | [1] | | [3] |
| 2 | 11 | × | | × | | | |

$t_{start}$ : planned start date WS : workstation

[1] : Kanban for workstation 1

11479e

**Fig. 20.3** Release list integrated with a Kanban board

[4] The problem is less obvious when orders are released periodically then with an event oriented release: At the start of the order release a number of WIP accounts are usually below the WIP limits, thus there is an opportunity for orders with a number of operations to be released for production.

ensured that Order 1 can be released as soon as a Kanban for WS 2 arrives. This method can in turn be applied when the WIP is measured in planned hours.

Chang and Yih simulated the method and could not find evidence of any significant advantages in comparison to a simple decentralized WIP control. Due to the simulated production – a production line – this is not surprising: All orders are routed through the same workstations. Correspondingly, a workstation specific load balance cannot be conducted (see Chap. 16).

**Workload Control with Different Release Conditions** The WIP control aspect of Workload Control can be limited to workstations with critical capacities by setting the WIP limits for workstations without critical capacities at a very high value. These workstations thus lose their influence on the order release and the order release then only considers the workstations with critical capacities. The Bottleneck Control (Chap. 19) can thus be understood as an extreme case of Workload Control in which the WIP limit of the non-bottleneck workstations are set at an unlimited high value so that only the bottleneck workstation controls the order release.

## 20.2 Discussion of the Procedural Rules

The workstation specific WIP control is a fundamental characteristic of Workload Control. It leads to the load being balanced when orders are released.

In this next section, we will focus on the procedural rules with regards to the guidelines for designing manufacturing control methods.

1. *A manufacturing control method should allow the WIP of a production or workstation to be regulated at a defined level.*

Workload control limits the WIP for each workstation between the order's release and the end of the order's processing on the workstation. Thus the shorter the order throughput times are from the order release to the arrival on the workstation, the more precisely the (direct) WIP of the workstation can be controlled. In addition, the method limits the total WIP in the production via the WIP levels on the workstations.

2. *A manufacturing control method should reduce WIP fluctuations to a minimum.*

The longer the throughput time (from the release) of the order up until its arrival at the workstation is in relation to the workstation throughput times, the more orders in the workstation's WIP account are found on preceding workstations. Accordingly, the WIP limit of the workstation generally clearly exceeds the planned direct WIP. A certain leeway thus arises for fluctuations in the direct WIP; the further the workstation is from the order release, the greater this leeway is.

Order spectrums in which the order throughput time from the order release up to the arrival at a workstation with critical capacities strongly varies are particularly problematic for this method. In intervals where the majority of orders reach the workstation very quickly the total WIP of the workstation does not significantly exceed its direct WIP. Consequently, a relatively low WIP limit would be sufficient. However, if orders with long throughput times up to their arrival at the workstation dominate in other intervals, a higher WIP limit would be required. In this case the total WIP will be clearly higher than the direct WIP on the workstation.

If the intervals with shorter and longer throughput times up to the observed workstation quickly alternate, the WIP limit then would have to be constantly recalculated. This however would be laborious. This problem can arise especially when the position of a workstation in the order throughput strongly varies. The simulation trials conducted by Cigolini and Portioli-Staudacher seem to confirm this [Cigo-02*]. When the workstation has a constant position in the material flow though, the problem should arise much less frequently if at all.

3. *A manufacturing control method should cause as little blocked WIP as possible.*
Workload control blocks orders at the point of release. In comparison to a pure Due Date Based Order Release the method can also block orders whose planned start dates have already been passed. It also releases orders early if the utilization of the workstation can be ensured by doing so. However, during the order's flow through the production, Workload Control does not block any WIP.

4. *A manufacturing control method should contribute to closely aligning the load to the capacities.*
Closely aligning the load to the capacities is a fundamental characteristic of Workload Control. The method holds back orders which are routed through workstations where the WIP limit is exceeded. However, it also releases orders early which (only) load workstations that have not yet reached their WIP limits. Thus with a comparable WIP this technique can attain a higher output rate than a manufacturing control method that does not have this type of load balance.

Workload control is based on the assumption that the WIP between the order release and completion on a workstation is an indication of how well a workstation is protected from breaks in the material flow. Obviously, the possibility of a break in the material flow is also dependent on the position of the WIP. In other words, direct WIP protects a workstation in the near future from breaks in the material flow more reliably than WIP that still has to pass through preceding workstations before it arrives at the observed workstation. The effectiveness of balancing the loads with Workload Control is dependent on the reliability of the predictions about the distribution of the WIP on the workstations (see also [Land-96*] and the comments regarding Load Oriented Order Release in Sect. 21.2).

5. *A manufacturing control method should cause as few sequencing interchanges as possible.*
Workload control causes sequencing interchanges when the orders are released. This allows the load to be balanced. Furthermore, some variations of the method have the tendency to delay orders that require many operations. It is thus practical to compensate for the resulting sequencing interchanges during the orders' throughput. It then

seems obvious to combine this method with a sequencing rule that takes into consideration the order's planned end date.

6. *A manufacturing control method should be able to balance schedule deviations within the context of backlog control.*

In the version from Tatsiopoulos, Kingsman and Hendry, Workload Control is integrated into a higher level, hierarchical system for planning and controlling make-to-order productions. Their planning concept includes among other aspects a decision about increasing the capacities. The authors recommend increasing the capacities of workstations if this makes it possible to promise delivery times that concur with the market [King-02*]. Thus the variation of Workload Control from Tatsiopoulos, Hendry and Kingsman is one of the few methods that integrate adjusting the capacities over the short term. An explicit backlog control, however, is not included in this version. It is thus practical to supplement this method with a backlog control in order to be able to ensure the delivery reliability even when there are schedule deviations.

7. *A manufacturing control method should take into consideration the bottleneck principle.*

Workload control considers the bottleneck principle at least implicitly. Usually it is the bottleneck workstations on which the WIP limits are (first) exceeded. If the WIP on the bottleneck workstation falls below the WIP limit when an order is completed, Workload Control releases another order. Exceeding the WIP limit prevents further orders from being released for that workstation. Moreover, the method is also able to control productions in which the bottleneck workstations alternate.

8. *A manufacturing control method should be as simple as possible.*

The procedural rules for Workload Control are relatively simple. However, most of the variations require feedback for each of the operations. Yih and Chang implement workstation specific Kanbans for this purpose (see also Sect. 20.1.2). In further variations of Workload Control, Tatsiopoulos [Tats-93*] as well as Land and Gaalman [Land-96b*] avoid the necessity of reporting the completion of operations and limit feedback instead to orders.

## 20.3 Setting the Method's Parameters

The WIP limits of the workstations are the most important parameters of Workload Control. The WIP limits determine the WIP levels and the (WIP dependent) utilization of the workstations as well as the throughput times of the orders. Furthermore, in case the early release of orders should be limited to a defined period of time, the advance release window has to be set. Setting the advance release window is extensively discussed for the Load Oriented Order Release (Sect. 21.3). Since this procedure can be directly transferred to Workload Control (without any changes) it will not be further discussed here.

As with all other WIP regulating control methods, the WIP limits can be determined in a gradual, continuous improvement process. In order to do so a value known to be sufficient is chosen to start with and subsequently reduced step-by-step up until the loss of utilization has reached a tolerable level. This procedure, however, does not release the enterprise from the necessity of positioning itself between the competing logistic objectives i.e., between low WIP levels and short throughput times, on the one hand, and a high utilization, on the other hand. Thus, in the following it is supplemented with a model based procedure which includes both a Logistic Positioning of the production's workstations and a calculation of the WIP limits for the workstations.

**Logistic Positioning** The Logistic Positioning generally does not differ with the various manufacturing control methods. It is comprised of two steps:

- calculating the Logistic Operating Curves for the workstations, and
- setting a planned WIP level for the workstations (Logistic Positioning).

Both of these steps are extensively discussed in Sect. 3.2.2.3. With the planned WIP the second step also sets the planned throughput and inter-operation times of the workstation. Therefore, based on a Logistic Positioning it is possible to determine planned throughput times for all of the orders.

**Calculating the WIP Limits** Manufacturing control is responsible for setting the WIP limits so that the planned WIP can be attained. This thus raises the question of the mathematical correlation between the WIP limit and the WIP.

Due to the dynamic of the manufacturing control alone, an exact rule of conversion cannot be determined. However, it is relatively simple to estimate the correlation between the WIP and the WIP limit. In order to simplify this process, it is first assumed that the total WIP of a workstation corresponds to its WIP limit:

$$WIP_{tot} \approx WIPL \tag{20.2}$$

where

| | |
|---|---|
| $WIP_{tot}$ | total WIP [hours], |
| WIPL | WIP limit [hours]. |

If the ratio between the total WIP and the direct WIP is known, the WIP limit can be set as follows:

$$WIPL = WIP_{m,plan} \cdot \frac{WIP_{tot,m,plan}}{WIP_{m,plan}} \tag{20.3}$$

where

| | |
|---|---|
| WIPL | WIP limit [hours], |
| $WIP_{m,plan}$ | mean planned (direct) WIP [hours], |
| $WIP_{tot,m,plan}$ | mean planned total WIP [hours]. |

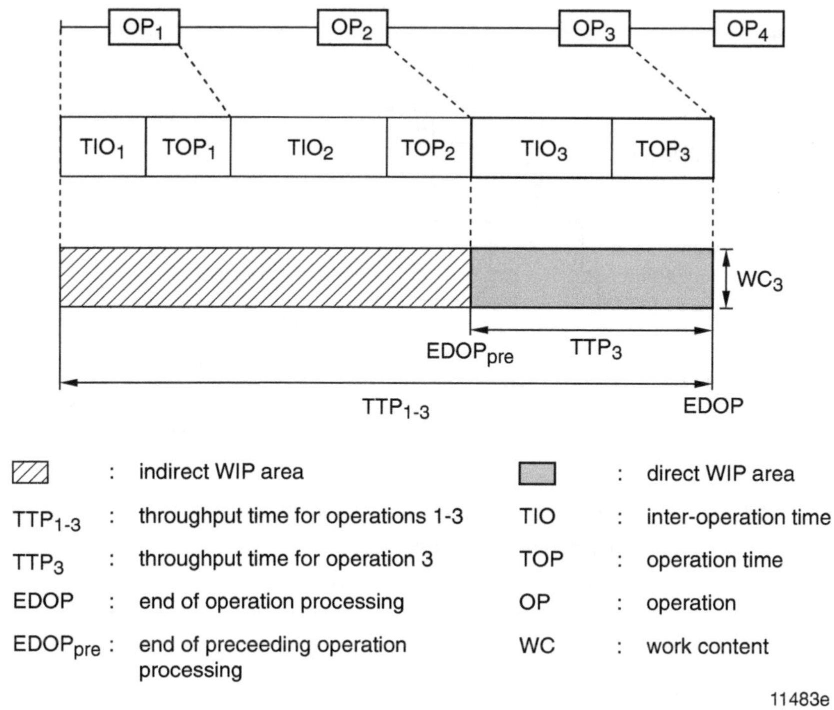


**Fig. 20.4** Representation of the direct and indirect WIP in extended throughput elements

The ratio of the mean planned total WIP and the planned direct WIP is relatively easily determined. In order to do so the throughput of a single order from its point of release up to its completion on a workstation is considered (Fig. 20.4).

From the order release up to its input on the workstation, the order is considered indirect WIP for the workstation and from its input till its processing is completed as direct WIP. The ratio of the total WIP area and the direct WIP area is calculated as:

$$\frac{AWIP_{tot}}{AWIP_{dir}} = \frac{TTP_{tot,i,j}}{TTP_{i,j}} \tag{20.4}$$

where

$AWIP_{tot}$ total WIP area [hours · SCD],
$AWIP_{dir}$ direct WIP area [hours · SCD],
$TTP_{tot,i,j}$ throughput time of order $i$ from its release till the end of processing on workstation $j$ [SCD],
$TTP_{i,j}$ throughput time of order $i$ on workstation $j$ [SCD].

This procedure can be extended to include a number of orders, in which case the throughput times of the orders need to be weighted (Fig. 20.5). Accordingly,

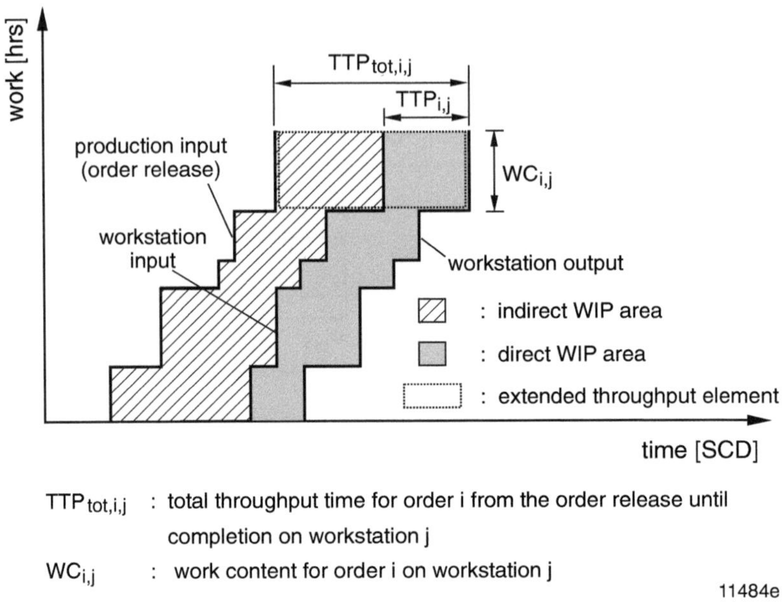


**Fig. 20.5** Representation of direct WIP and indirect WIP in a throughput diagram

$$\frac{AWIP_{tot}}{AWIP_{dir}} = \frac{\sum_{i=1}^{NO} \left(WC_{i,j} \cdot TTP_{tot,i,j}\right)}{\sum_{i=1}^{NO} \left(WC_{i,j} \cdot TTP_{i,j}\right)} \tag{20.5}$$

where

| | |
|---|---|
| $AWIP_{tot}$ | total WIP area [hours/SCD], |
| $AWIP_{dir}$ | direct WIP area [hours/SCD], |
| $WC_{i,j}$ | work content of order *i* on workstation *j* [hours], |
| $TTP_{tot,i,j}$ | throughput time of order *i* from its release till the end of processing on workstation *j* [SCD], |
| $TTP_{i,j}$ | throughput time of order *i* on workstation *j* [SCD], |
| NO | number of orders [−]. |

The ratio between the direct WIP and the total WIP corresponds to the ratio of their areas (see [Wien-97b]). Thus with Eq. 20.3:

$$WIPL_j = WIP_{m,plan,j} \cdot \frac{\sum_{i=1}^{NO} \left(WC_{i,j} \cdot TTP_{plan,tot,i,j}\right)}{\sum_{i=1}^{NO} \left(WC_{i,j} \cdot TTP_{plan,i,j}\right)} \tag{20.6}$$

where

| | |
|---|---|
| $WIPL_j$ | WIP limit on workstation *j* [hours], |
| $WIP_{m,plan,j}$ | planned WIP on workstation *j* [hours], |
| $WC_{i,j}$ | work content of order *i* on workstation *j* [hours], |
| $TTP_{plan,tot,i,j}$ | planned throughput time of order *i* from its release till the end of processing on workstation *j* [SCD], |
| $TTP_{plan,i,j}$ | planned throughput time of order *i* on workstation *j* [SCD], |
| NO | number of orders [−]. |

In order to set the WIP limits based on a model the planned throughput times for a representative quantity of orders needs to be determined. The WIP limit then results from the product of the planned WIP and the quotient of the total WIP and direct WIP on the workstation. The idea to use a representative quantity of orders for dimensioning the WIP limit is based on a method from Bertrand and Wortmann for dimensioning Workload Control [Bert-81*].

When orders are released periodically, WIP in the production between two release points is reduced by the amount of work completed during the period. Bertrand and Wortmann take this into consideration by increasing the WIP limit by the planned output for the period [Bert-81*]. Accordingly, with Eq. 20.6:

$$WIPL_{j,period} = WIP_{plan,j} \cdot \frac{\sum_{i=1}^{NO} \left(WC_{i,j} \cdot TTP_{plan,tot,i,j}\right)}{\sum_{i=1}^{NO} \left(WC_{i,j} \cdot TTP_{plan,i,j}\right)} + OUT_{plan,j} \tag{20.7}$$

where

| | |
|---|---|
| $WIPL_{j,period}$ | WIP limit on workstation *j* with periodic order release [hours], |
| $WIP_{plan,j}$ | planned WIP workstation *j* [hours], |
| $WC_{i,j}$ | work content of order *i* on workstation *j* [hours], |
| $TTP_{plan,i,j}$ | planned throughput time of order *i* on workstation *j* [SCD], |
| $TTP_{plan,tot,i,j}$ | planned throughput time of order *i* from its release till processing end on workstation *j* [SCD], |
| $OUT_{plan,j}$ | planned output of workstation *j* during release period [hours], |
| NO | number of orders [−]. |

Furthermore, Bertrand and Wortmann increase the WIP limit of minimally used workstations above the calculated amount. They thus prevent these workstations from thwarting the release of orders when an above average number of orders temporarily load the workstation.

## 20.4 Implementation Across the Supply Chain

Implementing Workload Control across the supply chain would be unusual as it would require the supplying and receiving enterprises to have the same sized production lots as well as a very close supply relationship in general: Such constellations are rarely found in the industry.

## 20.5 Application

The aim of Workload Control is a steady utilization of workstations. Accordingly, an enterprise would implement it when a steady utilization is an important logistic target that is not already ensured by the scheduling of orders. This is to be assumed for example if the orders are generated without taking the capacities into consideration.

Like all methods that have a workstation specific load balance, Workload Control generates sequencing interchanges with the release of orders. The method thus causes a large amount of variance in the input deviations. When orders are processed in the sequence of their input (FIFO) this leads to comparable variance in the output lateness. The method therefore only obtains a high schedule reliability when the enterprise accepts a large due date tolerance or the variance in the output lateness is reduced by implementing a suitable sequencing rule. The method should thus be combined with a sequencing rule that takes into consideration the planned completion date of the orders.

Based on the discussion of the procedural rules (Sect. 20.2) further conditions for implementing this method can be derived:

- A minimum complexity of the material flow is necessary for the method to conduct a workstation specific load balance.
- The position of the workstations in the order throughput should remain the same as far as possible.
- Operation feedback is required for at least the base method.

Furthermore the procedural rules should be modified or designed so that the release of orders with many operations is not delayed disproportionately long.

Bertrand and Wortmann describe a successful application of this method in the semi-conductor industry in one of Philips' factories located in the Netherlands [Bert-81*]. By applying Workload Control it was possible to reduce the mean throughput time of the lots from approximately 35 days to 25 days. At the same time the variance of the throughput times decreased, through which in turn the proportion of defective products could be strongly reduced.

# References

[Bert-81*] Bertrand, J.W.M., Wortmann, J.C.: Production Control and Information Systems for Component-Manufacturing Shops. Studies in Production and Engineering Economics, vol. 1. Elsevier, Amsterdam (1981)

[Chan-94*] Chang, T.M., Yih, Y.: Generic Kanban systems for dynamic environments. Int. J. Prod. Res. **32**(4), 889–902 (1994)

[Cigo-02*] Cigolini, R., Portioli-Staudacher, A.: An experimental investigation on workload limiting methods with ORR policies in a job shop environment. Prod. Plan. Control **12**(7), 602–612 (2002)

[Hend-91*] Hendry, L.C., Kingsman, B.G.: Job release: part of a hierarchical system to manage manufacturing lead times in make-to-order companies. J. Op. Res. Soc. **42**(10), 871–883 (1991)

[Jend-78] Jendralski, J.: Kapazitätsterminierung zur Bestandregelung in der Werkstattfertigung. Dissertation, TU Hannover (1978)

[King-89*] Kingsman, B.G., Tatsiopoulos, I.P., Hendry, L.C.: A structural methodology for managing manufacturing lead times in make-to-order companies. Eur. J. Oper. Res. **40**, 196–209 (1989)

[King-02*] Kingsman, B., Hendry, L.: The relative contributions of input and output control on the performance of a workload control system in make-to-order companies. Prod. Plan. Control **12**(7), 579–590 (2002)

[Land-96*] Land, M., Gaalman, G.: Workload control concepts in job shops. A critical assessment. Int. J. Prod. Res. **46–47**, 535–548 (1996)

[Land-96b*] Land, M., Gaalman, G.: Towards simple and robust workload norms. In: Proceedings of Workshop on Production Planning and Control, pp. 66–96. Mons (1996)

[Tats-93*] Tatsiopoulos, I.P.: Simplified production management software for the small manufacturing firm. Prod. Plan. Control **4**(1), 17–26 (1993)

[Wien-97b] Wiendahl, H.-P.: Fertigungsregelung. Logistische Beherrschung von Fertigungsabläufen auf Basis des Trichtermodells. Carl Hanser Verlag, Munich/Vienna (1997)

# Chapter 21
# Load Oriented Order Release

Among all of the manufacturing control methods there are, probably only a few have triggered as much scholarly debate as the Load Oriented Order Release (LOOR). Since it was introduced by Bechte in his dissertation at the Leibniz Universität Hannover's Institute of Production Systems and Logistics in 1980 [Bech-84 (reprint)], the discussion surrounding this method has not stopped. LOOR is the subject of numerous publications both in German language journals as well as international ones. In German speaking countries it is also taught at many universities and polytechnics. Due to the scholarly debate about supposed and actual strengths and weaknesses (see for example [Knol-91*]) it can be easily forgotten that LOOR has proven itself in industrial applications [Wien-91].

In addition to the dissertation from Bechte, Wiendahl's work [Wien-97b] is the most important and perhaps most cited source regarding the LOOR. The following comments deviate considerably from these traditional presentations in order to ensure comparability with the manufacturing control methods discussed here in this book. However, these works are emphatically recommended as more extensive reading.

In Sect. 21.1 we will consider the procedural rules for the Load Oriented Order Release using an example to illustrate them more clearly. Section 21.2 then discusses the procedural rules for LOOR based on the guidelines for designing manufacturing control methods, whereas, Sect. 21.3 is dedicated to setting the method's parameters. The method described here deviates considerably from Bechte. Without changing the procedural rules, the methods range of applicability can thus be greatly extended. Manufacturing control methods such as Workload Control, Bottleneck Control or CONWIP can then be emulated just by setting the method's parameters. Section 21.4 then focuses on applying LOOR across the supply chain and finally Sect. 21.5 describes experiences in applying it.

H. Lödding, *Handbook of Manufacturing Control*,
DOI 10.1007/978-3-642-24458-2_21, © Springer-Verlag Berlin Heidelberg 2013

## 21.1 Method Description

Figure 21.1 depicts the principle of the Load Oriented Order Release: A central PPC system generates a list of urgent orders using backwards scheduling. *An order is released when the WIP limit*[1] *is not exceeded on any of the workstations that the order is routed through.*

Bechte recommends releasing the orders periodically for this method. However, LOOR is generally suitable for event oriented order release as well.

Unlike Workload Control, LOOR does not always credit the workstations' WIP accounts with the full work content of the orders: If an order flows through other workstations before it arrives at a workstation, the booked work content is reduced by the so-called 'discount factor' of the preceding workstations.

Load Oriented Order Release is thus based on four elements:

- The *release list* contains all of the known, but not yet released orders whose planned start date falls within a specific advance release window.
- Each workstation is allocated a *WIP account*. Orders are credited to these accounts with a booked work content when the orders are released and debited from them upon completion. Orders that are part of the workstation's direct WIP are credited with their full work content. Orders for the workstation that are still on preceding workstations (indirect WIP) are taken into consideration with a discounted (reduced) booked work content in the WIP accounts.

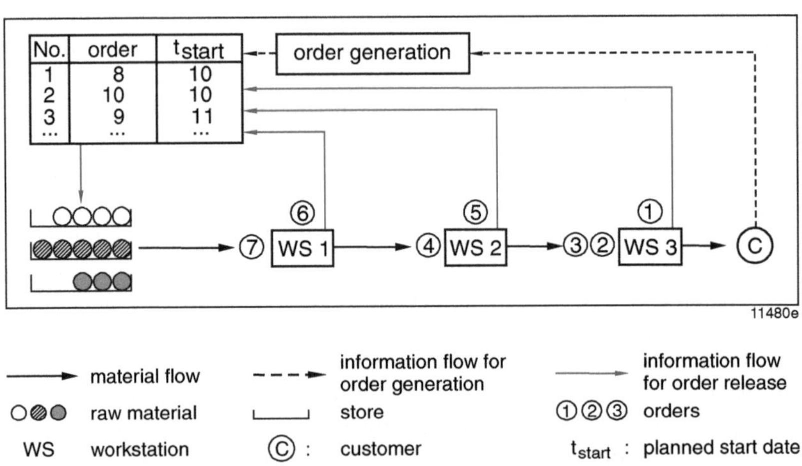


**Fig. 21.1** Principle of Load Oriented Order Release

[1] Bechte originally used the term load limit (*Belastungsschranke*) .

- *WIP limits* are defined for each workstation. If the WIP level in the account exceeds the WIP limit, the method blocks the release of all orders that would flow through the impacted workstation.
- Each workstation is allocated a *discount factor*. This determines the percentage of an order's work content with which it will be taken into consideration in the WIP accounts of the following workstations.

The discounted work content is a fundamental characteristic of the Load Oriented Order Release. The basic idea is that the longer an order is expected to require before it arrives at an observed workstation, the more its booked work content should be reduced. The booked work content of an order is calculated as the product of the discount factors of the preceding workstations and the work content on the workstation. Thus according to [Bech-84]:

$$WC_{WA,j} = \begin{cases} WC_j & \text{if CWS} = \text{j} \\ WC_j \cdot \prod_{i=CWS}^{j-1} \left(\frac{DFA_i}{100}\right) & \text{if CWS} < \text{j} \end{cases} \tag{21.1}$$

where

| | |
|---|---|
| $WC_{WA,j}$ | booked work content in WIP account from workstation *j* [hours], |
| $WC_j$ | work content on workstation *j* [hours], |
| $DFA_i$ | discount factor for workstation *i* [%], |
| CWS | index of current workstation [–]. |

The following example based on an order that flows through four workstations will allow us to take a closer look at this formula (Table 21.1).

The table contains the discount factor for each workstation. When the order is released is it allocated to the first workstation (index of current workstation = 1). The order is credited to the first workstation's WIP account with the full work content. On the *second* workstation the work content is reduced by the *first* workstation's discount factor ($WC_{WA,2} = 7.0 \cdot 0.5 = 3.5$), on the *third* workstation it is reduced by the *first two* workstations' discount factors ($WC_{WA,3} = 5.0 \cdot 0.5 \cdot 0.8 = 2.0$) and so on. The (percentual) booking factor of an order credited to the WIP account results from this (second last column in Table 21.1).

With the completion of the first operation, the order is debited from the WIP account of the first workstation. Moreover, the booked work contents of the orders

**Table 21.1** Discounting the work content when releasing orders

| OP [–] | WS [–] | WC [hours] | DFA [%] | Booking factor [%] | $WC_{WA}$ [hours] |
|---|---|---|---|---|---|
| 1 | 5,000 | 3.0 | 50 | 100 | 3.0 |
| 2 | 4,600 | 7.0 | 80 | 50 | 3.5 |
| 3 | 6,000 | 5.0 | 75 | 40 | 2.0 |
| 4 | 6,200 | 2.0 | 75 | 30 | 0.6 |

*OP* operation, *WC* work content, $WC_{WA}$ booked work content, *WS* workstation, *DFA* discount factor

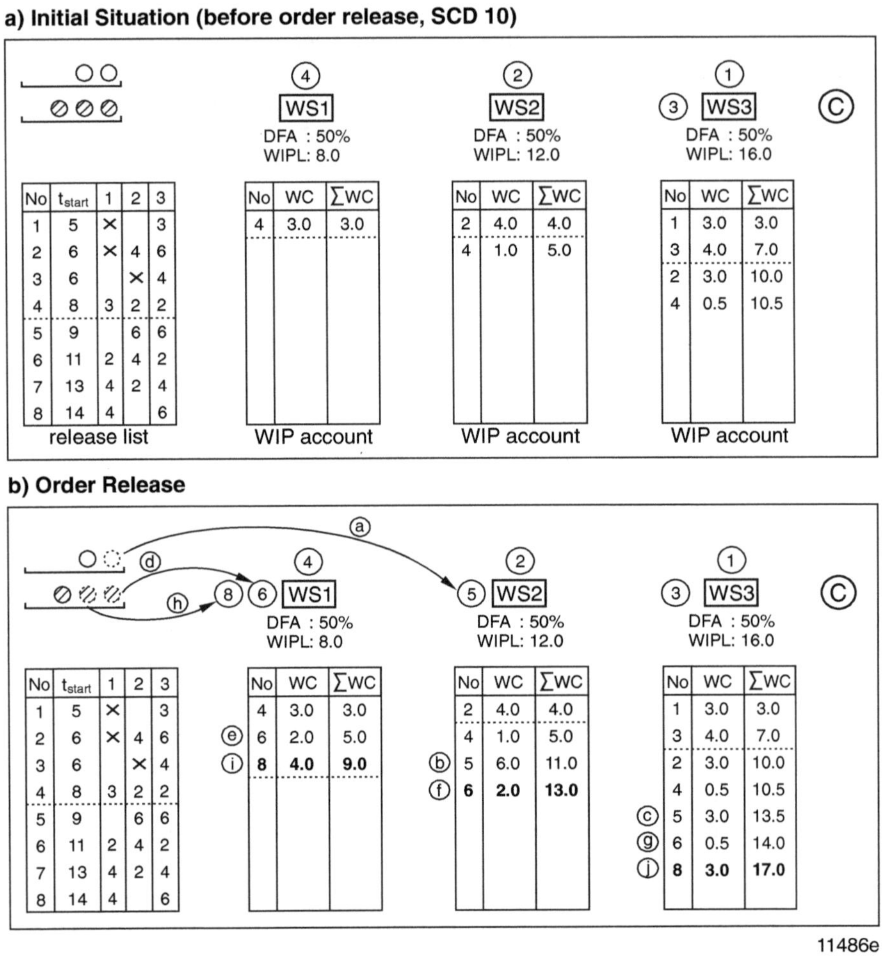


**Fig. 21.2** How load oriented order release functions

are re-calculated for the WIP accounts of the remaining workstations. The reduction by the workstation's discount factor is discontinued.

In order to further explain this and other aspects of the LOOR, we will consider another example based on a production with three workstations (Fig. 21.2).

The WIP account for each workstation is represented. In the WIP account, all of the orders that have been released but not yet completed on the workstation are credited with the booked work content. The direct WIP on the workstation and the indirect WIP on the preceding workstations are separated in the WIP account by a dashed line. The final column contains the sum of the booked work contents. If the WIP limit is exceeded the last row is bold-faced.

**c) Workstation 1 completes order 4.**

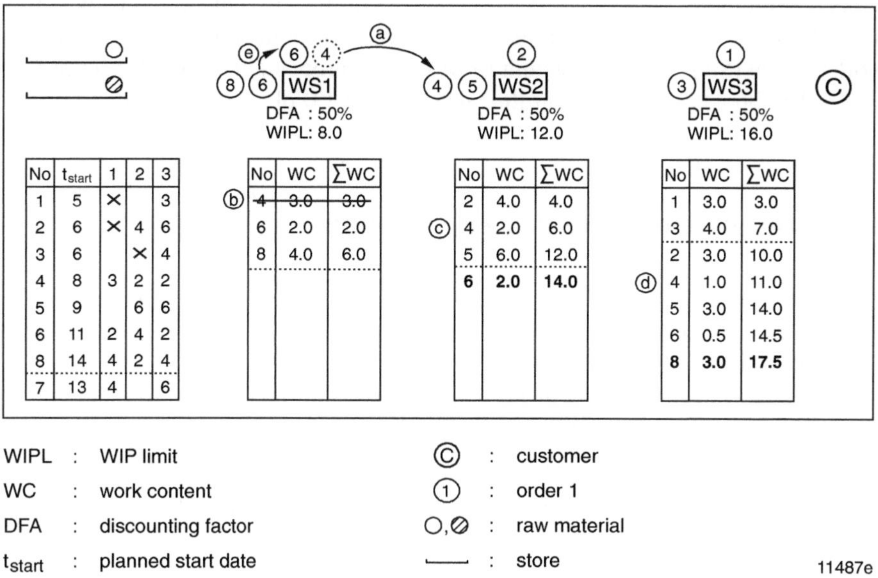


**Fig. 21.2** (continued)

The figure also depicts a list of orders that are either already released (above the line of dashes) or that are still to be released (below the dashed line). The list contains a job number for each of the orders, the planned start date, and the workstations that the order will flow through (or already has flown through). Completed operations are designated by a cross, whereas the work content on the workstation is entered for each of the remaining operations.

In the initial situation (Fig. 21.2a) the WIP levels on all of the workstations are below the WIP limits. At the start of the observed period, LOOR assesses the release of each order still waiting to enter production in the same sequence as their priority. It first releases Order 5 (Step A, Fig. 21.2b) and credits the WIP account of WS 2 with the order's full work content (Step B). In the WIP account from WS 3 only half of the work content is considered due to the discount on WS 2 (Step C). Releasing Order 5 does not cause any of the WIP limits to be exceeded on any of the workstations. LOOR thus also releases Order 6 (Step D). The order is credited to the WIP account of WS 1 (Step E) with the full work content and with the discounted work content in the WIP account of WS 2 and 3 (Steps F and G).

As a result the WIP limit on WS 2 is exceeded. The algorithm therefore does not release Order 7 because it would flow through WS 2. Instead, LOOR releases Order 8 (Step H) which only flows through WS 1 and WS 3. This in turn causes the WIP levels on both of these workstations to be exceeded (Steps I and J). No further orders can be released; the order release phase is thus completed.

As soon as WS 1 finishes processing Order 4 (Step A in Fig. 21.2c), the order is debited from the workstation's WIP account (Step B) and transported to WS 2. As direct WIP, Order 4 is now entered into the WIP account from WS 2 with its full work content (Step C). The discount from WS 1 is dropped in the WIP account from WS 3. The booked work content thus increases to 1.0 planned hour (Step D).

## 21.2 Discussion of the Procedural Rules

The workstation specific WIP control and the discounting of the work content are fundamental characteristics of the Load Oriented Order Release method. They allow the load to be balanced when releasing orders.

In this next section, we will once again take the opportunity to examine the procedural rules more extensively based on the guidelines for designing manufacturing control methods.

*1. A manufacturing control method should allow the WIP of a production or workstation to be regulated at a defined level.*

The Load Oriented Order Release limits the WIP for each workstation between the release of the order and the end of the orders' processing on the workstation. As a result the shorter the throughput time of the orders from their release to their arrival at the workstation, the more precisely the (direct) WIP can be controlled. The WIP of the entire production is thus also regulated by the method.

*2. A manufacturing control method should reduce WIP fluctuations to a minimum.*

Preventing WIP fluctuations is not one of the obvious strengths of the Load Oriented Order Release: When a workstation finishes processing an order, the *complete work content* of the order is debited from the workstation's WIP account. Newly released orders are however only credited to the workstation's WIP account with a *discounted booked work content* when they flow through other workstations first. Thus, completing *one* operation can trigger the release of *a number* of orders and thus cause the WIP to fluctuate. This is the other side of the improved load balance that results from the discounted work content (see the discussion of the fourth guideline).

Nevertheless, this disadvantage remains within bounds particularly when orders are released periodically as well as when there are no clearly defined bottlenecks in the production:

- A *periodic order release* is generally connected with WIP fluctuations (see Chap. 16), therefore, the additional WIP fluctuations caused by the discounting often make no difference.
- If the *bottleneck is not clearly defined*, a number of workstations restrict the order release. The release of more orders is then frequently hindered by the WIP limit on other workstations being exceeded.

Wiendahl recommends crediting the operations in the WIP account only when the work content exceeds a maximum value. He mentions 3% of the maximum permissible load as a possible value [Wien-97b]. With reference to corresponding applications in the industry, Breithaupt, Land and Nyhuis propose crediting the WIP accounts of the workstations with only the first four operations of an order [Brei-02b*]. Both of these suggestions prevent the work content from being discounted extremely. Obviously the number of operations that should be considered when balancing the load is dependent on the position of the workstations with the critical capacities in the material flow. Workstations with critical capacities that are located in the end part of the material flow are especially problematic.

*3. A manufacturing control method should cause as little blocked WIP as possible.*
The Load Oriented Order Release blocks WIP with the order release. Unlike when orders are released according to the schedule, it can block orders, whose planned start date has already passed. In exchange though, it also releases orders early when the utilization of the workstations can be ensured by doing so. Once the orders are released the Load Oriented Order Release does not cause any blocked WIP.

*4. A manufacturing control method should contribute to closely aligning the load to the capacities.*
The close alignment of the load to the capacities is a fundamental characteristic of the Load Oriented Order Release. The method holds back orders when the WIP limit on at least one of the workstations in the production flow has been exceeded. In exchange, it pulls orders forward that load only workstations where the WIP limit has not yet been reached. With a comparable WIP, the Load Oriented Order Release can thus attain a higher output rate than manufacturing methods that do not balance the loads in this manner.

In contrast to Workload Control, Load Oriented Order Release considers the current position of released orders when deciding upon the release of new orders. The shorter the time until an order arrives at a workstation, the more heavily this method weights the orders in the workstation's WIP account. The total WIP calculated in this way is therefore a better measure of the protection a workstation has against disrupted material flows than the total WIP used by Workload Control (see [Land-96*]).

In order to develop a better understanding of this let us consider an example with three workstations (Fig. 21.3).

Two different load situations are presented here. In Fig. 21.3a, the orders are evenly distributed on the production's workstations. Neither Workload Control nor LOOR release orders for Workstation 3 because it is not threatened by any break in the material flow.

In contrast, the orders in Fig. 21.3b accumulate on the first two workstations. Obviously Workstation 3 is affected by a break in the material flow. However, Workload Control would not release any orders because it cannot tell the difference between this load situation and that presented in Part A of the figure. The Load Oriented Order Release, however, would release orders that only load

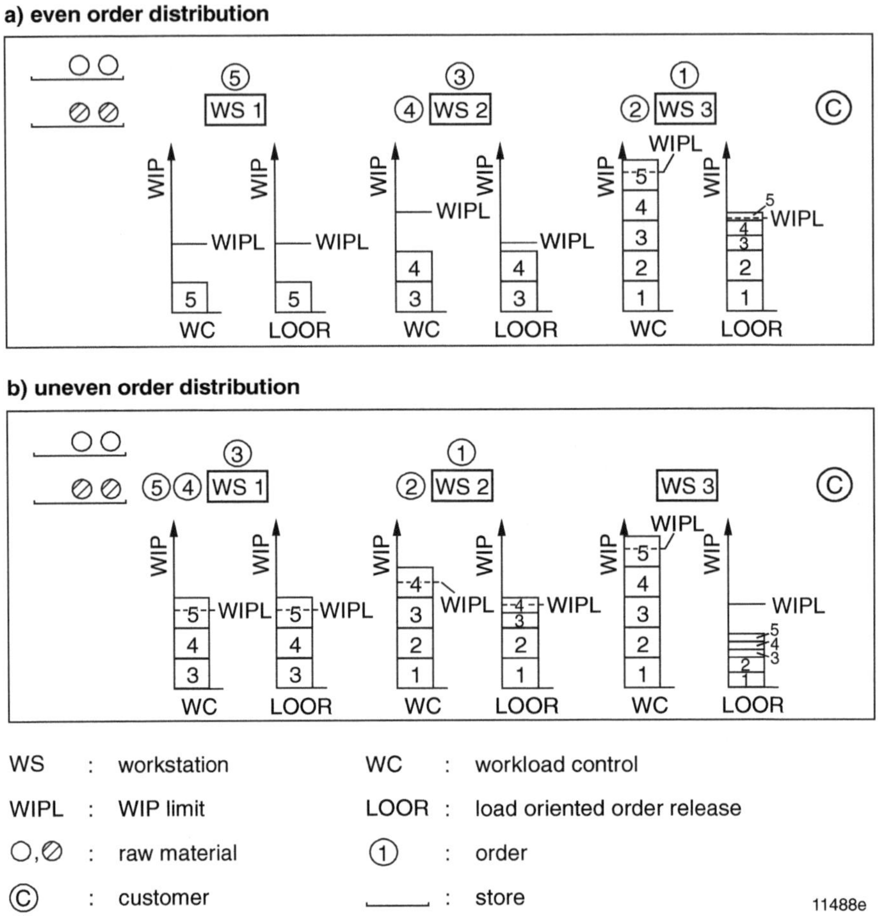


**Fig. 21.3** Comparison of Workload Control and Load Oriented Order Release with different order distributions on the production's workstations

Workstation 3. If jobs such as this are in the list of available orders LOOR then leads to an improved load balance in comparison to Workload Control.

Simulations conducted by Cigolini and Portioli-Staudacher seem to confirm this. In a production with a very complex material flow, releasing orders with discounted work contents attains better results than releasing orders without discounting the work content [Cigo-02*, Pero-96*].

Nevertheless, it is obvious that the load balance aspect of LOOR can still be improved because it is based on a heuristic principle.

5. *A manufacturing control method should cause as few sequencing interchanges as possible.*

The Load Oriented Order Release causes sequencing interchanges when releasing jobs in order to allow the load to be balanced. Moreover, the method has a tendency to release orders with many operations later than orders with fewer operations (see the corresponding discussion for Workload Control in Chap. 20).[2] It is thus practical to compensate for the impact of these sequencing interchanges of the scheduling while the orders flow through the production. Combining the method with a sequencing rule that considers the planned end date of the operations seems to be a reasonable solution.

6. *A manufacturing control method should be able to balance schedule deviations within the context of backlog control.*

It is the responsibility of manufacturing control to prevent a threatening backlog and to reduce a backlog once it has developed. The procedural rules of the Load Oriented Order Release do not consider a backlog control. The method should thus be combined with a backlog control in order to ensure a high delivery reliability even when there are schedule deviations. A basic concept for this is outlined in [Nyhu-09*].

7. *A manufacturing control method should take into consideration the bottleneck principle.*

The Load Oriented Order Release at least considers the bottleneck principle implicitly. Generally it is on the bottleneck workstations that the WIP limits are (first) exceeded. If the WIP falls below the limit on a workstation once an order is completed, the method also triggers the release of orders. When the WIP limit for a workstations is exceeded the release of orders for that workstation is hindered. The method is also able to control productions that have alternating bottlenecks.

8. *A manufacturing control method should be as simple as possible.*

The procedural rules of the Load Oriented Order Release require explaining first and foremost due to the discounting of the work content. This can cause considerable difficulties in convincing practitioners of the methods applicability. The method also requires feedback for the operations and thus an electronic system for compiling operation data is generally a prerequisite.

## 21.3 Setting the Method's Parameters

The Load Oriented Order Release is based on three parameters:

- the advance release window (Sect. 21.3.1),
- the workstations' discounting factors (Sect. 21.3.2), and
- the workstations' WIP limits (Sect. 21.3.3).

Setting the method's parameters based on a model is described in the indicated sections. Bechte recommends closely linking the WIP limits and the discount factors of the workstations. The corresponding procedure is described in Sect. 21.3.4.

Depending on how the parameters are set the Load Oriented Order Release can also emulate various other manufacturing control methods (for the respective

---

[2] This disadvantage of the method can be avoided by a minor change to the procedural rules (see Chap. 20).

parameter values see Sect. 21.3.5). LOOR is thus a generic manufacturing control method and is therefore particularly suitable for integrating into PPC software.

### *21.3.1 Advance Release Window*

The advance release window sets the maximum length of time within which an order can be released early i.e. before the planned start date has been reached. The order release is thus bound to the following condition:

$$DPS \leq TPL_0 + ARW \tag{21.2}$$

where

| | |
|---|---|
| DPS | planned start date [SCD], |
| $TPL_0$ | planning time point [SCD], |
| ARW | advance release window [SCD]. |

The advance release window indirectly influences the production's logistic objectives:

- The longer an advance release window is, the more orders are available for balancing the load. The effectiveness of the load balance thus has a tendency to increase with the length of the advance release window [Wien-97b]. Accordingly, a specific output rate can be achieved with a lower WIP level over a longer advance release window than with a shorter advance release window.
- The longer the advance release window is, the earlier orders can be released. There are three possible ways this can negatively impact the production: First, there is a tendency for the proportion of orders completed early to increase and with that the finished goods stores too. Second, there is an increased probability that the customer's desired changes to the orders will arrive *after* the orders have already been released. Third, there tends to be a greater proportion of non-urgent orders in the production which then compete with urgent orders for the available capacities [Wien-97b].

It can be assumed that the effectiveness of the load balance only climbs under-proportionally with the number of orders in the release list.[3] This should be taken into account when setting the advance release windows.

Generally, the advance release window can only compensate to a certain extent for fluctuations in the capacity requirements caused by changes in the order mix. When the capacity requirements strongly fluctuate the available capacities have to be adjusted.

[3] As clarification: If the release list includes only one order, the probability that a second order has a different order throughput is relatively high. However, if the list already contains a large number of orders, there is a greater probability that an additional order will be routed through the same workstations as a number of other orders in the list. The additional order can thus no longer improve the load balance.

### 21.3.2 Discounting Factor

Bechte recommends the discounting factor be tied to the WIP limits. The procedure for doing so is explained in Sect. 21.3.4 and has proven itself on the shop floor. Generally though, it is more practical for a number of reasons to set the discounting factor and WIP limits separately. One reason for this is that WIP frequently does not reach the WIP level on workstations minimally utilized. Accordingly, with the recommendation from Bechte there is a tendency to underestimate the probability of orders on this workstation being completed within a planning period [Pero-96*]. Perona and Portioli [Pero-96*] deliver a possible solution for this problem. A second reason is the possibility of WIP fluctuations with a high discount.

Generally, the following recommendations can be made for setting the discounting factor:

- When there is a clearly defined production bottleneck the discounting factor should be set rather high in order to avoid WIP fluctuations.[4] In contrast, when the capacities are well balanced, lower discounting factors can also be chosen. High WIP fluctuations are then prevented by exceeding the WIP limits on other workstations.
- The higher the throughput time on a workstation is, the lower the selected discounting factor should be. The planned throughput times of the workstations thus provide a suitable measure for coordinating the discounting factors of the various workstations.
- For workstations with critical capacities the work content used when booking released orders in the WIP accounts should be verified for a typical selection of orders. On the one hand, the discounting of orders with different throughput times should clearly differ up to the observed workstation. This is a critical condition for improving the load balance. On the other hand, the discounting should generally not be too high in order to limit the resulting WIP fluctuations.

It can be difficult to meet the conditions mentioned in the last point for all of the workstations, especially with orders that have a number of operations. One solution for this problem is to not tightly link the discounting factor with individual workstations; instead, the discounting factor can be set as a function of the (expected) throughput time of the order up until it arrives at the workstation.

Figure 21.4 depicts a possible curve of a function for the discounting factor. The discounting factor decreases from 100% for the direct WIP (planned throughput time up until arrival at the workstation = 0 SCD) to a minimum value (here: planned throughput time up until arrival on the workstation = 5 SCD). With a longer planned throughput time the discounting is continually conducted based on the minimum discounting factor.

[4] See also the discussion regarding the second guideline for designing manufacturing control methods in Sect. 21.2.

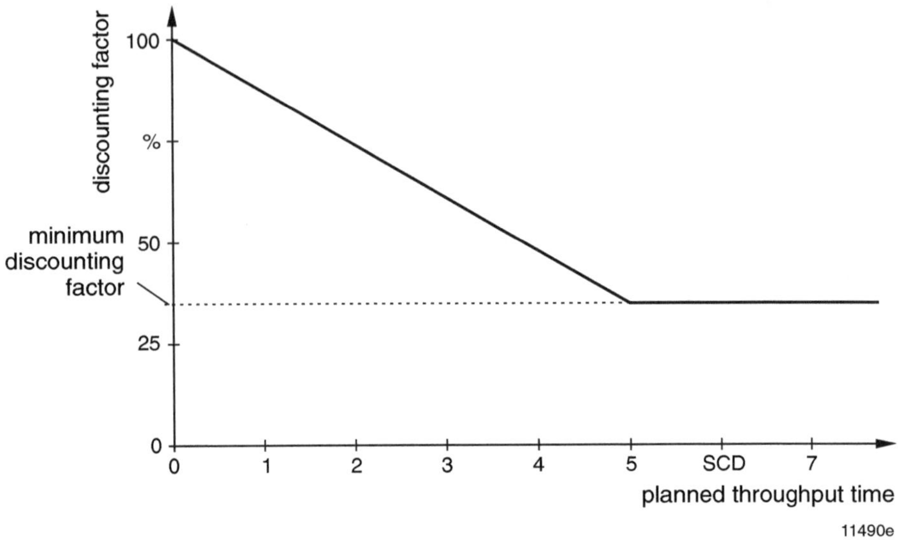


**Fig. 21.4** Setting the discounting factor as a function of the remaining planned throughput time up until the arrival at the workstation

## *21.3.3 WIP Limit*

The WIP limit sets the amount of booked WIP after which a workstation blocks the release of orders routed through it. The WIP limit of the workstation thus determines the WIP and utilization of the workstation as well as the throughput times of the orders. As with other manufacturing control methods it is practical to continually verify the WIP levels and, where possible, gradually reduce them in a continual improvement process.

The procedure for setting the WIP limits of workstations based on a model is similar to that previously explained for Workload Control (Sect. 20.3). It consists of two steps: a Logistic Positioning and the setting of the WIP limit based on the results of the positioning.

**Logistic Positioning** The Logistic Positioning is comprised of two steps:

- calculating the Logistic Operating Curves, and
- setting the planned WIP for the workstations.

Both steps are extensively discussed in Sect. 3.2.2.3.

**Calculating the WIP Limits** The WIP limit for a workstation should be set so that the resulting WIP corresponds as closely as possible to the planned WIP that was determined by the Logistic Positioning for the workstation. In order to estimate the correlation between the WIP and the WIP limit, an assumption is first made for the purpose of simplification. According to it, the mean booked

WIP (and thus the mean planned total WIP) corresponds approximately to the workstations WIP limit:

$$WIP_{bk,m} = WIP_{tot,m,plan} \approx WIPL \tag{21.3}$$

where

| | |
|---|---|
| $WIP_{bk,m}$ | mean booked WIP [hours], |
| $WIP_{tot,m,plan}$ | mean planned total WIP (booked WIP) [hours], |
| WIPL | WIP limit [hours]. |

The WIP limit can then be calculated from the planned WIP of the workstation and the ratio of the total WIP and direct WIP of the workstation:

$$WIPL = WIP_{m,plan} \frac{WIP_{tot,m,plan}}{WIP_{m,plan}} \tag{21.4}$$

where

| | |
|---|---|
| WIPL | WIP limit [hours], |
| $WIP_{tot,m,plan}$ | mean planned total WIP (booked WIP) [hours], |
| $WIP_{m,plan}$ | mean planned WIP [hours]. |

The mean planned WIP is known from the Logistic Positioning of the workstation. The ratio of the total WIP to the direct WIP can be determined relatively easily for each order. Figure 21.5 depicts the throughput of an order with four operations which are conducted on WS 1–4. The throughput elements are presented for the first three operations. Below these throughput elements the WIP areas with which the orders on WS 3 are considered, are mapped.

In comparison to Workload Control the discounting of the orders also needs to be considered here. After the order is released, the WIP is entered with a booking work content – which is reduced by the discounting factor of WS 1 and 2 (thus in the example with a quarter of the work content) – into the WIP account of WS 3. The product of the throughput time at WS1 and the booking work content results in the indirect WIP area of the order during its throughput on Workstation 1. Once the operation on WS1 is completed the booking work content increases, since the discounting which occurs due to WS 1 is no longer applicable. The order is now weighted with half of the work content. With its completion on WS 2, the order enters WS 3: For the throughput time on this workstation, the entire work content is taken into consideration in the WIP account.

The ratio of the total WIP area to the direct WIP area can thus be calculated for an order as:

$$\frac{AWIP_{tot}}{AWIP_{dir}} = \frac{\sum\limits_{n=1}^{j-1}\left(TTP_{i,n} \cdot \prod\limits_{k=n}^{j-1}\left(\frac{DFA_k}{100}\right)\right) + TTP_{i,j}}{TTP_{i,j}} \tag{21.5}$$

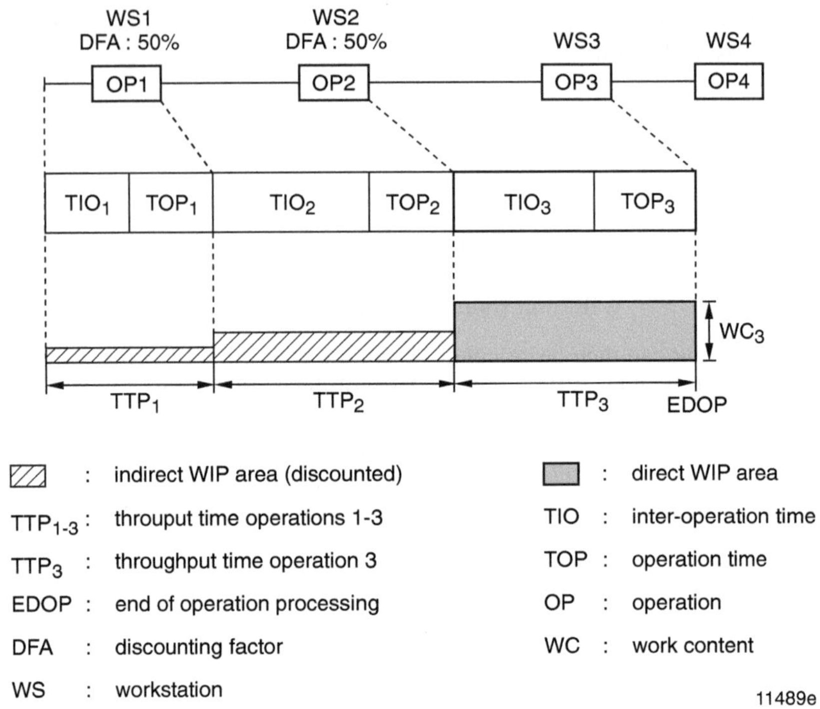


**Fig. 21.5** Representation of the direct and indirect WIP in throughput elements

where

| | |
|---|---|
| $AWIP_{tot}$ | total WIP area [hours·SCD], |
| $AWIP_{dir}$ | direct WIP area [hours·SCD], |
| $TTP_{i,n}$ | throughput time of Order *i* on Workstation *n* [SCD], |
| $DFA_k$ | discounting factor Workstation *k* [%], |
| j | index number of the observed workstation [–]. |

This procedure can also be expanded to a number of orders. In order to do so the ratio of the total WIP area to the direct WIP area of an order should be weighted with its work content. Accordingly,

$$\frac{AWIP_{tot}}{AWIP_{dir}} = \frac{\sum_{i=1}^{NO}\left(WC_{i,j}\cdot\left(\sum_{n=1}^{j-1}\left(TTP_{i,n}\cdot\prod_{k=n}^{j-1}\left(\frac{DFA_k}{100}\right)\right)+TTP_{i,j}\right)\right)}{\sum_{i=1}^{NO}\left(WC_{i,j}\cdot TTP_{i,j}\right)} \tag{21.6}$$

where

| | |
|---|---|
| $AWIP_{tot}$ | total WIP area [hours·SCD], |
| $AWIP_{dir}$ | direct WIP area [hours·SCD], |
| $WC_{i,j}$ | work content of Order *i* on Workstation *j* [hours], |

| | |
|---|---|
| $TTP_{i,n}$ | throughput time of Order *i* on Workstation *n* [SCD], |
| $DFA_k$ | discounting factor Workstation *k* [%], |
| j | index number of the observed workstation [–], |
| NO | number of orders [–]. |

In order to set the WIP limits based on a model, the planned throughput time for a typical quantity of orders needs to be determined. The WIP level results then from the product of the planned WIP and the ratio of the booking WIP and direct WIP of the workstation. The idea of using a typical quantity of orders for dimensioning WIP limits is based on a method from Bertrand and Wortmann for dimensioning Workload Control [Bert-81*].

With a periodic order release, WIP between two release points decreases by the work completed during the period. This should be taken into consideration when dimensioning the WIP limits (see [Bech-84, Bert-81*]):

$$WIPL_{j,period} = WIPL_{j,event} + OUT_{plan,j} \tag{21.7}$$

where

| | |
|---|---|
| $WIPL_{j,period}$ | WIP limit for Workstation *j* with periodic order release [hours], |
| $WIPL_{j,event}$ | WIP limit for Workstation *j* with event oriented order release [hours], |
| $OUT_{plan,j}$ | planned output from Workstation *j* [hours]. |

## 21.3.4 Setting the Discounting Factor According to Bechte

Bechte suggests a model based procedure for calculating the discounting factor of a workstation from the WIP limit [Bech-84]. The methodology for doing so is extensively discussed in [Wien-97b].

Bechte defines the loading percentage of a workstation as the ratio of the WIP limit to the period's planned output [Bech-84]:

$$LP = \frac{WIPL}{OUT_{plan}} \cdot 100 \tag{21.8}$$

where

| | |
|---|---|
| LP | loading percentage [%], |
| WIPL | WIP limit [hours], |
| $OUT_{plan}$ | planned output [hours]. |

Obviously the probability with which an order will be processed on a workstation within a release period is higher, the lower the loading percentage is. Bechte therefore sets the inverse value of the loading percentage as the discounting factor of a workstation [Wien-97b].

$$DFA = \frac{100}{LP} \cdot 100 \tag{21.9}$$

where

DFA discounting factor [%],
LP loading percentage [%].

This procedure has proven itself when applied on the shop floor. Nonetheless, methodically it seems more practical not to define the discounting factor as a function of the WIP limit, because the two parameters serve different purposes. The WIP limit determines the WIP level of a workstation, whereas the discounting factor serves to calculate a suitable load measure on subsequent workstations. Decoupling the WIP limit from the discounting factor also allows other order release methods to be emulated by LOOR. This is further explained in the following section.

### *21.3.5 Emulating Other Manufacturing Control Methods by Setting the Parameters*

The Load Oriented Order Release is able to emulate various manufacturing control methods simply through the selection of the method's parameters. This is particularly beneficial for the development of PPC software: If the Load Oriented Order Release is integrated in PPC software, it can then also simultaneously control a production using Workload Control, CONWIP, Bottleneck Control or a Due Date Oriented Order Release. We will now turn our discussion to how these parameters should be set in order to emulate these manufacturing control methods. Table 21.2 summarizes the required parameter settings.

**Table 21.2** Parameter settings for various manufacturing control methods

| Manufacturing control method | Parameter | | |
|---|---|---|---|
| | WIP limit [hours] or [–] | Discounting factor [%] | Advance release window [SCD] |
| Load oriented order release | * | * | * |
| Workload control | * | 100 | * |
| CONWIP | ∞, except last WS | 100 | * |
| Bottleneck control | ∞, except bottleneck WS | 100 | * |
| Due date order release | ∞ | omitted | 0 |

*WS* workstation
* Set by user

**Workload Control** The Workload Control (see Chap. 20) can be emulated by the Load Oriented Order Release when all discount factors are set at 100%. By doing so the discounting aspect is prevented and the orders are credited to the workstations' WIP accounts with their complete work content.

**CONWIP** CONWIP regulates a production line's WIP (see Chap. 18) thus only the last workstation of the production line is relevant for the order release. Accordingly, the WIP limits of the remaining workstations should be set to infinity in order to emulate CONWIP with the Load Oriented Order Release. Furthermore, the discounting factor for the workstations should be set at 100% in order to prevent the discounting. In order to emulate the original method, the WIP should be measured in either the number of orders or parts. Moreover, the order release should be event oriented.[5]

**Bottleneck Control** Bottleneck Control (see Chap. 19) regulates WIP in a production line up to and including the bottleneck workstation. In order to emulate Bottleneck Control with the Load Oriented Order Release the WIP limits of the remaining workstations should be set to infinity. Similar to CONWIP the discounting factor should be set at 100%. Moreover, it is practical to combine Bottleneck Control with an event oriented order release.

**Due Date Based Order Release** With Due Date Based Order Release, jobs are released once their planned start date has been reached (see Chap. 17). LOOR performs this type of order release when the workstations' WIP limits are set extremely high and the advance release window set at zero.

## 21.4 Implementation Across the Supply Chain

It would be unusual to implement the Load Oriented Order Release across supply chains. This would require production lot sizes to be coordinated and generally a very close supply relationship between the suppliers and the receiving enterprises. Such constellations are rarely found in the industry.

## 21.5 Application

The application range of LOOR corresponds with an important difference to Workload Control (see the extensive discussion in Sect. 20.5): LOOR can also be implemented when the location of workstations in the order throughput clearly

[5] When there is no clear final workstation, further adjustments are required.

varies. Since LOOR causes WIP fluctuations due to the discounting of the work content though it seems obvious that Workload Control should be implemented when the positions of the workstations in the order throughput are almost constant.

The Load Oriented Order Release has been/is applied in various industrial enterprises. Bechte reports of considerable WIP level and throughput time reductions attained with this method [Bech-88*]. A number of application accounts that are worth reading are found in [Holz-91, Knec-91, Sohn-91, Stor-91, Wien-91]. This method is currently being successfully implemented in one of Siemens' plants for controlling the production of electronic components.

Furthermore, there are also enterprises that have abandoned controlling the production with LOOR. On the one hand, there are reasons for this that are not related to the method, for example, the introduction of a new PPC system which does not emulate LOOR. On the other hand, introducing LOOR is to some extent associated with overstated expectations. LOOR can reliably control the WIP and throughput times of the production. However, it cannot revise wrong decisions made when accepting orders or planning the production program. If unrealistic delivery dates are promised or if loads are accepted that exceed available capacities the orders will obviously be completed late also when applying the Load Oriented Order Release.

## References

[Bech-84] Bechte, W.: Steuerung der Durchlaufzeit durch belastungsorientierte Auftragsfreigabe bei Werkstattfertigung. VDI Progress Reports, Series 2, No.70, Düsseldorf (1984)

[Bech-88*] Bechte, W.: Theory and practice of load-oriented manufacturing control. Int. J. Prod. Res. **26**(3), 375–395 (1988)

[Bert-81*] Bertrand, J.W.M., Wortmann, J.C.: Production Control and Information Systems for Component-Manufacturing Shops. Studies in Production and Engineering Economics, vol. 1. Elsevier, Amsterdam (1981)

[Brei-02b*] Breithaupt, J.-W., Land, M., Nyhuis, P.: The workload control concept: theory and practical extensions of load oriented order release. Prod. Plan. Control **12**(7), 625–638 (2002)

[Cigo-02*] Cigolini, R., Portioli-Staudacher, A.: An experimental investigation on workload limiting methods with ORR policies in a job shop environment. Prod. Plan. Control **12**(7), 602–612 (2002)

[Holz-91] Holzhüter, E., Friedrichs, E.: Erfahrungen mit dem System KPSF zur Belastungsorieniterten Fertigungssteuerung in einem Maschinenbaubetrieb mittlerer Größe. In: Wiendahl, H.-P. (ed.) Anwendung der Belastungsorientierten Fertigungssteuerung, pp. 139–152. Carl Hanser Verlag, Munich/Vienna (1991)

[Knec-91] Knecht, R.: Belastungsorientierte Auftragsfreigabe der Hilti AG. In: Wiendahl, H.-P. (ed.) Anwendung der Belastungsorientierten Fertigungssteuerung, pp. 179–193. Carl Hanser Verlag, Munich/Vienna (1991)

[Knol-91*] Knolmayer, G.: A widely acclaimed method of load-oriented job release and its conceptual deficiencies. In: Fandel, G., Zäpfel, G. (eds.) Modern Production Concepts, pp. 219–236. Springer, Berlin (1991)

[Land-96*] Land, M., Gaalman, G.: Workload control concepts in job shops. A critical assessment. Int. J. Prod. Res. **46–47**, 535–548 (1996)

[Nyhu-09*] Nyhuis, P., Wiendahl, H.-P.: Fundamentals of Production Logistics. Theory, Tools and Applications. Springer, Berlin (2009)

[Pero-96*] Perona, M., Portioli, A.: An enhanced loading model for the probabilistic workload control under workload imbalance. Prod. Plan. Control **7**(8), 68–78 (1996)

[Sohn-91] Sohn, H.: Erfahrungen bei der Implementierung der Belastungsorientierten Auftragsfreigabe. In: Wiendahl, H.-P. (ed.) Anwendung der Belastungsorientierten Fertigungssteuerung, pp. 267–274. Carl Hanser Verlag, Munich/Vienna (1991)

[Stor-91] Storfinger, R.: Die Belastungsorientierte Fertigungssteuerung für mechanische Fertigungen in der BMW AG. In: Wiendahl, H.P. (ed.) Anwendung der Belastungsorientierten Auftragsfreigabe. Carl Hanser Verlag, Munich/Vienna (1991)

[Wien-91] Wiendahl, H.-P. (ed.): Anwendung der Belastungsorientierten Auftragsfreigabe. Carl Hanser Verlag, Munich/Vienna (1991)

[Wien-97b] Wiendahl, H.-P.: Fertigungsregelung. Logistische Beherrschung von Fertigungsabläufen auf Basis des Trichtermodells. Carl Hanser Verlag, Munich/Vienna (1997)

# Chapter 22
# Order Release with Linear Programming

Releasing orders with linear programming was introduced by Irastorza and Deane in 1974 [Iras-74*] and is thus one of the oldest methods for releasing orders. Nonetheless, it combines not only essential aspects of methods developed later, but also contains further elements such as weighting logistic target criteria. However, this method receives little attention in publications about production control, perhaps because Irastorza and Deane themselves wrote few publications about the method. Moreover, the method requires a module that can solve linear equation systems. These requirements were still a substantial hurdle for many manufacturers in the 1970s.

In the following, the procedural rules for this method (which we will refer to here as "Order Release with Linear Programming") will first be described (Sect. 21.1). We will then examine these rules from the perspective of logistics (Sect. 22.2). Next, Sect. 21.3 is dedicated to setting the method's parameters, whereas the reasons for why applying this method to control production across a supply chain would be unusual are presented in Sect. 22.4. Our discussion of this method will then conclude with a discussion of areas where it can be applied (Sect. 22.5).

## 22.1 Method Description

Order Release with Linear Programming (ORLP) is simular in a number of aspects to Load Oriented Order Release or Workload Control (see Fig. 22.1 for a presentation of the principle behind it):

- A release list (or job pool) contains all known but not yet released orders. Unlike the other methods mentioned though the early release of orders is not restricted by an advance release window.
- The method tries to control the WIP on the workstations as close to the planned WIP as possible. The WIP is allocated to a workstation from its release up until its completion on the workstation. This corresponds to the Workload Control's definition of WIP.

H. Lödding, *Handbook of Manufacturing Control*,
DOI 10.1007/978-3-642-24458-2_22, © Springer-Verlag Berlin Heidelberg 2013

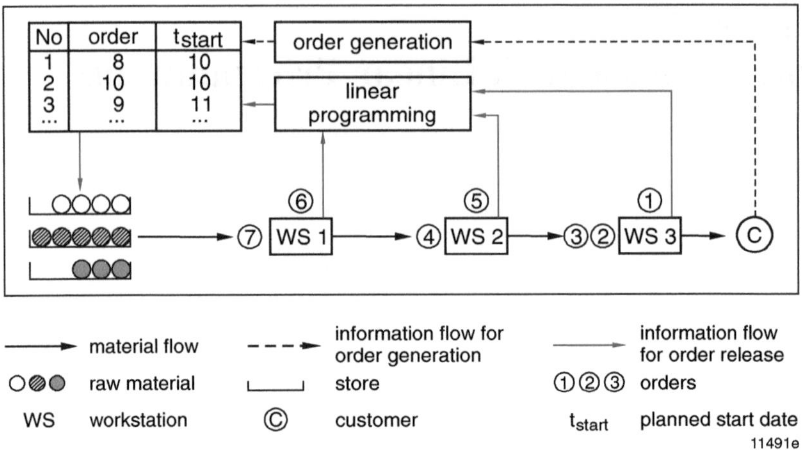


**Fig. 22.1** Principle of the order release with linear programming

In other aspects ORLP differs greatly from the remaining methods for releasing orders. It thus implements an optimization process in order to determine which jobs should be released from the release list and which should be held back. For this purpose a *target function* and *constraints* are defined.

## Target Function

Irastorza and Deane define a decision variable for each job in the release list [Iras-74*]. The value of a decision variable for an order is 1 if the order is to be released and 0 if it is to be held back:

$$X_j = \begin{cases} 0, \text{where Order j is not released} \\ 1, \text{where Order j is released} \end{cases} \tag{22.1}$$

where $X_j$ decision variable [−].

A decision to release the order is thus clearly defined by setting the decision variables for the jobs in the release list.

The target function is a function of the orders' decision variables. It calculates a desired value that is greater,

- the larger the deviations of the actual WIP from the planned WIP on the workstations are, and,
- the earlier the orders are released before the planned completion date.

The optimization algorithm sets the decision variables so that the target value is as small as possible.

Irastorza defines the target function as follows [Iras-74*]:

$$TV = \sum_{j=1}^{NWS} \left(WF_{j,WIP<WIPplan} \cdot \Delta WIP_{j,WIP<WIPplan}\right) + \sum_{j=1}^{NWS} \left(WF_{j,WIP>WIPplan} \cdot \Delta WIP_{j,WIP>WIPplan}\right) - \sum_{i=1}^{NO} \left(f\left(EDO_{plan,i}\right) \cdot X_i\right) \tag{22.2}$$

where

| | |
|---|---|
| TV | target value [−], |
| $WF_{j,WIP<WIPplan}$ | weighting factor Workstation *j* for the WIP difference, where actual WIP< planned WIP [−/hrs], |
| $\Delta WIP_{j,WIP<WIPplan}$ | WIP difference on Workstation *j*, where actual WIP < planned WIP [hrs], |
| $WF_{j,WIP>WIPplan}$ | weighting factor Workstation *j* for the WIP difference, where actual WIP > planned WIP [−/hrs], |
| $\Delta WIP_{j,WIP>WIPplan}$ | WIP difference on Workstation *j*, where actual WIP > planned WIP [hrs], |
| $EDO_{plan,i}$ | planned end of processing for Order *i* [SCD], |
| $X_i$ | decision variable [−], |
| NWS | number of workstations [−], |
| NO | number of orders [−]. |

The components of the target function will now be explained, however, the first two sums weight the deviation from the planned WIP on the workstation. Both an overload (actual WIP is greater than the planned WIP) and an underload (actual WIP is less than the planned WIP) are assessed as deviations from the target. The weighting factor allows overloads and underloads to be sanctioned to different extents. The greater the weighting factor is, the more weight is given to the WIP deviation on the workstation. The WIP differences are calculated as follows:

$$\Delta WIP_{j,WIP<WIPplan} = \begin{cases} WIP_{j,plan} - WIP_{j,act} & \text{where } \mathrm{WIP}_{j,plan} > WIP_{j,act} \\ 0 & \text{where } \mathrm{WIP}_{j,plan} \leq WIP_{j,act} \end{cases} \tag{22.3}$$

$$\Delta WIP_{j,WIP>WIPplan} = \begin{cases} WIP_{j,act} - WIP_{j,plan} & \text{where } \mathrm{WIP}_{j,act} > WIP_{j,plan} \\ 0 & \text{where } \mathrm{WIP}_{j,act} \leq WIP_{j,plan} \end{cases} \tag{22.4}$$

where

| | |
|---|---|
| $\Delta\ WIP_{j,WIP<WIPplan}$ | WIP difference on Workstation *j*, where actual WIP < planned WIP [hrs], |
| $WIP_{j,plan}$ | planned WIP Workstation *j* [hrs], |
| $WIP_{j,act}$ | actual WIP Workstation *j* [hrs], |
| $\Delta WIP_{j,WIP>WIPplan}$ | WIP difference on Workstation *j*, where actual WIP > planned WIP [hrs]. |

The planned WIP levels are a parameter that needs to be defined. The actual WIP of a workstation is calculated as the sum of the work contents for all orders that are released but not yet completed on the workstation. In this calculation, orders released during the planning runs are already considered. The setting of the decision variables thus impacts the calculated WIP difference. This will be further explained in a later example.

The last sum in Eq. 22.4 is a measure of the order's urgency. The due date function $f(EDO_{plan})$ should be defined so that the calculated due date weight is greater the earlier the end date of an order is. Irastorza and Deane generally leave it to the user to select a function suitable for the case it is being applied in. For their simulation elements they apply the following due date function [Iras-74*][1]:

$$f\left(EDO_{plan}\right) = \frac{C}{0.1 + (EDO_{plan} - TPL_0)} \tag{22.5}$$

where

| | |
|---|---|
| $EDO_{plan}$ | planned end of order processing [SCD], |
| C | constant [SCD], |
| $TPL_0$ | planning time point [SCD]. |

**Constraints**

Irastorza and Deane define constraints in order to control the work content that is being released for the entire production or for individual workstations. Where possible, the optimization algorithm sets the decision variable of the orders so that the constraints are met. Irastorza and Deane suggest the following constraints [Iras-74*].

*Upper and lower limits for the releasable work content for the entire production:* The sum of the work content released for all of the workstations should be within an upper and lower limit:

$$WCR_{LL} \leq \sum_{i=1}^{NO} \left( \left( \sum_{j=1}^{NWS} WC_{i,j} \right) \cdot X_i \right) \leq WCR_{UL} \tag{22.6}$$

where

| | |
|---|---|
| $WCR_{LL}$ | lower limit of releasable work content [hrs], |
| $WCR_{UL}$ | upper limit of releasable work content [hrs], |
| $WC_{i,j}$ | work content for Order *i* on Workstation j [hrs], |
| $X_i$ | decision variable for Order i [−], |
| NWS | number of workstations [−], |
| NO | number of orders [−]. |

[1] Irastorza and Deane measure time in hours and not in shop calendar days as is done here.

*Upper limit of the releasable work content for individual workstations:* The releasable work content for specific workstations should not exceed an upper limit:

$$\sum_{i=1}^{NO} \left(WC_{i,j} \cdot X_i\right) \leq WCR_{UL,j} \tag{22.7}$$

where

| | |
|---|---|
| $WC_{i,j}$ | work content for Order *i* on Workstation j [hrs], |
| $X_i$ | decision variable for Order *i* [−], |
| $WCR_{UL,j}$ | upper limit of releasable work content for Workstation *j* [hrs], |
| NO | number of orders [−]. |

We will now examine the procedural rules more closely based on the example of a production with three workstations (Fig. 22.2). The release list contains four jobs (Nr. 5–8). For each order the work content on the workstations is provided as well as the value of the decisions variable.

The planned WIP and WIP accounts are depicted for each workstation. A WIP account contains all released orders that are either on the workstation itself or on preceding workstations. The calculation of the target function and the constraints are framed by a box and depicted in different tables. In order to make decisions about the release of jobs the equation system (target function and constraints) is programmed into operations research software, which can solve the linear equations or binary integer optimization problems. The basis for the calculations is clearly presented by Hillier and Lieberman [Hill-90*] for example. For the (simplified) example Microsoft Excel's optimization function is used. The program calculates the combination of decision variables for which the target function in consideration of the constraints is (nearly) as small as possible. The result of the optimization calculation is depicted in Fig. 22.2b. Orders 5 and 7 are released (decision variable $X = 1$); Orders 6 and 8 are held back until the next release decision (decision variable $X = 0$).

With this release decision the planned WIP on Workstation 1 is not quite met, whereas on WS 2 it is attained and on WS 3 it is exceeded by 3 h (Table B). The sum of the due date weights for the releasable orders is 4.9 (Table C) and altogether the resulting target value is −0.9 (Table D). The constraints are met (Table E). All in all 22 h of work are released. The upper limit for WS 3 is not exceeded. Every other release decision would increase the target value or breech the constraints.

Once the release decision has been made the orders are released (Step A for Order 4 in Fig. 22.2c), credited to the workstations' WIP accounts (Steps B and C) and removed from the release list (Step D). Steps E to I describe the same procedure for Order 7. The two jobs that are held back (Order 6 and 8) remain on the release list.

## 22.2 Discussion of the Procedural Rules

Order Release with Linear Programming is characterized by two properties:

- the load balancing conducted via an optimization algorithm, and
- the multitude of parameters involved.

The latter complicates on the one hand, the assessment of the method in that the suitability of releasing orders with linear programming is even more closely linked to the selection of the method's parameters than with other methods. On the other hand, it is this variety of the parameters that makes it possible to adjust the method

**a) Initial situation (before order release, SCD 5).**

| No | EDO | 1 | 2 | 3 | X |
|---|---|---|---|---|---|
| 5 | 13 | | 6 | 6 | 0 |
| 6 | 17 | 2 | 4 | 2 | 0 |
| 7 | 19 | 4 | 2 | 4 | 0 |
| 8 | 20 | 4 | | 6 | 0 |

WS 1:

| No | WC | ΣWC |
|---|---|---|
| 4 | 3.0 | 3.0 |

WS 2:

| No | WC | ΣWC |
|---|---|---|
| 2 | 4.0 | 4.0 |
| 4 | 2.0 | 6.0 |

WS 3:

| No | WC | ΣWC |
|---|---|---|
| 1 | 3.0 | 3.0 |
| 3 | 4.0 | 7.0 |
| 2 | 6.0 | 13.0 |
| 4 | 2.0 | 15.0 |

a)

| | | X·WCij | | |
|---|---|---|---|---|
| No | X | 1 | 2 | 3 |
| 5 | 0 | 0 | 0 | 0 |
| 6 | 0 | 0 | 0 | 0 |
| 7 | 0 | 0 | 0 | 0 |
| 8 | 0 | 0 | 0 | 0 |
| Σ | 0 | 0 | 0 | 0 |
| total | 0 | | | |

b)

| | | | | underload | | overload | |
|---|---|---|---|---|---|---|---|
| WS | WIP | RE | WIP+RE | WF | ΔWIP | WF | ΔWIP |
| 1 | 3 | 0 | 3 | 1 | 5 | 2 | 0 |
| 2 | 6 | 0 | 6 | 1 | 8 | 2 | 0 |
| 3 | 15 | 0 | 15 | 2 | 7 | 1 | 0 |
| | | | | 27 | | 0 | |

c)

| | due dates | | | |
|---|---|---|---|---|
| No | EDO | f(EDO) | X | X·f(EDO) |
| 5 | 13 | 3.1 | 0 | 0.0 |
| 6 | 17 | 2.1 | 0 | 0.0 |
| 7 | 19 | 1.8 | 0 | 0.0 |
| 8 | 20 | 1.7 | 0 | 0.0 |
| Σ | - | - | 0 | 0.0 |

$$f\,(EDO) = \frac{25}{0.1 + (EDO - 5)}$$

d)

| | target function | | | target value |
|---|---|---|---|---|
| criterion | underload | overload | due dates | |
| calculation | Σ(ΔWIP·WF)j | Σ(ΔWIP·WF)j | ΣXj·f(TAEj) | |
| sign | + | + | - | |
| value | 27.0 | 0.0 | 0.0 | 27.0 |

e)

| | conditions | | | |
|---|---|---|---|---|
| criterion | released work | | | |
| | total | WS1 | WS2 | WS3 |
| act | 0.0 | 0.0 | 0.0 | 0.0 |
| max | 30.0 | ∞ | ∞ | 20 |
| min | 20.0 | - | - | - |

WS : workstation
WIP$_{tar}$ : target WIP
RE : released work content
ΔWIP : difference from target WIP ( ≥0)
○,⊘ : raw material
①,② : orders
⎵ : store

Ⓒ : customer
WIP : work-in-progress
WF : weighting factor
WC : work content
X : decision variable
EDO : end of order processing

11492e

**Fig. 22.2** How the order release with linear programming functions

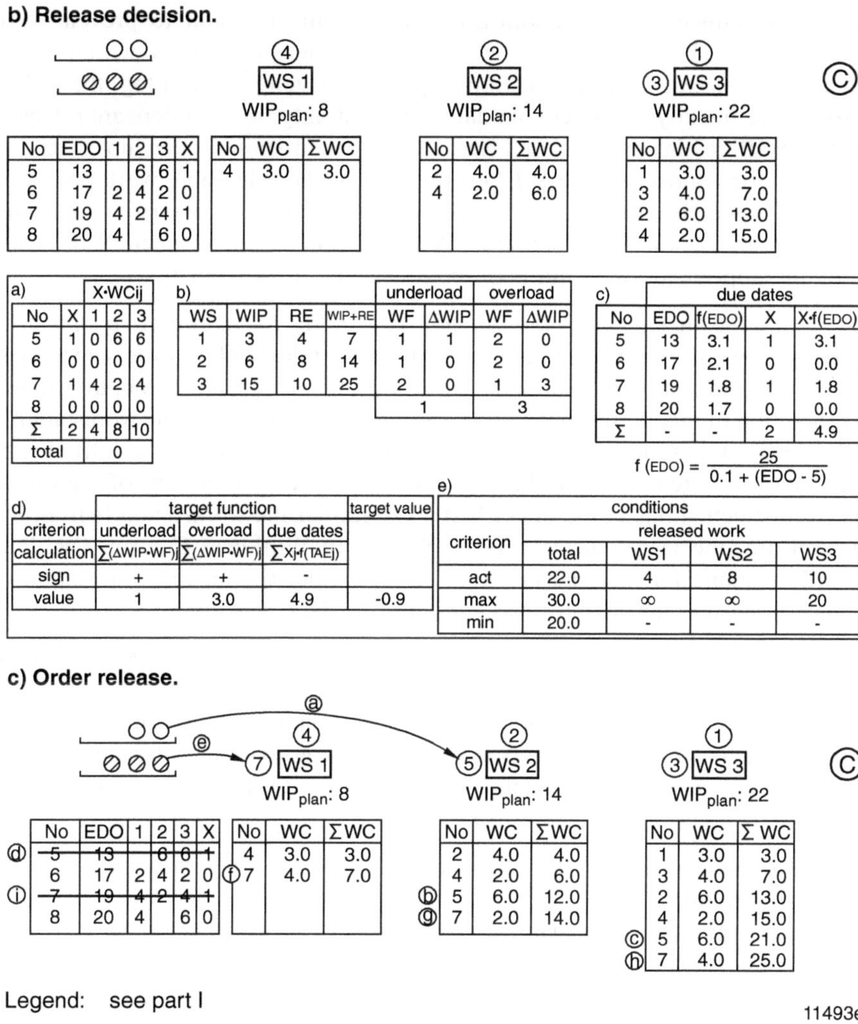


Fig. 22.2 (continued)

to fit different production environments. In the following, we will turn our focus to a discussion about the procedural rules for ORLP based on the guidelines for designing production control methods.

1. *A manufacturing control method should allow the WIP of a production or workstation to be regulated at a defined level.*

ORLP offers two possibilities for controlling the WIP of a production or workstation. First, the target function provides planned WIP levels for the individual workstations. WIP deviations increase the value of the target function and are extensively reduced by the optimization algorithm. The aim to prevent WIP deviations nonetheless competes with that of releasing orders with heavily weighted due dates. If the production in general develops a backlog, the orders receive more heavily weighted due dates. Accordingly, more orders are released and the WIP in the production increases. The constraints open up a second possibility for controlling the WIP in the form of the releasable work content's upper and lower limits. In order to do so though, it is necessary to take into consideration the current production WIP as well as the planned WIP when setting the upper and lower limits of the releasable work content.

2. *A manufacturing control method should reduce WIP fluctuations to a minimum.* The longer the orders' throughput times are from the release of the orders up to their arrival on a workstation in relation to the workstation throughput time, the greater the proportion of orders still found on preceding workstations is in the workstation's WIP account. Accordingly, the planned WIP of the workstations will usually clearly exceed the preferred direct WIP. Consequently, a certain amount of leeway is created for fluctuations in the direct WIP: The farther the workstation is from the order release, the larger the leeway is. Similar to Workload Control, ORLP is primarily suitable for material flows in which the position of the workstation in the material flow is for the most part constant (see the discussion in Sect. 20.2).

3. *A manufacturing control method should cause as little blocked WIP as possible.* ORLP blocks orders at the release point. Moreover, unlike a Due Date Based Order Release the method can also block orders whose planned start date has already passed. In exchange, it also releases orders when doing so ensures the utilization of a workstation. During the throughput of the orders ORLP does not cause any blocked WIP.

4. *A manufacturing control method should contribute to closely aligning the load to the capacities.*

Load balancing is the main characteristic of Order Release with Linear Programming. Unlike the rest of the methods with a load balance described here, ORLP weighs underloading workstations in comparison to overloading them. Furthermore, the optimization algorithm is presumably more effective than the simple heuristics used in Workload Control or Load Oriented Order Release.

However, it is not ensured that ORLP optimally releases the orders. Similar to Workload Control, the method is based on the assumption that WIP between the order release and its completion on a workstation serves as a degree of protection for the workstation from breaks in the material flow. Unlike Load Oriented Order Release, ORLP does not consider the position of WIP. The effectiveness of the load balance is thus dependent on the reliability with which the distribution of WIP on the workstations can be forecasted (see also the discussion on Load Oriented Order Release).

5. *A manufacturing control method should cause as few sequencing interchanges as possible.*

The Order Release with Linear Programming causes sequencing interchanges as the jobs are released. The method should thus be combined with a sequence which takes into consideration the planned due date of the orders. Doing so should again compensate for the negative impact of the sequencing interchange on the schedule reliability (at least to some extent). Once the orders are released the method does not cause any further sequence interchanges.

6. *A manufacturing control method should be able to balance schedule deviations within the context of backlog control.*

It is the responsibility of production control to prevent a threatening backlog and to reduce an existing backlog – in order to do so capacities are temporarily increased. The procedural rules for ORLP do not consider a backlog control. The method should thus be supplemented with a backlog control in order to be able to ensure a high level of schedule reliability even when there are schedule deviations.

7. *A manufacturing control method should take into consideration the bottleneck principle.*

Unlike Workload Control or Load Oriented Order Release, ORLP is not inherently rigid in its orientation on bottlenecks. Even when the WIP limit on the bottleneck workstation has been exceeded, the method continues to release orders for the bottleneck station when doing so reduces the underload on other workstations. Since a bottleneck workstation determines the output rate of a production, this has a tendency to lead to WIP escalations without increasing the output rate of the production on the whole. This applies at least when there is a clearly defined bottleneck.

It is however possible to orient the production more closely on the bottleneck by suitably defining the upper limit of the releasable work content for the bottleneck workstation.

8. *A manufacturing control method should be as simple as possible.*

The complexity of releasing orders with linear programming is probably its greatest weakness. The method not only requires feedback regarding the individual operations, ORLP also necessitates that numerous parameters are set for it. Without an experienced production controller who has a solid understanding of how ORLP functions, the latter can rarely be successfully managed. Furthermore, few PPC systems include the optimization methods required for executing this method.

## 22.3 Setting the Method's Parameters

Setting the parameters for the Order Release with Linear Programming determines both its character as well as its aptitude. The following sections present how the parameters of the target function and constraints can be set. Where possible, references to corresponding procedures for other production control methods will be made. The following parameters will be addressed:

- planned WIP of the workstations (Sect. 22.3.1),
- weighting factors for the underload and overload (Sect. 22.3.2),
- definition of the due date function (Sect. 22.3.3),
- upper and lower limits for the work content that is to be released for the production (Sect. 22.3.4), and
- upper limit for the work content that is to be released for individual workstations (Sect. 22.3.5).

### *22.3.1 Planned WIP for the Workstations*

According to the definition of Order Release with Linear Programming, the (booking) WIP of a workstation contains the work content of all released orders not yet completed on the workstation. It is thus comprised of both direct WIP on a workstation and indirect WIP on preceding workstations. This WIP definition corresponds exactly to that of Workload Control. The planned WIP can thus be set with the same procedure as the WIP limits for the workstations in Workload Control method (see Sect. 20.3).

### *22.3.2 Weighting Factor for Underload and Overload*

The weighting factors for the under and overload determine the weight with which the deviations from the planned WIP are evaluated within the target function. Their setting cannot be directly derived from the logistic objectives. It is however possible to formulate general guidelines:

- The weighting factors on the workstations with critical capacities should be selected at a relatively high level.
- The weighting factor for the underload on the workstations with non-critical capacities should be set low. Doing so will prevent an underload on these workstations from triggering the release of orders that load workstations with critical capacities.
- With a periodic order release, an overload should not be heavily weighted when the workstation has sufficient capacities to reduce the WIP during the release period.
- The weighting factor for the overload and underload should be coordinated with the due date function in order to prevent the undesired dominance of one of the target criteria.

### 22.3.3 Definition of the Weight Function

The definition of the weight function is critical to the effectiveness of the Order Release with Linear Programming. When releasing orders the method weighs between the quality of the load balancing and the schedule reliability. Thus, inappropriate due date weights endanger the schedule reliability of the production. Irastorza and Deane leave it up to the user of ORLP to find a function well suited to their needs. For their simulation trials they used a very simple due date function (Eq. 22.5). Due to a number of reasons, however, this choice of function is not suitable for most industrial applications: It implicitly requires constant planned throughput times since it allocates orders with the same planned completion date the same due date weight. If the planned throughput times of two orders with the same planned completion date differ, it is however practical to allocate the order with the longer planned throughput times a heavier weight. Moreover, they can also be negatively weighted when planned completion dates are passed. The optimization algorithm thus receives an incentive to hold orders such as these back. This is obviously not practical. A suitable weight function is therefore necessary for the majority of practical applications. Such a function should exhibit the following characteristics:

- The weights of orders with the same planned start date should be equal.[2]
- The weights of orders whose planned start dates have been clearly exceeded should be clearly greater than other orders.
- The weight of orders whose planned start dates have not yet been reached should be negative.

Figure 22.3 depicts a function that fulfils these requirements. Four different zones can be identified:

*Zone 1: Planning Time Point < Planned Start Date – Advance Release Window*
This zone contains all orders that lay outside of the advance release window; they receive a clearly negative weight. They are thus only released when the load of the production will then be balanced to a particular degree or the constraints can otherwise not be maintained.

*Zone 2: Planned Start Date – Advance Release Window ≤ Planning Time point < Planned Start Date – Release Tolerance*
The orders within this zone receive a slightly negative weight. They can be released when by doing so the load in the production is balanced. The slope of the due date function in this zone tends to be minimal.

[2] Alternatively, the slack of an order can also be applied as a criterion for allocating the due date weight.

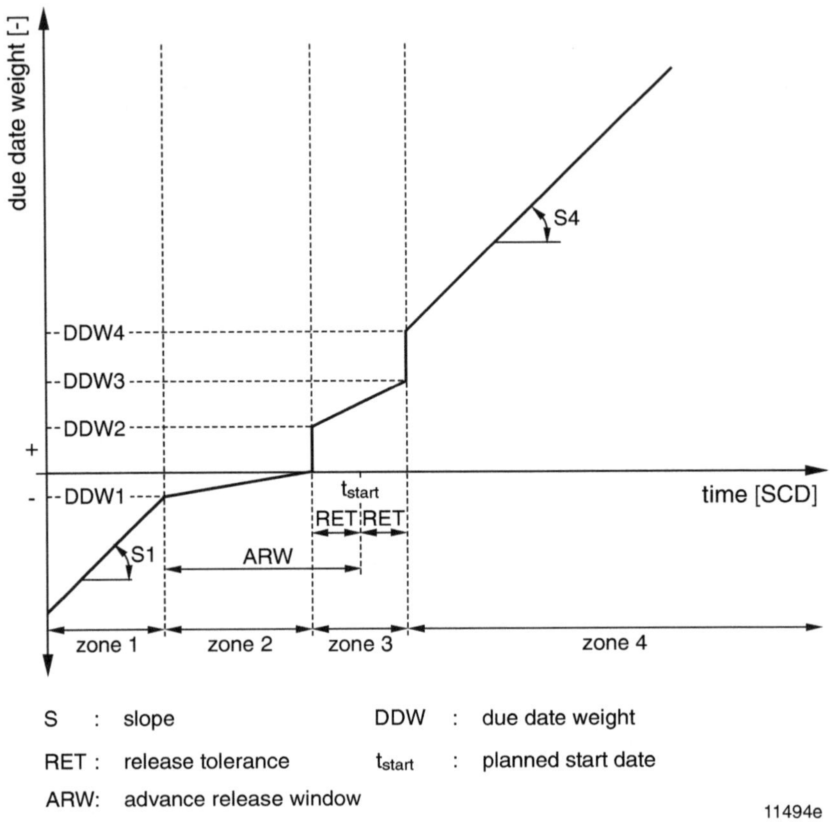


**Fig. 22.3** Curve and parameters for a weight function

*Zone 3: Planned Start Date – Release Tolerance ≤ Planning Time Point ≤ Planned Start Date + Release Tolerance*
Orders in this zone are weight more heavily. This is the zone in which the orders should be released according to plan and is defined by the release tolerance. Existing deviations from the planned start date can be relatively easily compensated while the orders flow through the production.

*Zone 4: Planned Start Date+ Release Tolerance < Planning Time Point*
Orders in this zone are already clearly delayed. They are thus very heavily weighted.

The due date function can be mathematically described as:

$$f(DPS)=\begin{cases}(TPL_0-DPS+ARW)\cdot S1+DDW1 & \text{for } TPL_0<DPS-ARW\\ (TPL_0-DPS+RT)\cdot\dfrac{-DDW1}{ARW-RT} & \text{for } DPS-ARW\leq TPL_0<DPS-RT\\ (TPL_0-DPS+RT)\cdot\dfrac{DDW3-DDW2}{2\cdot RT}+DDW2 & \text{for } DPS-RTL_0\leq TPL_0\leq DPS+RT\\ (TPL_0-DPS-RT)\cdot S4+DDW4 & \text{for } DPS+RT<TPL_0\end{cases}. \tag{22.8}$$

where

| | |
|---|---|
| f(DPS) | weighting function, |
| $TPL_0$ | planning time point [SCD], |
| DPS | planned start date [SCD], |
| ARW | advance release window [SCD], |
| RT | release tolerance [SCD], |
| S1 | slope of due date function in Zone 1 [−/SCD], |
| S4 | slope of due date function in Zone 4 [−/SCD], |
| DDW1 | weight for $TPL_0$= DPS - ARW [−], |
| DDW2 | weight for $TPL_0$= DPS - RT [−], |
| DDW3 | weight for $TPL_0$= DPS + RT [−], |
| DDW4 | weight on lower boundary of Zone 4 [−]. |

The function can be configured the most easily in a practical test. It should be verified whether or not the method makes practical release decisions for various WIP situations and various combinations of the release list.

### *22.3.4 Upper and Lower Limits for the Work Content to be Released for the Entire Production*

The Order Release with Linear Programming concept allows an upper and lower limit to be set for the work content being released for the entire production (Eq. 22.6). It is possible to control the WIP via the upper and lower limits of the releasable work content. In order to do so limits should be set as a function of the production's planned WIP and actual WIP:

$$WCR_{UL} = WIP_{tot,plan} - WIP_{tot,act} + WIPT \tag{22.9}$$

$$WCR_{LL} = WIP_{tot,plan} - WIP_{tot,act} - WIPT \tag{22.10}$$

where

| | |
|---|---|
| $WCR_{UL}$ | upper limit for work content to be released [hrs], |
| $WCR_{LL}$ | lower limit for work content to be released [hrs], |
| $WIP_{tot,plan}$ | planned total WIP of the production [hrs], |
| $WIP_{tot,act}$ | actual total WIP of the production [hrs], |
| WIPT | WIP tolerance [hrs]. |

The planned WIP can be determined based on a Logistic Positioning of the production's workstations in five steps. In the first step, the Logistic Operating Curves are calculated for the workstations in the production. The basis for these calculations is explained in Sect. 3.2.2.3. In the second step, the planned WIP for the workstations are set within a Logistic Positioning (see Sect. 3.2.2.3). In order to determine the total WIP of the production the planned WIP levels are first converted into the unit 'number of orders' (third step). Then (see Eq. 18.4 as well as [Nyhu-09*]):

$$WIPO_{plan} = \frac{WIP_{plan}}{WC_m} - \frac{U_m \cdot WC_v^2}{100} \tag{22.11}$$

where

| | |
|---|---|
| $WIPO_{plan}$ | planned WIP (in number of orders) [−], |
| $WIP_{plan}$ | planned WIP (in standard hours) [hrs], |
| $WC_m$ | mean work content [hrs], |
| $U_m$ | mean WIP dependent utilization [%], |
| $WC_v$ | variation coefficient of the work content [−]. |

The planned WIP of the individual workstations can then be added together for the planned WIP of the production (fourth step):

$$WIPO_{plan,tot} = \sum_{j=1}^{NWS} WIPO_{plan,j} \tag{22.12}$$

where

| | |
|---|---|
| $WIPO_{plan,tot}$ | planned WIP of the production (in number of orders) [−], |
| $WIPO_{plan,j}$ | planned WIP of Workstation *j* (in number of orders) [−], |
| NWS | number of workstations [−]. |

In the last step, the planned WIP is again converted into the unit of standard hours. In order to do so the planned throughput time for a representative quantity of orders has to be determined (see Sect. 20.3):

$$WIP_{plan,tot} = WIPO_{plan,tot} \cdot \frac{\sum_{i=1}^{NO}\left(TTP_{i,plan} \cdot \sum_{j=1}^{NWS} WC_{i,j}\right)}{\sum_{i=1}^{NO} TTP_{i,plan}} \tag{22.13}$$

where

| | |
|---|---|
| $WIP_{plan,tot}$ | planned WIP of the production [hrs], |
| $WIPO_{plan,tot}$ | planned WIP of the production (in number of orders) [−], |
| $TTP_{i,plan}$ | planned throughput time of Order *i* [SCD], |
| $WC_{i,j}$ | work content of Order *i* on Workstation *j* [hrs], |
| NO | number of orders [−], |
| NWS | number of workstations [−]. |

With a periodic order release the planned WIP should be increased by an amount equal to the planned output during the release period (see Sect. 20.3).

The WIP tolerance determines the allowable deviation from the planned WIP. If a very narrow WIP tolerance is selected it may not be possible to make a valid release decision. In any case the leeway for the release decision would be unnecessarily restricted. As a typical value, the WIP tolerance can be determined according to the following formula:

$$WIPT = WC_{tot,m} + 3 \cdot WC_{tot,s} \tag{22.14}$$

where

| | |
|---|---|
| WIPT | WIP tolerance [hrs], |
| $WC_{tot,m}$ | mean total work content [hrs], |
| $WC_{tot,s}$ | standard deviation of the total work content [hrs]. |

The throughput time deviation which results from the WIP tolerance can be estimated as:

$$\Delta TTP \leq \frac{WIPT}{ROUT_m} \tag{22.15}$$

where

| | |
|---|---|
| ΔTTP | throughput time deviation [SCD], |
| WIPT | WIP tolerance [hrs], |
| $ROUT_m$ | mean output rate of the production [hrs/SCD]. |

### 22.3.5 *Upper Limit for the Work Content to be Released for Individual Workstations*

The Order Release with Linear Programming method allows an upper limit for the work content being released for individual workstations to be set (Eq. 22.7). If the upper limit is linked with the actual WIP of the workstation, a WIP control can be achieved. Thus, for the upper limit of work content being released:

$$WCR_{UL,j} = WIP_{plan,j} - WIP_{act,j} \tag{22.16}$$

where

| | |
|---|---|
| $WCR_{UL,j}$ | upper limit of work content to be released for Workstation *j* [hrs], |
| $WIP_{plan,j}$ | planned WIP on Workstation *j* [hrs], |
| $WIP_{act,j}$ | actual WIP on Workstation *j* [hrs]. |

With a periodic order release this value should be increased by the planned output during the next release period. Furthermore, it should also be ensured that the work content of individual orders is never higher than the planned WIP of a workstation in order to prevent a complete blockage of orders of this type. Setting the planned WIP is explained in Sect. 20.3 for the Workload Control.

## 22.4 Implementation Across the Supply Chain

It would be unusual to implement the Order Release with Linear Programming method across a supply chain as it would require the production lot sizes to be coordinated and a very close delivery relationship between the supplying and receiving enterprises. Such configurations are rarely found in the industrial practice.

## 22.5 Application

The same comments apply for the application range of ORLP method as for Workload Control (for a more extensive discussion see Sect. 20.5):

- The steady utilization of the workstations should be an important objective and is not already ensured by the scheduling of orders.
- The enterprise accepts a higher due date tolerance or can compensate for sequence interchanges in the order input during the order's throughput.
- The positions of the workstations in the order throughput should be as constant as possible.
- The enterprise provides feedback with regards to the completion of operations.

Whereas Workload Control can balance the load via different order throughputs, ORLP can balance the load via different order throughputs or with different order times. ORLP thus poses fewer demands on the complexity of the production's material flow. In exchange though, greater expert knowledge is necessary when setting the method's numerous parameters. Furthermore, it is everything but simple to forecast the method's release decision. Therefore, the impact of the method is more difficult to estimate than Workload Control.

There is no available information about applying this method.

## References

[Hill-90*] Hillier, F.S., Lieberman, G.J.: Introduction to Operations Research, 5th edn. McGraw-Hill, New York (1990)

[Iras-74*] Irastorza, J., Deane, R.: A loading and balancing methodology for job-shop control. AIIE T. **6**, 302–307 (1974)

[Nyhu-09*] Nyhuis, P., Wiendahl, H.-P.: Fundamentals of Production Logistics. Theory, Tools and Applications. Springer, Berlin (2009)

# Chapter 23
# POLCA Control

POLCA was developed by Suri in the 1990s and positioned as an alternative to both the MRP II concept and Kanban [Suri-98*]. The method is a component of the comprehensive Quick Response concept aimed in particular at organizing production into manufacturing cells. In the ideal case (but not inevitably), the manufacturing cells operate according to the one-piece-flow principle. POLCA stands for Paired-Cell Overlapping Loops of Cards with Authorization. The abbreviation thus describes the localized WIP control loops between manufacturing cells in a production.

In order to control the WIP each pair of manufacturing cells is allocated POLCA cards which rotate between the manufacturing cells and authorize the production of orders. In addition, POLCA implements an order release according to due dates: A manufacturing cell is only allowed to process an order when the assigned release date has been exceeded. The direct connection between POLCA and a production organized in manufacturing cells seems to make comparing it to other production control methods difficult. It is, however, possible to implement POLCA also in conventional production environments in which case it then controls the material flow between pairs of workstations. Moreover, other production control methods can also be easily transferred to productions in which a number of workstations are combined together into manufacturing cells. The following discussion therefore refers more to the procedure's algorithm than to the organization of the production.

In Sect. 23.1 we will consider the procedural rules of POLCA and expand upon them using an example. Subsequently, we will discuss the procedural rules based on the guidelines for designing production control methods (Sect. 23.2). Setting the method's parameters is the subject of Sect. 23.3, while the reasons why applying POLCA across the supply chain would be unusual are addressed following that (Sect. 23.4). Our discussion of POLCA then concludes in Sect. 23.5 with a review of reported experiences applying it.

H. Lödding, *Handbook of Manufacturing Control*,
DOI 10.1007/978-3-642-24458-2_23, 

## 23.1 Method Description

Figure 23.1 depicts the principle behind POLCA. It is based on a local WIP control which regulates the material flow *between* manufacturing cells; generally, *within* the manufacturing cells any (other) suitable production control method can be implemented [Suri-98*]. In addition, POLCA is centered on a schedule of orders generated by a higher level PPC system.

Implementing POLCA is based on three elements:

- *Release list:* The list contains all orders that can be released. It is generated by a higher level PPC system e.g., through backwards scheduling.
- *POLCA cards:* A POLCA card (Fig. 23.2) is always allocated to a specific pair of manufacturing cells. It contains the abbreviations for the manufacturing cells as well as their designations. The POLCA cards for a pair of manufacturing cells are numbered making it easier to control the number of POLCA cards. POLCA cards authorize the processing of an order that is meant for a specific destination cell (F2 in Fig. 23.2) on the originating cell (in the example from Fig. 23.2: P1).
- *Release Date:* A release date for each order is determined by a higher level PPC system for every manufacturing cell that they are routed through. A manufacturing cell is allowed to process an order only when the release date has been attained or exceeded. Release dates are calculated with the aid of backwards scheduling, based on the planned end of processing date. The planned throughput times of the manufacturing cells form the basis for the calculation. This calculation does not differ from conventional PPC systems [Suri-98*] (see also Sect. 23.3).

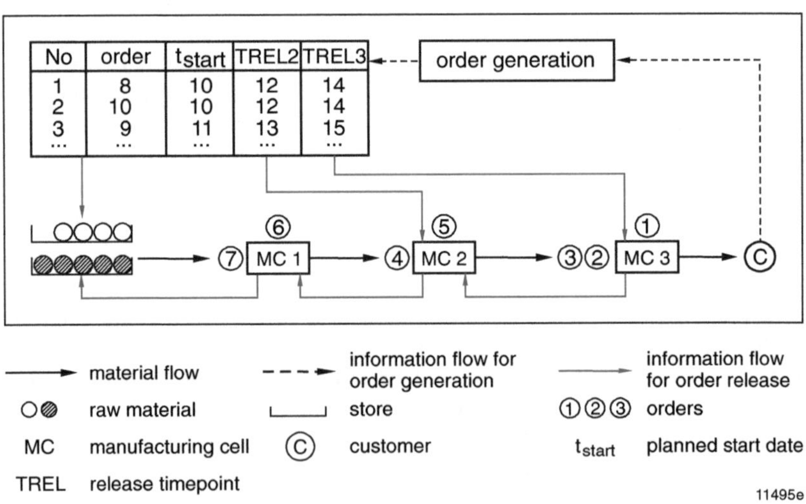


**Fig. 23.1** Principle of POLCA

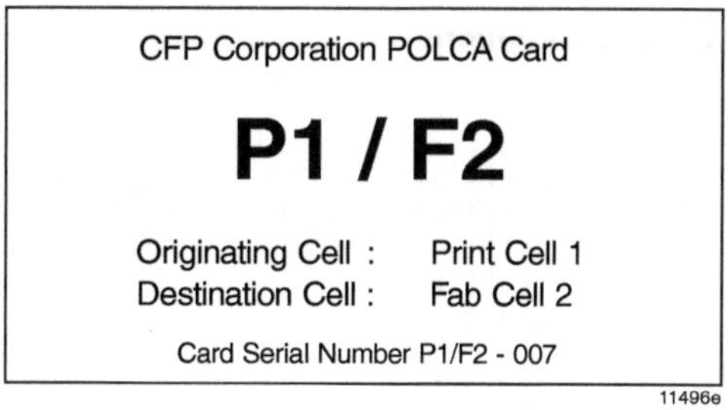


**Fig. 23.2** Sample POLCA card (Suri)

Based on these elements, POLCA can be summarized with a few rules:

*Rule 1:* An originating cell is only allowed to process an order when the release date of the order has been reached and a free POLCA card that is allocated to the originating cell and destination cell is available. If one of these conditions are not met, the order is blocked. (The first and last cells in the material flow are exceptions here: The first manufacturing cell uses the planned start date of the order as a release date, the last manufacturing cell can begin processing an order without a POLCA card.[1])

*Rule 2:* If an order is blocked, the manufacturing cell verifies whether or not the release conditions are met for other orders in the WIP.

*Rule 3:* With the start of the processing on the originating cell a POLCA card is firmly allocated to an order. It remains with the order up to the end of processing on the destination cell. After the processing has ended on the destination cell, the POLCA card is again free and can authorize the processing of a further order on the originating cell.

We will now take a closer look at the procedural rules for POLCA based on the example of a production with five manufacturing cells (Fig. 23.3). The orders flow through three manufacturing cells: They are first processed on either Cell 1 or Cell 2, then on either Cell 3 or Cell 4. Finally, all orders flow through Cell 5. With the exception of Cell 5, each cell is allocated a POLCA board on which the available cards can be placed. Cell 5 is allowed to process all of the orders whose release date has been reached or exceeded. The orders' release dates are provided above the orders whereas the current shop calendar day is stated in the lower right corner of each section of the figure.

The release list is divided according to the cells on which the orders are first processed. In addition to the order number, it contains the planned start date of the orders as well as the sequence of the production throughput. Every possible pair of manufacturing cells is allocated two POLCA cards.

[1] The POLCA card from the control loop with the preceding cell passes through the last manufacturing cell, however, this is inevitable since it authorizes the processing on the preceding cell.

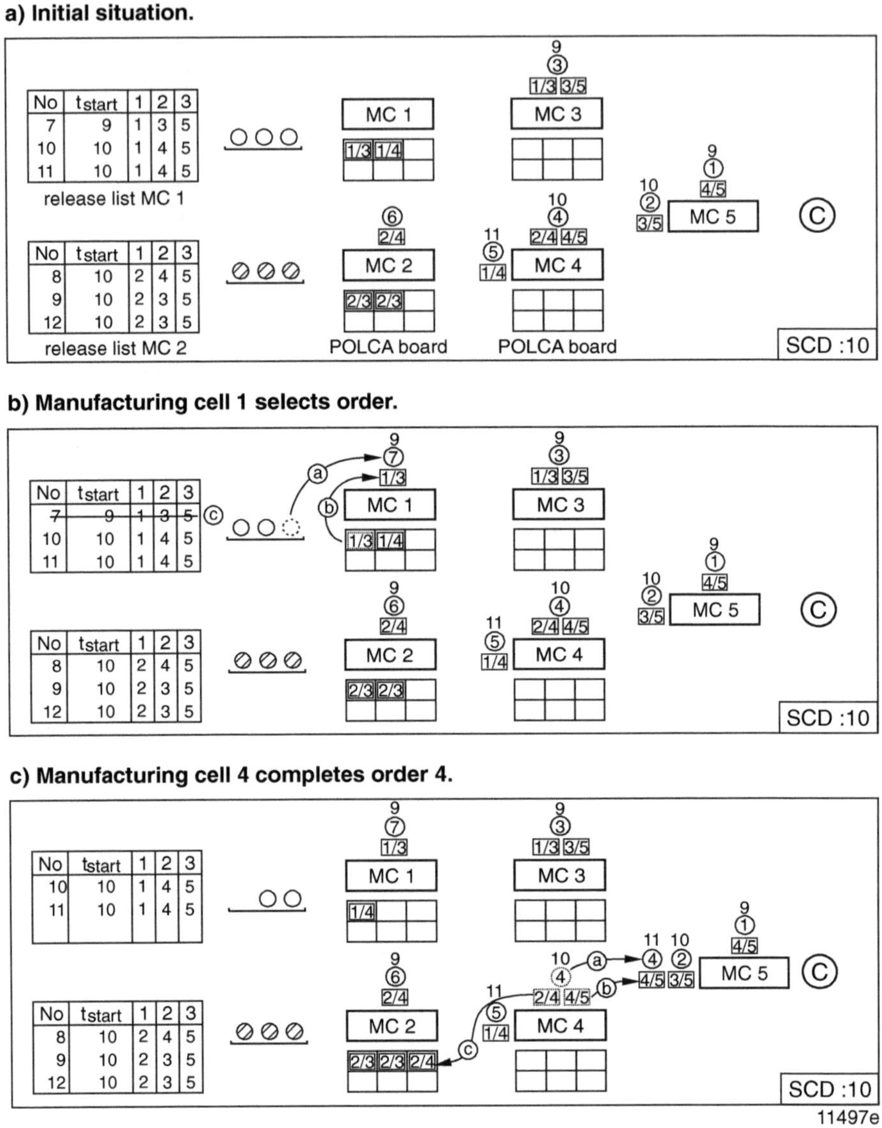


**Fig. 23.3** How POLCA functions

In the initial situation (Fig. 23.3a) Cell 1 has just completed processing an order. There are two free POLCA cards on the cell: At the top of the release list is Order 7, which will be processed after Cell 1 on Cell 3.

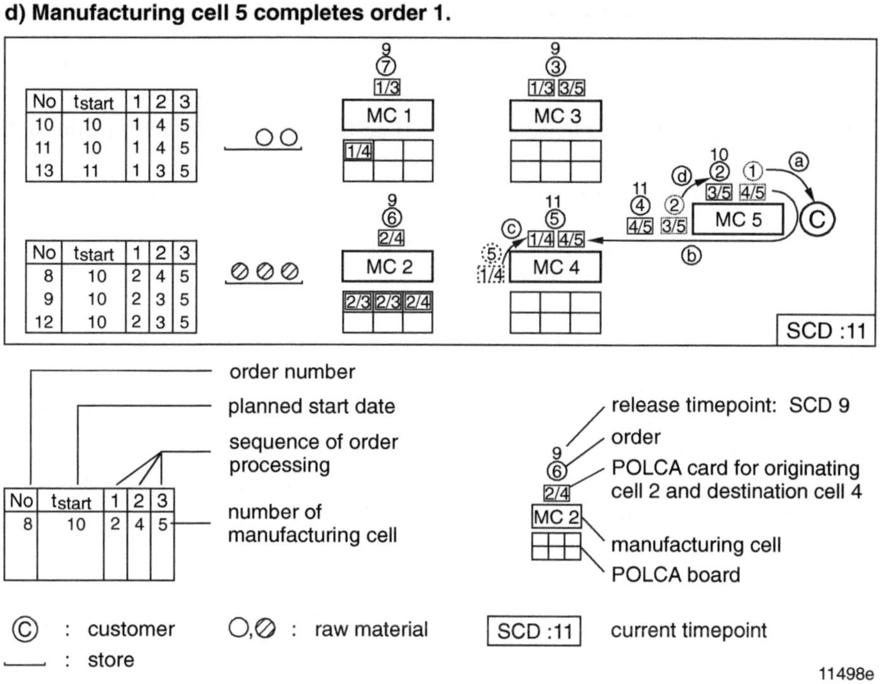


**Fig. 23.3** (continued)

A corresponding POLCA card is available so that the order is released (Step A in Fig. 23.3b). Cell 1 begins with the processing and allocates the POLCA card to Order 7 (Step B). The order is struck from the job list (Step C).

As soon as Cell 4 completes processing Order 4 (Fig. 23.3c) the order (Step A) is transported with POLCA Card 4/5 (Step B) to Cell 5. The order is allowed to be processed here on SCD 11 at the earliest, thus it is still blocked.

POLCA Card 2/4 is free and enters the POLCA board for Cell 2, where it can authorize the processing of a new order (Step C). The only order in the WIP from Cell 4 is Order 5. It cannot be processed because the required POLCA card is not available nor has the release date been reached. Order 5 is thus blocked until both release conditions are met. This is the case the next day when Cell 5 completes Order 1 (Fig. 23.3d).

Order 1 is then delivered to the customer (Step A). POLCA Card 4/5, which is then freed, authorizes the processing of Order 5 on Cell 4 (Step B). The cell begins processing it (Step C). While it is being processed Order 5 is allocated two POLCA cards. (It should be pointed out here that Order 5 would have been blocked despite the available POLCA cards when Cell 5 had completed processing Order 1 already on SCD 10 because its release date SCD 11 had not yet been reached.) Cell 5 begins processing Order 2 (Step D). Since Cell 5 conducts the last production stage, it only has to verify that the release date has been reached.

## 23.2 Discussion of the Procedural Rules

POLCA is characterized by:

- The calculation of the release date for all cells which an order flows through, and,
- A local WIP control between the production's manufacturing cells.

We will now examine the procedural rules more closely based on the guidelines for designing production control methods.

*1. A manufacturing control method should allow the WIP of a production or workstation to be regulated at a defined level.*
POLCA regulates the WIP between a pair of manufacturing cells, whereby the WIP is limited by the number of POLCA cards. The fewer the number of originating cells are and the better the intervals between a POLCA Card becoming free and the arrival of the order triggered by it at the destination cell can be estimated, the more precisely a cell's WIP can be regulated (see Sect. 23.3). Via the manufacturing cells' WIP POLCA also limits the production's WIP on the whole.

*2. A manufacturing control method should reduce WIP fluctuations to a minimum.*
The leeway for a manufacturing cell's WIP fluctuations is limited by the number of POLCA cards for which the manufacturing cell a destination cell is. It is thus usually less than with a centralized WIP control (such as that conducted by CONWIP). The leeway increases with the number of possible originating cells. The reason for this is that the number of POLCA cards is usually rounded up to a whole number from the calculated value (see Sect. 23.3).

*3. A manufacturing control method should cause as little blocked WIP as possible.*
Blocked WIP is one of POLCA's weaknesses: Compared to centralized order release methods it causes more blockades because it can refuse orders from being released for processing even during the orders' throughput. Compared to other decentralized methods – such as the Decentralized WIP Oriented Manufacturing Control – the fact that it does not regulate the WIP of a manufacturing cell or workstation as a whole is also disadvantageous. Instead, POLCA regulates the WIP from different originating cells in different control loops. It thus makes release decisions based on information regarding a part of the total WIP at a manufacturing cell. This reduces the method's efficiency.

When there are backflows in the material flow, POLCA cannot exclude deadlocked orders or manufacturing cells. A simple example with two variants A and B (Fig. 23.4) illustrates this.

Variant A is first routed through Cell 2, then Cell 1 and finally Cell 2 once more. Variant B follows the opposing material flow and is routed through the manufacturing cells according to the sequence C 1 – C 2 – C 1.

In the situation presented here, none of the manufacturing cells can process one of the orders. Cell 1 does not have a POLCA card for releasing orders of type 1/2; Variant B has been allocated these. Conversely, Cell 2 requires a POLCA card from type 2/1 in order to be able to process the orders of Variant B. These however

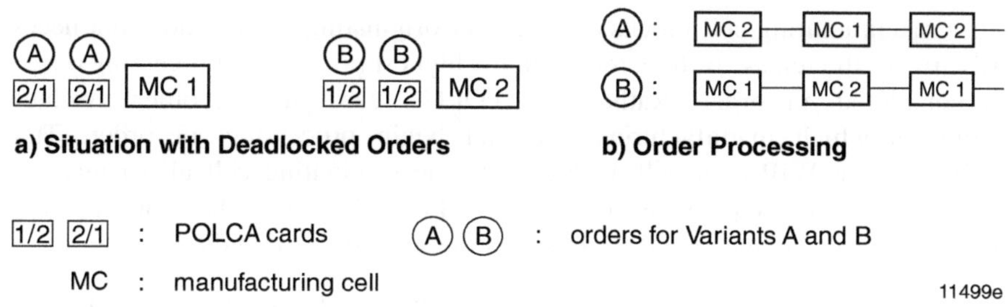


**Fig. 23.4** Deadlocked orders with a material backflow

are allocated to Variant A on Cell 1. Consequently, the orders are permanently blocked on both manufacturing cells. Thus when implementing POLCA it is critical to check for possible deadlocks which can arise when there are material back flows.

4. *A manufacturing control method should contribute to closely aligning the load to the capacities.*

POLCA supports a load balance between the manufacturing cells. The limited number of POLCA cards prevents more orders from being released for processing than a cell can immediately process (load balance by delaying order release). If a manufacturing cell is not allowed to release an order for processing due to a lack of a POLCA Card, it tries to pull forward another order. If it succeeds, losses of utilization on a non-overloaded cell can be prevented (load balance by accelerating order release). POLCA thus reduces the WIP buffer required for ensuring the utilization. The effectiveness of this mechanism is nonetheless decreased because POLCA divides the total WIP of a manufacturing cell according to the originating cell and controls it separately. Consequently, it makes decisions on the basis of information about a part of the WIP. Moreover, the release date restricts the leeway for a load balance. Unlike for instance the Load Oriented Order Release, POLCA also does not support a centralized load balance.

5. *A manufacturing control method should cause as few sequencing interchanges as possible.*

POLCA causes sequence interchanges during the orders' throughput. They arise when a manufacturing cell pulls the processing of an order forward because an order with a higher priority is blocked. However, this does not inevitably impact the schedule reliability negatively: An order that is blocked due to the lack of a POLCA card would frequently encounter a high WIP level on the following cell when it is processed according to the sequence.[2] As a result of the sequencing interchange the WIP can decrease, so that the order enters the following cell at a lower WIP level and is probably processed no later than without the change in sequence.

[2] Among other factors, this depends on the number of POLCA cards.

If a manufacturing cell blocks a number of originating cells, it does not necessarily trigger the release of the order with the highest priority for processing once it has completed an order. Rather, the POLCA card that becomes available determines which manufacturing cell can begin processing an order. Thus distributing the WIP of a cell according to the originating cell also negatively impacts the caused sequencing interchanges. POLCA should, therefore, be combined with a sequencing rule that considers the operations' due dates.

6. *A manufacturing control method should be able to balance schedule deviations within the context of backlog control.*

It is the responsibility of production control to prevent threatening backlogs and to reduce backlogs that have already developed. In order to do so capacities are temporarily increased. The procedural rules of POLCA do not take into consideration a backlog control. It should thus be combined with a backlog control in order to be able to ensure a high delivery reliability even when there are deviations from the production plan.

7. *A manufacturing control method should take into consideration the bottleneck principle.*

POLCA takes into consideration the bottleneck principle at least implicitly: Orders and POLCA cards accumulate automatically at the bottleneck cell, as a result, orders on the originating cell are blocked due to a lack of free POLCA cards. This mechanism prevents the originating cell from producing more orders than the bottleneck cell can immediately process. If the bottleneck cell completes an order, a POLCA card becomes available thus triggering the production of an order at the originating cell. POLCA therefore ensures that the bottleneck is supplied with orders in the future.

8. *A manufacturing control method should be as simple as possible.*

The procedural rules for POLCA are quite simple. Moreover, with the POLCA cards the control method also has a clear visual aid. These can also be implemented to control transportation: If there is an order in the outgoing area of a manufacturing cell the transporter can read the destination cell from the POLCA card [Suri-98*]. Nonetheless, when the material flow is more complex it also becomes more difficult to implement POLCA. The number of POLCA cards then has to be set for a great many different pairs of manufacturing cells. This can require a considerable effort and influences the extent to which the logistic targets are met. However, POLCA was not developed for such a production environment.

## 23.3 Setting the Method's Parameters

The two essential parameters of POLCA are the number of POLCA cards (Sect. 23.3.1) and the release date of the orders at the manufacturing cells (Sect. 23.3.2). In the following section we will examine how they can be set based on the logistic targets. In addition, it is also possible to define the advance release window. This sets the length of time in which the orders can be processed before their planned start date in order to support balancing the production's load. A load balance of this type is, however, not Suri's intention; he refers instead to the WIP increase related to completing orders early [Suri-98*].

### *23.3.1 Number of POLCA Cards*

Suri mentions the following equation which is based on Little's Law and can be used to calculate the number of POLCA cards for two manufacturing cells *i* and *j* [Suri-98*]:

$$NPC_{ij} = \left(TTP_{m,i} + TTP_{m,j}\right) \cdot \frac{NO_{ij}}{P} \tag{23.1}$$

where

| | |
|---|---|
| $NPC_{ij}$ | number of POLCA cards with Cell *i* as originating cell and Cell *j* as destination cell [−], |
| $TTP_{m,i}$ | mean throughput time of the (originating) Cell *i* [SCD], |
| $NO_{ij}$ | number of orders which flow through Cells *i* and *j* consecutively during the period P [−], |
| P | length of the reference period [SCD]. |

This equation should be used cautiously because the throughput times are not homogenously defined: An order only requires a POLCA card for the throughput *through* Cell *i*, but not for the waiting time *before* Cell *i*.[3] However, the POLCA card remains with the order while it is queued and during its processing on the destination cell *j*. This difference is critical especially when the waiting time before the originating cell *i* is relatively long. In this case the mean throughput time on the destination cell *j* can clearly exceed the target value when the waiting time on the originating cell *i* is included in the calculation of the number of POLCA cards.

Moreover, the question of which throughput times can practically be attained remains open. For this, Suri refers to a special modeling software [Suri-98*]. In the following – as with the other production control methods – we will turn to the funnel model. The procedure for setting the number of POLCA cards is comprised

[3] Naturally, it is possible to allocate a POLCA card to an order already in the queue before the originating cell, so that the waiting time is also considered. However, this would influence the sequence in which the orders are processed at the manufacturing cell. Furthermore, it does not prevent an order from reaching the originating cell when there is no POLCA card available for the next control loop.

of a Logistic Positioning of the manufacturing cells, on the one hand, and the calculation of the number of POLCA cards based on that, on the other hand.

**Logistic Positioning of the Manufacturing Cells** The goal of the Logistic Positioning is to determine the planned WIP and throughput times for the individual manufacturing cells. In order to weigh between the logistic objectives (WIP, throughput time and output rate) the Logistic Operating Curves, which depict the mutual correlations of the objectives, are calculated. In the following it will be assumed that a manufacturing cell consists of a number of workstations. According to the type of part flow, two different procedures can then be distinguished:

1. *Lot-wise transport:* Within the manufacturing cell each workstation processes a complete lot which is transported to the next workstation only after it is completed. In this case a Production Operating Curve can be calculated for each workstation and a target WIP determined. This should then be converted into the unit 'number of orders' and added together for all of the workstations. The corresponding procedure is explained in Sect. 18.3 for the CONWIP Control and can be similarly transferred to POLCA.
2. *One-pieceflow:* Within a manufacturing cell a part is transported to the next workstations directly after being processed. A single Manufacturing Cell Operating Curve should then be calculated for the entire cell. The basis for the calculation is extensively discussed in [Lödd-01b].

In both cases a planned throughput time can be calculated from the planned WIP and planned output rate for the manufacturing cell (Little´s Law):

$$TTP_{plan,j} = \frac{WIPO_{plan,j}}{ROUTO_{plan,j}} \quad (23.2)$$

where

| | |
|---|---|
| $TTP_{plan,j}$ | planned throughput time for Cell $j$ [SCD], |
| $WIPO_{plan,j}$ | planned WIP for Cell $j$ (in number of orders) [−], |
| $ROUTO_{plan,j}$ | planned output rate for Cell $j$ (in number of orders) [−/SCD]. |

The planned throughput time contains both the waiting time before the cell as well as the routing time through the manufacturing cell. The net throughput time *through* the manufacturing cell results from subtracting the waiting time before the cell from the planned throughput time:

$$TTP_{net,plan,j} = TTP_{plan,j} - WT_{plan,j} \quad (23.3)$$

where

| | |
|---|---|
| $TTP_{net,plan,j}$ | planned net throughput time for Cell $j$ [SCD], |
| $TTP_{plan,j}$ | planned throughput time for Cell $j$ [SCD], |
| $WT_{plan,j}$ | planned waiting time *before* Cell $j$ [SCD]. |

**Calculating the Number of POLCA Cards** If the enterprise has made a decision about the planned WIP of the manufacturing cell it is the responsibility of production control to set the number of POLCA cards so that the planned WIP can be attained. An exact mathematical correlation between the number of POLCA cards and the WIP levels on the manufacturing cells cannot be determined. In order to estimate this correlation an assumption which simplifies the situation is made. According to this assumption, all POLCA cards are always allocated to an order. If this is intermittently not the case in reality, the actual WIP will be lower than the planned WIP. Under this assumption a simple procedure for calculating the number of POLCA cards can be derived. It is comprised of three steps:

1. *Determining the number of POLCA cards for the direct WIP of the destination cell.*

The number of POLCA cards for the direct WIP is determined straight from the planned WIP of the manufacturing cell. This should be proportionately allocated to the POLCA cards for the possible originating cells. Accordingly:

$$NPC_{ij,dir} = WIPO_{plan,j} \cdot \frac{NO_{ij}}{\sum_{l=1}^{m} NO_{lj}} \tag{23.4}$$

where

| | |
|---|---|
| $NPC_{ij,dir}$ | number of POLCA cards for the direct WIP with Cell $i$ as originating cell and Cell $j$ as destination cell [−], |
| $WIPO_{plan,j}$ | planned WIP for Cell $j$ (in number of orders) [−], |
| $NO_{ij}$ | number of orders which are routed through Cell $i$ and Cell $j$ one after the other during reference period $P$ [−], |
| m | number of manufacturing cells [−]. |

2. *Determining the number of POLCA cards for the indirect WIP of the destination cell.*

The planned WIP of the destination cell alone is insufficient for determining the number of POLCA cards, instead, the indirect WIP that is already being processed at the originating cell also has to be considered. It can be modeled with the aid of a Throughput Diagram (Fig. 23.5).

In addition to the input curve of the destination cell, an indirect input curve is also plotted. It is formed similarly to the (conventional direct) input curve, with the difference that the orders are already included from the start of processing at the originating cell.

The area between the curve of the indirect input and the direct input curve corresponds to the indirect WIP area of the destination cell. The mean indirect WIP can be determined from it (see [Wien-97b]):

$$WIPO_{m,ind} = \frac{AWIP_{ind}}{P} \tag{23.5}$$

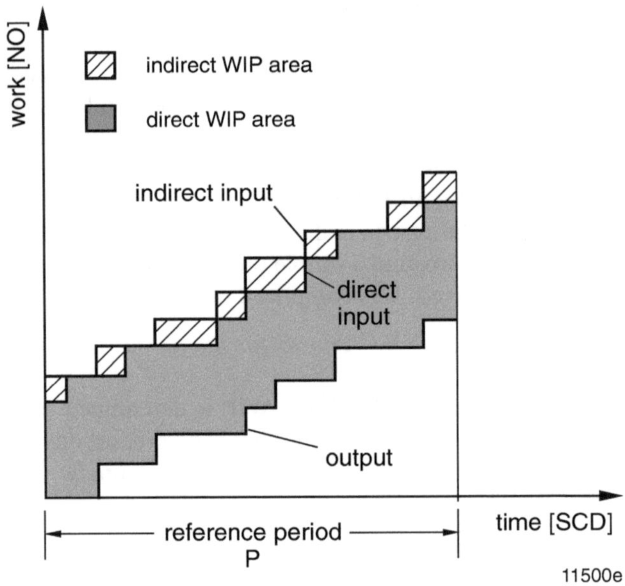


**Fig. 23.5** Representation of indirect WIP in the extended throughput diagram

where

| | |
|---|---|
| $WIPO_{m,ind}$ | mean indirect WIP (in number of orders) [−], |
| $AWIP_{ind}$ | indirect WIP area [SCD], |
| P | length of reference period [SCD]. |

The longer the net throughput times of the orders on the originating cell are and the greater the number of orders that should be processed in the reference period, the larger the indirect WIP area. After a number of conversions the following relation results from Eq. 23.5:

$$NPC_{ij,ind} = TTP_{m,net,i} \cdot \frac{NO_{ij}}{P} \tag{23.6}$$

where

| | |
|---|---|
| $NPC_{ij,ind}$ | number of POLCA cards for the indirect WIP with Cell *i* as originating cell and Cell *j* as destination cell [−], |
| $TTP_{m,net,i}$ | mean net throughput time Cell *i* [SCD], |
| $NO_{ij}$ | number of orders which are routed through Cell *i* and Cell *j* consecutively during reference period *P* [−], |
| P | length of the reference period [SCD]. |

*3. Determining the Total Number of POLCA Cards*

In the last step, the number of POLCA cards for the direct WIP and indirect WIP of a manufacturing cell are added together. The result – when necessary – is rounded up to the next whole number.

$$NPC_{ij} = \textit{rounded}\ \text{up}\ \left(NPC_{ij,dir} + NPC_{ij,ind}\right) \tag{23.7}$$

where

| | |
|---|---|
| $NPC_{ij}$ | number of POLCA cards with Cell $i$ as originating cell and Cell $j$ as destination cell [−], |
| $NPC_{ij,dir}$ | number of POLCA cards for the direct WIP with Cell $i$ as originating cell and Cell $j$ as destination cell [−], |
| $NPC_{ij,ind}$ | number of POLCA cards for the indirect WIP with Cell $i$ as originating cell and Cell $j$ as destination cell [−]. |

When a manufacturing cell has a number of possible originating cells, the following problems can arise:

- *The number of POLCA cards in the control loops is very low:* In many cases a POLCA control loop is allocated the minimum number of POLCA cards. When an above average number of orders for a specific sequence of manufacturing cell occasionally arises, blocked WIP and losses of utilization can arise.
- *The total number of POLCA cards for all of the control loops is high:* This is a direct consequence of rounding up the number of POLCA cards. Consequently, a leeway for WIP increases and WIP fluctuations arises.

Both effects could be proven in simulation trials with POLCA and highly complex material flows [Lödd-01]. A very simple material flow is investigated by Luh, Zhou and Tomastik [Luh-00*]. Under these conditions POLCA attained good results.

## 23.3.2 Release Date

The release date determines the earliest time point after which a manufacturing cell is allowed to start processing an order. POLCA is the only release method that can block an order because of a due date criterion after it has initially been released. Suri suggests setting the release date on a manufacturing cell with conventional backward scheduling [Suri-98*]. It seems appropriate to select the planned end of operation for an order on the originating cell as the release date: [4]

[4] An alternative would be the planned start of operation on the observed cell. This however would leave no leeway for balancing the load (see discussion below).

$$ReD_i = EDOP_{plan,i-1} \tag{23.8}$$

where

| | |
|---|---|
| $ReD_i$ | release date on manufacturing Cell *i* [SCD], |
| $EDOP_{plan,i-1}$ | planned end of operation on originating cell [SCD]. |

Basically, the release date determines the leeway of the manufacturing cell for balancing the load. It sets an advance release window that other production control methods only determine for the entire production or entire order. The earlier the release date of the orders on the manufacturing cell can be placed, the greater the leeway is for a load balance and therefore the higher the WIP increase is that needs to be consciously accepted for this purpose.

In practice, the setting of the release dates can often be omitted. This is especially true when the order throughput times are short.

## 23.4 Implementation Across the Supply Chain

It would be very unusual to implement POLCA across a supply chain as it would require the production lots to be coordinated between the supplying and receiving enterprises. POLCA's control loops, which are specific to the material flow, are nonetheless generally well suited for controlling production across a supply chain. They allow control loops to be developed between two enterprises without having to directly consider the supply relationship to other enterprises.

## 23.5 Application

The range of application for POLCA is comparably large. Possible reasons for *not* applying the method include:

- When the order generation is not coordinated and the material flow is complex, an order release with workstation specific load balances will attain a higher output rate with a comparable WIP level.
- When implementing finite scheduling for the orders it is possible that the planned WIP of a manufacturing cell is temporarily higher than POLCA's WIP limits allow, which can in turn not implement the targets set by production planning. This tends to occur more with POLCA's tight control loops than with a centralized WIP control (such as the CONWIP control).
- With highly complex material flows the method also leads to blocked WIP and loses its efficiency.

In a technical report, Suri and Krishnamurthy give an account of introducing POLCA at two enterprises in the USA [Suri-03*]. The first of these, *Olsen*

*Engineering*, a metal processing enterprise in Iowa, was able to reduce the throughput times of various products by 22–68% by implementing POLCA. Furthermore, the stock in the production and in the stores could be reduced by 75–90%. Suri and Krishnamurthy also report an increased transparency, improved employee moral and various improvements in the area of quality. On the whole, introducing POLCA stimulated a continual improvement mindset in the enterprise.

The second, *Rockwell Automation* introduced POLCA in one of their divisions which produces motor controls. The throughput time there was able to be reduced by 25% and the delivery reliability increased to 92% with a very strict due date tolerance.

## References


[Lödd-01] Lödding, H.: Dezentrale Bestandsorientierte Fertigungsregelung. VDI Progress Reports, Series 2, No. 587, Düsseldorf (2001)

[Lödd-01b] Lödding, H., Hernández, R., Wiendahl, H.-P.: Dimensionierung und Bewertung von Fertigungsinseln mit logistischen Kennlinien. Io Manage. **70**(1–2), 25–32 (2001)

[Luh-00*] Luh, P. B., Zhou, X., Tomastik, R.N.: The performance of a new material control and replenishment system: a simulation and comparative study. In: Proceedings of Quick Response Manufacturing 2000, pp. 805–826. 7–9 June 2000

[Suri-03*] Suri, R., Krishnamurthy, A.: How to plan and implement POLCA: a material control system for high-variety or custom-engineered products. Technical Report. Center for Quick Response Manufacturing, University of Wisconsin-Madison, pp. 1–17. (2003)

[Suri-98*] Suri, R.: Quick Response Manufacturing. A Companywide Approach to Reducing Lead Times. Productivity Press, Portland (1998)

[Wien-97b] Wiendahl, H.-P.: Fertigungsregelung. Logistische Beherrschung von Fertigungsabläufen auf Basis des Trichtermodells. Carl Hanser Verlag, Munich/Vienna (1997)

# Chapter 24
# Decentralized WIP Oriented Manufacturing Control

Decentralized WIP Oriented Manufacturing Control (DEWIP) is based on decentralized WIP control loops between the production's workstations. It thus allows the production's workforce being integrated in the responsibility of attaining logistic targets. The method was developed by Lödding at the Leibniz University of Hanover's Institute of Production Systems and Logistics in a project sponsored by the German Research Foundation [Lödd-01, Wien-01].

Below, Sect. 24.1 presents the procedural rules for DEWIP, following that Sect. 24.2 examines these rules based on the guidelines for designing manufacturing control methods. Section 24.3 then describes how the parameters for DEWIP can be set based on the logistic objectives. Finally, the reasons for why it would be unusual to implement DEWIP across enterprises in a supply chain are presented in Sect. 24.4, whereas areas for applying it are outlined in Sect. 24.5.

## 24.1 Method Description

In the following section, the basic version of Decentralized WIP Oriented Manufacturing Control is introduced. Subsequent to that, Sect. 24.1.2 describes the WIP Regulated Throughput Control – a predecessor of DEWIP.

### *24.1.1 Basic Method*

Figure 24.1 depicts the basic principle of Decentralized WIP Oriented Manufacturing Control. A centralized PPC system generates a list of urgent orders based on information about the market and/or customers' orders. In order to do so the PPC system can for example conduct an MRP run. It is however possible to combine DEWIP with other methods for generating orders. The orders in the list are released for production via the decentralized WIP control loops. DEWIP releases orders

H. Lödding, *Handbook of Manufacturing Control*,
DOI 10.1007/978-3-642-24458-2_24, © Springer-Verlag Berlin Heidelberg 2013

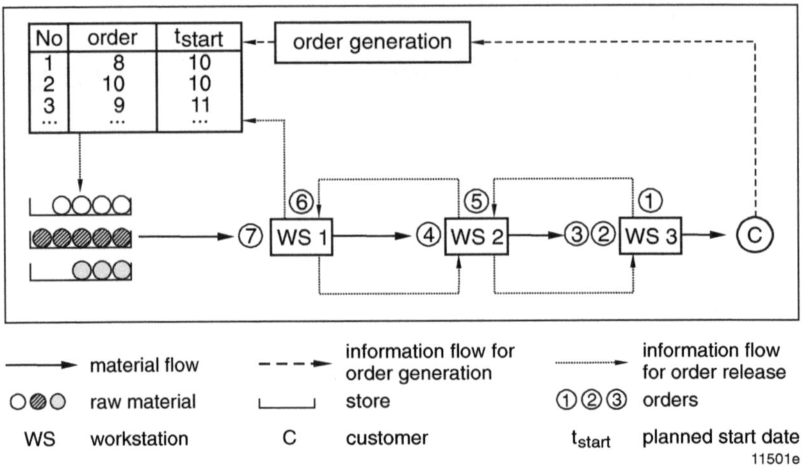


**Fig. 24.1** Principle of the decentralized WIP oriented manufacturing control

for each operation separately. The release decision is made based on the WIP on the next workstation. The corresponding request for a 'go-ahead' from the next workstations creates an information flow in the same direction as the material flow. The decision about whether or not the order is allowed to be processed and the receipt of this 'go-ahead' however results in an information flow opposing the material flow.

Controlling production with DEWIP is based on six procedural rules [Lödd-01]:

1. *Before an operator begins working on an order they request a 'go-ahead' from the workstation that will process the order next. The last operation of an order can always be processed.*
   Through this rule the subsequent workstation learns of a possible load due to an order. This information proves being particularly helpful when an overlapped production is being conducted. As a result, the attainable WIP levels and throughput times can be reduced. A corresponding extension of the method is described in [Lödd-01].
2. *The operator on the subsequent workstation decides whether or not to give a 'go-ahead' based on the WIP limit. If the WIP on the workstation (direct WIP) and the WIP being processed on the preceding workstation (indirect WIP) exceeds the workstation's WIP limit the 'go-ahead' for the order is refused. Otherwise the 'go-ahead' is authorized and its work content is added to the WIP account.*
   This rule allows a workstation to regulate the influx of work. Processing releases ensure that the workstations are supplied with orders. If a workstation refuses the 'go-ahead' it prevents more orders from flowing to it, than it can immediately process. It thus prevents a WIP build-up.

3. *If a workstation receives a 'go-ahead' for an order, it begins processing it. Otherwise, in the case where other orders in its queued WIP are processed on another subsequent station, it requests a 'go-ahead' for these.*
   The effort to obtain the 'go-ahead' for other queued orders is aimed at ensuring the workstation's utilization. If no releases are acquired the workstation is blocked. It is thus usually not a bottleneck, since otherwise it would not be able to cause a WIP increase on subsequent workstations.
4. *Once an order is processed the operator strikes it from the workstation's total WIP. If the WIP account falls below the WIP limit the operator authorizes 'go-aheads' for orders that were previously refused up until the WIP limit is exceeded.*
   This ensures not only that the workstation will be supplied with orders in the future but also that the previously blocked workstation will learn that the blockade is lifted.
5. *Each workstation is allocated a position number which identifies the relative position of a workstation in the material flow. Independent of the WIP limit being exceeded, a workstation authorizes a 'go-ahead' for orders when the workstation requesting it has a higher position number* i.e., *where there is a backflow.*
   This rule prevents a deadlock of the workstations. It also annuls the strived for WIP control to some extent. The position numbers should thus be assigned so that material flows to workstations with lower numbers (backflows) are kept to a minimal.
6. *The production planning generates a list of urgent orders. These orders are released when the WIP on the initial workstation falls below the WIP limit. As they are released they are added to the initial systems WIP. If the material provision is very efficient, the orders can also be released directly before the processing starts (c.f.* [Lödd-01]*).*
   The sixth rule describes the connection between DEWIP and the production planning. As with all methods for releasing orders it is practical to define an advance release window. This sets how many days in advance an order is allowed being released.

We will now consider three examples in order to explain these procedural rules. The first example clarifies the basic mechanism of DEWIP. The second example presents the problem of deadlocked workstation, while the third example explains the two alternatives to releasing orders with DEWIP.

### Example 1: The Basic Mechanism of DEWIP

The first example depicts the throughput of seven orders through three workstations in a production (Fig. 24.2).

Each workstation is allocated a WIP account. It contains the work content of the orders in the WIP and indicates the WIP limits. Orders that are processed on preceding workstations (indirect WIP) are circled with a dashed line.

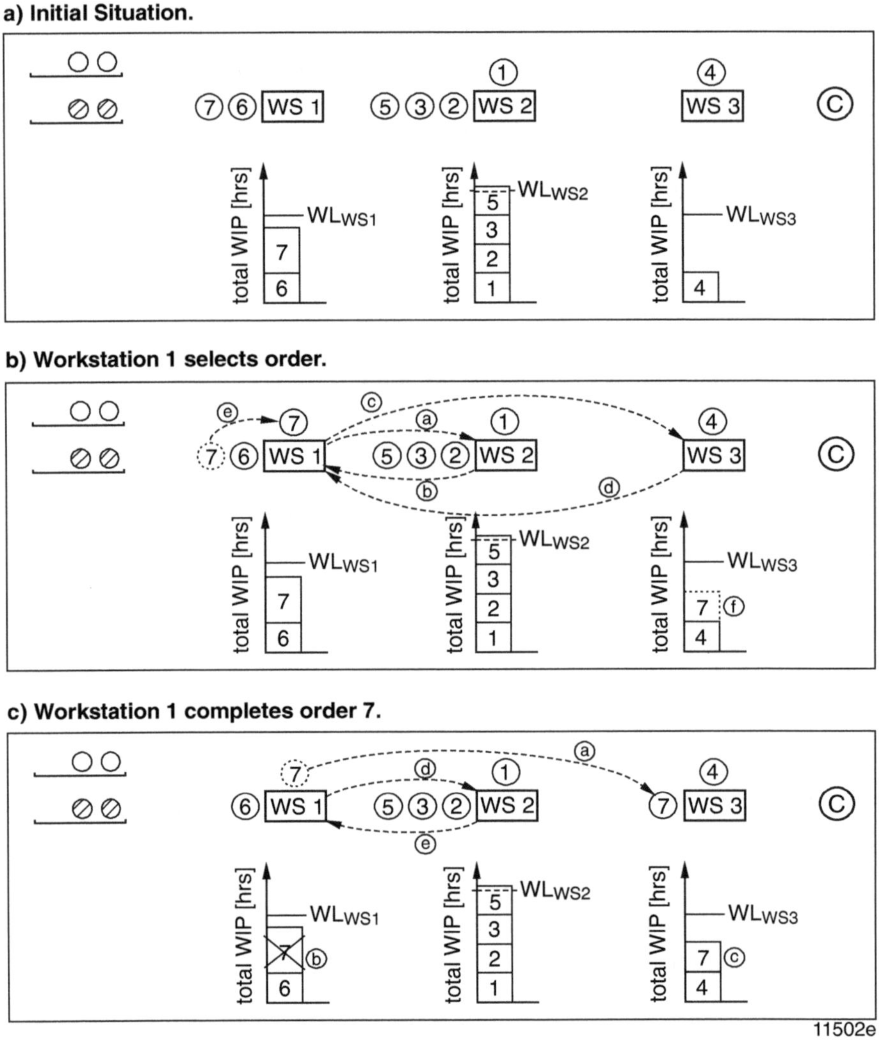


**Fig. 24.2** How the decentralized WIP oriented manufacturing control functions

In the initial situation, the WIP limit on Workstation 2 is exceeded and Workstation 1 just finished processing Order 5 (Fig. 24.2a).

Workstation 1 would now like to begin processing Order 6 (Fig. 24.2b). It thus asks Workstation 2, which processes the next operation for Order 6, for a 'go-ahead' (Step A). However, at Workstation 2 the sum of the work content for the orders in the WIP already exceeds the WIP limit. It therefore refuses the 'go-ahead' (Step B).

Workstation 1 then asks Workstation 3 – which processes the next operation for Order 7 – for a 'go-ahead' for Order 7 (Step C). At Workstation 3 only Order 4 is in

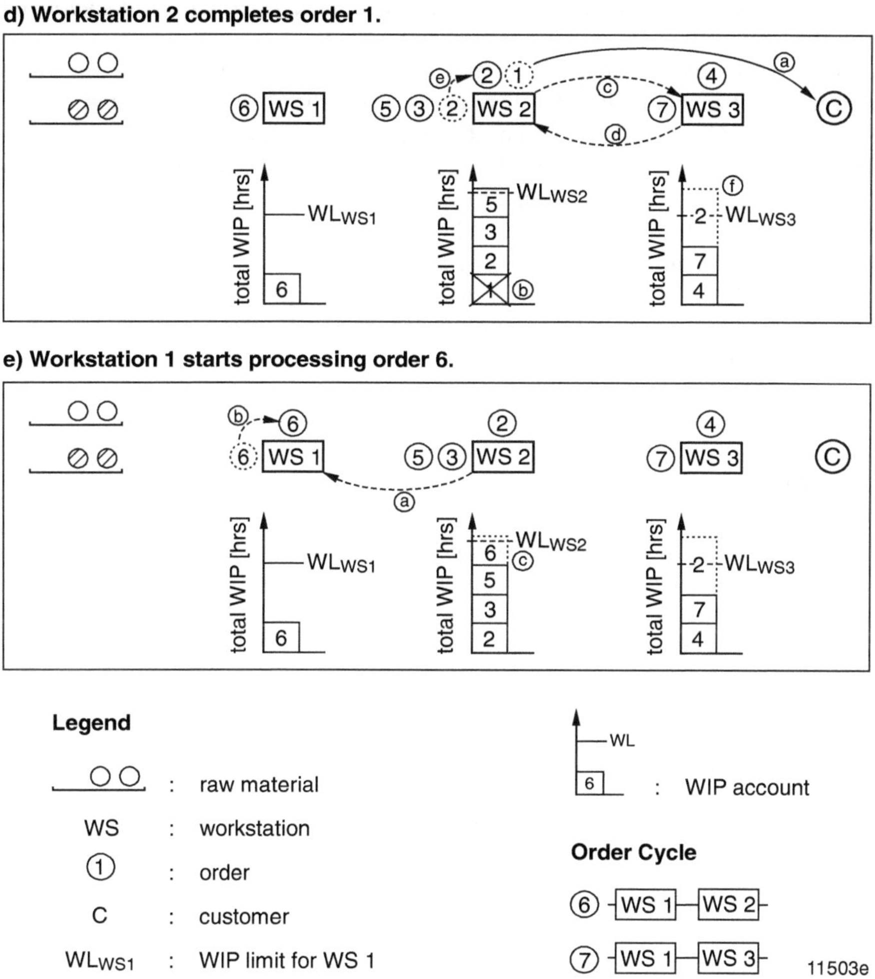


**Fig. 24.2** (continued)

its WIP and is currently being processed. The WIP level is thus not yet exceeded. Workstation 3 thus authorizes the 'go-ahead' (Step D). Workstation 1 begins with the processing (Step E), upon which Order 7 is credited to the WIP account of Workstation 3 (Step F). While it is being processed Order 7 counts as direct WIP in the WIP account of Workstation 1 and as indirect WIP for the WIP account of Workstation 3. On both workstations Order 7 is assessed with the work content specific to each of the operations. There is a clear difference between them in this case.

Workstation 1 finishes processing Order 7 (Step A in Fig. 24.2c) and removes it from the WIP account (Step B). The order is transported to Workstation 3, where it is now managed as direct WIP (Step C). Workstation 1 once again requests a 'go-ahead' for Order 6 from Workstation 2 (Step D). Nevertheless, the WIP limit is still exceeded here so the 'go-ahead' is refused once more (Step E). Workstation 1 is thus blocked as there are no more orders in the WIP.

With the completion of Order 1 (Step A in Fig. 24.2d) Workstation 2 removes it from the WIP account (Step B). It requests a 'go-ahead' from Workstation 3 for Order 2 (Step C), is granted it (Step D) and begins with the processing (Step E). Order 2 is credited at Workstation 3 as indirect WIP, whereby the WIP limit of the workstation is exceeded for the first time (Step F).

At the same time Workstation 2 authorizes the 'go-ahead' of Order 6 at Workstation 1 (Step A in Fig. 24.2e), since the WIP level falls below the limit upon completing Order 1. Workstation 1 begins processing it immediately (Step B). As a result of the indirect WIP from Order 6, the WIP limit for Workstation 2 is again exceeded (Step C).

### Example 2

Without considering the position numbers that occurs with Rule 5, workstations could be deadlocked when there is a complex material flow. A situation such as this is presented in Fig. 24.3 using the example of two workstations.

The orders on Workstation 1 are all intended for Workstation 2 and vise versa. Furthermore, the WIP limits for both workstations are exceeded due to the orders already in the WIP. In a case such as this, neither Workstation 1 nor Workstation 2 can process an order, since each of the workstations have to refuse a 'go-ahead' as a result of their WIP limits already being exceeded. The workstations are therefore deadlocked.

In order to prevent a situation such as this a position number is assigned to each workstation. As mentioned above, the position number identifies a workstation's relative position in the material flow (see Sect. 24.3.2 for determining the position number). Even when the WIP limit has been exceeded a workstation will grant

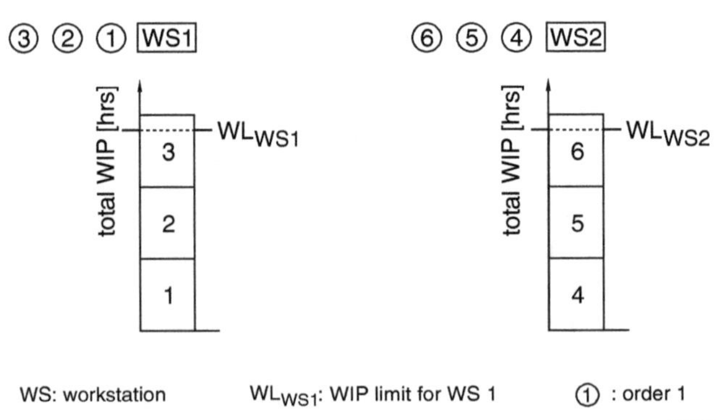


**Fig. 24.3** Deadlocked workstations

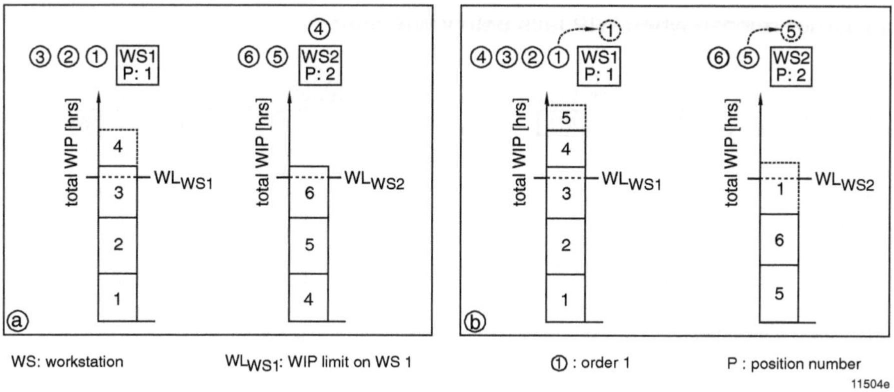


**Fig. 24.4** Preventing deadlocks with position numbers

a 'go-ahead' for an order, when the requesting workstation's position number is higher than its own.

If in the above example, Workstation 1 has the position number 1 and Workstation 2 position number 2, a deadlock cannot arise (Fig. 24.4).

Workstation 1 grants the 'go-ahead' for Order 4, since Workstation 2 has a higher position number than it does. As a result the WIP account at Workstation 1 increases further (Fig. 24.4a). Once Order 4 has been processed the WIP at Workstation 2 falls below the WIP limit (Fig. 24.4b). It thus grants Workstation 1 the 'go-ahead' for Order 1. Furthermore, due to the lower position number Workstation 2 receives the 'go-ahead' for Order 5 from Workstation 1 and immediately begins processing it. The WIP account at Workstation 2 thus increases further.

### Example 3: Releasing Orders with DEWIP

There are basically two different possibilities to control the release of an order for production with DEWIP (Fig. 24.5):

1. *When the WIP on the first workstation falls below the WIP limit the release of an order is triggered* (Fig. 24.5a). The most urgent order on the release list is authorized for production. A procedure such as this does not take into consideration the WIP situation of the subsequent workstation.
2. *The orders are only released directly before the start of processing on the first workstation* (Fig. 24.5b). This requires however that the workstation has already received the 'go-ahead' *and* that the raw materials can be quickly supplied. The main advantage of this method is that it takes into consideration the load situation on the subsequent workstation. Moreover, it reduces the throughput times on the first workstation to the operation time. This method is particularly well suited for productions with standardized raw materials.

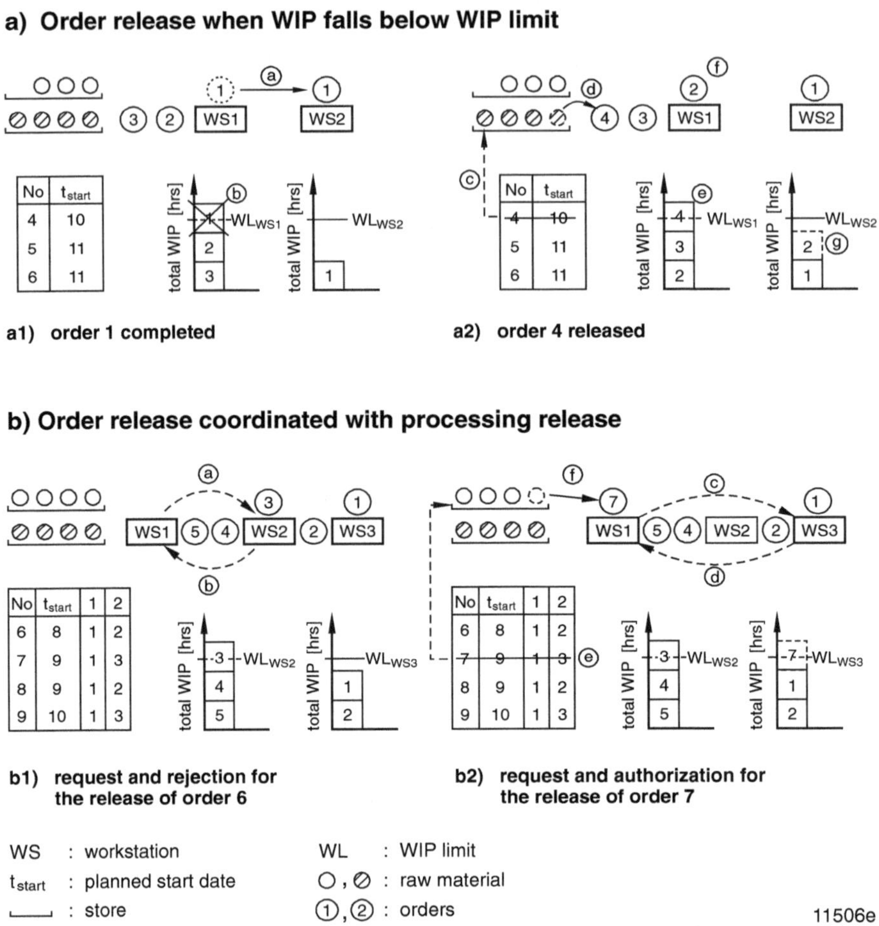


**Fig. 24.5** Releasing orders with DEWIP

## 24.1.2 *WIP Regulated Throughput Control*

Already in the late 1980s Busch suggested a decentralized WIP control as part of a more comprehensive concept for production planning and control [Busc-87]. According to it, a workstation could process an order when the WIP on the subsequent workstation did not fall below a defined WIP level.

This form of the WIP control differs from DEWIP in two particular ways: First, it does not consider the indirect WIP that is already being processed on the preceding workstation. When there is a complex material flow this leads to increased WIP fluctuations (see the discussion in [Lödd-01]). Second, when there are backflows in the material flow it can lead to workstations being deadlocked.

Despite these disadvantages, Busch probably earns the credit for developing the first decentralized method for releasing orders. It can also be successfully implemented in linear (or at least simple and directed) material flows.

## 24.2 Discussion of the Procedural Rules

DEWIP is characterized by the decentralized control loops between the production's workstations. In this next section we will examine the procedural rules more closely in light of the guidelines for designing manufacturing control methods.

1. *A manufacturing control method should allow the WIP of a production or workstation being regulated at a defined level.*

Regulating the WIP at a workstation at a defined level is a fundamental idea of the Decentralized WIP Oriented Manufacturing Control. As a result the WIP of the entire production is limited.

2. *A manufacturing control method should reduce WIP fluctuations to a minimum.*

DEWIP is able to limit WIP fluctuations. This is especially true when the workstations are (able to be) assigned position numbers so that there is little backflow from workstations with higher position numbers to those with lower position numbers.

3. *A manufacturing control method should cause as little blocked WIP as possible.*

DEWIP is not able to fully prevent blocked WIP during the order's throughput. Quite the contrary, blocking orders is an integral aspect of the method when otherwise a WIP build-up would arise on the subsequent workstation. This is a property that DEWIP shares with other decentralized manufacturing control methods.

4. *A manufacturing control method should contribute to closely aligning the load to the capacities.*

Load balancing accelerates or delays the release or throughput of an order with the aim of preventing breaks in the material flow (acceleration) or preventing the WIP from increasing (delay).

*Balancing the Load when Releasing Orders:* DEWIP releases orders early when the WIP would otherwise fall below the planned WIP. In order to do so, an advance release window is defined which determines how many days in advance an order is allowed being released. Inversely, DEWIP holds back orders whose planned start date has already been attained when releasing them would cause the planned WIP being exceeded. With both of these characteristics DEWIP supports a load balance when releasing orders. Unlike, for example, the Load Oriented Order Release, the Decentralized WIP Oriented Manufacturing Control does not however – with the exception of the initial workstation – consider the load situation on the individual workstations. DEWIP therefore does not specifically hold back those orders which would flow through workstations that are already overloaded.

*Balancing the Load during the Order Throughput:* During the order throughput the WIP limits prevent more orders than can be immediately processed by a workstation from receiving a 'go-ahead' (balancing the load via a load shift). A workstation whose request for a 'go-ahead' for an order is refused, tries to receive authorization for other orders in its WIP. If the workstation is successful an order is consequently pulled forward for a workstation that is not overloaded. It is thus possible that a loss of utilization can also be prevented for this workstation (balancing the load by pulling the load forward).

5. *A manufacturing control method should cause as few sequencing interchanges as possible.*

As a result of orders being blocked, DEWIP can cause sequence interchanges. It is, nonetheless, questionable if these sequence interchanges actually lead to a reduction in the schedule reliability: An order whose 'go-ahead' was denied, would meet a high WIP level on the subsequent workstation when it was processed according to sequence. Due to the sequence interchange the WIP level can decrease so that the order then generally meets a lower WIP level on the subsequent workstation. It is therefore frequently processed no later than it would have been without the sequence interchange.

However, the sequence interchanges that result due to attempts at preventing workstations from being deadlocked, should be assessed more critically. Orders whose next operation is on a workstation with a lower position number are given preference when orders with a higher priority are blocked.

Generally, DEWIP should be combined with a sequencing rule that considers the planned end of processing date for the orders or operations. As a result the impact of the sequence interchanges caused by this method can at least be reduced.

6. *A manufacturing control method should be able to balance schedule deviations within the context of backlog control.*

A backlog control is not actually part of basic DEWIP, however, the WIP control aspect delivers valuable information about current bottlenecks in the production. If an enterprise temporarily increases the capacities of a bottleneck system – e.g. via overtime – it can increase the output rate of the entire production thus reducing an existing backlog and improving the schedule reliability. This is the basis for an extension of DEWIP, comprehensively described in [Lödd-01, Lödd-02].

7. *A manufacturing control method should take into consideration the bottleneck principle.*

DEWIP at least implicitly considers the bottleneck principle: With a high utilization of the production, the WIP at the bottleneck workstation approaches the WIP limit or exceeds it. The reason for this is that the workstations before a bottleneck can attain a higher output rate. They thus cause the WIP to build-up at the bottleneck until the WIP limit is exceeded and the bottleneck denies the release of other orders. As a result DEWIP prevents an uncontrolled WIP increase on the bottleneck workstation. Every time the WIP drops below the limit due to the bottleneck workstation completing an order, the bottleneck triggers the start of another order's processing on the preceding workstation. DEWIP therefore does not generally have to maintain a particularly high WIP for bottlenecks. Rather the utilization is ensured because the bottlenecks function like a metronome i.e. regulate the release rhythm.

*8. A manufacturing control method should be as simple as possible.*
The procedural rules for DEWIP were consciously kept simple in order to support implementing it on the shop floor. Nonetheless, suitable aids should be utilized in order support its application. Card systems such as those implemented with the Kanban or POLCA can serve here as an example.

## 24.3 Setting the Method's Parameters

The two parameters of DEWIP that are specific to the method are the *WIP limits* and *position numbers*. In addition, the advance release window should be set when orders are allowed being released before their planned start date in order to balance the load. Selecting the advance release window is explained in §18.3 for the CONWIP control and can be transferred to DEWIP.

Let us now turn our attention to how the level of the WIP limit can be set from the logistic targets with the aid of the Funnel Model (Sect. 24.3.1). Moreover, since the position numbers of the workstations should be set so that no backflows arise we will consider how this can be achieved in Sect. 24.3.2.

### *24.3.1 WIP Limits*

The WIP limits can be derived from a Logistic Positioning with the aid of the Logistic Operating Curves Theory (Sect. 24.3.1.1). Alternatively or supplemental to this they can be set by operators on the shop floor.

#### 24.3.1.1 Setting Based on a Model

The dimensioning of the WIP limits can be divided into two components: target WIP and indirect WIP, in which case [Lödd-01]:

$$WIPL = WIP_{tar} + WIP_{m,ind} \tag{24.1}$$

where

| | |
|---|---|
| WIPL | WIP limit [hrs], |
| $WIP_{tar}$ | target WIP [hrs], |
| $WIP_{m,ind}$ | mean indirect WIP [hrs]. |

The following comments describe how both components can be set with the aid of the Logistic Operating Curves Theory. Moreover, they explain why and under which circumstances the actual WIP on a workstation can deviate from the target WIP.

**Target WIP** The target WIP corresponds to the direct WIP that should develop on a workstation. The mean WIP of a workstation determines both its (WIP dependent) utilization as well as its throughput time. The value of the target WIP levels is therefore critical. It needs being set within the frame of a Logistic Positioning for every workstation. A suitable aid for this is Production Operating Curves for the workstations, which present the utilization and throughput time curves as a function of the workstation's mean WIP. Deriving and calculating the Production Operating Curves are described more extensively in Sect. 3.2.2.3.

**Indirect WIP** The target WIP alone is not sufficient for setting the WIP limits. Rather, the indirect WIP being processed on the preceding workstation also needs being considered.

The indirect WIP of a workstation can be modeled with the aid of a Throughput Diagram (Fig. 24.6). In order to do so, the times at which the orders began being processed on the preceding workstation are measured and the corresponding work contents are plotted cumulatively over the time. The work contents are assessed with their work content on the observed workstation and *not* with the work content on the preceding workstations.

In the Funnel Model, the time at which an order's processing ends on a preceding workstation is defined as the input time for the subsequent workstation. Therefore, in a Throughput Diagram, the area between the curve for the indirect and direct input to a workstation corresponds to the WIP area of the indirect WIP. Thus, for the mean value of the indirect WIP [Lödd-01, Wien-97b]:

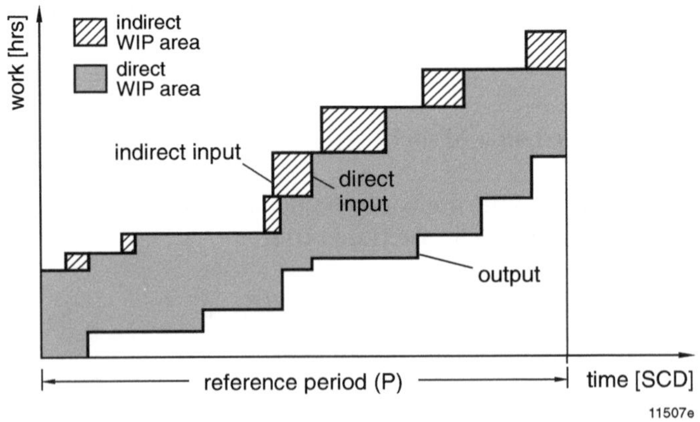


**Fig. 24.6** Extended throughput diagram of a workstation

$$WIP_{m,ind} = \frac{AWIP_{ind}}{P} \tag{24.2}$$

where

| | |
|---|---|
| $WIP_{m,ind}$ | mean indirect WIP [hrs], |
| $AWIP_{ind}$ | indirect WIP area [hrs·SCD], |
| P | length of the reference period [SCD]. |

In order to calculate the indirect WIP, the indirect WIP area and the length of the reference period need being determined. The longer the operation times on the preceding workstations are and the greater the work contents are on the observed workstation, the greater the indirect WIP area is. According to [Lödd-01], it is calculated as:

$$AWIP_{ind} = \sum_{i=1}^{NOP} \left(WC_i \cdot TOP_{i,Pr\ WS}\right) \tag{24.3}$$

where

| | |
|---|---|
| $AWIP_{ind}$ | indirect WIP area [hrs·SCD], |
| $WC_i$ | work content of Order *i* [hrs], |
| $TOP_{i,PrWS}$ | operation time for Order *i* on preceding workstation [SCD], |
| NOP | number of operations during reference period. |

With a fully utilized workstation that has only one work center, the length of the reference period can be replaced with sum of the operation times for the orders. If the WIP dependent utilization as well as the number of work centers are also considered, the following equation results [Lödd-01]:

$$P = \frac{\sum_{i=1}^{NOP} TOP_i}{\frac{U}{100} \cdot NW} \tag{24.4}$$

where

| | |
|---|---|
| P | length of the reference period [SCD], |
| $TOP_i$ | operation time for Order *i* [SCD], |
| U | WIP dependent utilization [%], |
| NW | number of work centers [−], |
| NOP | number of operations during reference period [−]. |

Accordingly, for the mean indirect WIP [Lödd-01]:

$$WIP_{m,ind} = \frac{\sum_{i=1}^{NOP} \left(WC_i \cdot TOP_{i,PrWS}\right)}{\sum_{i=1}^{NOP} TOP_i} \cdot NW \cdot \frac{U}{100} \tag{24.5}$$

where

| | |
|---|---|
| $WIP_{m,ind}$ | mean indirect WIP [hrs], |
| $TOP_{i,PrWS}$ | operation time for Order *i* on preceding workstation [SCD], |
| $WC_i$ | work content of Order *i* [hrs], |
| $TOP_i$ | operation time for Order *i* [SCD], |
| NW | number of work centers [−], |
| U | WIP dependent utilization [%], |
| NOP | number of operations during reference period [−]. |

**Deviation of the Actual WIP from the Target WIP** The actual attained WIP will deviate more or less greatly from the target WIP. The reasons for this vary [Lödd-01]:

- The WIP limits on bottleneck workstations are frequently exceeded, consequently, the mean WIP is higher than planned.
- The workstations grant 'go-ahead's even when the WIP limits are exceeded if the position numbers of the requesting workstation is higher than their own. A WIP level higher than that planned can thus arise.
- The start of processing on the preceding workstation can be delayed. This is the case primarily when the preceding workstation pulls an order forward and can process it because the 'go-ahead' for another order has been denied. If, while processing this order, it receives a 'go-ahead' for the previously blocked order, it will first complete the current order. When the processing start on the preceding workstation is delayed it leads to a lower WIP on the observed station.
- Different work times on the workstations (e.g., one shift operation on preceding workstation, double shift operation on observed workstation) can cause WIP reductions. This should be considered when determining the WIP limits.
- With a lower utilization of the production the target WIP levels are generally not met when the lower utilization is not already considered in calculating the WIP limits. The latter is however not recommendable. Large orders could exceed the resulting low WIP limits and thus cause a greater degree of blocked WIP [Lödd-01].

#### 24.3.1.2 Setting Parameters in the Practice

Alternatively or supplemental to setting parameters based on a model, employees can also set the WIP limits on the shop floor. In order to so they are first set at a value known being sufficient. In a continuous improvement process the operators then gradually reduce the WIP limits up until the loss of utilization exceeds a tolerable degree. A procedure such as this can motivate employees to search for possibilities to continually reduce the attainable WIP level and throughput times. It is thus especially well suited for localized manufacturing control methods. The WIP limits determined with the Funnel Model can then serve as benchmark figures.

### 24.3.2 Position Numbers

The position numbers identify the relative position of a workstation in the material flow of a production. The higher the position number of a workstation is, the farther down in the material flow it is on average for all of the orders. Position numbers are implemented in order to prevent workstations from being deadlocked (see Fig. 24.3). When the workstation requesting a 'go-ahead' has a higher position number, a workstation will grant the release for an order even when the WIP limits are exceeded. Since this (partially) annuls the WIP control, the position numbers should be allotted so that as few backflows as possible arise. A backflow is defined as a material flow from a workstation with a higher position number to a workstation with a lower position number. The number of backflows can be determined with the aid of a material flow matrix in which the workstations are organized according to the sequence of their position numbers. The material flow matrix results from the orders and the corresponding routings for the planning period. In the material flow matrix the element $a_{ij}$ identifies the number of orders that run from a workstation with the position number $i$ to a workstation with the position number $j$. The production's backflow are mapped by the elements below the diagonal of the material flow matrix (Fig. 24.7a).

The number of backflows are calculated as [Lödd-01]:

$$NBF = \sum_{i=1}^{n} \sum_{j=1}^{n} a_{ij} \text{ where } i > j \tag{24.6}$$

where

| | |
|---|---|
| NBF | number of backflows [−], |
| $a_{ij}$ | number of orders that run from Workstation $i$ to Workstation $j$ [−], |
| n | number of workstations [−], |
| i | row index [−], |
| j | column index [−]. |

An algorithm for minimizing the number of backflows – based on heuristically determined initial values of the position numbers – was originally developed for simulation trials. The basic idea of it is to reduce the backflows by shifting a workstation in the material flow matrix to a lower position number. The shifted workstation is thus allocated a lower position number. The position numbers of the workstations located between the old and new position of the shifted workstation increase by one.

For a possible shift, the number of backflows prevented by the shift is compared to the number of additional backflows. The number of backflows prevented by shifting a workstation from position number x to the lower position number y amounts to (see

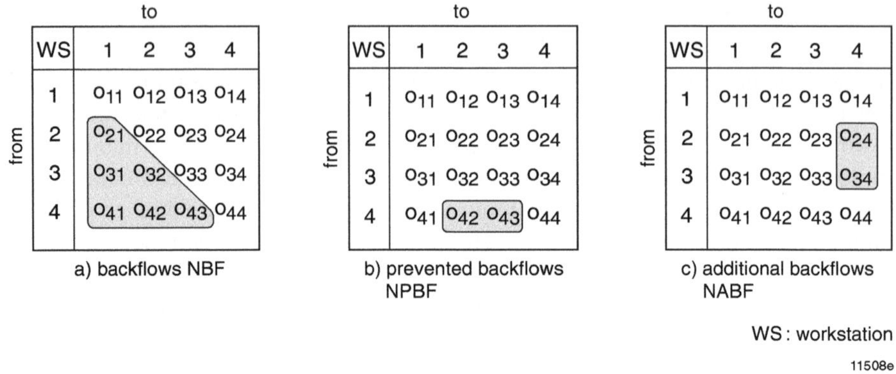


**Fig. 24.7** Representation and calculation of backflows in the material flow matrix

the cells marked with a grey background in Fig. 24.7b for a shift of the workstation with position number x = 4 to position number y = 2) [Lödd-01]:

$$NPBF = \sum_{j=y}^{x-1} a_{xj} \tag{24.7}$$

where

| | |
|---|---|
| NPBF | number of prevented backflows [−], |
| $a_{xj}$ | number of orders that run from Workstation $x$ to Workstation $j$ [−], |
| j | column index [−], |
| x | original position number of the observed workstation [−], |
| y | new position number of the observed workstation [−]. |

The number of additional backflows (see the cells marked with a grey background in Fig. 24.7c for a shift of the workstation with position number x = 4 to position number y = 2) [Lödd-01]:

$$NABF = \sum_{i=y}^{x-1} a_{ix} \tag{24.8}$$

where

| | |
|---|---|
| NABF | number of additional backflows [−], |
| $a_{ix}$ | number of orders that run from Workstation $i$ to Workstation $x$ [−], |
| i | row index [−], |
| x | original position number of the observed workstation [−], |
| y | new position number of the observed workstation [−]. |

If the number of prevented backflows is greater than the number of additional backflows (NPBF > NABF), a shift reduces the number of backflows. Based on Eqs. 24.7 and 24.8 a simple algorithm which minimizes the number of backflows by shifting the position numbers can be programmed. The algorithm verifies for each workstation if the number of backflows can be reduced by assigning it a lower position number. It then executes the shift with which the number of backflows can be reduced the most and then repeats this process until no more backflows can be prevented by shifting a workstation.

It is recommendable to employ a heuristic method for determining a solution for the initial allocation of the position numbers to the workstations. This can be gained for example, from the mean position of the workstations in the routings.

## 24.4 Implementation Across the Supply Chain

It would be unusual to implement DEWIP across a supply chain as it would require both a very close supply relationship between the enterprises as well as coordinating the production lot sizes. Moreover, the transportation needs being taken into consideration or even regarded as a workstation of its own when determining the WIP limits.

## 24.5 Application

The discussion concerning the application range of POLCA generally also applies for Decentralized WIP Oriented Manufacturing Control (see Sect. 23.5). Unlike POLCA, however, DEWIP can also be implemented in complex material flows. Up until now the method has not been applied on the shop floor. Experiments with simulation models of real production areas nonetheless prove the suitability of DEWIP for application in the industry [Lödd-01]. The successful implementation of related methods such as POLCA or WIP Regulated Throughput Control support this.

The Decentralized WIP Oriented Manufacturing Control concludes our discussion about methods for releasing orders. The next part of the book is dedicated to sequencing.

## References

[Busc-87] Busch, U.: Bestandsgeregelte Durchfluß-Steuerung (BGD). CIM Management **1**, 18–23 (1987)

[Lödd-01] Lödding, H.: Dezentrale Bestandsorientierte Fertigungsregelung. VDI Progress Reports, Series 2, No. 587, Düsseldorf (2001)

[Lödd-02] Lödding, H., Lopitzsch, J., Begemann, C.: Rückstandsregelung erhöht die Termintreue. Eine Erweiterung der Dezentralen Bestandsorientierten Fertigungsregelung (DBF). wt Werkstattstechnik online **92**(5), 248–252 (2002)

[Wien-97b] Wiendahl, H.-P.: Fertigungsregelung. Logistische Beherrschung von Fertigungsabläufen auf Basis des Trichtermodells. Carl Hanser Verlag, Munich/Vienna (1997)

[Wien-01] Wiendahl, H.-P., Lödding, H.: Dezentrale Bestandsorientierte Fertigungsregelung. Grundlagen und Anwendung. Werkstattstechnik **91**(4), 202–207 (2001)

# Part D
# Sequencing

# Chapter 25
# Sequencing

The job of sequencing is simple to describe: It determines which order in a workstation's queue should be processed next. In order to do so it allocates every order a priority level based on defined criteria. The order with the greatest priority is the most urgent and is – when the operator at the workstation maintains the sequence – processed first. The primary task of sequencing is to adjust the actual sequence in which the orders are completed to the planned sequence in order to ensure a high schedule reliability and service level (Fig. 25.1).

The higher the WIP level on the workstation is and the more orders queued, the greater the impact of the sequencing on the attainment of logistic targets [Wien-97b]. Therefore, with the aim to reduce the WIP, the significance of sequencing has the tendency to decrease. Even with an appropriate WIP level however, individual orders can be greatly accelerated or delayed due to sequence interchanges [Yu-01]. The impact on the mean though is rather small [Wien-97b].

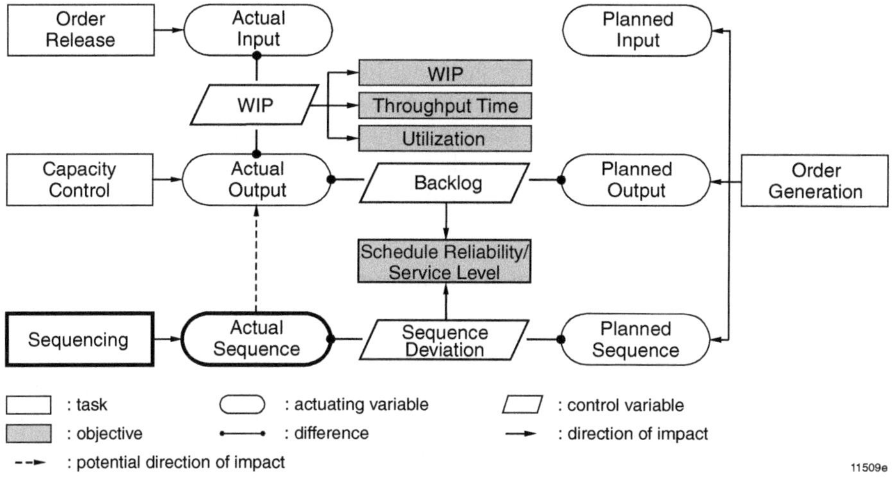


**Fig. 25.1** Location of sequencing within the manufacturing control model

H. Lödding, *Handbook of Manufacturing Control*,
DOI 10.1007/978-3-642-24458-2_25, © Springer-Verlag Berlin Heidelberg 2013

At the latest since the pioneering work by Conway, Maxwell and Miller [Conw-67*], sequencing has been a traditional part of manufacturing control. Since then a vast number of sequencing rules have been developed, published and discussed. The practical significance of these sequencing rules is rather questionable. Ultimately there are only a few possibilities for selecting a sequencing rule. Moreover, a sequencing rule can only compensate to a certain degree for mistakes made in previous steps of the production planning and control. To overstate it: On the shop floor, preventing a random sequence with its completely negative impact on the delivery reliability is frequently a greater and more worthwhile challenge than finding a special sequencing rule X that fulfils a not always understandable target criterion Y better than sequencing rule Z that is mostly known only by experts. In view of this critique, however, it should be kept in mind that many authors have developed sequences not for the actual sequencing in the production, but rather as heuristic methods for finite scheduling orders (see Chap. 4). This application of sequencing in the production though will be explicitly excluded here.

There will thus be no attempt made here to systematize and evaluate the hundreds of sequencing rules. Instead, a small selection of rules that appear to be practical will be discussed in relation to the logistic objectives. By far the greatest influence the sequence has is on the schedule reliability of a production. Sequencing rules that support the schedule reliability or at least do not negatively influence it are described in Sect. 25.1. In a make-to-stock production the service level is the most decisive logistic objective from the customers' perspective. Sequencing rules that take into consideration the demand after orders are generated can positively influence the service level (Sect 25.2). In specific cases, the sequencing can increase the actual output and with that the output rate (see the dashed arrow in Fig. 25.1). Such sequencing rules are the subject of Sect. 25.3. When applying them the target conflict with the schedule reliability generally has to be monitored.

Since the work by Conway, Maxwell and Miller the shortest processing time rule is known as a simple sequence that positively influences the WIP, throughput time and schedule reliability. Section 25.4 discusses the assumptions that this assessment are based on and demonstrates why applying this sequence is practical in a few exceptional cases at the most. The chapter ends with a discussion on sequencing compliance (Sect. 25.5). In particular we will consider variables with which the sequencing compliance can be measured.

## 25.1 Sequencing Rules for Increasing the Delivery Reliability

The delivery reliability is the logistic objective that the sequencing influences the most. This influence can be exercised in both directions. A random sequence decreases the delivery reliability, whereas sequencing rules that consider the planned sequence can compensate for plan deviations and thus increase the delivery reliability. In the following, four sequencing rules that positively influence the delivery reliability will be discussed:

- First In – First Out (FIFO), (Sect. 25.1.1),
- Earliest Planned Start Date (ESD) (Sect. 25.1.2),
- Earliest-Operation-Due-Date (EODD) (Sect. 25.1.3),
- Least Slack (Sect. 25.1.4).

### 25.1.1 First In – First Out (FIFO)

When applying the FIFO rule, orders are processed in the sequence that they arrive at the workstation. The FIFO rule avoids sequencing interchanges and is therefore referred to as a natural sequence [Nyhu-09*, Wien-97b]. The FIFO rule has two advantages: First, it leads to minimal variance among the production's throughput times [Wien-97b]. Consequently, the production planning can estimate the throughput times well, which in turn positively impacts the delivery reliability. Second, the FIFO rule can be more easily implemented than other sequences. There are physical aids – for example a flow rack – that enforces the processing of orders in the sequence of their arrival.

The advantage of the low throughput time variance is also accompanied by disadvantages. The FIFO rule is thus not able to correct plan deviations in that it cannot for example accelerate the throughput of orders that are released late. It thus does not use the available potential of sequencing. Moreover, it enforces standard throughput times even when the production planning has slated different planned throughput times. It should thus not be implemented together with scheduling methods that include sequencing interchanges.

### 25.1.2 Earliest Planned Start Date (ESD)

The ESD rule assigns the highest priority to the order with the earliest planned start date of the operation on the workstation. The sequence aims to process the orders in the planned sequence. It can compensate for plan deviations of individual orders (within limits) and in order to do so accelerates, for example, the throughput of orders that are released late due to a lack of materials. In these cases, it then causes sequence interchanges. Physical measures for ensuring the sequence compliance are thus not implemented as easily as when processing according to the FIFO rule.

The ESD rule requires production planning to determine planned start dates for each operation and to communicate these to the operators; this is frequently not the case in enterprises. The order's accompanying documents frequently only contain a planned end date for the operations, which can also be used for sequencing (see the following section).

### 25.1.3 Earliest-Operation-Due-Date (EODD)

The EODD rule[1] allocates the highest priority to the order with the earliest planned due date for the operation on the workstation. It only differs from the ESD rule in certain cases:

- With finite scheduling the EODD rule on the individual workstations does not differ from the ESD rule: Orders with an earlier planned start date also have an earlier planned due date. With workstations that have a number of parallel placed work centers, the sequences differ when the work contents vary.
- When scheduling with unlimited capacities, two orders can be allocated the same planned due date also on single workstations. In this case, according to the ESD rule, the order with the largest work content would have the higher priority, whereas the EODD rule assigns both the same priority.

The EODD rule can frequently be implemented more easily than the ESD rule, since the planned due dates of the operations are often already noted in the accompanying order documentation. From a theoretical perspective, the ESD rule leads to (slightly) better results with regards to the schedule reliability.

### 25.1.4 Least Slack

The least slack rule assigns the orders with the least slack the highest priority. The slack refers to the amount of time up until the order's planned completion date that is not required for processing or possibly the minimum inter-operation time. It is calculated as:

$$Slack = EDO_{plan} - TPL_0 - \sum_{i=COP}^{NOP} TOP_i - \sum_{i=COP+1}^{NOP} TIO_{\min,i} \qquad (25.1)$$

where

| | |
|---|---|
| slack | slack time [SCD], |
| $EDO_{plan}$ | end of order processing [SCD], |
| $TPL_0$ | planning time point [SCD], |
| $TOP_i$ | operation time of the $i^{th}$ operation [SCD], |
| $TIO_{min,i}$ | minimum inter-operation time of the $i^{th}$ operation [SCD], |
| COP | index of the current operation [−], |
| NOP | number of operations for the order [−]. |

[1] This is to be differentiated from the frequently implemented Earliest-Due-Date Rule, which determines the priority based on the due date of the entire order.

When the slack is negative, the order will be completed late if the order cannot be processed using overlapped production. The delay will then correspond to at least the amount of slack. When there is zero slack, the order can still be completed punctually when it is processed directly after arriving at each workstation. On the shop floor, however, this is improbable because workstations are generally already occupied with another order when the order arrives. If the slack is positive, the order can then also be punctually completed when it has to wait at the workstations to be processed. The sum of these waiting times however is not allowed to exceed the slack time.

The basic idea behind the slack rule is that delaying an order that has less slack more often leads to a late completion than with orders that have more slack. The slack time rule thus makes it possible to take future delays or disruptions into consideration in the sequencing.

In comparison to sequencing with the earliest planned start date (Sect. 25.1.2) the slack rule has one disadvantage: It can cause sequence interchanges also when there are no plan deviations. After an order is released it usually has a lot of slack and therefore has a low priority. Other orders are therefore pulled ahead also when they are not delayed in comparison to the plan. During the course of time the order then ends up being delayed compared to the plan and gains at the same time priority. Near the end of the throughput the order's processing is accelerated because there is less slack. A good portion of the slack rule's sequence interchanges thus only serve to compensate again for previous sequence interchanges. The more complex the material flow, the greater the impact of this. The slack rule is therefore not suitable for precisely implementing the production plan.

Variations of the slack rule set the slack in relation to the number of the remaining operations or the remaining operation times (see Eq. 25.4). They thus approximate sequencing according to the planned start date.

## 25.2 Sequencing Rules for Increasing the Service Level

Sequencing rules can increase the service level in make-to-stock productions. In order to do so, they take into consideration the demand that occurs after the stock falls below the order point. The basic idea is to accelerate orders for which the demand during the replenishment time exceeds the planned demand. As a result a negative correlation between the demand and the replenishment time can be achieved. This means there is a tendency that the higher the actual demand is in comparison to the planned demand, the shorter the replenishment times. Consequently, a higher service level can be attained than with sequences that do not consider the demand during the replenishment time.

A suitable figure for setting the priority can be directly derived from Schönsleben's Capacity Oriented Material Management (CORMA, see Chap. 11) and the included probability scheduling. Here, the length of time in which the stock level is expected to fall below the safety stock level is calculated and set in relation to the sum of the order's remaining operation times. The priority figure and the sequence will be referred to in the following as Flow Rate Probability Scheduling.[2]

$$FR_{PS1} = \frac{\dfrac{SL - SSL}{RDMD_m}}{\sum\limits_{i=COP}^{NOP} TOP_i} \qquad (25.2)$$

where

| | |
|---|---|
| $FR_{PS1}$ | Flow Rate Probability Scheduling (Variant 1) [–] |
| SL | stock level [units] |
| SSL | safety stock level [units], |
| $RDMD_m$ | mean demand rate [units/SCD], |
| $TOP_i$ | operation time of the $i^{th}$ operation [SCD], |
| COP | index of the current operation [–], |
| NOP | number of operations for the order [–]. |

The order with the lowest priority figure has the highest priority and is accordingly processed first. Alternatively, the following priority figure can be utilized (see Chap. 11):

$$FR_{PS2} = \frac{\dfrac{SL}{RDMD_m}}{\sum\limits_{i=COP}^{NOP} TOP_i} \qquad (25.3)$$

where

| | |
|---|---|
| $FR_{PS2}$ | Flow Rate Probability Scheduling (Variant 2) [–], |
| SL | stock level [units], |
| $RDMD_m$ | mean demand rate [units/SCD], |
| $TOP_i$ | operation time of the $i^{th}$ operation [SCD], |
| COP | index of the current operation [–], |
| NOP | number of operations for the order [–]. |

Both priority figures consider the actual stock of a variant and the production's progress regarding the replenishment order. Implementing the sequence requires that the current production progress and stock are compiled in detail. When there are a large number of variants this can only be ensured via an electronic information system.

In a mixed production, in which a part of the variants are made-to-stock and another part are made-to-order, the stock and customer orders have to be

[2] The equation requires there to be no additional replenishment orders for any of the variants in the production. Otherwise, the input of those orders closer to completion has to be taken into consideration.

coordinated. For the customer orders the following priority figure can be calculated and is referred to as the 'critical ratio' in publications:

$$CR = \frac{EDO_{plan} - TPL_0}{\sum_{i=COP}^{NOP} TOP_i} \tag{25.4}$$

where

| | |
|---|---|
| CR | critical ratio [−], |
| $EDO_{plan}$ | end of order processing [SCD], |
| $TPL_0$ | planning time point [SCD], |
| $TOP_i$ | operation time of the $i^{th}$ operation [SCD], |
| COP | index of the current operation [−], |
| NOP | number of operations for the order [−]. |

If customer orders enjoy a higher priority compared to stock orders, this can be taken into consideration, for example, by introducing a factor into Eq. 25.4.

## 25.3 Sequencing Rules for Increasing the Output Rate

Generally, expectations about the possible positive impact of the sequence on the output rate should not be exaggerated. Nevertheless, at least two sequences have a positive influence:

- The best possibility results when the setup times are sequence dependent in which case a setup time optimizing sequence suggests itself (Sect. 25.3.1).
- The Extended Work in Next Queue (XWINQ) sequence, which was already the focus of research in the 1960s, uses the possibility of a decentralized load balance (Sect. 25.3.2).

### *25.3.1 Setup Time Optimizing Sequencing*

The aim of setup time optimizing sequencing is to reduce the setup times and with that the setup costs through a suitable sequence. One of the conditions for this is that the setup times of at least part of the order are sequence dependent i.e., dependent on the item number of the previously processed order. Setup time optimizing sequencing can increase the available capacities of a workstation or even of the entire production when the workstation is a throughput bottleneck. In this case setup time optimizing sequencing can then even positively impact the schedule reliability when it reduces the production's backlog.

On the shop floor, sequence dependent setup times are primarily relevant when a group of variants can be produced with the same setup. An example is circuit board production in which enterprises frequently manufacture different circuit board formats. Some machines have to change setups when the format is changed, however, within a format various orders can be processed without further setups; related examples are found in many industries.

The danger of setup time optimizing sequencing is that orders requiring time consuming setups can be disproportionately delayed. This has to be prevented with suitable measures. Possibilities here include:

- *Fixed Sequence Cycles*: The enterprise sets a fixed cycle in which the production repeats the different setup groups (e.g., A-B-A-C-A-B).
- *Minimal Setup Frequency*: The frequency with which each setup group has to at least be setup (e.g., daily or every other day) is determined.
- *Load Shift Restrictions:* The enterprise sets an advance release window within which orders are allowed to be processed early. Similarly, a maximum delay can be set for an order. Should the maximum delay be exceeded the order receives a higher priority than all other orders for which this is not yet the case.[3]

In addition to the target conflict with the delivery reliability, setup time optimal sequencing also counteracts the goal of reducing the WIP. A sufficiently large enough supply of orders is required to exhaust the potential of sequence dependent setup times. In order to prevent an uncontrolled WIP increase, enterprises should precisely set the required WIP level. In each case the enterprise should weigh whether or not the advantages of setup time optimizing sequencing exceed the disadvantages of the greater effort required for implementing the control, the higher WIP levels, and the longer, more varied throughput times on a workstation. When possible, reducing the setup times is even more effective than setup optimizing sequencing. This aspect is especially emphasized by the Japanese production philosophy. If setup times are successfully reduced, setup time optimizing sequencing is usually no longer required.

## 25.3.2 Extended Work in Next Queue (XWINQ)

The XWINQ rule was suggested by Conway, Maxwell and Miller in the 1960s and evaluated in simulation trials [Conw-67*]. One of the aims of this sequencing rule is to prevent breaks in the material flow on downstream workstations. The criterion for assigning priority is the current WIP on the following workstation including

[3] This can be problematic when the workstation develops a backlog in comparison to the planned output rate. In a situation such as this, the setup time optimizing could contribute to reducing the backlog, however, this is prevented by orders delayed due to a non-setup-optimized production.

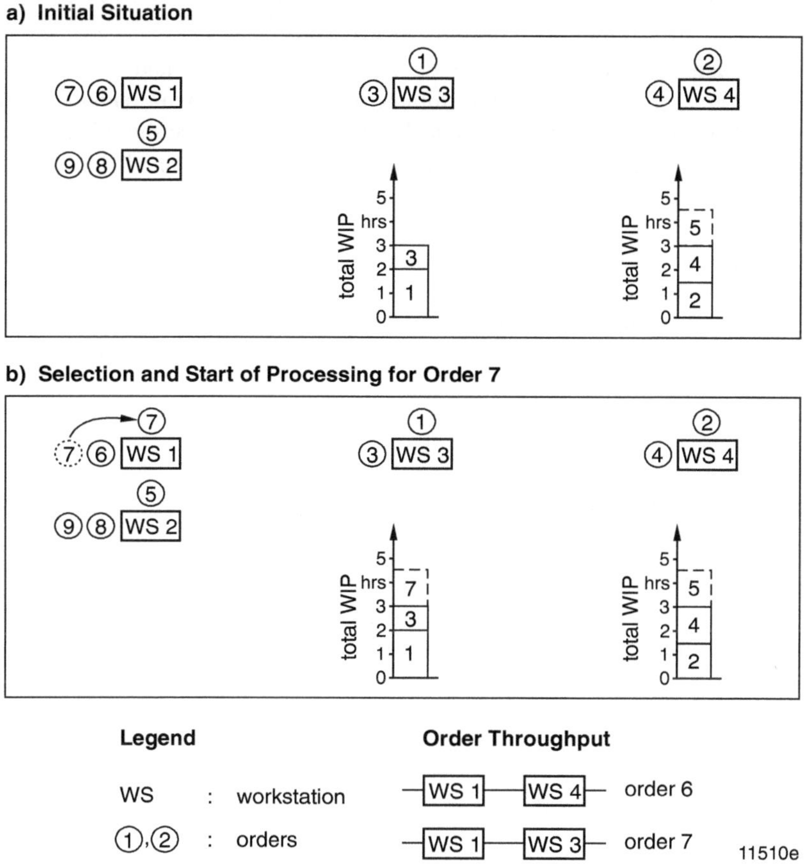


**Fig. 25.2** How the XWINQ sequencing rule functions (example)

WIP that is already being processed on the preceding workstations. The lower the WIP is, the higher the priority of the corresponding orders is.

We will take a closer look at this sequence by considering an example (Fig. 25.2). In a production with four workstations, WS 1 has just completed processing an order. In the queue are Orders 6 and 7. After being completed on WS 1 these are processed on WS 3 (Order 7) and WS 4 (Order 6). The WIP on WS 3 is 3 h. Accordingly, Order 7 is assigned the priority figure 3. The total WIP at WS 4 consists of Orders 2 and 4 as direct WIP and Order 5 as indirect WIP.

Order 6 is assigned the priority figure 4.5. Consequently, Order 7 has the higher priority and is processed first. The sequence leads to a load balance and can thus positively impact the production's utilization and WIP.

Possible disadvantages of the sequence include:

- When there are a different number of machines per workstation and the workstations have different operating times (e.g., single/double shift operation), the WIP is not an appropriate criterion for estimating the risk of a break in the material flow. Instead of WIP, the WIP range should be used in these cases (see Chap. 3 regarding the definition of the range).

- The sequence does not differentiate between bottleneck and non-bottleneck workstations. It is thus primarily suitable for productions with balanced utilization. When there is a defined bottleneck system, the sequence should be adjusted.
- The XWINQ sequence does not consider the orders' planned due dates and should therefore be combined with a schedule criterion.
- The sequence is generally only suitable for complex material flows. In a linear material flow it assigns all orders the same priority.
- The XWINQ sequence requires production operators to be informed about the current WIP on the possible downstream workstations as well as about the orders that are being processed on their preceding workstations. It thus poses relatively high demands on the compilation of operating data.

## 25.4 Shortest Processing Time (SPT)

The SPT rule was extensively analyzed by Conway, Maxwell and Miller [Conw-67*]. It assigns the order with the smallest work content the highest priority. Simulation experiments demonstrate that in comparison to other sequences, the SPT rule helps to attain: low WIP levels, short mean throughput times, low mean delay of orders and a high delivery reliability [Conw-67*]. The trio of authors identified the variance of the throughput times which causes extreme delays for individual orders as a disadvantage of the sequence.[4] Nevertheless, the authors emphatically recommend the SPT rule. Recognized textbooks also join in this generally positive evaluation of the SPT rule.

In contrast, Wiendahl and Nyhuis are critical of the SPT rule [Nyhu-09*, Wien-97b]. They point out for instance that the claimed positive impact of the sequence is strongly dependent on the WIP level. With the low WIP levels that are strived for in the industry, the sequence loses its effectiveness. They derived an approximation equation for the impact of the work content dependent sequence on the throughput time and WIP (see Fig. 25.3).

Moreover, they criticize the validity of the mean (unweighted) throughput time parameter that attaches the same importance to small orders – which at least also have a tendency to be less critical – as to large orders. If in making the calculation, the orders' are weighted with their work content, the differences in the mean throughput times that are targeted with the various sequencing rules disappear.

I would like to further develop Wiendahl and Nyhuis' criticism here: The throughput time is a very important logistic objective, however, its significance is largely due to its influence on the delivery times. With given planned delivery times

[4] This disadvantage can also possibly be reduced by assigning orders a very high priority after they have waited a specific amount of time.

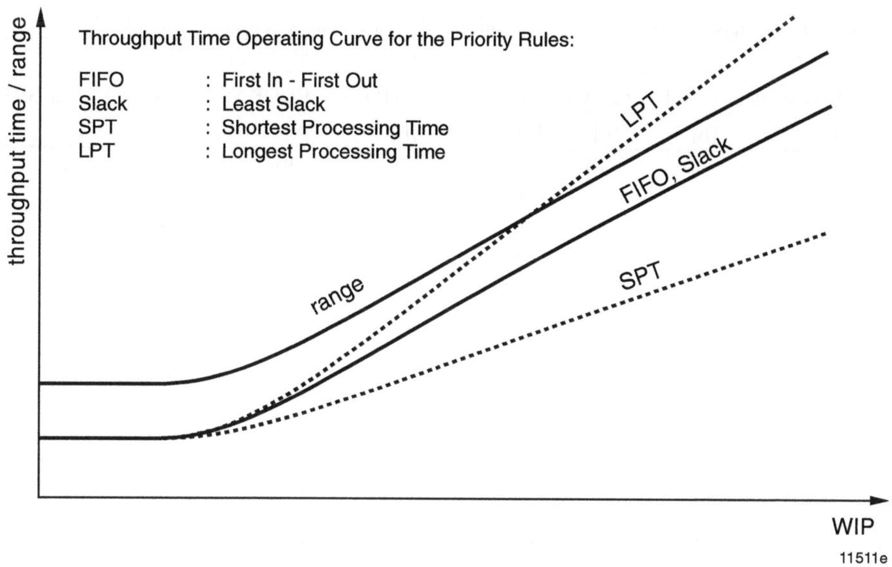


**Fig. 25.3** Influence of work content dependent sequencing rules on the mean throughput time (Nyhuis/Wiendahl)

(this is to be assumed given the order release) the throughput time thus loses its significance.

After the order is placed, the customer only perceives whether or not the order is punctually delivered or not. Early deliveries are often frowned upon and are frequently even refused. The measured shorter throughput times are then only significant when the enterprise can promise shorter delivery times. Due to the large variance among the throughput times though, this is barely possible. They are thus usually meaningless with regards to the economical success of the enterprise.

The supposed logistic advantageousness of the SPT rules is extremely questionable when a workstation or production's backlog is included in the analysis. Two cases can be distinguished here: A production with a backlog and a production without a backlog.

**Production with a Backlog** If the production or a workstation has a backlog, the SPT rules holds back larger orders. The backlog is then distributed across a few larger orders, and many small orders in comparison can be shipped punctually or even early. As with the throughput times, the wisdom of distributing the lateness in this way can be debated. In the end, the SPT rule positively impacts the measured delivery compliance parameter or mean lateness.

This is clear in Table 25.1 which depicts the results of processing orders according to the earliest planned start date and according to the shortest processing time rule.

**Table 25.1** Comparison of the ESD rule and the SPT rule at a workstation with a backlog

| Input data for simulation | | | | Simulation ESD | | Simulation SPT | | |
|---|---|---|---|---|---|---|---|---|
| No. | Work content | $EDOP_{pre}$ | EDOP | EDOP ESD | TTP ESD | EDOP SPT | TTP SPT | Seq. SPT |
| [–] | [h] | [h] | [h] | [h] | [h] | [h] | [h] | [–] |
| 1 | 5 | 0 | 15 | 30 | 30 | 30 | 30 | 1 |
| 2 | 10 | 0 | 25 | 40 | 40 | 45 | 45 | 3 |
| 3 | 15 | 0 | 40 | 55 | 55 | 60 | 60 | 4 |
| 4 | 5 | 0 | 45 | 60 | 60 | 35 | 35 | 2 |
| 5 | 25 | 20 | 70 | 85 | 65 | 135 | 115 | 9 |
| 6 | 20 | 40 | 90 | 105 | 65 | 110 | 70 | 8 |
| 7 | 5 | 45 | 95 | 110 | 65 | 65 | 20 | 5 |
| 8 | 10 | 55 | 105 | 120 | 65 | 75 | 20 | 6 |
| 9 | 15 | 70 | 120 | 135 | 65 | 90 | 20 | 7 |
| 10 | 30 | 100 | 150 | 165 | 65 | 165 | 65 | 10 |
| 11 | 40 | 140 | 190 | 205 | 65 | 245 | 105 | 15 |
| 12 | 5 | 145 | 195 | 210 | 65 | 170 | 25 | 11 |
| 13 | 20 | 165 | 215 | 230 | 65 | 195 | 30 | 13 |
| 14 | 5 | 170 | 220 | 235 | 65 | 175 | 5 | 12 |
| 15 | 10 | 180 | 230 | 245 | 65 | 205 | 25 | 14 |
| 16 | 35 | 215 | 265 | 280 | 65 | 325 | 110 | 20 |
| 17 | 20 | 235 | 285 | 300 | 65 | 270 | 35 | 17 |
| 18 | 5 | 240 | 290 | 305 | 65 | 250 | 10 | 16 |
| 19 | 15 | 255 | 305 | 320 | 65 | 290 | 35 | 19 |
| 20 | 5 | 260 | 310 | 325 | 65 | 275 | 15 | 18 |
| Number of late orders | | | | 20 | | 8 | | |
| Mean throughput time | | | | | 61 | | 44 | |
| Start of simulation: hour 25 | | | | | | | | |

*$EDOP_{pre}$* end of preceding operation processing (time of input), *EDOP* end of operation processing (time of output), *TTP* throughput time, *ESD* earliest planned start date, *SPT* shortest processing time

Orders that are completed late are underlined. The simulation was conducted manually and began at Hour 25. The ESD rule did not complete any orders punctually; the SPT rule at least managed 12 of 20. Both sequences, however, failed to attain satisfactory results.

**Production Without Backlog** The same simulation trial was conducted for a production without a backlog. Here the simulation was already initiated at Hour 10. Given otherwise similar conditions, when applying the ESD rule all orders were punctually completed, whereas with the SPT rule only 14 were (Table 25.2).

The reason for the delayed shipment of six orders when applying the SPT rule is the large variance among the throughput times.

Thus, for the SPT rule, it can in general be said: With a positive backlog it usually delivers better in comparison to other sequences, but still unsatisfactory in absolute terms. In a backlog free production i.e. the strived for target state, it delivers comparably poor results. A very high schedule reliability thus cannot be attained

**Table 25.2** Comparison of the ESD rule and SPT rule at a workstation without a backlog

| Input data for simulation | | | | Simulation ESD | | Simulation SPT | | |
|---|---|---|---|---|---|---|---|---|
| No. | Work content | $EDOP_{pre}$ | EDOP | EDOP ESD | TTP ESD | EDOP SPT | TTP SPT | Seq. SPT |
| [–] | [h] | [h] | [h] | [h] | [h] | [–] | [h] | [–] |
| 1 | 5 | 0 | 15 | 15 | 15 | 15 | 15 | 1 |
| 2 | 10 | 0 | 25 | 25 | 25 | 30 | 30 | 3 |
| 3 | 15 | 0 | 40 | 40 | 40 | 45 | 45 | 4 |
| 4 | 5 | 0 | 45 | 45 | 45 | 20 | 20 | 2 |
| 5 | 25 | 20 | 70 | 70 | 50 | 120 | 100 | 9 |
| 6 | 20 | 40 | 90 | 90 | 50 | 70 | 30 | 6 |
| 7 | 5 | 45 | 95 | 95 | 50 | 50 | 5 | 5 |
| 8 | 10 | 55 | 105 | 105 | 50 | 80 | 25 | 7 |
| 9 | 15 | 70 | 120 | 120 | 50 | 95 | 25 | 8 |
| 10 | 30 | 100 | 150 | 150 | 50 | 150 | 50 | 10 |
| 11 | 40 | 140 | 190 | 190 | 50 | 195 | 55 | 12 |
| 12 | 5 | 145 | 195 | 195 | 50 | 155 | 10 | 11 |
| 13 | 20 | 165 | 215 | 215 | 50 | 230 | 65 | 15 |
| 14 | 5 | 170 | 220 | 220 | 50 | 200 | 30 | 13 |
| 15 | 10 | 180 | 230 | 230 | 50 | 210 | 30 | 14 |
| 16 | 35 | 215 | 265 | 265 | 50 | 265 | 50 | 16 |
| 17 | 20 | 235 | 285 | 285 | 50 | 310 | 75 | 20 |
| 18 | 5 | 240 | 290 | 290 | 50 | 270 | 30 | 17 |
| 19 | 15 | 255 | 305 | 305 | 50 | 290 | 35 | 19 |
| 20 | 5 | 260 | 310 | 310 | 50 | 275 | 15 | 18 |
| Number of late orders | | | | 0 | | 6 | | |
| Mean throughput time | | | | | 46 | | 37 | |
| Start of simulation: hour 10 | | | | | | | | |

*$EDOP_{pre}$* end of pre-operation processing (time of input), *EDOP* end of operation processing (time of output), *TTP* throughput time, *ESD* earliest planned start date, *SPT* shortest processing time

in either of the two cases with the SPT rule. For most enterprises the SPT rule is therefore not suitable.

When evaluating the simulation results, the trend of the backlog should generally be constantly indicated alongside. Practically interpreting the attained schedule reliability or lateness is otherwise not possible. Nonetheless, up until now this is not typically the case.

## 25.5 Sequencing Compliance

Implementing the sequencing rule is decisive when it comes to whether or not the desired impact is also actually attained. Many enterprises are not able to implement the necessary sequencing compliance. To some degree operators make decisions about the processing sequence based on irrelevant criteria. The impact on the delivery reliability and therefore on the economic success of the enterprise can be dramatic. The variance of the lateness has a particularly drastic impact when the production supplies an assembly and the delay of one of many components can already lead to an assembly stand still for a complex final product.

Communicating the significance of the sequencing compliance is therefore a critical task. The rule according to which the processing sequence should be set has to be clearly defined for each workstation. Furthermore, the information required for complying with the sequence has to be provided. Systems which force the compliance with the sequence are ideal – these are especially simple when applying the First-in-first-out rule. The sequencing compliance can usually only be derived from operational feedback data with a great deal of effort. A simple possibility of changing this would be to save the number of orders in the workstation's queue which have a higher priority and are thus passed over, for every operation. Figure 25.4 depicts various (fictive) distributions of this parameters with decreasing sequence compliance.

The sequencing compliance describes the percentage of all sequencing decisions that conformed with the priority of the orders.

$$SeqC = \frac{\text{NOP with NSkO} = 0}{\text{NOP}} \cdot 100 \qquad (25.5)$$

where

SeqC sequencing compliance [%],
NOP number of operations [−],
NSkO number of skipped orders [−].

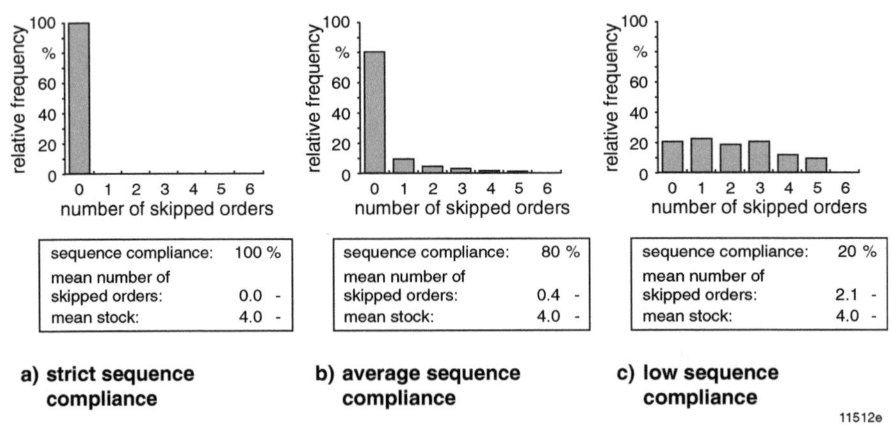


**Fig. 25.4** Visualization of sequencing compliance

The average number of skipped orders is calculated as:

$$NskO_m = \frac{\sum_{i=1}^{NOP} NSkO_i}{\text{NOP}} \tag{25.6}$$

where

| | |
|---|---|
| $NSkO_m$ | mean number of skipped orders [−], |
| $NSkO_i$ | number of skipped orders at $i^{th}$ operation [−], |
| NOP | number of operations [−]. |

When interpreting this variable the average number of orders in the queue during the investigation period should be taken into consideration. When there is a very long queue, the same value for the mean number of skipped orders is assessed more positively than with a very short queue.

Compiling this parameter makes it possible to measure the production and workstation or operator's scheduling compliance. It thus creates an important condition for localizing and eliminating the causes for an unsatisfactory delivery reliability.

## References


[Conw-67*] Conway, R.W., Maxwell, W.L., Miller, L.W.: Theory of Scheduling. Addison-Wesley, Reading (1967)

[Nyhu-09*] Nyhuis, P., Wiendahl, H.-P.: Fundamentals of Production Logistics. Theory, Tools and Applications. Springer, Berlin (2009)

[Wien-97b] Wiendahl, H.-P.: Fertigungsregelung. Logistische Beherrschung von Fertigungsabläufen auf Basis des Trichtermodells. Carl Hanser Verlag, Munich/Vienna (1997)

[Yu-01] Yu, K.-W.: Terminkennlinie. Eine Beschreibungsmethodik für die Terminabweichung im Produktionsbereich. VDI Progress Reports, Series 2, No. 576, Düsseldorf (2001)

# Part E
# Controlling Capacities

# Chapter 26
# Fundamentals of Capacity Control

Whereas capacity planning determines in advance the capacities required to implement a production program, capacity control determines the actual capacities implemented shortly beforehand. The capacity control generally decides about the work hours and about which workstation cross-trained operators are allocated to. In particular it thus determines when overtime, shortened work hours and special measures related to the capacity flexibility are implemented.

Figure 26.1 illustrates how the capacity control fits into the manufacturing control model; the tasks and corresponding actuating variable are outlined in bold.

Via the actual output, the capacity control impacts the backlog and consequently the schedule reliability of a production. Ensuring high schedule reliability is the primary logistic target of the capacity control. Furthermore, one of the essential economical aims is to efficiently implement the capacity flexibility.

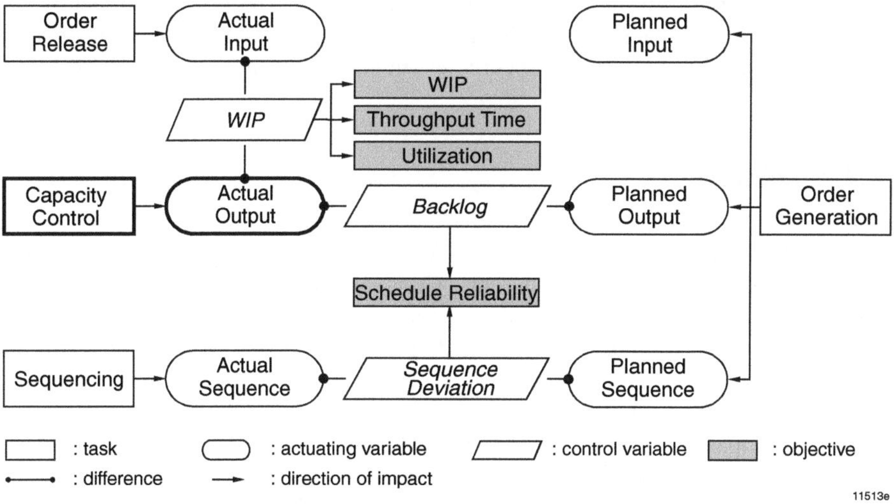


**Fig. 26.1** Positioning of the capacity control within the manufacturing control model

H. Lödding, *Handbook of Manufacturing Control*,
DOI 10.1007/978-3-642-24458-2_26, © Springer-Verlag Berlin Heidelberg 2013

In the industry, there is a wide array of possibilities for designing the capacity control. In order to organize the design process, Sect. 26.1 summarizes the general characteristics which the enterprise needs to make decisions about.

The basis of the capacity control is the flexibility of the capacities (Sect. 26.2). On the one hand, the capacity flexibility describes the degree to which the capacities can be adjusted, and on the other hand, it determines how quickly the capacity changes are effective.

In order to implement the capacity flexibility as effectively and efficiently as possible the principles of capacity control should be taken into consideration (Sect. 26.3). It should help ensure that the capacity increases have the desired impact on the production's output and do not lead to an undesired build-up of WIP in the production.

Section 26.4 is dedicated to workforce planning, whereby the question of which workstation an operator should be assigned to when they are cross-trained is raised.

## 26.1 Characteristics for Classifying Capacity Controls

Three characteristics for classifying capacity controls can be distinguished:

- The capacity control's *criterion* determines the characteristic according to which the enterprise makes decisions about adjusting the capacities (Sect. 26.1.1).
- The capacity control's *degree of detail* (Sect. 26.1.2) determines whether the capacities are adjusted for the entire production (low degree of detail) or for only part of the production (high degree of detail).
- Periodic and an event oriented capacity control are differentiated from one another according to the *trigger logic* (Sect. 26.1.3). The first makes decisions about adjusting capacities at defined intervals, while the latter makes decisions following the occurrence of a specific event e.g., exceeding a backlog limit.

### *26.1.1 Criteria*

The capacity control determines the work hours of the production workers and therefore can directly give rise to costs. Consequently, it impacts the social and financial aspects more immediately than the remaining manufacturing control tasks. As a result, it is all the more important to define suitable criteria that can trigger the capacities to be adjusted on short notice. Two types of criteria can be roughly differentiated: Criteria 1–5 are aimed at fulfilling customer demand or planning values. In comparison, criteria 6 and 7 should lead to the alignment of the capacities of a production's workstations. In many cases, an enterprise has to consider a number of criteria in a capacity control in order to attain both the logistic objective of high schedule reliability as well as the economical target of efficiently implementing the capacity flexibility.

1. *Fulfilling Customer Demand:* The basic task of capacity planning and control is to orient the capacities on the customer demand. The more successful they are, the more punctual orders can be completed (make-to-order productions) or the less stock is required to cover the demand (make-to-stock productions). In (almost) every case, orienting the capacities on the customer demand is a component of capacity planning. Especially when the customer demand is difficult to forecast and fluctuates it can also be the main criteria for the capacity control. A simple example here would be the proprietor of a banquet hall who flexibly adjusts the work hours of the personnel to the duration of the celebration.
2. *Stock:* In a make-to-stock production the stock of a product or semi-finished products can serve as a criterion for the capacity control. Enterprises increase the capacities when the stock has fallen below a lower limit and reduce the capacities if it exceeds an upper limit. The capacity control is thus indirectly coupled to the customer demand through the stock (see also Sect. 28.4).

In many cases, enterprises do not orient the capacity control directly on the customer demand, but rather on the production planning. Criteria for the capacity control are then deviations from the planned output (backlog) or from the planned capacities. The implicit assumption of such a method, is that the production planning ensures the coupling with the customer demand.

3. *Backlog*: With a backlog control, enterprises increase the capacities when the actual output falls below the planned output (backlog $> 0$). In comparison, if the production's actual output surpasses the planned output the capacities are decreased (backlog $< 0$). The aim of the backlog control is to realize the due dates and output level provided by production planning as precisely as possible even when there are disruptions. Due to the significance of the backlog control in the industry an entire chapter is dedicated to it (Chap. 27).
4. *Deviations from the Planned Capacity:* With this form of the capacity control, enterprises increase the actual capacity if it is lower than the planned capacity (the reverse case of a capacity reduction when the planned capacities are exceeded rarely occurs on the shop floor). A simple example is when one of the machines in a machine group breaks down and the production management extends the operating hours of the remaining machines. Unlike with backlog control, the capacity measure can already be introduced before a backlog develops. However, since other reasons for the development of a backlog remain unconsidered, the deviation from the planned capacity generally does not suffice as the only criterion for the capacity control (see also Sect. 28.1). If the deviation from the planned capacity only impacts part of the workstations, it inevitably results in different capacity profiles in the production. Thus, taking into consideration this criterion contributes to a better alignment of the capacities.

5. *Probable Lateness of Orders:* Begemann designed a method for controlling capacities that increases the capacities when one or more orders would otherwise be late [Bege-05]. In order to do so, the method continually (re)calculates the probable completion date of the orders via forwards scheduling (for an extensive discussion see Sect. 28.2). Unlike the backlog control, by increasing capacities the method can also reduce lateness resulting from sequence interchanges or a late input.

A number of criteria for controlling capacities are aimed at aligning the capacities of a production's workstations. When the capacities are not aligned there is a risk of breaks in the material flow on workstations with higher capacities and a WIP build-up on workstations with lower capacities. It is the task of capacity planning to keep the unavoidable capacity differences to a minimum with a well adjusted work hours and personnel planning. Frequently though, it is also necessary to implement additional control measures as a supplement. Possible criteria for such a capacity control include:

6. *WIP Before or After the Workstation:* The basic idea here is to reduce capacities when the WIP builds-up after a workstation and exceeds an upper limit. This indicates that the capacities on the following workstations are lower. *Increasing* the capacities is reasonable when the WIP falls below a lower limit after a workstation and the following workstations are threatened by a break in the material flow. On the other hand, WIP *before* the workstation should be sufficient for protecting the workstation from breaks in the material flow once the capacities are increased.
   Especially with manual work places, there is the possibility that differences in capacities are compensated not by changing the work hours, but rather by adjusting the work speed. Reducing the work speed causes losses of productivity when operators are paid hourly. Conversely, increasing the work speed beyond a reasonable degree can lead to stress and exhaustion as well as a lack of quality.
7. *Active and Passive Blockings:* If the WIP before or after a workstation is limited (e.g. due to the number of storage spaces), the resulting blockings are a possible criteria for the capacity control (for an extensive discussion see [Lödd-01]). A *passive* blocking arises when a workstation is unable to produce anymore because the upper limit after the workstation is exceeded (the workstation *is* blocked). With an active blocking WIP exceeds the upper stock limit before the workstation and thus blocks the preceding workstation. Similar to the comments about the WIP criterion, it is reasonable to reduce the capacities for a workstation that is blocked. In comparison, it is possible to increase the capacities for those workstations that are actively blocking other workstations, without themselves being blocked.

According to the criterion of the capacity control, the methods for controlling capacities that are described in the following chapters can be classified as follows (Table 26.1):

**Table 26.1** Methods of controlling capacities classified according to the capacity control criteria

| Chapter/section | Method of capacity control | Criterion |
|---|---|---|
| 27 | Backlog control | Backlog |
| 28.1 | Plan oriented capacity control | Deviation from the planned capacity |
| 28.2 | Due date oriented capacity control | Probable lateness |
| 28.3 | Output rate maximizing capacity control | Depending on bottleneck status of workstation |
| 28.4 | Inventory based capacity control | Stock level |

**Table 26.2** Methods of controlling capacities classified according to degree of detail

| Chapter/section | Method of capacity control | Degree of detail |
|---|---|---|
| 27 | Backlog control | Low/High |
| 28.1 | Plan oriented capacity control | High |
| 28.2 | Due date oriented capacity control | High |
| 28.3 | Output rate maximizing capacity control | High |
| 28.4 | Inventory based capacity control | Low |

### *26.1.2 Degree of Detail*

The capacity adjustment can apply to the entire production (low degree of detail) or for the individual workstations (high degree of detail). The degree of detail required depends on the reasons for the capacity adjustment.

Generally, changes in the customer demand require the capacities to be adjusted for the entire production. In comparison, measures that compensate for differences in the capacities should be set for each of the workstations, whereas both forms can be useful for a backlog control. Table 26.2 summarizes the reasonable degree of detail for the capacity control techniques described in this book.

Typical measures which impact the entire production include additional shifts, days closed, or collective overtime. The following measures allow capacities to be adjusted individually:

- *Adjusting the Work Speed:* Especially with manual operations it is possible to some extent to increase the work tempo above the usual on the bottleneck workstations. Possible negative consequences include poor quality or overwhelming the operator.
- *Adjusting the Work Allocation to Different Operators and Workstations:* Occasionally, enterprises can lighten the load on bottleneck workstations by allocating a portion of the work to other operators or workstations.
- *Temporarily Assigning Operators from Production Areas with Overcapacity to Bottleneck Workstations:* This is a possibility when the capacity of the bottleneck can be increased with additional operators and is not limited by technical restrictions. Naturally, this depends on operators having the corresponding qualifications.

- *Individually Adjusting Work Hours:* Increasing the work hours on bottleneck stations and reducing them on workstations with over capacities are effective measures for balancing differences in capacities. The resulting difference in the work hours, however, requires a decoupling via WIP.

### *26.1.3 Trigger Logic*

Capacity adjustments can be triggered in defined intervals (periodic capacity controls) or by specific events such as exceeding a backlog limit (event oriented backlog control).

**Periodic Capacity Control** With the periodic capacity control an enterprise makes decisions about measures for adjusting capacities at predetermined time points; a possible interval could be e.g., 1 week. The reaction of the periodic capacity control to events that trigger a backlog (e.g., machine disruptions, production of rejects) is inherently delayed. Especially when the planning periods are longer, it is frequently recommendable to supplement the periodic capacity control with event oriented components. In exchange, the periodic capacity control can integrate measures that require time to react in order to increase capacities (e.g., weekend shifts) more easily.

The periodic capacity control frequently resembles re-planning the capacities, particularly with long planning periods.

**Event Oriented Capacity Control** The event oriented capacity control changes the capacities of a workstation or production when a defined event occurs, examples here include exceeding a backlog limit, the breakdown of machinery or falling below an inventory limit. Since it triggers a measure as soon as the event occurs, the event oriented capacity control reacts more quickly than a periodic capacity control which always has to wait till the next decision point.

## 26.2 Capacity Flexibility

The greater an enterprise's capacity flexibility is, the greater possibilities it has to compensate for backlogs by increasing capacities on short notice, and thus preventing delivery delays or at least keeping delays within limits. The following section defines the concept of capacity flexibility and shows how it can be visualized in the form of Capacity Envelope Curves (Sect. 26.2.1). The capacity flexibility is influenced by both the flexibility of the resources (Sect. 26.2.2) as well as, in particular, the flexibility of the workforce (Sect. 26.2.3). The resulting capacity flexibility is determined by the bottleneck principle and is thus constantly limited by the element with the least capacity flexibility (Sect. 26.2.4).

### 26.2.1 *Definitions and Fundamentals*

The concept of flexibility is the subject of numerous publications in the field [Beac-00*, deTo-90*, Seth-90*]. The Duden dictionary defines it as the ability to adjust. The greater the number of possibilities to make adjustments and the less the costs and time required to make them, the greater the flexibility.

*Capacity Flexibility* is the ability of the production to adjust their capacities quickly, cost-effectively and to the greatest degree possible. Breithaupt and Wiendahl employ Capacity Envelope Curves to describe the capacity flexibility of a workstation (Fig. 26.2) [Brei-01].

The Capacity Envelope Curve plots the possible additional (and/or diminished) capacities that a workstation can supply over the reaction time. In the example from Fig. 26.2 the enterprise can increase the capacities of the observed workstation by 4 h/SCD with a reaction time of 5 days. In addition, the possibility of reducing the capacities below the normal value is depicted. It follows the same logic as increasing the capacities and should be considered in a backlog control.

Most enterprises use a portion of the available capacities already during the production planning to chase expected demand fluctuations (Fig. 26.3a). This portion is then no longer available for correcting plan deviations.

The capacity control utilizes only the capacities not yet budgeted (Fig. 26.3b). The possibilities of the backlog control tend to be greater, the lower the customer demand is. When the possible demand exceeds the capacity limits, it is reasonable to abstain from a backlog control under certain circumstances and to orient the capacity control on a maximum throughput. This however requires a certain degree of load flexibility and the production planning to be coordinated with the production progress. An example of a capacity control configuration such as this is described Sect. 28.3.

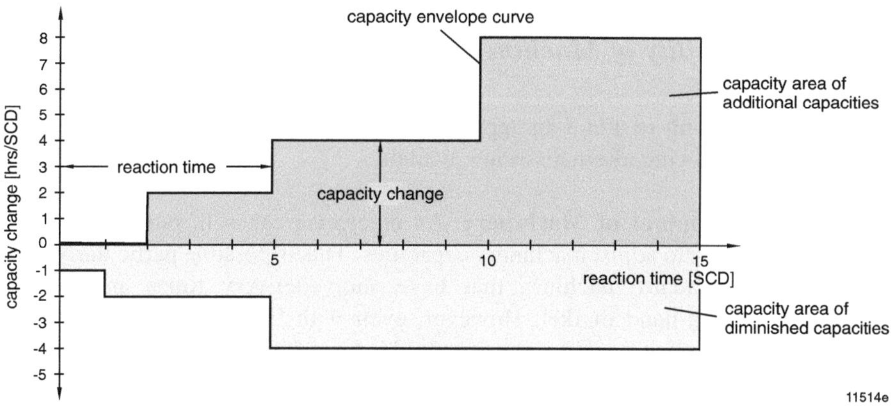


**Fig. 26.2** Sample capacity envelope curve (according to Breithaupt)

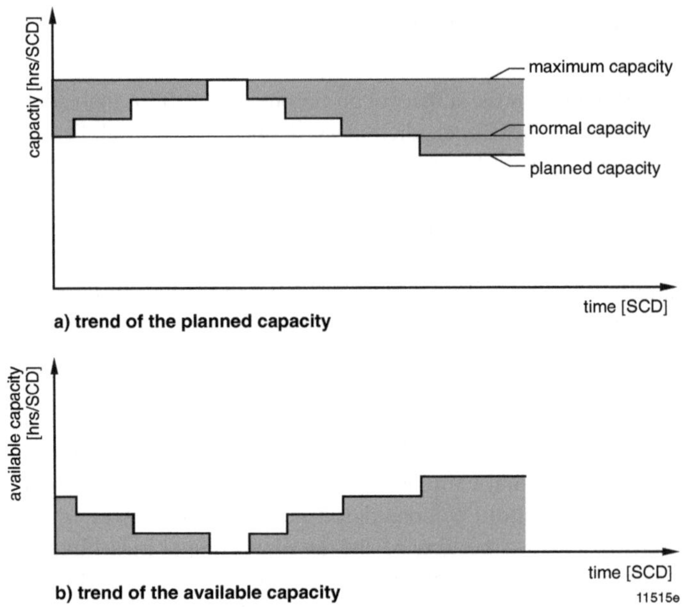


**Fig. 26.3** Determining the capacity available for the backlog control

If a backlog arises, the enterprise should be able to react quickly in order to prevent delivery delays. Those possibilities for increasing capacities that require little to no reaction time to be effective should be left to the backlog control.

The capacity flexibility of a workstation results from the capacity flexibility of both the resources and the workforce. We will turn our attention to these in the next sub-sections.

## *26.2.2 Flexibility of Machinery Capacities*

Usually it is difficult to attain an appreciable flexibility of machinery. Generally, however, the following alternatives are available.

**Changing the Amount of Machinery** An enterprise can sell, purchase or rent equipment in order to adjust machinery capacities. This is possible particularly with cost-effective standard machines that have short delivery times and a well established second-hand market. However, even with favorable conditions such as these, an enterprise usually implements this measure only under exceptional circumstances e.g., when there is a threat of large penalties for delayed orders and the production has a considerable backlog.

**Changing the Intensity of Machinery** Sometimes it is possible to increase the processing speed of the resource. A known example of this is lathes with which the tool-wear increases along with the cutting speed. The cutting speed with the optimal costs is usually below the cutting speed at the maximum capacity. The cutting speed can then be increased at the cost of stronger tool wear in order to increase the capacities and reduce the existing backlog. The possibilities for increasing the intensity of machinery are nevertheless frequently very restricted in the industrial practice.

**Subcontracting Orders or Operations** An enterprise can also subcontract orders or specific operations in order to reduce a backlog. This is not a capacity increase in the narrower sense, but rather a form of load flexibility. The impact on the backlog however is comparable. Windt intensively analyses subcontracting orders in [Wind-01]. One of the key focuses of her observations there is the common use of redundant capacities in production networks.

**Shifting Maintenance Measures** One strategy which carries considerable risks with it is the shifting of maintenance measures in order to use the operating time of a machine as much as possible for productive purposes. The resulting possibilities for temporarily increasing the capacities are generally limited, particularly as the maintenance work must then be caught up at a later time.

**Shifting to Alternative Machinery** Sometimes it is possible to shift specific operations to alternative machinery that is not completely utilized e.g., using machines in the prototype or training workshop. The possible disadvantages of utilizing such an emergency solution over the long-term are obvious. They range from additional costs and quality problems up to a negative impact on the quality of the training or the schedule reliability of the prototype workshop.

### *26.2.3 Flexibility of Workforce Capacities*

The flexibility of the workforce is comprised of four components: the flexibility of work hours, the flexibility of the working speed, the flexibility in engaging or dismissing the workforce and their cross training.

**Flexibility of Work Hours** The flexibility of work hours is all the more greater, the lower the costs are, the shorter the reaction times are with which the enterprise can change the standard work hours, and the larger the possible change is. Overtime, weekend shifts and reduced hours are typical measures for changing the standard work hours. The costs and reaction time with which they can be implanted are determined by legislators, management/collective bargaining agents and operational agreements. The same applies to the maximum number of work hours per day or week. Many enterprises arrange with their employees work hour models that are specifically aimed at increasing the flexibility of the work hours.

**Flexibility of the Work Speed** Particularly at manual work centers it is possible to change the work speed. Due to the possible strain on operators and the negative impact on the quality, many enterprises refrain from actively using this potential. Nevertheless, in numerous enterprises the correlation between the work speed and WIP on subsequent workstations needs to be monitored.

**Flexibility of Hiring and Dismissing Employees** The flexibility in engaging and dismissing employees is determined by the time and money involved in hiring and dismissing employees. This is similar to the logic of purchasing and selling machines, but is based on the social and economical situation of actual and potential employees. Decisions about engaging and dismissing employees thus also have to be made from a moral viewpoint.

The flexibility of engaging employees is influenced by the availability of the sought qualifications on the job market. If a qualification is not available, the additional time and costs required for the training need to be taken into consideration.

The flexibility of dismissing employees is essentially determined by legal requirements and the employer's fiduciary duty towards employees. The costs involved in dismissing employees, however, also impact decisions about hiring employees. The latter is usually influenced by uncertainty about the future of the enterprise's economic situation. If the costs for dismissing a yet to be hired employee are very high, enterprises are more likely to try and absorb additional demand with the flexibility of the existing employees' work hours rather than hiring new employees. This can then lead to a situation in which employees are working overtime although there are qualified people in the job market who are looking for work.

One possibility for increasing the flexibility in hiring and dismissing employees is temporary employment agencies which procure employees for enterprises for a limited time. They can be implemented for a backlog control when the training periods for a sought job are short or when subcontracted worker(s) with the required qualification(s) are available.

**Cross-Training of Employees** Employees who are cross-trained can carry out various tasks and can, for example, operate different machines. Cross-training is a highly significant form of staff qualifications. It is not only required for modern production concepts such as manufacturing cells, but also opens up the possibility of increasing the capacity flexibility of individual workstations. Generally speaking, cross-training is associated with the following advantages:

1. *Provides protection for the enterprise against fluctuations in qualifications demands:* Usually, the demand for individual qualifications fluctuates with the customer demand for the enterprise's different products. If employees are cross-trained, the enterprise can correct these fluctuations by preferentially

implementing employees at tasks which are particularly in demand. This is frequently a condition in order for the enterprise's production program to be highly flexible.

2. *Possibility to quickly increase the production's capacity:* In many cases the production's capacity cannot be quickly increased because the required professional qualifications are not available on the job market and the time to train new employees is too long. If employees are cross-trained they can change to tasks requiring longer training times when new employees are hired. The new employees then initially take over tasks that only require a short training time. In this way the production's capacity can be quickly increased (see [Suri-98*]).
3. *Protects an enterprise against the absence or loss of employees:* If all of the tasks can be completed by a number of employees, the enterprise is less susceptible to unexpected losses or absences of employees e.g. as a result of dismissals or illness.
4. *Protects employees against the loss of qualifications:* If the qualification of an employee becomes redundant e.g., due to a new technology, the employee can fall back on their remaining qualifications.
5. *Understanding of the entire system:* When employees are involved in more than one task, they gain a better overview of the entire production system and learn to understand the dependencies between the processes. This is critical for example in continual improvement processes.

The first two advantages are the most important for controlling capacities. Nevertheless, cross-training is also related to the following disadvantages which should be weighed against the advantages if there are any doubts:

1. *Expenditures and time involved with cross-training:* Cross-training employees incurs costs and requires time which may then be lacking for productive purposes. It can thus temporarily reduce the available capacities.
2. *Higher wages for cross-trained employees:* There is at least a tendency for cross-trained employees to earn more. This aspect is emphasized in particular by Suri [Suri-98*].
3. *Loss of advantages related to being a specialist:* If an employee is involved in a number of tasks, they do not attain the level skill for an individual task that they would gain if they specialized in one task.

A qualification matrix is an important aid in analyzing the qualification profile of an employee (Table 26.3).

The qualifications matrix compares employees with regards to existing tasks. Three levels of qualification are frequently differentiated within professional literature:

1. The employee can partially carry out task.
2. The employee can completely carry out task.
3. The employee can carry out task completely and is able to train other employees to do it as well.

**Table 26.3** Qualifications matrix (sample)

| Employee | Workstation | | | | |
|---|---|---|---|---|---|
| | Turning | Milling | Drilling | Cutting | Grinding |
| Müller | ■ | ○ | ■ | ■ | |
| Meier | | | | ● | |
| Schmidt | | | ■ | ● | ● |
| Konrad | ○ | ■ | ● | ● | |

■ employee can carry out task completely and is able to train other employees to do it as well,
● employee can carry out task completely,
○ employee can partially carry out task

The qualification matrix should be analyzed at least according to the following viewpoints:

1. *With regards to which task is there a threat of an 'employee bottleneck'?* In the above sample matrix, this applies primarily to grinding.
2. *Which employee can only complete a few tasks?* In the above sample matrix, this applies to the employee named Meier.

## *26.2.4 Resulting Capacity Flexibility of a Production*

Generally, the bottleneck principle applies to the capacity flexibility. On the one hand, it can be applied to the overlapping of the capacity flexibility for both the machinery and workforce. That means, the usable capacity of a workstation results from the minimum machinery and work force capacity. On the other hand, however, it also applies to deriving the capacity flexibility of an entire production from the capacity flexibility of the individual workstations in the production. When only one workstation is unable to increase its capacities, then the entire production is unable to increase its throughput.

In order to be able to compare the capacity flexibility of the workstations with one another, the capacity should be set in relation to the typical capacity requirement of a workstation:

$$CAP_{rel} = \frac{CAP}{CAP_{req,typ}} \cdot 100 \qquad (26.1)$$

where

| | |
|---|---|
| $CAP_{rel}$ | relative capacity [%], |
| CAP | capacity of a workstation [h/SCD], |
| $CAP_{req,typ}$ | typical capacity requirements of a workstation [h/SCD]. |

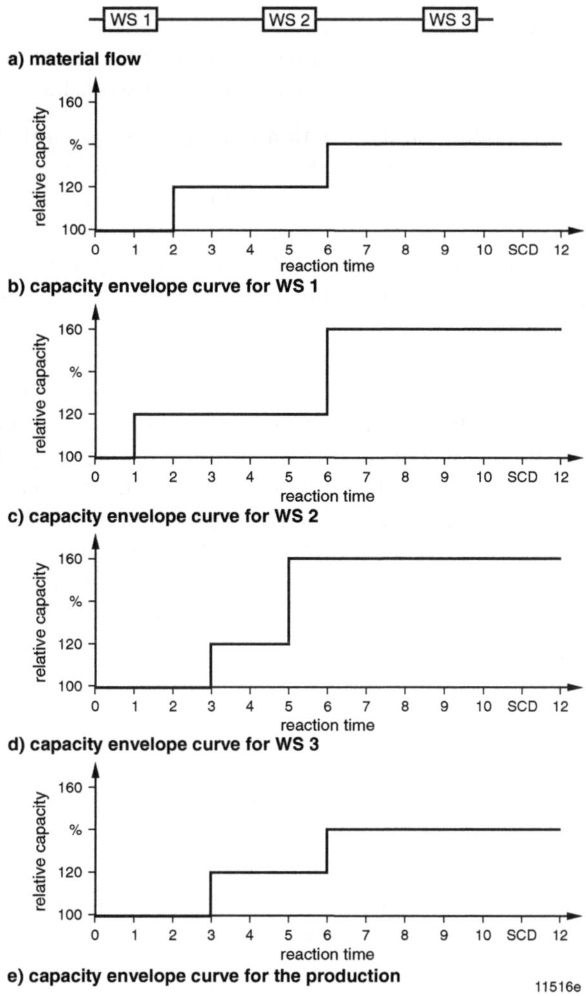


**Fig. 26.4** Overlapping the capacity envelope curves of a production's three workstations into a capacity envelope curve for the production

If the relative capacity is less than 100%, then the planned output rate, even with typical capacity requirements, can only be achieved by increasing the capacity. In comparison, when the value is above 100% there is a loss of utilization on the workstation when there are typical capacity requirements.

Capacity envelopes, which instead of the absolute additional capacity plot the relative capacity of the workstation over the reaction time, can be directly compared with one another (Fig. 26.4). According to the bottleneck principle, the relative capacity of the production corresponds to the minimum relative flexibility of the workstations.

$$CAP_{rel,ent}(t) = Minimum\big(CAP_{rel,1}(t); CAP_{rel,2}(t); \ldots; CAP_{rel,n}(t)\big) \quad (26.2)$$

where

$CAP_{rel,ent}$ (t) relative capacity of the entire production [%],

$CAP_{rel,i}$ (t) relative capacity of workstation *i* [%].

Depending on the reaction time it is also possible for different workstations to limit the throughput of the production. In the example from Fig. 26.4, Workstation 3 limits the capacity flexibility of the production up until a reaction time of 5 days, to some degree also with other workstations. Between 5 and 6 days of reaction time Workstation 2 is the throughput bottleneck, whereas from a reaction time of 6 days on Workstation 1 limits the relative capacity of the production.

When there is a complex material flow it should be verified whether or not a significant part of the orders does *not* flow through the throughput bottleneck. If this is the case, then a separate backlog control can be calculated for these orders on the remaining workstations. By increasing the capacities accordingly on these workstations it is then possible to utilize the production more homogenously and limit delays to orders routed through the bottleneck.

Occasionally, it is difficult to determine the throughput bottleneck of a production. This is especially so when the relative capacities of the workstations do not significantly differ and the bottleneck varies due to changes in the product mix.

## 26.3 Principles of Capacity Control

There are two fundamental principles that enterprises should consider when controlling the capacity: The orientation on the planned output as well as the consideration of the bottleneck principle.

1. *The capacity control should orient itself on the planned output (or the customer demand).*

The first principle provides the target reference for the capacity control. Deviations from the planned output lead to the delayed completion of orders (actual output < planned output) or to the early completion of orders and increased stock of finished products (actual output > planned output). A capacity control that is oriented on the planned output thus supports a higher schedule reliability and compliance with planned stock levels.

2. *The capacity control should take into consideration the bottleneck principle, according to which the bottleneck of the production determines the yield of the production.*

The second principle concerns the economic efficiency of the capacity control and at the same time, defines the frame within which a backlog control can act: If an enterprise does not align the capacity increase on all of the workstations with the possible capacity increase on the throughput bottleneck, the capacity increase will not achieve its purpose. Moreover, undesired side effects frequently arise.

In the following, we will discuss this under the assumption that the maximum capacity for the throughput bottleneck of a production is completely exhausted. If in this situation an enterprise increases the capacity on the remaining workstations it leads to an asymmetric capacity profile for the production's workstations. The following effects result in an idealized case:

- The throughput of the production does not change (with the exception of the temporary effects described below), since the throughput bottleneck does not manage more than previously. Accordingly, the production's backlogs and lateness continue essentially unchanged. Generally, the hoped for positive impact of the capacity increase fails to occur.
- In the part of the production before the manufacturing bottleneck the production's WIP climbs. This WIP increase is paid for dearly through the additional costs for overtime or additional shifts. The WIP build-up is limited by either a WIP control or with a Due Date Oriented Order Release because to a lack of urgent orders. In both cases the workstations are threatened by a break in the material flow.
- In the part of the production after the throughput bottleneck the WIP is reduced. As a result the measured output rate is temporarily increased and the lateness as well as the throughput times are reduced. After the WIP is reduced there is the threat of breaks in the material flow.[1]

Thus, even when there is a production backlog, it is usually not reasonable to increase the capacities of a workstation more than the workstation with the smallest reserve of capacities can additionally manage. In particular, the decision about increasing the capacities on a workstation cannot be made isolated from the decisions on other workstations. This makes a decentralized backlog control conducted by the operators of the workstations more difficult.

## 26.4 Employee Allocation

With the cross-trained employees described in Sect. 26.2 comes the necessity of deciding which task or workstation an operator should be allocated to.

- Section 26.4.1 addresses the criteria for making a decision and with that the questions of *how* and with *which* objectives the decisions about allocating an employee to a workstation is met.

[1] Breithaupt installed a special controller in order to increase the capacities only when there is sufficient WIP available [Brei-01].

- Section 26.4.2 clarifies the trigger logic and with that the question of *when* decisions about the employee allocation have to be made.
- Section 26.4.3 discusses the locality of the decision and with that the question about *who* makes the decision about the employee allocation.

### 26.4.1 Criteria for Allocating Employees

At least four different logistic criteria can be identified: the priority of the orders on the workstation, the backlog of the workstation, the bottleneck status of the workstation and the efficiency of employees for various tasks:

- *Order Priority:* An enterprise allocates an employee to the available workstation at which the order with the highest priority is waiting. A prerequisite for this is that the priorities of orders on different workstations are comparable. An example would be to select the orders with the earliest planned start date (or the earliest planned end date, lowest critical ratio etc.). All of the criteria mentioned are aimed at increasing the schedule reliability.
- *Workstation Backlog:* The enterprise allocates the employee to the available workstation with the highest workstation backlog (or the greatest–expected–lateness). Doing so aims at ensuring the schedule reliability.
- *Bottleneck Status:* When an enterprise is working at its capacity limit, the bottleneck determines the output rate of the production. Accordingly, in these situations the bottleneck workstations should always be allocated an operator. Similarly, provisions should be made which extend the operating hours of the bottleneck workstation. This can be attained for example with overtime or special shifts on the bottleneck workstation. An additional possibility is to introduce a measure which ensures the workstation is operated even during regular breaks.
- *Employee Efficiency:* If there is a difference in the efficiency with which the employees can conduct specific tasks, it is beneficial to allocate employees to the type of tasks that they can most efficiently accomplish. The aim here is to reduce the production costs. An approach such as this can also positively influence the output rate and with that the backlog and mean lateness of orders. It should however be verified (and carefully considered) whether or not such an allocation of employees leads to increased variance in the output lateness.

### 26.4.2 Trigger Logic

Two types of employee allocation can be generally identified: periodic and event oriented.

**Periodic Employee Allocation** Periodic employee allocation makes decisions about which resources employees will be assigned to at defined times. An example here is an enterprise which allocates employees at the start of a shift. One of the advantages of periodic employee allocation is the minimal effort required to make decisions. It can however lead to situations in which employees work on orders whose planned start date is not yet reached while orders whose planned start date has already been exceeded are waiting at unoccupied workstations.

**Event Oriented Employee Allocation** Event oriented employee allocation is triggered by specific events. A reasonable event is the completion of an order by an employee. Two different strategies can in turn be identified: With the first strategy, the employee always remains at the workstation when there is another order waiting in the workstation's queue. Similar to periodic employee allocation this is related to the disadvantage that an employee can also end up processing orders that are not urgent. A method such as this can nevertheless be necessary when starting up or shutting down a workstation is connected with (considerable) costs. The second strategy is that employees consider changing workstations after completing each order. This requires that the employee be able to quickly grasp which workstations are not allocated employees and which orders at these workstations have the highest priority. In this case, this strategy will usually lead to better results than the first strategy. It is also superior to periodic employee allocation. Nevertheless, with event oriented employee allocation the times at which decisions are made cannot be coordinated.

### *26.4.3 Locality of the Employee Allocation*

Generally, it can be differentiated between whether employees are allocated to machines from a centralized planning authority or if the employees themselves make the decisions about which workstations they work at. In both cases the criteria described in Sect. 26.4.1 are to be considered. Due to the uncoordinated decision points (see previous section) a (logistically advantageous) event oriented employee allocation can generally be more easily realized when the responsibility for the employee allocation is delegated to the employees. The lack of motivation or possibility of delegating the responsibility for the employee allocation probably contributes to the wide spread implementation of periodic employee allocation that is inherently connected to disadvantages. In comparison, a centralized employee allocation is advantageous when the efficiency of employees for specific tasks differs greatly among the employees.

## References

[Beac-00*] Beach, R., et al.: Manufacturing operations and strategic flexibility survey and cases. Int. J. Oper. Prod. Man. **20**(1), 7–30 (2000)

[Bege-05] Begemann, C.: Terminorientierte Kapazitätssteuerung. Reports from the IFA, Hannover, No. 2/2005

[Brei-01] Breithaupt, J.-W.: Rückstandsorientierte Produktionsregelung von Fertigungsbereichen. Grundlagen und Anwendung. VDI Progress Reports, Series 2, No. 571, Düsseldorf (2001)

[deTo-90*] de Toni, A., Tonchia, S.: Manufacturing flexibility: a literature review. Int. J. Prod. Res. **36**(6), 1587–1617 (1990)

[Lödd-01] Lödding, H.: Dezentrale Bestandsorientierte Fertigungsregelung. VDI Progress Reports, Series 2, No. 587, Düsseldorf (2001)

[Seth-90*] Sethi, A.K., Sethi, S.P.: Flexibility in manufacturing: a survey. Int. J. Flex. Manuf. Syst. **2**(4), 289–328 (1990)

[Suri-98*] Suri, R.: Quick Response Manufacturing. A Companywide Approach to Reducing Lead Times. Productivity Press, Portland (1998)

[Wind-01] Windt, K.: Engpassorientierte Fremdvergabe in Produktionsnetzen. VDI Progress Reports, Series 2, No. 579, Düsseldorf (2001)

# Chapter 27
# Backlog Control

The basic idea of backlog control is to temporarily adjust the capacities so that the production attains the planned output despite disruptions. The most important goal is ensuring high schedule reliability.

Unlike methods for generating or releasing orders the procedural rules are not clearly defined. This is not only because individual enterprises tend to control their capacity flexibility differently, but also because of the diverse conditions they are faced with which necessitate customized solutions being defined for backlog control even when their capacity flexibility might be the same.

What is common among the solutions is that they arrange capacitive measures as a function of the measured backlog in order to correct deviations in the actual output from the planned output. The basic way in which they function is described in Sect. 27.1. The most important components of backlog control are the measuring of the backlog (Sect. 27.2) and selection of the measure for adjusting the capacities (Sect. 27.3).

We will discuss backlog control based on the guidelines for designing manufacturing control methods in Sect. 27.4, whereas, how enterprises can set the parameters for backlog control is the subject of Sect. 27.5. The chapter will conclude with an example of applying backlog control in the aviation industry (Sect. 27.6).

## 27.1 Method Description

Backlog control can be generally described with three rules:

1. Measure the backlog of the production or its workstations at set times (backlog measurement).
2. Select measures for increasing capacities in order to reduce backlogs if the backlog exceeds a defined upper backlog limit (selection of measures).
3. Select measures for reducing capacities in order to correct negative backlogs in case the backlog falls below a defined lower backlog limit.

H. Lödding, *Handbook of Manufacturing Control*,
DOI 10.1007/978-3-642-24458-2_27, © Springer-Verlag Berlin Heidelberg 2013

The principle of backlog control is depicted in Fig. 27.1.

The upper part of the figure depicts the cumulated planned and actual output of a production over time, whereas, the middle part presents the backlog trend and the lower part the capacity trend. The actual output soon falls below the planned values so that a backlog arises which first exceeds the upper backlog limit at time $T_0$. As a result, the enterprise opts for a capacity increase which goes into effect once the response time has passed at time $T_1$. From this point on, the backlog is reduced by the capacity increase and the slope of the output curve becomes steeper. Shortly after the backlog is completely eliminated the enterprise retracts the capacity increase at time T2.

Figure 27.1b also depicts from time $T_1$ on the backlog trend that would have resulted if the capacities had not been increased. The distance between it and the backlog trend with the capacity increase steadily grows over the duration of the capacity increase. From time T2 on the curves run parallel to one another.

Before we finish this discussion, let us examine a simple example which will clarify the basic idea of backlog control (Fig. 27.2). The object of our considerations is a production with three workstations and a linear material flow. All of the workstations are operating with a normal capacity of 8 h per day (single shift operation). The capacity flexibility of the enterprise allows up to 2 h of overtime without advance notice. Each workstation is allocated one employee. The upper and lower backlog limits are both zero.

The enterprise has defined backlog control as follows: At the end of a shift the backlog is measured on each workstation. If the planned output is not attained (backlog $> 0$), it arranges overtime on the impacted workstation equal to the extent of the backlog. The maximum overtime is 2 h. If the workstation attains the planned output before the end of the shift, the allocated employee stops working.

Due to a disruption that has already been resolved WS 2 has a backlog of 2 h in the initial situation (Fig. 27.2a). The distruption also impacts WS 3 due to a break in the part supply, it thus has a backlog of 1 h. In comparison, WS 1 does not have a backlog at all. As a result of the disruption, three orders accumulate at WS 2, while the queue before WS 3 is empty.

The enterprise reacts to the backlog with 2 h of overtime on WS 2 and 1 h of overtime on WS 3. The employee on WS 1 has fullfilled the daily target by the end of the shift at 2 pm and stops working (Step A in Fig. 27.2a).

At 3 pm the employee at WS 3 completes Order 1, which can still be delivered to the customer at the planned date (Step A in Fig. 27.2b). The backlog at the workstation is thus eliminated (Step B) so the employee stops his work (Step C).

The employee on WS 2 completed Order 2 during the first hour of over overtime and began working on Order 3. The backlog on WS 2 is thus reduced to 1 h (Step D).

After another hour, the employee on WS 2 completes Order 3 (Step A in Fig. 27.2c). The output thus increases to the planned value of 8 h so that the backlog is completely caught up (Step B). The employee leaves the production (Step C).

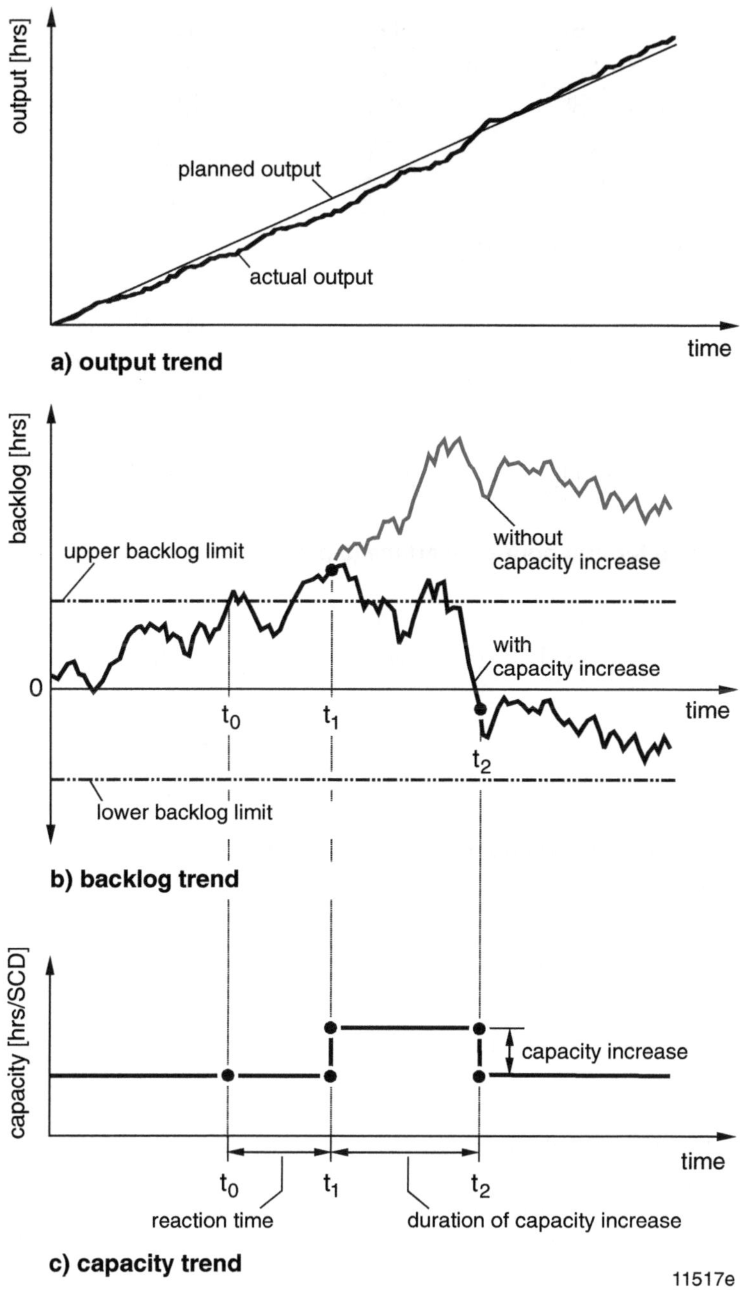


**Fig. 27.1** Principle of backlog control

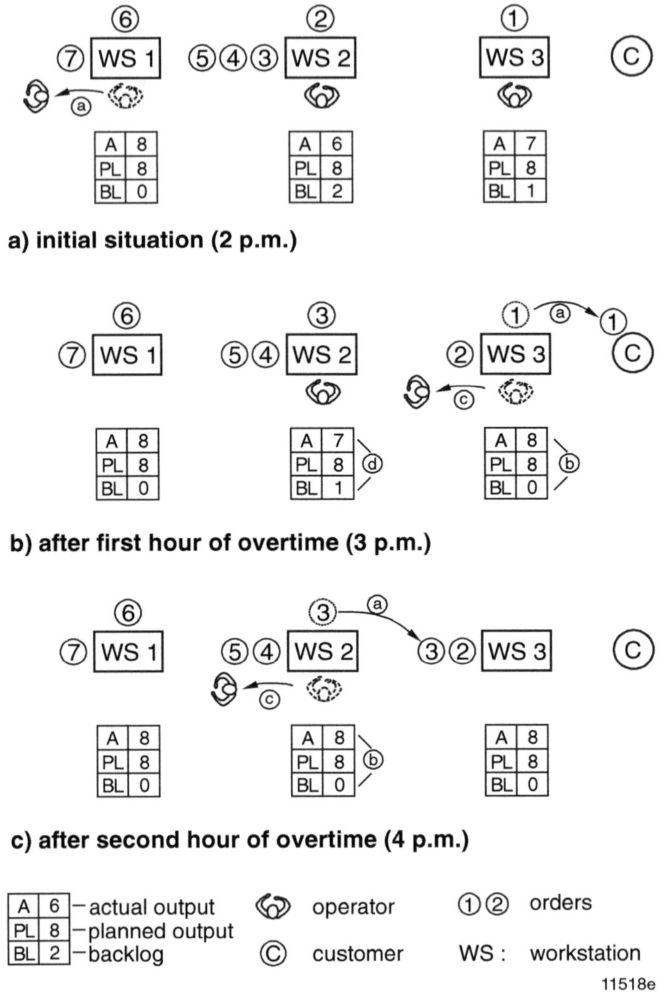


**Fig. 27.2** How backlog control functions (example)

Due to backlog control, the enterprise has suceeded in converting the production back to the planned state. In comparison to the initial situation, the WIP is evenly distributed across the workstations in the production and protects them from sudden breaks in the material flow.

In the industrial practice though, backlog control is frequently not so easily realized as in this simplified example. The reason for this is that the response times are longer and there is less capacity flexibility.

## 27.2 Measuring the Backlog

Backlog measuring should measure the backlog of a workstation or production as promptly, precisely and efficiently as possible. According to Eq. 3.21, the backlog at a specific time is defined as the difference between the planned output and the actual output [Pete-96]:

$$BL(t) = OUT_{plan}(t) - OUT_{act}(t) \tag{27.1}$$

where

| | |
|---|---|
| BL(t) | backlog [h], |
| $OUT_{plan}(t)$ | planned output [h], |
| $OUT_{act}(t)$ | actual output [h]. |

Both the planned output as well as the actual output should always be measured in standard hours. Otherwise, if the actual work content deviates from the planned work content, it is no longer possible to determine whether or not the planned output was attained.

As described in Chap. 3, the backlog can take on both positive (planned output is not attained) as well as negative (planned output is exceeded) values. The following Section first describes the fundamentals of measuring backlogs independent of the method selected (Sect. 27.2.1). How enterprises can measure the backlog with different manufacturing control methods is then the subject of Sect. 27.2.2.

### *27.2.1 Method Independent Fundamentals of Measuring the Backlog*

The measuring of the backlog can be classified according to five criteria:

- degree of detail,
- precision of measurement,
- means of measurement,
- measured variable,
- time of measurement.

**Degree of Detail** The backlog can be determined from the orders' planned end-date for the entire production (low degree of detail). In order to do so the cumulated planned output at the time of the measurement is compared with the cumulated actual output.

However, the backlog can also be calculated for each of the workstations in the production (high degree of detail). In this case, the cumulated planned output of a workstation at the time of the measurement is compared with the actual output. In order to calculate the planned output it is necessary to determine due dates for the individual operations during the scheduling.

Generally, for a backlog control both the backlog for each workstation as well as for the entire production should be determined. This is necessary because the backlog situation of the individual workstations can siginificantly differ from the entire production.

**Precision of Measurement** The precision of the measurement is basically dependent on whether the actual output is compiled continuously or discretely. Discretely measuring the actual output takes an order into consideration the moment its processing is completed. In comparison, a continuous backlog measurement already considers the part of the order's work content that is being processed before it is completed.

The discrete backlog measurement is less precise than a continuous backlog measurement. This point is emphasized by Breithaupt [Brei-01]. Particulary with large work contents the discrete backlog measurement can distort the results of the measurement and lead to false conclusions. This can be explained with a simple example: At a specific time the planned output on a workstation is 300 h. The already completed orders correspond to an actual output of 285 h. A discrete backlog measurement identifies a backlog of 15 h or two shifts. This would justify temporarily increasing the capacities. However, if there is a larger order of 20 h being processed it is possible that the workstation has a lower or even no backlog. This depends on how many hours of work on the order has already been completed.

The disadvantage of the continuous backlog measurement is that it requires more effort. Furthermore, especially with small work contents the imprecision of a discrete backlog control is frequently negligible so that an enterprise can refrain from a continuous backlog control.

**Means of Measurement** In many cases the backlog can be measured via an electronic monitoring system. Its precision depends primarily on whether the system requires a manual and thus faulty and frequently delayed data entry or whether the data collection can be automated. There is also a possibility that enterprises can also refrain from an electronic data collection. This applies when the backlog is able to be visually identified. In this case a large degree of transparency is required e.g., free storage places in a visual Kanban system.

**Measured Variables** Usually the actual output is measured and compared with the planned output in order to determine the backlog. However, to some degree it is also possible to calculate the backlog with a – possibly simpler to measure – subsidiary variable. Thus with a Due Date Based Order Release the backlog can be deducted by comparing the actual WIP and the planned WIP (see Eq. 16.1).

**Time of Measurement** According to the trigger logic the following alternatives can be distinguished, which can also be implemented together.

- *Periodic Backlog Measurement*: The backlog is measured at a defined time (e.g. end of a shift or day).
- *Event Oriented Backlog Measurement*: The backlog is measured after defined events, e.g., the end of a disruption or following a time limited measure for adjusting the capacity.
- *Continuous Backlog Measurement*: Especially in mass productions with manufacturing or assembly lines the backlog is so important that enterprises frequently measure it continuously and display it on electronic monitors. The reason for this is not always a backlog control: Frequently important productivity and cost targets depend on attaining the planned output.

### 27.2.2 Method Dependent Backlog Measurement

It is often easiest to measure the backlog as the difference between the planned and actual output, however, this requires defined lot sizes and planned due dates. Not all manufacturing control methods schedule orders though. A number of methods also leave the lot sizes open. Even under such conditions though, a backlog control is still necessary. In this next Section, I would thus like to describe how the backlog can be measured when applying specific methods for generating orders. Moreover, we will briefly discuss measuring the backlog when applying techniques for releasing orders.

**Make-to-Order Productions/MRP II** Since orders are scheduled there are at least planned end dates for the orders as well as frequently for the operations. The backlog can be determined by measuring the actual output and comparing it with the planned output (see Eq. 27.1).

**Order Point System** There are two possibilities for estimating the backlog with the Order Point System: by scheduling the replenishment orders and by analyzing the stock levels.

*Scheduling Orders:* Many enterprises schedule the replenishment orders and set due dates for the operations and the entire order. This allows the backlog to be determined by comparing the actual output with the planned output both on a workstation level as well as for the entire production (see Eq. 27.1).

*Analyzing Stock Levels:* An approximate indication of a production's backlog can be derived by analyzing the stock level. If the demand during the replenishment time coincides with that which was planned and supplies the production punctually then for the mean stock (see Fig. 8.1):

$$SL_m = \frac{LS}{2} + SSL \tag{27.2}$$

where

| | |
|---|---|
| $SL_m$ | mean stock level [units], |
| LS | lot size [units], |
| SSL | safety stock level [units]. |

When a production develops a backlog, the stock levels sink. In the reverse situation, the stock levels climb above the planned value. A measure for the backlog is thus the sum of the deviations from the planned stock for all of the variants.

$$BL = \sum_{k=1}^{NVa} \left(SL_{plan,m,k} - SL_{act,k}\right) \tag{27.3}$$

where

| | |
|---|---|
| BL | backlog [units], |
| $SL_{plan,m,k}$ | mean planned stock for Variant *k* [units], |
| $SL_{act,k}$ | actual stock for Variant *k* [units], |
| NVa | number of variants [–]. |

Generally it is practical to convert the backlog into the unit hours, then:

$$BL = \sum_{k=1}^{NVa} \left(\frac{SL_{plan,m,k} - SL_{act,k}}{LS_k} \cdot WC_{total,k}\right) \tag{27.4}$$

where

| | |
|---|---|
| BL | backlog [units], |
| $SL_{plan,m,k}$ | mean planned stock for Variant *k* [units], |
| $SL_{act,k}$ | actual stock for Variant *k* [units], |
| $WC_{tot,k}$ | total work content for Variant *k* [h], |
| $LS_k$ | lot size for Variant *k* [units], |
| NVa | number of variants [–]. |

**Kanban** Kanban does not generate any (explicit) planned end dates for replenishment orders. Thus Eq. 27.1 cannot be directly applied. Nevertheless, there are two approaches to measuring the backlog:

When there is a detailed production program, the backlog results from the difference between the actual output and the production program's target value (see the following discussion regarding Cumulative Production Figures for the equation).

The second approach analyses the stock in the outgoing store of a workstation or the distribution of the kanbans. If the workstation has a backlog, the stock in the outgoing store sinks below the safety stock level and there are more kanbans on the board than planned. If the workstation completes the replenishment orders ahead of time, the stock in the outgoing store climbs and the number of kanbans sink. The backlog of the workstation can therefore be estimated either from the stock in the outgoing store or from the number of kanbans on the workstation. It thus seems obvious to estimate the backlog similarly to the procedure with the Order Point System via Eq. 27.4.

On the shop floor, it is frequently practical to take additional information also into consideration in order to estimate the danger of a supply shortage; included here in particular is the progress of the work on the orders being processed (see Sect. 27.2.1).

Formally calculating the backlog is frequently not necessary when the employees keep working until, depending on the delivery tolerance, the production program for the day or week is accomplished.

**Capacity Oriented Material Management (CORMA)** CORMA is a manufacturing control method that was developed especially for mixed productions (see Chap. 10). In a mixed production, make-to-stock and customer orders are manufactured on the same machinery.

On the one hand, CORMA provides good conditions for measuring the backlog, since all of the orders are scheduled. On the other hand, the constant re-scheduling of make-to-stock orders allows backlogs for the make-to-stock production only to be identified by analyzing the stretch factor, if at all: The method shortens the planned inter-operation time when the processing of an order falls behind schedule. As a result, the planned processing dates and thus also the planned output of the workstations are shifted into future (with the exception of the last operation). The backlog with regards to the original plan is no longer identifiable due to the re-planning. There are two possibilities for solving this problem:

- The backlog is calculated in comparison to the planned output, which results from a scheduling with the originally planned inter-operation times.
- Based on the new scheduling the capacity requirements are calculated for each workstation. The capacities are then re-set based on these requirements. In this case the backlog measurement is eliminated.

**Synchro MRP** The comments made above regarding Kanban also apply to measuring the backlog with Synchro MRP. Nevertheless, fulfilling the daily production program is generally preferred over the analysis of the stock.

**Hybrid Kanban/CONWIP Control** This manufacturing control method is similar to the Order Point System combined with the decentralized control loop of Kanban (see Chap 12). Accordingly, the comments for these methods also apply for this hybrid application. The backlog can be derived by:

- scheduling the replenishment orders (see Order Point System) or
- analyzing the stock (see Order Point System).

**Cumulative Production Figures** With Cumulative Production Figures (see Chap 13), the backlog can be calculated from the comparison of the actual and planned Cumulative Production Figures of the variants:

$$BL = \frac{\sum_{k=1}^{NVa}\left(\left(CPF_{plan,k} - CPF_{act,k}\right)\cdot T_{p,k}\right)}{60} \tag{27.5}$$

where

| | |
|---|---|
| BL | backlog [h], |
| $CPF_{plan,k}$ | planned Cumulative Production Figure of Variant *k* [units], |
| $CPF_{act,k}$ | actual Cumulative Production Figure of Variant *k* [units], |
| $T_{p,k}$ | processing time per unit of Variant *k* [min/ME], |
| NVa | number of variants [−]. |

The equation does not consider any setup times. Where setup times and lot sizes are known, the following applies for the backlog:

$$BL = \sum_{k=1}^{NVa}\left(\frac{CPF_{plan,k} - CPF_{act,k}}{LS_k}\cdot WC_k\right) \tag{27.6}$$

where

| | |
|---|---|
| BL | backlog [h], |
| NVa | number of variants [−], |
| $CPF_{plan,k}$ | planned Cumulative Production Figure of Variant *k* [units], |
| $CPF_{act,k}$ | actual Cumulative Production Figure of Variant *k* [units], |
| $LS_k$ | lot size of Variant *k* [units], |
| $WC_k$ | work content of Variant *k* [h]. |

**Base Stock System** The Base Stock System can be interpreted as a special case of Cumulative Production Figures in which the control CPF of a control block corresponds to the sum of the cumulated demand and base stock of the control block (see Chap 14).

Accordingly, for the planned CPF of a variant in the Base Stock System (see Chap 14):

$$CPF_{plan} = CPF_{DMD} + BS - TRP_{plan}\cdot RDMD_m \tag{27.7}$$

where

| | |
|---|---|
| $CPF_{plan}$ | planned Cumulative Production Figure for the control block [units], |
| $CPF_{DMD}$ | Cumulative Production Figure for the demand [units], |
| BS | base stock [units], |
| $TRP_{plan}$ | planned replenishment time for the control block [SCD], |
| $RDMD_m$ | mean demand rate [units/SCD]. |

The backlog of the Base Stock System can then be calculated according to Eq. 27.5 and 27.6.

**Production Authorization with Cards (PAC)** The backlog measurement is largely dependent on the selected configuration of parameters. A backlog can be derived from the following parameters:

- number of PA cards on the PA card board of a workstation,
- number of order tags in the outgoing goods store of a workstation,
- the stock in the outgoing goods store of a workstation.

The backlog can be described in more detail for specific configurations of the PAC control.

**Order Release Methods** Order release methods can help to determine the backlog of a production. With a *Due Date Based Order Release* the backlog of the production corresponds to the difference of the actual WIP and the planned WIP (see Eq. 16.1 in Sect. 16.1.2). With a *WIP Regulating Order Release* (see Sect. 16.1.3) a backlog can be measured not only in the output, but also in the input of a production. This can provide an early indication of threatening lateness in the production's output and can be significant for timely adjusting capacities.

## 27.3 Selecting Measures

Despite its significance for the delivery reliability, backlog control is still a quite new field of research. In comparison, basic ideas of backlog control have been implemented in the industry for quite some time. It is also understandable even without a theoretical background that capacities should be increased when the actual output falls behind the planned output. However, when a backlog control is not appropriately applied, the desired results may not be attained and in the worst case even new problems can arise e.g., increased WIP or not using the supplied capacities. The most problematic aspects are delayed or over/understated reactions to backlogs.

In the next Section, we will turn our attentions to how enterprises can determine the duration and magnitude of the capacity adjustment (Sect. 27.3.1) as well as consider possibilities for selecting measures to adjust the capacities (Sect. 27.3.2).

### *27.3.1 Determining the Duration and Magnitude of the Capacity Adjustment*

The goal of increasing capacities is to reduce the backlog of a workstation or entire production. Therefore, the product of the duration and magnitude of a capacity adjustment should generally correspond to the backlog. Accordingly,

$$\Delta CAP \cdot T_{impl} = BL \tag{27.8}$$

where

| | |
|---|---|
| ΔCAP | capacity adjustment [h/SCD], |
| $T_{impl}$ | duration of capacity adjustment [SCD], |
| BL | backlog [h]. |

Generally formulated, the integral of the capacity increase over the duration of the capacity adjustment should correspond to the backlog. This formulation also covers a gradual capacity increase [Brei-01]. The duration and magnitude of the capacity adjustment can be determined in various ways.

### Calculating the Magnitude of the Capacity Adjustment from the Duration of the Capacity Adjustment

Particularly, with a periodic backlog control the capacity of a workstation or a production is frequently set for the entire next period. The duration of the capacity adjustment corresponds then to the length of the period and backlog control takes on the character of re-planning the capacities. The production planning does not explicitly provide the capacity increase, but rather the (mean) capacity for the next planning period. The greater the backlog is from the preceding period, the higher this mean capacity is. Scholtissek describes two alternatives for the planned capacity target (Fig. 27.3).

In Alternative 1 (Fig. 27.3a), the planned capacities are calculated from the difference of the values for the planned output at the planned period end and the actual output at the beginning of the planned period as well as the length of the planned period [Schö-95*]:

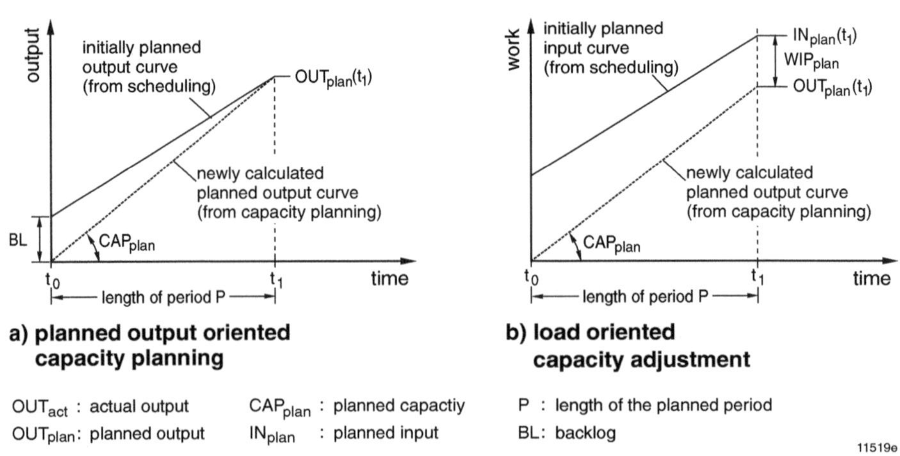


**Fig. 27.3** Alternatives for determining planned capacities (according to Scholtissek)

$$CAP_{plan} = \frac{OUT_{plan}(t_1) - OUT_{act}(t_0)}{P} \tag{27.9}$$

where

| | |
|---|---|
| $CAP_{plan}$ | planned capacity [h/SCD], |
| $OUT_{plan}$ $(t_1)$ | planned output at period end [h], |
| $OUT_{act}$ $(t_0)$ | actual output at period start [h], |
| P | length of planned period [SCD]. |

Alternative 2 follows a suggestion from Scholtissek for a Load Oriented Capacity Adjustment (Fig. 27.3b). Scholtissek calculates the planned output by subtracting a (fixed) planned WIP as well as the value of the actual output at the beginning of the period from the value of the planned input at the end of the period [Scho-95].

$$CAP_{plan} = \frac{IN_{plan}(t_1) - WIP_{plan} - OUT_{act}(t_0)}{P} \tag{27.10}$$

where

| | |
|---|---|
| $CAP_{plan}$ | planned capacity [h/SCD], |
| $IN_{plan}$ $(t_1)$ | planned input at period end [h], |
| $WIP_{plan}$ | planned WIP [h], |
| $OUT_{act}$ $(t_0)$ | actual output at period start [h], |
| P | length of planned period [SCD]. |

Calculating the capacity adjustment from the length of the period requires that all of the workstations have sufficiently flexible capacities. If this is not the case, the capacity increase should be aligned with the bottleneck workstation of the production (see also below). The capacity adjustment must then also be sustained during subsequent periods.

### Calculating the Duration of the Capacity Adjustment from the Maximum Possible Capacity Adjustment

The inverse way is to first determine the maximum possible capacity adjustment. In doing so the capacity restrictions for all of the production's workstations are taken into consideration. A simple procedure for doing so is to determine the maximum relative capacity of the production according to Eq. 26.2 and from that derive the maximum capacity increase of the workstations. Then, the duration which the capacity adjustment should be implemented is:

$$T_{\Delta CAP\,\max} = \frac{BL}{\Delta CAP\,\max} \tag{27.11}$$

where

| | |
|---|---|
| $T_{\Delta CAPmax}$ | duration of capacity adjustment [SCD], |
| BL | backlog [h], |
| $\Delta CAPmax$ | maximum possible capacity adjustment [h/SCD]. |

At times it is practical to increase the capacities beyond this degree on individual workstations for a limited time. This applies for example when individual workstations are impacted by disruptions – once the disruption is resolved the capacities can be increased in order to catch up on the backlog.[1] Furthermore, specific measures e.g., weekend shifts cannot be dispensed as precisely as preferred. Deviations from the calculated capacity increase can then not be prevented. It is critical that capacities are not systematically increased or implemented for extended periods beyond the limit of that which the production's throughput bottleneck can manage.

### *27.3.2 Deriving Measures for Adjusting Capacities*

Frequently, enterprises have a variety of measures for changing the capacities: possible alternatives include e.g., implementing overtime, weekend shifts or temporary workers. It is backlog control's task to determine which of these alternative measures should be selected and possibly combined in order to reduce the backlog.

Breithaupt introduced a sequence in which different measures should be activated [Brei-01]. To do so he sorted the measures according to the reaction time. If the first measure is not sufficient for reducing the backlog, the next measure is activated and so on. In doing so he also considered the minimum time required for implementing measures in order to prevent an overreaction. Thus, for example, it is not practical to introduce an additional shift for a minimal backlog when this measure has to be sustained for a longer duration.

Generally, the aim is to select the measure so that the backlog can be quickly and cost-effectively reduced. The load on the operators can also be another criteria that is considered. When there are a large number of alternative measures with different reaction times and costs, it may be practical to use optimization methods in order to help make the decision; usually though, this is not necessary.

In each case, enterprises should define clear criteria which trigger the individual measures for increasing capacities. These can, for example, be documented in the form of a decision table [Brei-02] (see Table 27.1 as well as the practical example in Sect. 27.6).

---

[1] With workstations before the throughput bottleneck this is obviously only practical to the extent required to supply the bottleneck.

**Table 27.1** Decision table for determining measures for increasing capacities (example)

| Backlog BL | Measure | Duration |
|---|---|---|
| $0\ h < BL \leq 3\ h$ | 1 h of overtime per SCD | Until $BL \leq 0\ h$ |
| $3\ h < BL \leq 6\ h$ | 2 h of overtime per SCD | Until $BL \leq 1\ h$ |
| $6\ h < BL \leq 8\ h$ | 1 weekend shift per week | Once |
| $8\ h < BL \leq 11\ h$ | 1 weekend shift per week + 1 h of overtime per SCD | Overtime: until $BL < 7\ h$ |
| $BL > 11\ h$ | 1 weekend shift per week +2 h of overtime per SCD | Overtime: until $BL < 8\ h$ |

The assessment of different alternatives for controlling the backlog should be oriented on two essential criteria: The effectiveness in reducing the backlog (and reducing the lateness/increasing the schedule reliability) as well as the efficiency with which this reduction is attained. The costs related to the measure for increasing the capacities or the number of implemented hours of overtime are examples of measures of this efficiency.

## 27.4 Discussion of the Procedural Rules

Backlog control is the most important method for regulating capacities. In the following, we will examine it more closely based on the guidelines for designing manufacturing control methods (see Sect. 3.3.2).

*1. A manufacturing control method should allow the WIP of a production or workstation to be regulated at a defined level.*

Backlog control is not aimed at regulating the WIP of a production or individual workstations, nevertheless, when the order release does not regulate the WIP (e.g., with a Due Date Based Order Release, see Chap. 17), backlog control still contributes to regulating WIP. As long as the input to the production corresponds to the planned value, it provides a WIP control by attaining the planned output.

However, regulating WIP via backlog control is neither as easy nor effective as with order release methods. One of the reasons for this is the reaction times required for changing the capacities, another is that the magnitude of the capacity adjustment is also limited.

In comparison, the order release can generally react without a delay. Moreover, it can vary the supply of orders between the extremes of completely stopping the release of orders and authorizing all of the orders not yet released.

*2. A manufacturing control method should reduce WIP fluctuations to a minimum.*

A backlog control prevents or reduces WIP fluctuations that result from a deviation of the actual output from the planned output. Otherwise, it tends to have minimal influence on WIP fluctuations.

*3. A manufacturing control method should cause as little blocked WIP as possible.*
Backlog control does not cause any blocked WIP.

*4. A manufacturing control method should contribute to closely aligning the load to the capacities.*
Backlog control does not contribute to adjusting the load, in fact, the basic idea is the opposite i.e., adjusting the capacities to the planned load. Supplementing a backlog control with a WIP regulating order release method is particularly practical when the flexibility of the capacities is insufficient for quickly and effectively eliminating the backlog.

*5. A manufacturing control method should cause as few sequencing interchanges as possible.*
A backlog control does not cause any sequencing interchanges.

*6. A manufacturing control method should be able to balance schedule deviations within the context of backlog control.*
This guideline emphasizes the significance of backlog control. In addition to the flexibility of the capacities, the degree to which backlog control can compensate for plan deviations is dependent on the extent of the expected plan deviations.

*7. A manufacturing control method should take into consideration the bottleneck principle.*
In particular, simple forms of backlog control can violate the bottleneck principle, e.g., when it increases the capacities of a workstation despite the subsequent workstation already working at its maximum capacity. Enterprises should therefore design the procedural rules so that the bottleneck principle is taken into consideration.

*8. A manufacturing control method should be as simple as possible.*
The idea of backlog control is straightforward, simple to grasp and thus easy to convey. Nonetheless, designing and implementing a backlog control can be difficult. The complexity of backlog control is dependent on the following factors:

- *Reaction time of the capacity measure:* With long reaction times, measures for adjusting capacities that have already been introduced but not yet completed need to be taken into consideration when making decisions about additional measures. This usually leads to more complex procedural rules than with shorter reaction times.
- *Number of alternative capacity measures:* Implementing a number of different measures for adjusting the capacities can increase the total flexibility, however, this requires selecting a number of measures which in turn has the tendency to increase the complexity of the methods.
- *Differences in the capacity flexibility of the workstations:* In the most simple of cases, the capacity flexibility of all the workstations corresponds with one another. The greater the difference in the (relative) capacities, the more important it is to consider the bottleneck principle in the procedural rules.

## 27.5 Setting the Method's Parameters

The basic parameters for backlog control are the upper and lower backlog limits. They determine the backlog levels after which an enterprise triggers a capacity adjustment. A lower backlog limit leads to a quicker reaction, nevertheless, with fluctuating backlog levels it can cause a nervous control behavior where measures for increasing and decreasing capacities follow one another in quick succession.

The following factors should be considered by enterprises when setting the backlog limits:

- *Distance between steps in adjusting the capacities:* If enterprises are unable to finely adjust capacities (e.g., are only able to implement additional shifts) the backlog limit should be set relatively high in order to prevent overreactions.
- *Delivery time buffer/safety stock:* The higher the backlog limit is, the higher the expected backlogs of the production are and the longer the orders are delayed. Accordingly, backlog limits and delivery time buffers or safety stock levels should be coordinated.
- *Stability of the processes:* The greater the variability of a workstation's output, the less exceeding the backlog limit conveys. In order to restrict the nervousness of backlog control, the backlog limit should therefore be set higher than with stable processes with fewer output fluctuations.

## 27.6 Application

*Please Note: The exemplary application of backlog control described below was conducted by Lufthansa Technik – without the collaboration of the author* [Brei-02].

Lufthansa Technik Logistik (a company in Hamburg also known as LTL), stores, transports and lends replacement parts for airplanes. The length of time that a plane has to remain on the ground is dependent on the service level of the enterprise. Due to the high costs related to the downtime of airplanes, short delivery times and a high degree of delivery reliability are critical. Usually, LTL promises their customers delivery within a day. In order to reliably keep this promise despite the strongly fluctuating order input, the enterprise relies on very flexible capacities [Brei-02]:

- *Overtime/flex-time:* Employees can increase their work time up to 10 h per day and can reduce their overtime hours during periods where capacity requirements are lower.
- *Cross-training:* The employees are trained on similar jobs within the department. Moreover, the enterprise has set a goal of training 25% of the employees as floaters (or stand-ins) who are able to be implemented in various departments.

- *Employing students:* For short term capacity peaks during off-hours (evenings, weekends, holidays) students are employed on-call.
- *Employing qualified subcontract workers:* The enterprise has developed a group of qualified subcontract workers. The subcontracted workers can be employed on a short and midterm basis for tasks which require higher qualifications.

In order to use the capacity flexibility effectively and efficiently, the enterprise estimates the backlog twice daily. As a subsidiary variable it measures the inventory of open orders and derives the backlog from that.[2] With the first measurement at 11:00 the planned WIP according to definition is 80 orders [Brei-02]. Therefore, the backlog can be estimated according to the following formula:

$$BL = WIPO_{act} - WIPO_{plan} = WIPO_{act} - 80 \tag{27.12}$$

where

| | |
|---|---|
| BL | backlog (in number of orders) [−], |
| $WIPO_{act}$ | actual WIP (in number of orders) [−], |
| $WIPO_{plan}$ | planned WIP (in number of orders) [−]. |

From this LTL calculates the expected output lateness with a planned output of 15 orders per hour [Brei-02]:

$$L_{out,\exp} = \frac{WIPO_{act} - WIPO_{plan}}{ROUTO_m} = \frac{WIPO_{act} - 80}{15} \tag{27.13}$$

where

| | |
|---|---|
| $L_{out,\,exp}$ | expected output lateness (in hours) [h], |
| $WIPO_{act}$ | actual WIP (in number of orders) [−], |
| $WIPO_{plan}$ | planned WIP (in number of orders) [−], |
| $ROUTO_m$ | output rate (in orders/hour) [−/hours]. |

From the calculated backlog and the expected lateness, LTL directly derives the required capacity adjustment. If the expected lateness is 1 h, then the shift is extended by 1 h, if it is 2 h then by 2 h etc.

The enterprise drew up a decision table that precisely determines which measure should be used with which backlog (Table 27.2).

Lufthansa Technik Logistik has had very positive experiences with this backlog control. The schedule reliability from the customer's perspective increased greatly with the introduction of this system, so that the enterprise regularly exceeded the target value of 95% [Brei-02].

[2] With a Due Date Based Order Release there is a direct correlation between the WIP and the backlog (see Eq. 16.1).

**Table 27.2** Decision table for the capacity control at Lufthansa Technik Logistik (Breithaupt/Panten)

| **Orders on Hand** [-] | **Expected Output Lateness** [hrs] | **Capacity Change** Early Shift [hrs] | Mid Shift [hrs] | Late Shift [hrs] | **Further Measures** |
|---|---|---|---|---|---|
| 5 | -5.0 | -2 | -2 | -1 | Increased use of flex-days |
| 20 | -4.0 | -2 | -2 | | |
| 35 | -3.0 | -2 | -1 | | |
| 50 | -2.0 | -2 | | | |
| 65 | -1.0 | -1 | | | |
| 80 | 0.0 | | | | |
| 95 | 1.0 | 1 | | | |
| 110 | 2.0 | 2 | | | Need to employ floaters or subcontractors is examined. |
| 125 | 3.0 | 2 | 1 | | |
| 140 | 4.0 | 2 | 2 | | |
| ≥155 | ≥5.0 | 2 | 2 | 1 | |

For the implementation of a backlog control, two aspects are exemplary here:

- *Capacity Flexibility:* The enterprise is not content with the available possibilities and develops the capacity flexibility systematically.
- *Formalization:* Lufthansa Technik Logistik clearly states which measures are to be used under which circumstances. This is decisive for implementing backlog control.

# References


[Brei-01] Breithaupt, J.-W.: Rückstandsorientierte Produktionsregelung von Fertigungsbereichen. Grundlagen und Anwendung. VDI Progress Reports, Series 2, No. 571, Düsseldorf (2001)

[Brei-02] Breithaupt, J.-W., Panten, M.: Maximierung der Termintreue durch Produktionsregelung. Ind. Manage. **18**(2), 25–28 (2002)

[Pete-96] Petermann, D.: Modellbasierte Produktionsregelung. VDI Progress Reports, Series 20, No. 193, Düsseldorf (1996)

[Scho-95] Scholtissek, P.: Simulationsprüfstand für Logistikkonzepte der Produktion. VDI Progress Reports, Series 2, No. 377, Düsseldorf, 1995

# Chapter 28
# Further Methods for Controlling Capacity

In addition to Backlog Control there are other methods for controlling capacity. These methods can be used as an alternative to Backlog Control but also can almost always be implemented as a supplement to the Backlog Control.

Both Plan Oriented Capacity Control (Sect. 28.1) as well as the Due Date Oriented Capacity Control (Sect. 28.2) developed by Begemann require a minimum of capacity flexibility on all workstations. The first reacts to deviations in the actual capacity from the planned capacity; the latter, attempts to optimize the schedule reliability of a production. To do so, it continuously calculates the expected lateness of the orders on a workstation triggering a capacity increase when it will prevent an order from being late or will at least reduce the delay.

In comparison, the Output Rate Maximizing Capacity Control (Sect. 28.3) is used when the capacity flexibility on the capacity bottleneck of the production is depleted. The objective here is to maximize the output rate of the entire production and at the same time prevent undesired side effects such as excessive WIP or breaks in the material flow after the bottleneck.

The Inventory Based Capacity Control (INCAP) is suitable only for make-to-stock productions (Sect. 28.4). It uses the capacity flexibility to control the inventory within defined stock limits.

## 28.1 Plan Oriented Capacity Control

Plan Oriented Capacity Control reacts to deviations of the actual capacity from the planned capacity. Its goal is to adjust the capacity so that they correspond to the planned capacity. It serves primarily to control the personnel capacity and makes decisions regarding measures such as implementing floaters, redistributing work content to members of a manufacturing cell and other capacitive measures which require short notice. It thus belongs to the daily activities of supervisors and production managers.

H. Lödding, *Handbook of Manufacturing Control*,

DOI 10.1007/978-3-642-24458-2_28, 

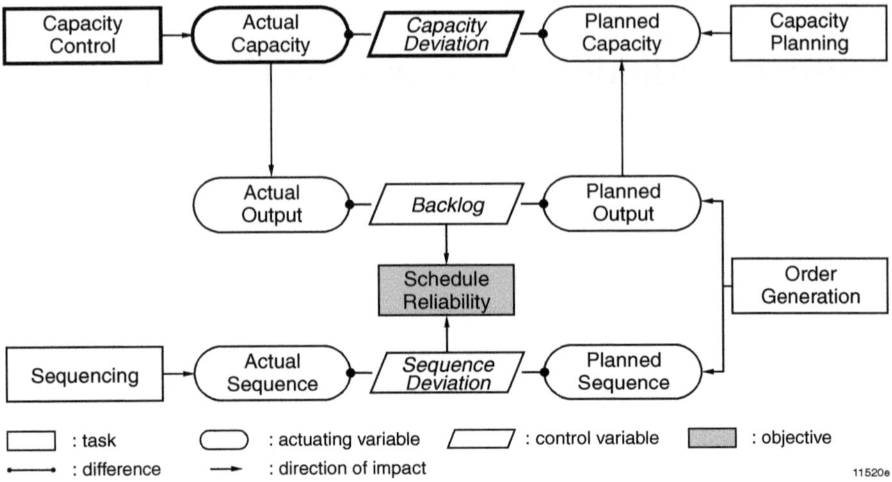


**Fig. 28.1** Location of plan oriented capacity control in the modified manufacturing control model

Figure 28.1 depicts the location of Plan Oriented Capacity Control in the manufacturing control model that has been modified for this purpose (see also Fig. 1.2). In addition to the basic model, it contains the actual and planned capacity as supplementary actuating variables and capacity planning as an independent task. The order release along with its logistic objectives and the actuating/control variables allocated to it are not depicted.

The capacity control sets the actual capacity which influences the backlog via the actual output and thus the schedule reliability. Similarly on the right side of the model the capacity planning is shown. In consideration of the planned output, it sets the planned capacity of the production and its workstations.

The capacity deviation, which is the control variable of Plan Oriented Capacity Control, results from the difference between the planned and actual capacity. The basic idea of the method is to keep the capacity deviations to a minimum even when there are disruptions. To do so, it implements measures which can be quickly implemented to adjust the capacity.

### *28.1.1 Method Description*

Plan Oriented Capacity Control corresponds to the long proven and good practice of many enterprises. It can thus not be traced back to a specific developer. Accordingly, the formulation of the procedural rules is not fixed. Nevertheless, Plan Oriented Capacity Control can be generally described by the following rules:

1. measure the capacity deviation.
2. increase the actual capacity when it is lower than the planned capacity.
3. decrease the actual capacity when it is higher than the planned capacity.

Unlike Backlog Control, Plan Oriented Capacity Control works preventively against backlogs and reacts even before backlogs develop. The following example, based on an assembly line, should help to illustrate how it functions (see Fig. 28.2).

The assembly line consists of seven stations (S1–S7). In the planned state, four operators are working on the assembly line, each of whom are allocated one to two workstations. The operators' routes are marked by the lines between stations.

In the initial situation one of the operators is suddenly unavailable due to illness (Step A in Fig. 28.2a). The enterprise can react to this with various measures.

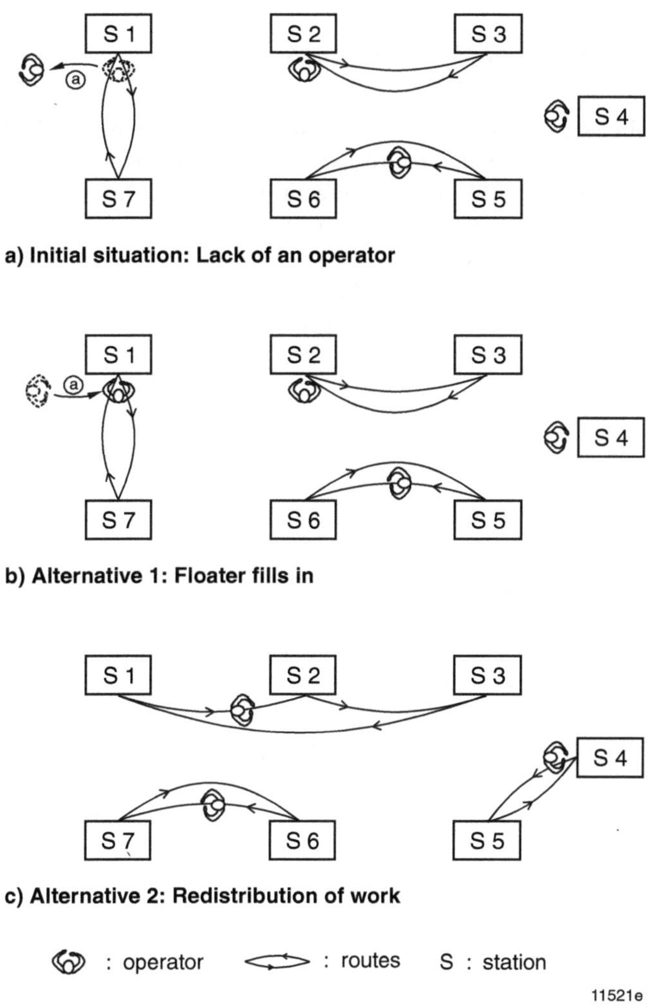


Fig. 28.2 Variations of plan oriented capacity control (example)

Alternative 1 (Fig. 28.2b): The enterprise can employ a floater to replace the ill employee. This requires constantly maintaining reserve capacity with the required qualifications; to some extent team leaders take on this task.

Alternative 2 (Fig. 28.2c): The enterprise re-distributes the work among the remaining employees. One operator takes on Stations 1–3, another operator takes on Stations 4 and 5, and a third takes on Stations 6 and 7. Frequently, the assembly line will yield less than the planned state, since the range of the work has increased for each of the operators. To some degree though the yield can be maintained at a constant for a limited time by increasing the work speed.

### *28.1.2 Discussion of the Procedural Rules*

We will refrain here from the usual reflections about this method with respect to the guidelines for designing manufacturing control methods as it is obvious that this method alone cannot fulfill the manufacturing control tasks or even that of controlling the capacity. Instead, it is much more practical to implement this method together with a Backlog Control (or other capacity control method). Compared to Backlog Control, however, Plan Oriented Capacity Control has a decisive advantage: it reacts immediately after the capacity is disrupted. Backlog Control alone would in contrast first intervene after a backlog developed due to the capacity disruption.

### *28.1.3 Application*

Almost every manufacturing enterprise and many service providers use the Plan Oriented Capacity Control, even when it is only rarely called this. It is clear that when an employee is absent or there is an obvious capacity disruption, there has to be a quick reaction; otherwise there is a threat of backlogs and with that the delayed completion of orders. List of employees who can at least from time to time replace an absent co-worker are frequently a component of practical applications.

## 28.2 Due Date Oriented Capacity Control

With Due Date Oriented Capacity Control, Begemann designed a method that is directly aimed at increasing the schedule reliability [Bege-05]. In order to do so, it continuously calculates the expected lateness of orders on a workstation and increases the capacity of the workstation when by doing so the lateness of at least one order can be reduced.

Section 28.2.1 describes the method and explains it based on an example. The discussion of the procedural rules (Sect. 28.2.2) and a comparison with Backlog Control (Sect. 28.2.3) provide insight into application areas for which the method is well-suited. Finally, Sect. 28.2.4 presents factors which should be taken into consideration when setting the lateness limits that are the method's parameters.

### 28.2.1 Method Description

Due Date Oriented Capacity Control can be described with five procedural rules which are based on a more extensive and algorithmic representation from Begemann [Bege-05].

*1. Calculating the expected lateness of the orders.*
The method calculates the expected lateness of all orders on the workstation for each of the workstations when there is an order input or output. In order to do so, it first arranges the orders in the sequence of their priority and then calculates the expected due date of all orders based on their work content and the workstation's capacity. Comparing the expected due date with the planned due date results in the expected lateness of each order on the workstation.

As a particular procedural rule, Begemann sets the condition that the selected sequencing rule should also support the schedule reliability [Bege-05], e.g., earliest planned start date or end date (see Chap. 25).

*2. Verifying prerequisites for a capacity increase.*
The Due Date Oriented Capacity Control then verifies whether or not the three conditions for increasing the capacity are fulfilled:

- At least one order is expected to be completed late and exceeds the defined lateness limit. The lateness limit is a parameter of the method which should prevent a minimum amount of lateness from causing capacity changes.
- Increasing capacity reduces or prevents at least one order from being delayed.
- The workstation can quickly increase its capacity.

*3. Calculating the required capacity.*
If all the conditions for a capacity increase are fulfilled, the method calculates the capacity required to prevent the delayed completion of the order with the highest expected lateness.

To do so it calculates the amount of work which, when the processing complies with the given sequence, remains before the order is completed. This is comprised of three components:

- The work which still needs to be completed on orders being processed,
- The work content of the order with the highest expected lateness,
- The sum of the work content for all of the orders that are queued before the order with the greatest expected lateness.

This work content is then set in relation to the time remaining until the planned due date. When the reaction time is negligible, then for the capacity increase [Bege-05]:

$$CAP_{req} = \frac{WC_{n+1} + \sum_{i=1}^{n} WC_i - WC_{comp}}{EDOP_{plan} - T_0} \tag{28.1}$$

where

| | |
|---|---|
| $CAP_{req}$ | required capacity [h/SCD], |
| $WC_{n+1}$ | work content of the order $n + 1$ [h], |
| n | number of orders in queue before the order [–], |
| $WC_i$ | work content of the $i^{st}$ order [h], |
| $WC_{comp}$ | already completed work content for orders being processed [h], |
| $EDOP_{plan}$ | end of operation processing for the order [SCD], |
| $T_0$ | current time [SCD]. |

Similar to a continuous backlog measurement, this equation takes into consideration the work content of orders that are already being processed (see Sect. 27.2.1). If the reaction time required to increase the capacity is not negligible, the later intervention time of the capacity increase needs to be considered:

$$CAP_{req} = \frac{WC_{n+1} + \sum_{i=1}^{n} WC_i - WC_{comp} - T_R \cdot CAP_{cur}}{EDOP_{plan} - T_0 - T_R} \tag{28.2}$$

where

| | |
|---|---|
| $CAP_{req}$ | required capacity [h/SCD], |
| $WC_{n+1}$ | work content of the order $n + 1$ [h], |
| n | number of orders in queue before the order [–], |
| $WC_i$ | work content of the $i^{st}$ order [h], |
| $WC_{comp}$ | already completed work content for orders being processed [h], |
| $T_R$ | reaction time for the capacity increase [SCD], |
| $CAP_{cur}$ | current capacity before the capacity increase [h/SCD], |
| $EDOP_{plan}$ | end of operation processing for the order [SCD], |
| $T_0$ | current time [SCD]. |

The calculation then only leads to practical results when an enterprise is able to increase the capacity before the planned due date of the orders. If this is not the case, enterprises should adjust the capacity to their maximum in order to minimize the resulting delays.

*4. Setting the capacity increase.*
The method increases the capacity to the required value as long as this does not exceed the maximum capacity; otherwise it adjusts the capacity to its maximum.

The latter also applies when the planned due date has already passed or the capacity increase does not go into effect before the planned due date.

*5. Lowering the capacity.*
After the output of an order that has triggered a capacity increase, the capacity is re-determined according to procedural rules 1–4. For the calculation of the expected lateness, the standard capacity (from the earliest possible time point) is applied.

In addition to these procedural rules Begemann sets a limit for the capacity increases on workstations before the bottleneck workstation in order to prevent the bottleneck from being overloaded. The capacity planning also has to coordinate the available flexibility of the workstations over the short term. Moreover, he also requires scheduling to set the operation due dates and to consider the available capacity flexibility.

Next, we will examine the procedural rules for the Due Date Oriented Capacity Control on the basis of an example. An enterprise manufactures products in a make-to-order production in three steps on workstations WS1–WS3 (Fig. 28.3). Our focus is the capacity control on WS 2 with a standard capacity of 8 h/SCD. The reaction time for increasing or decreasing the capacity is negligible. Given a backlog, the enterprise can immediately implement 2 h of overtime at the end of a shift (maximum capacity = 10 h/SCD). The lateness limit is 0 SCD. The enterprise thus reacts with a capacity adjustment as soon as a small delay is expected. The criteria for the sequencing is the operation due date of the orders on the respective workstation.

In the initial situation (Fig. 28.3a), Orders 2 and 5 are on WS 2. Order 2 has 4 h of work content remaining, Order 5 has 8 h. Accordingly, the workstation, which is operating with one shift, is expected to complete the orders in 0.5/1.5 SCD respectively, and thus, punctually or a day too early (see table below WS 2).

The situation changes with the completion of Order 4 on WS 1. With the earlier planned due date Order 4 is queued before Order 5 in the waiting line in front of WS 2 (Step A in Fig. 28.3b). The Due Date Oriented Capacity Control calculates the new expected completion dates and output lateness for all of the orders (Step B, first procedural rule). For Order 4 there is an expected delay of 0.5 SCD.

As a result, all of the conditions for a capacity increase are met (second procedural rule): At least one order is late (Order 4), the workstation can reduce the delay by increasing the capacity and its capacity flexibility is not yet exhausted. (Please note, with a lateness limit of 0.5 SCD or more there would be no increase in capacity.)

In the next step, the required capacity is calculated (see third procedural rule). According to Eq. 28.1 this is derived from the ratio of the work content that is to be completed before finishing Order 4 (12 h) and the length of time remaining until the planned due date (1 SCD) i.e., 12 h/SCD (Step D). Since this exceeds the maximum capacity, the enterprise increases the capacity to its maximum of 10 h/SCD (Step E, see fourth procedural rule) and makes arrangements for 2 h of overtime.

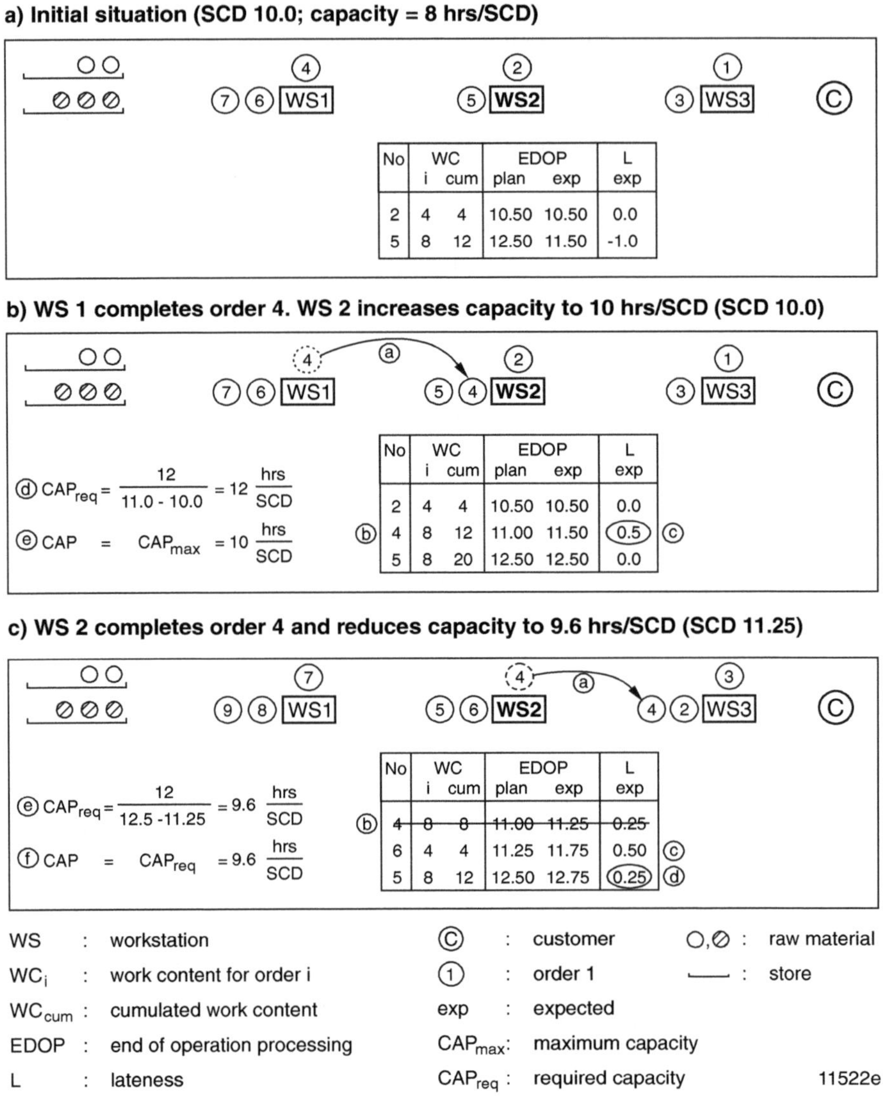


**Fig. 28.3** How the due date oriented capacity control functions (example)

The next time point considered is the completion of Order 4 on WS 2 (Step A in Fig. 28.3c, procedure corresponds with fifth procedural rule). The order is struck from the list of orders (Step B), the expected due dates for Orders 6 and 5 are recalculated (Steps C and D); an expected lateness of 0.5 SCD results for Order 6 and 0.25 SCD for Order 5. Since the workstation will still complete Order 6 on SCD 11, increasing the capacity via overtime would not decrease the delay: The overtime hours are first effective after the completion of the order. Correspondingly, according to the second procedural rule, Order 6 cannot trigger a capacity increase. The calculation of the required capacity is therefore based on Order 5, for which all of the conditions are met. The calculated capacity of 9.6 h/SCD is less than the maximum capacity and thus can be implemented (Step F).

### 28.2.2 Discussion of the Procedural Rules

We will now consider the Due Date Oriented Capacity Control from the perspective of the guidelines for designing manufacturing controls (see Sect. 3.3.2).

1. *A manufacturing control method should allow the WIP of a production or workstation to be regulated at a defined level.*

The procedural rules for the Due Date Oriented Capacity Control do not allow it to regulate the WIP in a production or on a workstation. It should thus be supplemented by other methods of manufacturing control e.g., with a WIP regulating order release.

2. *A manufacturing control method should reduce WIP fluctuations to a minimum.*

The Due Date Oriented Capacity Control has minimal influence on the WIP fluctuations of a workstation.

3. *A manufacturing control method should cause as little blocked WIP as possible.*

The method does not cause any blocked WIP.

4. *A manufacturing control method should contribute to closely aligning the load to the capacities.*

The Due Date Oriented Capacity Control does not contribute to balancing the load. On the contrary, the method is aimed at adjusting the capacity to the planned due date of the orders.

5. *A manufacturing control method should cause as few sequencing interchanges as possible.*

The Due Date Oriented Capacity Control does not cause any sequencing interchanges. Nevertheless, it should sensibly only be combined with sequencing rules that support increasing the schedule reliability. Otherwise, the enterprise would implement the capacity control for a goal that it could more effectively and more efficiently attain through sequencing. In his dissertation Begemann considers this to be the first procedural rule.

6. *A manufacturing control method should be able to balance schedule deviations within the context of Backlog Control.*

In normal situations, the method directly ensures that no large backlogs arise. A comparison with Backlog Control however shows that under certain circumstances the method's reaction to backlogs is delayed (see Sect. 28.2.3). In the simulations conducted by Begemann however, the method tends towards a negative backlog so that the actual output is slightly ahead of the planned output at the cost of a higher finished goods store [Bege-05]. The reason for this behavior is that the method is always oriented on the order with the greatest expected lateness. As a result the orders in the queue before this order tend to be accelerated more than necessary [Bege-05].

7. *A manufacturing control method should take into consideration the bottleneck principle.*

Begemann explicitly states that the bottleneck principle needs to be considered when applying this method. This prevents capacity increases on non-bottleneck workstations which would remain ineffective due to the limited bottleneck capacity for the entire order throughput [Bege-05].

8. *A manufacturing control method should be as simple as possible.*

The idea behind the Due Date Oriented Capacity Control is simple: Set the capacity on a workstation so that it can punctually complete as many of the orders as possible. Nonetheless, implementing it on the shop floor requires quite a bit of effort: It requires all of the orders on all of the workstations to be continually rescheduled. Moreover, on the workstations that precede the bottleneck workstation, it is instructed to continually consider the remaining bottleneck flexibility. It is thus primarily well suited for application cases in which the more simply implemented Backlog Control does not alone lead to the required schedule reliability. A comparison with Backlog Control will be the center of our focus in the next section.

### 28.2.3 Comparison with the Backlog Control

A simple basic configuration for ensuring a high schedule reliability can be derived from the model of manufacturing control (see Fig. 1.2). This consists of a combination of a Backlog Control with a sequencing rule that takes into consideration the planned due date. The aim of Backlog Control is to orient the actual output on the planned output. The sequencing rule ensures that the orders are in the right succession for high schedule reliability.

A configuration such as this is both effective and efficient for many practical cases: The laborious capacity control is activated when the planned output is not attained. In comparison, the easy to implement sequencing compensates for deviations in the planned sequence. Even when the influence of the sequencing on the mean throughput time tends to be minimal, it accelerates individual orders much more strongly than the capacity control is able to.

Selecting a configuration other than this basic one is justified especially when it is no longer adequate to ensure sufficiently high schedule reliability. This is particularly true when the deviations from the planned sequence are so severe that they can no longer be corrected by the sequencing rule alone. Here the Due Date Oriented Capacity Control exploits further possibilities in that it implements the capacity control in addition to the sequencing.

Nevertheless, the calculation of the expected lateness is limited to one workstation. This simplifies the implementation of the method, but, it cannot make a prognosis about whether or not the order's lateness could be corrected by the sequencing on the following workstation. In this case, implementing the capacity flexibility would be excessive. The more workstations an order has to flow through, the greater influence the sequencing has. Because of this, the danger of implementing the capacity flexibility unnecessarily is particularly high at the start of the material flow. For this reason, Begemann recommends dimensioning the lateness limits here particularly generously [Bege-05].

The strengths and weakness of Backlog Control and Due Date Oriented Capacity Control are visible in different situations:

Backlog Control is in danger of completing orders later when a workstation finishes a couple of orders too early e.g., because the input of more urgent orders is not punctual. Then when the more urgent orders do arrive, it is possible that they cannot be completed punctually even though the workstation does not have a backlog. Here, Due Date Oriented Capacity Control can ensure higher schedule reliability by increasing the capacity. The greater the capacity increase and the faster it goes into effect, the greater its influence. According to the Funnel Formula, doubling the output rate therefore halves the throughput time of the impacted order. With less capacity flexibility, the impact is correspondingly not as strong. Generally, Due Date Oriented Capacity Control accepts exceeding the planned output and thus increasing the WIP on the subsequent workstation in order to reduce the lateness of a couple of orders.

Conversely, Due Date Oriented Capacity Control cannot always detect a backlog in time. The reason for this is that the calculation of the expected lateness is limited to the orders on the workstation. If a more urgent order is late entering a workstation, the method does not calculate any delay for the queued orders despite the possibility that there is a backlog. Instead, this is first visible when the order enters the workstation.

It thus seems obvious to combine Due Date Oriented Capacity Control with Backlog Control, in order to use the advantages of both. Implementing Due Date Oriented Capacity Control only in the last part of the material flow can be practical especially when there are a large number of operations [Bege-05]: There is less danger of unnecessary capacity increases here because they directly impact the schedule reliability of the production.

### *28.2.4 Determining the Lateness Limits*

The lateness limit is the basic parameter of Due Date Oriented Capacity Control. It defines the value under which the lateness of the orders remains unconsidered. Only orders whose lateness exceeds this limit trigger a capacity increase [Bege-05].

The lateness limit thus sets the sensitivity with which Due Date Oriented Capacity Control reacts to threatening delays. A low value tends to lead to frequent capacity adjustments and therefore less lateness. With a higher lateness limit the number of capacity changes is correspondingly lower. This can make it easier for employees to accept the method, however, the enterprise then has to consciously accept delays [Bege-05].

Begemann mentions three criteria for setting the lateness limit [Bege-05]:

- *Market requirements:* The lateness limit should be selected so that the resulting lateness meets the market's requirements. At the same time, the length of the delivery time buffer and the safety stock should also be taken into consideration. The greater the delay accepted by the market and the greater the planned delivery time buffer, the greater the selected lateness can be.
- *Position of the workstation in the material flow:* The lateness limits should be oriented on the position of the workstation in the material flow. Since lateness at the beginning of the material flow can be corrected via the sequencing during the later stages of the order throughput, the lateness limits can generally be more generously dimensioned at the start of the material flow than on workstations at the end of the material flow.
- *Frequency of capacity adjustments:* The lower the lateness limits and the stronger the backlog fluctuations on the workstations are, the more frequently Due Date Oriented Capacity Control adjusts the capacity. Frequent capacity adjustments can cause costs as well as lower the acceptance of the method by employees. These aspects thus also need to be considered when selecting the lateness limit.

On the shop floor, enterprises can set an initial value for the lateness limit in consideration of these aspects and then adjust it more finely over time.

### 28.2.5 *Application*

There are no known practical applications of this method. Nevertheless, simulations support the general appropriateness of Due Date Oriented Capacity Control.

## 28.3 Output Rate Maximizing Capacity Control

The aim of the Output Rate Maximizing Capacity Control is to maximize the output rate of the production while preventing waste due to excessive WIP build-ups or unused capacity in the production. Both forms of waste can arise with unsteady capacity profiles, when no countermeasures are taken by the capacity control.

As in Goldratt's bottleneck philosophy [Gold-84*] the production is logically divided into three parts: the area before the bottleneck, the bottleneck itself, and the area after the bottleneck. The bottleneck workstation is usually operated with maximum capacity. It is the responsibility of the capacity control in the area before the bottleneck to ensure that the bottleneck is supplied. In comparison, after the bottleneck the most important criteria of the Output Rate Maximizing Capacity Control are a Backlog Control and the prevention of breaks in the material flow.

The method is suitable in bottleneck situations in which the capacity is not sufficient for satisfying the market demand. Moreover, enterprises who, due to high investment costs or technological restrictions, generally operate their production at the capacity limits can also apply it. They frequently try to influence the demand and coordinate it with the production's capacity by adjusting the prices.

Section 28.3.1 describes the control mechanism with which a method such as this can be realized. This in turn will be discussed in Sect. 28.3.2 based on the guidelines for designing manufacturing control methods.

## 28.3.1 Method Description

In comparison to other methods, the Output Rate Maximizing Capacity Control cannot regulate the capacity on the workstations according to uniform criteria or logic. Instead, the differences in the areas before the bottleneck, the bottleneck itself and the area after the bottleneck determine the properties of the capacity control (see Fig. 28.4).

*1. Capacity control in the area before the bottleneck*
The aim of the capacity control in the area before the bottleneck is to ensure that the bottleneck workstation is supplied. At the same time though, an excessive build-up of WIP on the bottleneck workstation should also be prevented. It thus seems obvious to use WIP as the criterion for the capacity adjustment.

Three alternatives for this can be distinguished. In Fig. 28.5 the same situation of a production with three workstations is depicted for each of these. WS 3 is the bottleneck workstation. Due to an (in the meantime resolved) disruption on WS 2, the WIP of the bottleneck workstation has been reduced to only two orders, whereas the WIP on WS 2 has climbed to six orders. Depending on the criteria used for the capacity control, different decisions result for the capacity adjustment.

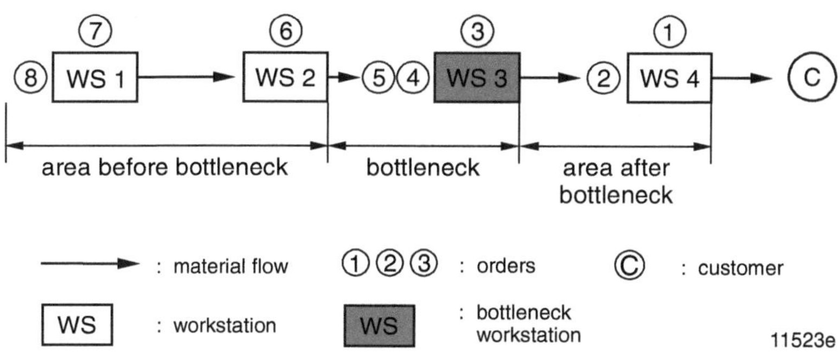


**Fig. 28.4** Structure of the production according to the bottleneck workstation

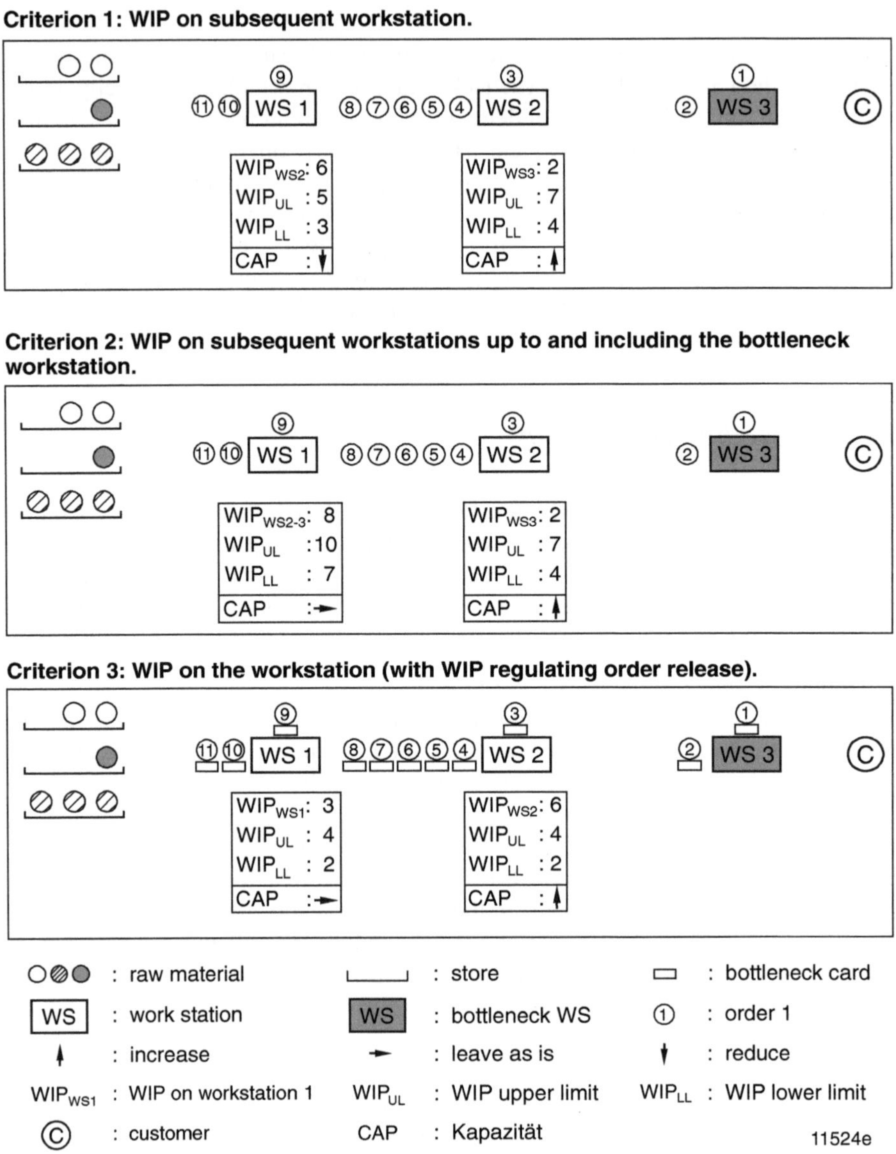


**Fig. 28.5** Criteria for the capacity control in the area before the bottleneck

*(a) WIP on the subsequent workstation (Criterion 1)*

Each workstation in the area before the bottleneck adjusts the capacity so that the WIP on the workstation immediately following it fluctuates only between the upper and lower WIP limits as much as possible. If the WIP of the following workstation climbs above the upper WIP limit it reduces the capacity in order to prevent a further build-up of WIP. If the WIP on the following workstations falls below the lower WIP limit, it increases the capacity in order to prevent a slack phase on the subsequent workstation.

In the example from Fig. 28.5 WS 2 is followed by the bottleneck WS 3. On this workstation the WIP is below the lower limit of four orders. The enterprise thus increases the capacity on WS 2. In comparison, on WS 2 the upper WIP limit is exceeded by one order so that WS 1 reduces its capacity.

A capacity control such as this requires a simple, linear material flow. If, for example, a workstation has a number of preceding workstations, the capacity increases on these need to be coordinated with one another. For more complex material flows the procedural rules therefore need to be extended.

*(b) WIP on subsequent workstations up to and including the bottleneck workstation (Criterion 2)*

This variation of the Output Rate Maximizing Capacity Control takes into consideration the entire WIP on the subsequent workstations up to and including the bottleneck workstation as the criterion for the capacity control. The definition of the WIP thus generally includes a number of workstations. Only on the workstation preceding the bottleneck is the WIP for one workstation considered i.e., for the bottleneck workstation. If the so defined WIP exceeds the *upper WIP limit*, the method reduces the capacity; if it falls below the *lower WIP limit*, the capacity is increased. This variation has two advantages: On the one hand, the completion of an order on the bottleneck workstation (and generally its output rate) directly impacts the capacity control on the preceding workstations. The connection to the bottleneck is thus quicker and more immediate. On the other hand, the WIP control range can be dimensioned rather larger. The capacity control, therefore, intervenes less frequently than with Criterion 1.

In the example from Fig. 28.5 these differences are visible on WS 1. For the decision about adjusting the capacity, all eight orders on WS 2 and 3 are taken into consideration. This WIP is still below the upper WIP limit of ten orders. Unlike with Criterion 1 the capacity is thus not reduced. Moreover, a capacity increase is also not an option as the WIP is above the lower limit. In comparison, the situation for WS 2 does not change because the subsequent workstation is the bottleneck workstation and thus the WIP and WIP limits are defined the same as with Criterion 1.

Criterion 2 also requires a linear material flow. For a more complex material flow the procedural rules need to be extended.

*(c) WIP on the workstation (Criterion 3)*

In a bottleneck situation it is particularly practical to implement a WIP regulating order release. In this case, as many orders as required to utilize the production bottleneck are released. At the same time it prevents an uncontrolled WIP build-up in the production. An example of such a release is the Bottleneck Control with bottleneck cards that is indicated in Fig. 28.5 (see Chap. 19).

Nevertheless, the WIP is not always distributed in the production as it is designated in the calculation of the planned WIP. If, for example, the capacity of a workstation before the bottleneck are insufficient due to a capacity disruption (WS 2 in the above figure) then a WIP build-up develops before the workstation. At the same time the WIP on the bottleneck workstation does not reach the planned level and there is a threat of breaks in the material flow.

With a constant total WIP, exceeding the planned WIP on a workstation is therefore a warning signal that can indicate a threat of a break in the material flow. Accordingly, an enterprise can regulate the capacity on the workstations before the bottleneck so that the workstations' WIP remain in the planned range. Thus when the upper WIP limit is exceeded the capacity should be increased, or conversely, when the WIP falls below the defined lower limit, the capacity should be reduced.

In the example from Fig. 28.5 (Criterion 3) WS 1 has a planned WIP of two to four orders. The actual WIP of three orders is within this range, thus, the capacity does not have to be adjusted. In comparison, the WIP on WS 2 clearly exceeds the upper limit of four orders. The enterprise should thus increase the capacity on this workstation.

Unlike with Criterion 1 and 2, this procedure can also easily be applied with complex material flows. It can also generally be combined with all of the centralized WIP regulating order release methods.

Nevertheless, the relationship to the bottleneck is not as close and clear as with Criterion 1 and 2. There are situations in which a bottleneck workstation is threatened by a break in the material flow even though the WIP on the preceding workstation is in the planned range.

*2. Capacity control on the bottleneck workstation*

The capacity control on a bottleneck workstation is very simple: The workstation is generally operated at maximum capacity in order to maximize the output rate of the entire production. Exceptions from this rule are:

- *Threat of a break in the material flow on the bottleneck WS:* When there is a threat of a break in the material flow due to e.g., disruptions on preceding workstations, it is not very practical to provide operators for a shift model that cannot be implemented. In this case it makes sense to only provide as much capacity as the work available requires.
- *"Virtual" Bottleneck operation:* With a "virtual" bottleneck, the production is quite capable of covering the customer demand i.e., the maximum bottleneck capacity is greater than the required capacity. Nevertheless, in comparison to the other workstations the bottleneck has such an outstanding position that it is practical to orient these directly on the bottleneck. In this case the bottleneck capacity can be regulated via a Backlog Control for example.

*3. Capacity control in the area following the bottleneck*

The capacity control in the area following the bottleneck is frequently characterized by a dilemma: Usually there is sufficient capacity flexibility available to catch-up with backlogs, however, the WIP is so low that it leads to a break in the material flow. As a result, additional costs arise for providing personnel who, due to the lack of available orders, cannot work.

Enterprises can then orient the personnel capacity at these workstations on the bottleneck capacity. Moreover, it is practical to define an upper and lower WIP limit for the capacity control. If the WIP at the workstation exceeds the upper WIP limit, the enterprise conducts a Backlog Control. The goal of the capacity control is to increase the schedule reliability.

Below the lower WIP limit though preventing a break in the material flow is first and foremost. To do so the enterprise reduces the capacity – even when the workstation has a backlog. The cause for the backlog is the lack of capacity flexibility on the bottleneck.

In the zone between the upper and lower limits no interventions in the capacity is required; rather, the workstation follows the bottleneck tact. The aim here is to prevent the control behavior from being too hectic, which would be especially probable when there is a backlog: In the first step, Backlog Control increases the capacity in order to reduce the backlog. As a result the WIP on the workstation is reduced until it reaches the lower limit. In the second step, the enterprise reduces the capacity in order to prevent a break in the material flow. The WIP and the backlog then grow again so that the capacity is once again increased when the upper WIP limit is exceeded, etc. . .

### 28.3.2 Discussion of the Procedural Rules

In this next section we will once again turn our attention to a discussion of the Output Rate Maximizing Capacity Control based on the guidelines for designing manufacturing control methods (see Sect. 3.3.2).

*1. A manufacturing control method should allow the WIP of a production or workstation to be regulated at a defined level.*

By coupling the capacity control to the WIP level on the workstations, the method attains a WIP control in the area leading up to the bottleneck. The quality of the WIP control increases with the capacity flexibility. Nevertheless, this WIP control remains ineffective with regards to the entire production because the order release inputs orders into the production independent of the capacity control. In order to attain a WIP control it is therefore practical to supplement it with a WIP regulating order release method.

*2. A manufacturing control method should reduce WIP fluctuations to a minimum.*

The impact of the method on the WIP fluctuations depends on how the procedural rules are designed. In particular, when the capacity control is oriented on the planned WIP of the workstations (see Criterion 3 in Fig. 28.5), the Output Rate

Maximizing Capacity Control impacts the WIP fluctuations positively in the area up to the bottleneck. In the area after the bottleneck the method prevents the WIP from falling and staying below the defined WIP level for an extended period.

3. *A manufacturing control method should cause as little blocked WIP as possible.*
The method does not block any WIP.

4. *A manufacturing control method should contribute to closely aligning the load to the capacities.*
The method adjusts the capacity on the workstation in the area before the bottleneck to the bottleneck capacity and thus aligns the load with the bottleneck capacity. However, this mechanism is generally not sufficient enough for attaining the logistic targets. Instead, when applying the Output Rate Maximizing Capacity Control the production planning itself should already be oriented on the bottleneck capacity in order to prevent overloading the capacity and subsequently missing deadlines as much as possible. Furthermore, in the most cases a WIP regulating order release is necessary in order to allow an effective load balance.

5. *A manufacturing control method should cause as few sequencing interchanges as possible.*
The method does not cause any sequence interchanges in the production.

6. *A manufacturing control method should be able to balance schedule deviations within the context of backlog control.*
Since the bottleneck according to definition is operated with the maximum capacity, a Backlog Control is not possible (with the exception of the virtual bottleneck operation). Thus high schedule reliability is difficult to attain with this method especially when there is minimal load flexibility. This is the essential reason why this method should only be implemented over the long term in exceptional cases; instead it is typically implemented temporarily in bottleneck situations.

7. *A manufacturing control method should take into consideration the bottleneck principle.*
The basic idea behind the method is to take into consideration the bottleneck principle. The aim is to attain the production's maximum output without squandering capacity on non-bottleneck workstations or developing a high WIP level in the production area before the bottleneck.

8. *A manufacturing control method should be as simple as possible.*
Even when the fundamental idea of the method is simple, implementing it on the shop floor frequently gives rise to problems. Basic causes for this include:

- The different logic for the capacity control in the three areas of the production (before bottleneck, bottleneck, after bottleneck) leads to organizational challenges. This is particularly true for the change from a Backlog Control to an Output Rate Maximizing Capacity Control.

- It is to some degree difficult to convey to customers and employees why in a bottleneck situation parts of the production are not working at their capacity limits although there is an obvious delivery backlog.

## 28.4 Inventory Based Capacity Control (INCAP)

Hermann Lödding and Steffen Lohmann

The basic idea of Inventory Based Capacity Control (INCAP) is to regulate the stock (and thus indirectly the service level) through the short-term capacity flexibility [Lödd-12*]. In order to do so, stock limits are defined which, when fallen below or exceeded, trigger a capacity adjustment. The aim is to regulate the inventory within defined limits and thus attain both stock and service level targets.

Figure 28.6 depicts the location of the Inventory Based Capacity Control in the expanded manufacturing control model; the tasks as well as the corresponding actuating and control variables are framed in bold.

INCAP regulates the store input via the actual output of the production. The inventory results from the difference between the cumulated store input and store output. The inventory (or stock level) is the control variable for INCAP and impacts the logistic objective 'service level'. At the same time the stock is also a logistic objective.

In the next sections, we will take a look at the procedural rules for the Inventory Based Capacity Control (Sect. 28.4.1) and discuss these based on the guidelines for designing manufacturing control methods (Sect. 28.4.2). Following that we will

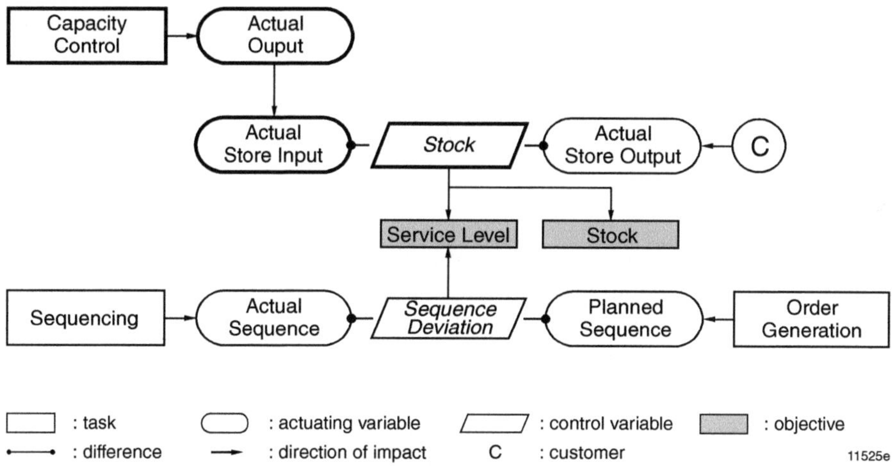


**Fig. 28.6** Location of inventory based capacity control in the manufacturing control model

consider under which conditions enterprises can particularly successfully apply the Inventory Based Capacity Control (Sect. 28.4.3).

### 28.4.1 Method Description

INCAP's procedural rules can be described as follows:

1. Increase the capacity when the stock falls below the lower limit.
2. Reduce the capacity when the stock exceeds the upper limit.
3. Orient the extent and duration of the capacity adjustment on the deviation from the planned stock.

The aim of the first procedural rule is to ensure the delivery capability of the enterprise. Causes for an inventory reduction could include both an increased customer demand (and thus a climbing store output) as well as a decreased output rate of the production (and thus a falling store input).

The second procedural rule should prevent an excessive WIP build-up and orient the enterprise's capacity on the customers' consumption. The cause for a climb in inventory is generally a fall in the customer demand. At least theoretically though, it may also be traced back to an increased production output and through that a higher store input.

The first two procedural rules determine the time of the capacity adjustment. Moreover, enterprises have to determine the extent and duration of the capacity adjustment (third procedural rule). One possibility of establishing the required capacity change is to derive it from the difference between the measured stock level and the planned stock level.

Similar to Eq. 27.9 when the processing times per unit are the same for every variant:

$$\Delta CAP \cdot T_{impl} = \left(SL_{plan} - SL_{act}\right) \cdot \frac{t_p}{60} \qquad (28.3)$$

where

| | |
|---|---|
| $\Delta CAP$ | capacity adjustment [h/SCD], |
| $T_{impl}$ | duration of capacity adjustment [SCD], |
| $SL_{plan}$ | planned stock level [unit], |
| $SL_{act}$ | actual stock level [unit], |
| $t_p$ | processing time per unit [min/unit]. |

The right side of the equation can be interpreted as the backlog of the production as compared to the planned state. The Inventory Based Capacity Control can thus be traced back to Backlog Control (with regards to Backlog Control see Chap. 27). The equation ignores setup times, however, these can be taken into consideration by adding a time allowance to the unit processing times (alternatively see Eq. 27.4 for a more precise calculation of the backlog).

If the unit processing times differ from variant to variant, a more detailed consideration is required. In this case, the total inventory does not allow a valid conclusion about the required capacity adjustment. Instead, it seems obvious to calculate a backlog for each variant by comparing the planned stock with the actual stock and to add all of these together (see Eq. 27.4). With the backlog calculated in this way, the enterprise can then conduct a Backlog Control .

Figure 28.7 depicts the principle of the Inventory Based Capacity Control. In the example, the stock falls below the lower stock limit at time $T_0$ (Fig. 28.7a), so that the enterprise increases the capacity.

Following the reaction time, this capacity increase becomes effective at time $T_1$ (Fig. 28.7b) and is maintained until $T_2$. During the duration of the capacity increase the stock climbs again to a level within the stock limits. In order to illustrate this Figure 28.7a includes a depiction of what the stock trend would have looked like

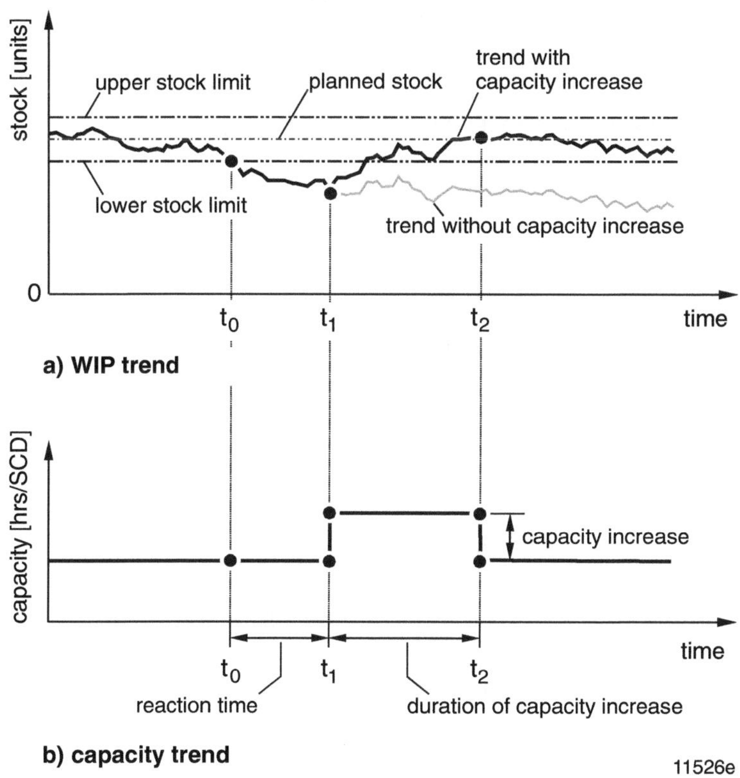


Fig. 28.7 Principle of inventory based capacity control

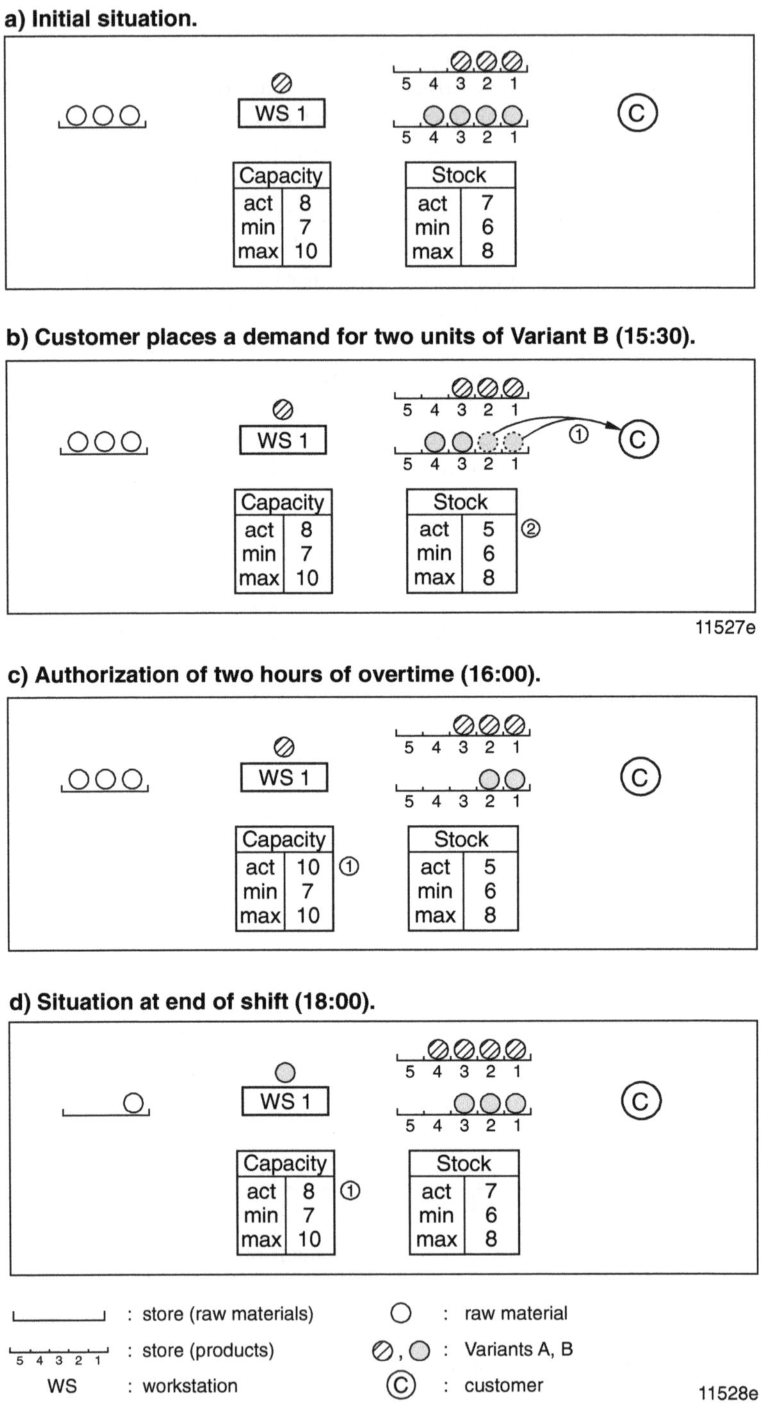


**Fig. 28.8** How inventory based capacity control functions

without a capacity increase. This is continuously located below the lower stock limit which negatively impacts on the enterprise's delivery capability.

Finally, in order to more fully explain the Inventory Based Capacity Control let us consider a simplified example in which an enterprise produces two variants on a production line as products from a common raw material (Fig. 28.8). The manufacturing time for one unit is 1 h. The planned stock level of the store is seven units i.e., mathematically 3.5 units per variant. The enterprise extends the operation time up to 2 h if the stock falls below the lower stock limit of six units of quantity before the end of the shift at 4 p.m. i.e., the capacity increase goes into effect on the same day (reaction time = 0 SCD).

Similarly, the enterprise reduces the hours of operation by an hour if the stock level at 3 p.m. exceeds the upper stock limit of eight units of quantity. The capacity change is maintained until the planned stock of seven units is once again attained.

The enterprise generates the orders via Visual Kanban (see Sect. 9.1.3) with five storage places per variant. In the initial situation the stock of seven units is at the planned level (Fig. 28.8a). With the customer demand for two units of Variant B (Step 1 in Fig. 28.8b) the lower stock limit is fallen below (Step 2). With the regular shift end at 4 p.m. the enterprise thus authorizes 2 h of overtime (Step 1 in Fig. 28.8c). Before the actual end of the shift at 18:00 it completes two units so that the stock once again attains the planned stock of seven units. The standard capacity of 8 h then applies for the next days (Step 1 in Fig. 28.8d).

### 28.4.2 Discussion of the Procedural Rules

Now let us examine the Inventory Based Capacity Control with regards to the guidelines for designing manufacturing controls (see Sect. 3.3.2). In doing so, it will become clear that the method can only fulfill a small number of the guidelines. It therefore has to be combined with other manufacturing control methods. When the manufacturing control is configured in such away though, the method can fundamentally improve the attainment of logistic targets.

1. *A manufacturing control method should allow the WIP of a production or workstation to be regulated at a defined level.*

The method is not suitable for regulating the production's WIP, but rather can only regulate the (total) warehouse inventory. It may therefore be supplemented with a WIP regulating order release method.

2. *A manufacturing control method should reduce WIP fluctuations to a minimum.*

Inventory Based Capacity Control only limits the stock fluctuations in the stores and not those in the production or at individual workstations.

3. *A manufacturing control method should cause as little blocked WIP as possible.*

INCAP method does not cause any blocked WIP.

4. *A manufacturing control method should contribute to closely aligning the load to the capacities.*

Inventory Based Capacity Control adjusts the capacity to the customer demand or its consumption. It however does not adjust the load to the available capacity though.

5. *A manufacturing control method should cause as few sequencing interchanges as possible.*

INCAP does not cause any sequencing interchanges in the production.

6. *A manufacturing control method should be able to balance schedule deviations within the context of Backlog Control.*

Compensating for plan deviations in the stock via a capacity control is the basic idea of the method. Ultimately, it can be considered a special form of a Backlog Control and as such fulfils the demands of this guideline in make-to-stock productions.

7. *A manufacturing control method should take into consideration the bottleneck principle.*

Inventory Based Capacity Control does not take into consideration bottlenecks in the production and thus requires sufficient capacity flexibility. The method is therefore not well suited to bottleneck situations. Different capacity profiles in the production should also be taken into consideration in the detailing of the procedural rules for more complex productions. The aim here is to prevent not only a WIP build-up in the production but also the arrangement of overtime for which there are no materials available for.

8. *A manufacturing control method should be as simple as possible.*

The simplicity of this method is an important advantage of Inventory Based Capacity Control. It orients an enterprise's capacity via the stock indirectly on the customer demand and allows the logistic objectives 'stock' and 'service level' to be regulated. The basic idea is easy to convey and – with flexible capacity and simply structured production – easy to implement.

### 28.4.3 Application

Inventory Based Capacity Control is primarily well suited for simple make-to-stock productions. Typical applications include controlling capacity in manufacturing or assembly lines i.e., producing standard variants to stock. In complex material flows with lot-wise transports between the workstations and strong variance among the unit processing times of the variants in the production, the procedural rules described here cannot be applied. The method does not consider the differentiated capacity profiles in the production nor does it provide different values for the capacity adjustment according to the workstations. For conditions such as these a Backlog Control is generally better suited.

In a multi-stage production with a number of storage echelons, enterprises can control the capacity of the production stages via decentralized control loops. Figure 28.9a depicts the resulting information flow for the Inventory Based

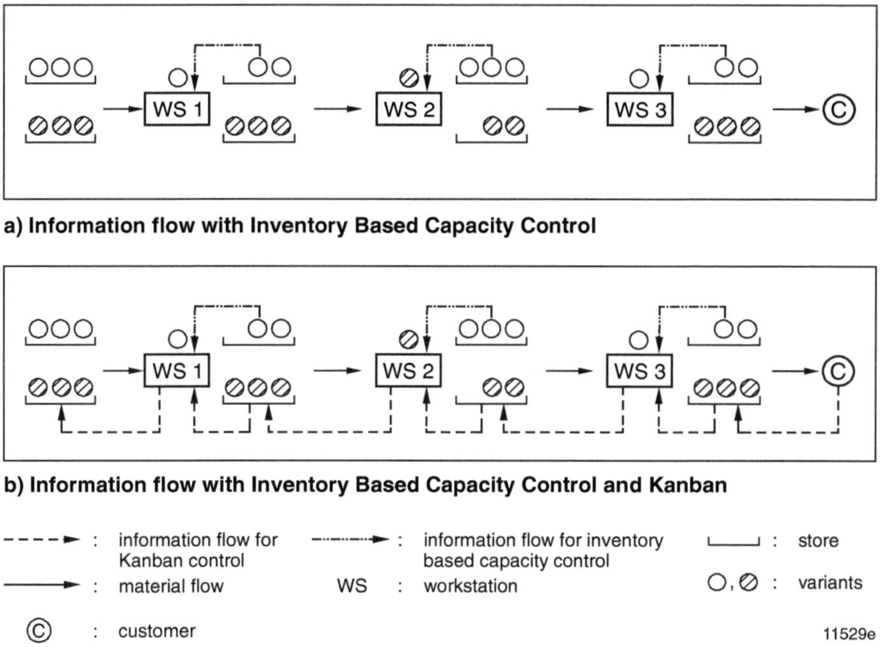


**Fig. 28.9** Information flow for inventory based capacity control and order generation with Kanbans in a multi-staged production

Capacity Control based on the example of a three stage production. Each of the three workstations control their capacity so that the stock in the following storage echelon is as close to the defined planned stock as possible. The flow of information runs in opposition to the material flow.

It seems obvious to combine Inventory Based Capacity Control with Kanban (or another method of order generation). Figure 28.9b supplements the additional flow of information for Kanban. Here too the flow of information opposes the direction of the material flow. Inventory Based Capacity Control regulates the total inventory in a configuration such as this, while Kanban ensures a well balanced mix of inventory. Especially when fluctuations in the store input (e.g., due to instable processes) or in the store output (e.g., due to fluctuating customer demand) arise, INCAP can fundamentally improve the enterprise's attainment of logistic targets. Nonetheless, this requires sufficient capacity flexibility.

# References

[Bege-05] Begemann, C.: Terminorientierte Kapazitätssteuerung. Reports from the IFA, Hannover, No. 2/2005

[Gold-84*] Goldratt, E.M., Cox, J.: The Goal. A Process of Ongoing Improvement. Gower, Aldershot (1984)

[Lödd-12*] Lödding, H., Lohmann, S.: INCAP. Applying short-term flexibility to control inventories. In: International Journal of Production Research, **50**(3), S. 909–919 (2012)

# Part F
# Configuring the Manufacturing Control

# Chapter 29
# Configuring the Manufacturing Control

In order for an enterprise to attain its logistic goals it is not enough to achieve outstanding results in a specific section of the manufacturing control. Instead, all of the aspects of the manufacturing control need to be coordinated with one another. Furthermore, they should take into consideration targets from the production planning as well as the structure of the product and production. Selecting suitable manufacturing control methods for a specific application case is not only difficult for practitioners; configuring the manufacturing control is also an underdeveloped field of research.

The following final chapter is for the most part limited to the manufacturing control tasks and methods described in the previous chapters. A more extensive discussion with regards to configuring production planning and control can be found in both Eversheim [Ever-99] as well as H.-H. Wiendahl [Wien-02]. Here, our discussion is focused on the four tasks dealt with in this book: order generation, order release, sequencing and capacity control (Fig. 29.1).

The aim is to attain a consistent configuration of the manufacturing control by choosing a coordinated selection of methods:

- Section 29.1 introduces the configuration of manufacturing control.
- Section 29.2 describes the criteria in selecting an order generation method.
- Section 29.3 is dedicated to selecting an order release method.
- Section 29.4 summarizes which criteria should be taken into consideration with the sequencing.
- Section 29.5 explains the selection of a method for the capacity control.
- Section 29.6 presents typical configurations of the manufacturing control in the overall context using examples from various enterprises.
- Section 29.7 discusses aspects which should be taken into consideration when implementing the configuration.

H. Lödding, *Handbook of Manufacturing Control*,
DOI 10.1007/978-3-642-24458-2_29, 

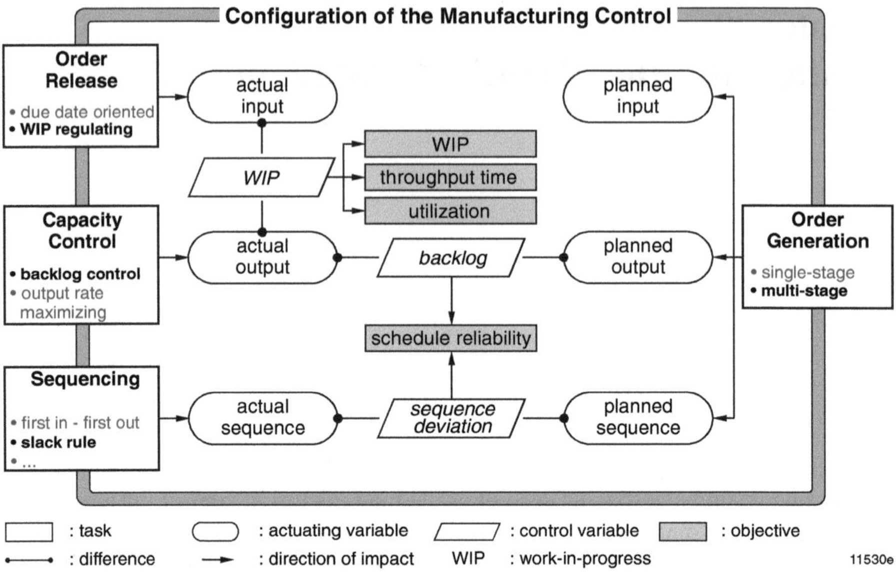


**Fig. 29.1** Configuration of the manufacturing control

## 29.1 Introduction

In the ideal case the configuration of the manufacturing control should be conducted together with the configuration of the production planning. Due to costs and the time required this is, however, not always possible. In the case described here the production planning will be considered as a given and the manufacturing control will be aligned with it.

Generally speaking, it is almost impossible to determine an *optimal* configuration of the manufacturing control or even to define it. This is also not the goal of the process described below. The problems of many enterprises in attaining logistic targets can rarely be traced back to them having decided on the wrong configuration during a systematic configuration process. By far the greatest cause of problems seems to be that many enterprises do not consciously implement a structured process to configure the manufacturing control and thus select imprecise and incomplete standards for executing manufacturing control tasks. If this assessment is accurate, a *systematic and consistent* configuration of the manufacturing control and the rigorous implementation of these targets should be sufficient for improving the logistic target attainment considerably. Similarly, it is more important to consider *all of the manufacturing control tasks* than to select particularly 'good' or 'modern' methods for a specific task while neglecting others.

The configuration of the individual manufacturing control tasks should be based on the same logistic targets and the same assumptions about the logistic conditions. Otherwise, they can for the most part be conducted independent of one another. An important exception from this rule applies to order release methods which cause sequence interchanges when releasing orders or while the orders flow through the production. These are the methods that balance the loads specifically for each workstation (Workload Control, LOOR, ORLP) as well as the decentralized order release methods (POLCA, DEWIP). They should be combined with a sequencing rule that considers the earliest due date of the operations or orders.

In order to configure the manufacturing control the following steps need to be conducted for each of the manufacturing control tasks: selecting the method, parameterizing the method and implementing the method. Our focus here will be on selecting the method, thus our procedure will entail:

1. selecting an order generation method,
2. selecting an order release method,
3. selecting the sequencing rule for the production's workstations,
4. selecting a capacity control method,
5. implementing the configuration.

The last step ensures that a configuration of the manufacturing control is not only developed as a concept but can also be realized on the enterprise's shop floor.

## 29.2 Selecting an Order Generation Method

The configuration of the order generation is an extremely important task. It should not be limited to the make-to-stock methods presented here in this book, but should also include methods that can be applied in make-to-order productions as well. In the following we will turn our attention to the conditions in which the selected make-to-stock methods can be practically implemented.

The key question of under which conditions a product should be manufactured according to the make-to-stock or make-to-order principles will not be discussed here because it is not part of the manufacturing control. For more information on this please refer to [Gläß-95] for example. Generally however, a make-to-stock production is necessary when the required delivery time is less than the feasible throughput time. If an enterprise has decided to implement a make-to-stock production then independent of the method: minimal variance in the demand rate helps to maintain smaller inventories.

With regards to the entire enterprise, the number of variants is a key criterion: The greater the number of variants, the greater the inventory. Moreover, as with make-to-order productions, a sufficient degree of capacity flexibility is advantageous.

In selecting a specific method for generating orders, an enterprise should at least analyze the following three main criteria:

- *Predictability of Demand:* This criterion states with which precision, how much in advance and with which effort the demand of a product can be forecasted and planned. The predictability of the demand generally declines when a product is supplied to a number of different customers and the order is not that valuable for the customer. Predictability can frequently be increased in cooperation with the customer and by implementing information and communication technology. A particularly high level of predictability is given when the customer establishes a reliable, precise and very detailed assembly plan (or more general: requirements plan) – this is for example common to some degree in the automobile industry.

- *Planning Necessity:* The higher the costs that result from wrongly estimating the requirements are, the greater the planning necessity is. It thus usually increases along with the value of the product. Moreover, frequent product changes and higher costs for failing to deliver increase the planning necessity.
- *Suitable Generation Scope:* Generally, a single stage order generation can be differentiated from a multi-stage order generation (see extensively Chap. 7). Single stage methods generate orders for every item number in separate control loops. Multi-stage methods generate orders for dependent demands and thus increase their predictability. Generally a multi-stage order generation is advantageous: It communicates the demands more quickly and usually also with less distortion than single stage methods. In selecting the appropriate scope for the generation, the effort required for the order generation should be considered in addition to the planning necessity. The effort is frequently (but not always) greater for the multi-stage order generation than for the single stage.

For the order generation methods described in Part B of this book, the following application areas can be derived:

*Order Point System:* The construction of the Order Point System is very simple and it is barely able to consider information about future demands, nevertheless, by changing the safety stock (and possibly the order lot-size) it is possible to react to changes in the demand. It is thus suitable when the predictability is not so good or when planning is not necessary. An exception here though is those products whose demand rate is almost constant. In this case the Order Point System can be practically implemented even when there is a high level of predictability and a great need for planning (e.g., also when the product value is high). The effort required for generating orders with this method is minimal. It is suitable only for single stage order generation.

*Kanban:* Kanban is a special form of the Order Point System. In the industrial application the method differs mainly in the way that it is implemented (see Sect. 9.2). When Kanban is integrated into a thoroughly conducted production planning and the demand rate fluctuations minimized, it can also control products with high value. Since the products are usually stored in the production, the available space limits the possible number of variants. With very short replenishment times the method triggers the early replenishment of seldom required variants and thus leads to a build-up of inventory. It is thus primarily suitable for products with a high demand rate. Kanban is a single stage method that is frequently combined with a multi-staged, higher level production planning.

*CORMA:* Capacity Oriented Materials Management is a supplement or extension of the Order Point System. Accordingly, the discussion regarding the Order Point System also applies here. Logistic targets can, however, be better attained by considering the demand during the replenishment time in the sequencing and by balancing the load. CORMA also coordinates the make-to-stock orders and the make-to-order jobs in the production.

*Synchro MRP:* This method supplements Kanban with a detailed and multi-staged, higher level production planning, making it simpler to efficiently manage technical changes and greater variance in the demand rate. Through the target values from the daily production program it is possible, for example, to prevent the seldom demanded products from being replenished too early. Synchro MRP thus requires the predictability of the (customer) requirements and at the same time supports the planning of dependent requirements.

*Hybrid Kanban/CONWIP Control:* Generally, the comments made about Kanban as well as for the Order Point System apply here as well. The method generates the orders, nevertheless, to some degree as multi-staged (via the finished goods kanbans) and to some degree as single staged (via conventional kanbans).

*Cumulative Production Figures:* This multi-stage method is particularly well suited for communicating information about requirements quickly and without distortion. It is thus reasonable especially when planning is vital and the customers' requirements are highly predictable. Similar to Synchro MRP, when the method is designed appropriately it can prevent the early replenishment of seldom demanded products. It is thus well suited for products with varied or low demand rates and increases the predictability of dependent requirements. A further advantage of Cumulative Production Figures is its flexibility. By adjusting the parameters it can emulate existing as well as new methods for generating orders. Moreover, when applied appropriately the method leaves room for decentralized decisions.

*Base Stock:* Base Stock can be interpreted as a special form of Cumulative Production Figures. It is particularly well suited for the quick and undistorted communication of demand information across the supply chain. Unlike Cumulative Production Figures, however, the method does not necessarily communicate the future development of requirements since, by definition, it conducts a WIP offset of the customer demand. Seen from a more positive perspective though, it can be applied even when the demand tends to be unpredictable. The method also requires that the replenishment times can be reduced when the demand increases (see Chap. 14), otherwise it is necessary to plan safety stock or regularly adjust the planned WIP of the control blocks.

*PA Cards:* The application range of this method is very dependent on how the parameters are set (see Chap. 15). Generalized statements about its application can thus not be easily made.

## 29.3 Selecting an Order Release Method

A two-staged process for selecting an order release method is recommended: First, a suitable class of methods should be selected (Sect. 29.3.1), then a decision can be made about a method within this chosen class (Sect. 29.3.2). In some cases, it makes sense to combine different order release methods with each other, thus a few reasonable combinations are discussed in Sect. 29.3.3.

### *29.3.1 Selecting a Method Class*

Three different classes of order releases can be distinguished: immediate order releases, due date oriented order releases and WIP oriented order releases. WIP oriented order releases can be further sub-divided into a WIP control with or without workstation specific load balancing (see Chap. 16). In this next section, we will consider under which conditions the three classes can be practically implemented.

**Immediate Order Release** Releasing the orders immediately after they have been generated usually increases the WIP in the production. Moreover, the method is unable to differentiate between urgent and non-urgent orders and is therefore only suitable in defined cases.

An example of this is when a single manufacturing cell or production line manufactures a semi-finished or end product from raw material. In this case the material is usually taken from the store just before it is consumed and there is usually only one order in the manufacturing cell or on the production line. Even when an enterprise formally releases other orders before, this does not lead to a build-up of WIP in the production. A constellation such as this is found in the industry especially when Kanban is implemented.

**Due Date Based Order Release** The order release according to planned start dates is aimed at directly implementing the production plan (see Chap. 16). It is therefore reasonable to apply it when the due date and capacity planning is thoroughly conducted, possibly with finite scheduling. If, in contrast, the orders are generated in a rather uncoordinated fashion e.g., by methods for a make-to-stock production and are not scheduled with limited capacities, the planned input date is less significant. Correspondingly, exactly implementing the planned input is also less important.

Furthermore, Due Date Based Order Releases require that the production planning targets can and should be realized. This is, however, not always the case. When there are disruptions in the production input (e.g. a lack of material) releasing orders according to due dates can cause breaks in the material flow within the production. When there are disruptions in the output it can lead to a build-up of WIP in the production.

Due Date Based Order Releases should not be used when the due date and capacity planning consciously does not plan the full capacities in order to leave some leeway for disruptions. At the same time, it is expected that the production – if possible – exceeds the planned output. In this case the manufacturing control should not comply exactly with the production planning. Due Date Based Order Release is therefore also not suitable for such a configuration of the production planning.

**WIP Regulating Order Release** If an immediate or due date based order release is related to considerable disadvantages, a WIP Regulating Order Release should be considered. This couples the production's input to its output and can thus specifically regulate the throughput times, WIP and utilization of the production. It requires that a portion of the orders can be released before their planned start date. Otherwise it can only prevent a production from being overloaded and cannot release any orders early in order to prevent a break in the material flow.

A WIP Regulating Order Release is particularly reasonable when the due date and capacity planning only determines approximate target values. This can be assumed, for example, when the orders (or a part of the orders) are generated by the simple methods used in make-to-stock productions. With a make-to-order production, a WIP Regulating Order Release is recommended in conjunction with infinite scheduling. Moreover, it is advantageous when the production's input is regularly influenced by (considerable) disruptions. The same applies for the output when the Backlog Control cannot or cannot quickly enough compensate for the deviations. Finally, a WIP Regulating Order Release is required when the due date and capacity planning only plans a part of the capacities in order to attain a high schedule reliability, but the production is supposed to produce more than planned as often as possible (see Sect. 29.6.1 for an example of such a configuration).

As already mentioned, WIP Regulating Order Releases are sub-divided into a release with or without a load balancing (Fig. 29.2). A workstation specific load balance is reasonable when the steady utilization of a production workstation is an important logistic target and the following conditions for it are given:

- *The production provides feedback regarding the completion of an operation:* The basic method for all three methods with workstation specific load balances requires feedback about the completion of operations. Only a few variations of Workload Control limit themselves to feedback about complete orders.

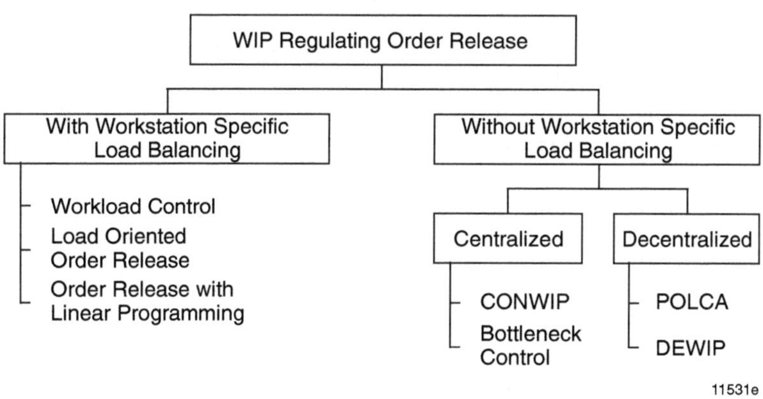


**Fig. 29.2** Classification of WIP regulating order releases

- *Sequence interchanges in the order release can be compensated for during the order throughput or the due date tolerance is relatively high:* Otherwise the sequence interchanges in the order release can lead to low schedule reliability.
- *The complexity of the material flow is great enough to allow an effective load balance:* The methods use differences in the order throughput to balance the load. The Order Release with Linear Programming is the only other method that can attain a load balance via the differences in work content.
- *At least part of the orders can be released before the planned start-date:* The effectiveness of the load balance increases with the number of jobs in the release list. If all of the orders can only be released at the planned start date, the list is frequently empty or contains only very few orders. Accordingly, the methods cannot attain their goal of balancing the load.

WIP Regulating Order Releases without workstation specific load balances are characterized by the following properties:

- The effectiveness of the method does not increase with the number of orders in the release list. As long as there is at least one order in the list the WIP of the production can be regulated. Accordingly, a shorter advance release window (or generally a lower load flexibility) than required by methods with a workstation specific load balance is sufficient.
- With complex material flows, changing bottlenecks and an uncoordinated order generation, the order release without a workstation specific load balance attains a somewhat lower output rate at comparable WIP levels than the order release with a workstation specific load balance (see Sect. 16.1.4).
- A WIP Regulating Order Release without workstation specific load balances leads to comparably few sequence interchanges (decentralized WIP control) or none at all when the orders are released (centralized WIP control). Accordingly, it is more agreeable with finite scheduling than a WIP Regulating Order Release with workstation specific load balancing. This combination is particularly reasonable when disruptions occur in the input or output of the production. On the other hand, methods without a workstation specific load balance do not lead to a target conflict with the schedule reliability.

WIP Regulating Order Releases without workstation specific load balancing can be further sub-divided into a centralized and decentralized WIP control. The advantages and disadvantages of a centralized or decentralized WIP control are extensively discussed in Chap. 16. Decisive criteria are frequently the preferred integration of the operators in the manufacturing control and the simplicity of the method. We will take a quick look at both of these criteria once more here.

*Integration of Operators:* A decentralized WIP control essentially offers greater possibilities for integrating operators in the manufacturing control and with that in the responsibility for attaining logistic targets. The closely meshed control loops convey information about the WIP situation on subsequent workstations. In principle it is possible to transfer the responsibility for the WIP level and thus also the throughput times of a workstation to the operators. With a centralized WIP control this is not possible. The operators have only minimal possibility to impact the WIP and throughput time. Moreover, they are not automatically informed about the WIP situation on other workstations [Hopp-96*]. The decentralized WIP control methods' better integration of the operators is not inevitably advantageous. It requires the operators to have sound knowledge of the method. If just one operator does not implement the method properly the entire production can end up blocked.

*Simplicity of the Method:* Centralized WIP control methods are very simple (especially when they do not balance the loads on each of the workstations). Since there are generally fewer method parameters to set than with a decentralized method, the danger of making an error in the dimensioning is also less.

### 29.3.2 Selecting a Method

In this next section on selecting an order release method we will first turn our attention to those methods which include a workstation specific load balance. Following that our discussion on selecting order release methods without a workstation specific load balance will consider both centralized and decentralized methods.

**Order Release with Workstation Specific Load Balance** There are three order release methods which include a workstation specific load balance: Workload Control (WC), Load Oriented Order Release (LOOR) and the Order Release with Linear Programming (ORLP). The complexity of the method and the quality of the load balance increase in this succession.

Setting the parameters for the ORLP seems so complex that only experts can estimate the impact of the method. This is particularly true regarding the method's influence on the schedule reliability. Due to the laboriousness of the calculations it is also of limited use for an event oriented order release: After each completed operation it has to conduct a new optimization run. The advantage of ORLP though, is that when releasing the orders it can use not only the different order throughput times for the load balancing, but also the different work contents of the orders. Neither LOOR nor Workload Control is able to satisfactorily do this.

The quality of the load balance with both LOOR as well as Workload Control is lower than ORLP due to the lack of optimization. However, in exchange, the impact on the schedule reliability and sequence interchanges can be better assessed and more easily limited than with ORLP. This is particularly valid when the procedural rules are designed so that orders with numerous operations are not disadvantaged by the order release (see Sect. 20.1.2).

LOOR is the only one of the three methods that takes into consideration the position of the released orders in the production and the sequence in which an order that is to be released flows through the workstations in a production by discounting the work content. In comparison to Workload Control or ORLP it is thus particularly well suited when the position of the workstations in the throughput of variants strongly differs (see Chap. 21). Compared to Workload Control, this advantage has to be weighed against the disadvantage of the greater amount of effort required to set the parameters (discounting factors) and compared to Workload Control and ORLP against the disadvantage of potentially greater WIP fluctuations.

The greatest advantage of Workload Control is its simplicity. It can be implemented with an acceptable amount of effort even without EDP solutions by utilizing so-called generic kanbans (see [Chan-94*]). This does not seem to be possible with either LOOR or ORLP. In the variation of the method developed by Tatsiopoulos, Workload Control even refrains from workstation specific feedback about orders. This may, however, negatively impact the quality of the load balance.

**Centralized WIP Control** CONWIP and the Bottleneck Control conduct a centralized WIP control without changing the sequence of the orders at their release. Bottleneck Control is suitable when:

- the bottleneck workstation is clearly defined
- the bottleneck does not change
- the position of the bottleneck system in the order throughput does not vary greatly

In the remaining cases, CONWIP should be favored.

**Decentralized WIP Control** As alternatives for a decentralized WIP control there is POLCA and the Decentralized WIP Oriented Manufacturing Control (DEWIP). POLCA can be practically implemented with simple material flows. Material flows in which the workstations have only one or very few possible workstations preceding them are especially advantageous. Conversely, POLCA is not well suited to complex material flows.

DEWIP can be implemented with both simple and complex material flows. However, setting the parameters and implementing it is easier with directed material flows without backflows – in which case position numbers do not need to be assigned or considered in the order release.

### *29.3.3 Combinations of Methods for Releasing Orders*

It is possible to combine different methods for releasing orders. Nevertheless, this is only reasonable when the methods belong to different classes. Thus the LOOR can be practically combined with a CONWIP control but not with a Workload Control. These types of combinations are for the most part new territory both in theory and in practice. The following combinations seem particularly reasonable:

**Combination of a Periodic Order Release with an Event Oriented Order Release** Such a combination (see Fig. 29.3) is particularly reasonable when the periodic order release is connected to a workstation specific load balance. It then sets the sequence in which the orders should be released for the production, whereas

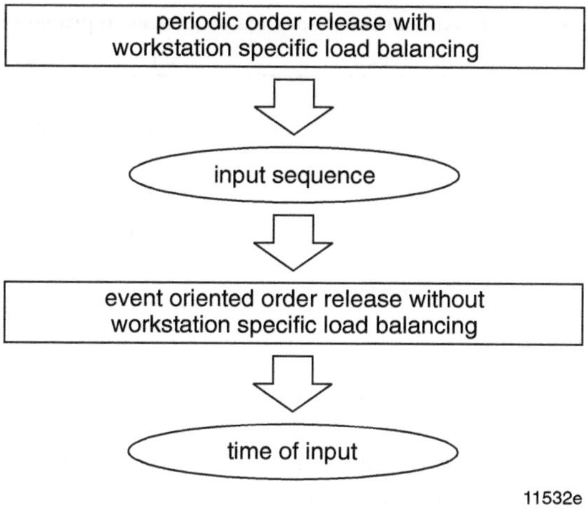


**Fig. 29.3** Combination of different methods for releasing orders

the event oriented order release is responsible for determining the exact time of the release. It thus prevents the unavoidable WIP fluctuations that occur in the production with the periodic order release especially when the planned periods are longer.

An example would be to generate a release list with Load Oriented Order Release or Order Release with Linear Programming (in the first stage) and to release these orders event oriented via a CONWIP or DEWIP.

**Hybrid WIP Control** A hybrid WIP control combines a centralized WIP control (with or without a workstation specific load balance) with a decentralized WIP control (see Fig. 16.3). One example of this would be the combination of a CONWIP Control with a DEWIP control. An order is then only released for production when both the WIP of the entire production as well as the WIP of the first workstation falls below the respective WIP limits. The resulting WIP control would be more restrictive than the order release with only one method. Simulations of the hybrid Kanban/CONWIP control with only one variant indicate that a hybrid WIP control promises to be advantageous [Bonv-97*]. It could attain a specific output rate with a lower WIP than the CONWIP control or the DEWIP when applied alone. This however needs to be confirmed in other simulation experiments. Furthermore, there are no approaches for setting the parameters of a hybrid WIP control based on a model. The main disadvantage of a hybrid WIP control is the larger amount of effort required to apply it.

## 29.4 Selecting Sequencing Rules

One of the key tasks in configuring the manufacturing control is setting a clearly defined sequencing rule for each of the workstations.

The most important criterion for making a decision is the *primary logistic objective*, that is, whether an enterprise wishes to improve the schedule reliability,

**Table 29.1** Allocation of selected sequences to primary logistic objectives

| Primary logistic objective | Sequencing rules |
|---|---|
| Schedule reliability | • Earliest planned start date<br>• Earliest operation due date<br>• Least slack/critical ratio<br>• First in – first out |
| Service level | • Flow rate probability scheduling |
| Output rate | • Setup time optimizing sequence<br>• Extended work in next queue (XWINQ) |

service level or output rate with a suitable sequencing rule. Due to the impact on the schedule reliability it is not reasonable to apply sequencing rules to improve WIP levels and throughput times. We will thus not consider these logistic objectives in the rest of our discussion on sequencing. The impact of selected sequencing rules is summarized in Table 29.1. The advantages and disadvantages of the individual sequences and their application areas are more extensively reviewed in Chap. 25.

The second decision criterion is the potential of the sequence on a workstation to influence the desired logistic objective. This criterion is particularly important for the sequence rules aimed at increasing the output rate. Whereas the XWINQ sequence requires a complex material flow, a setup time optimizing sequence is only effective when the setup times are noteworthy and dependent on the processing sequence of the orders. There is no doubt that the sequence has a considerable impact on the scheduling reliability (or in a make-to-stock production on the service level). Thus in the majority of cases, this logistic objective tends to be decisive in determining the sequence.

The third criterion for a decision is the *extent to which it can be realized*. Thus, for example, the sequence described in Sect. 25.2 for increasing the service level requires the priority numbers for the orders to be constantly re-calculated and communicated. While combined sequences can overwhelm the operators in the production, the FIFO rule can be implemented partially through physical aids such as flow through racks.

The fourth criterion for making a decision is the *level of detail and quality of the order scheduling*. If scheduling fixes the planned start point of every operation for all orders in a finite scheduling and also considers conditions such as sequence dependent setup times, it is often the responsibility of sequencing to implement the planned values as precisely as possible. It then seems obvious, independent of the logistic target, to process the order with the earliest planned start date on the workstation first.

Combining sequencing rules cannot always be prevented. This is particularly true when different logistic objectives are to be considered at the same time. Two cases will be discussed here:

1. *Competition between Output Rate and Service Level or Schedule Reliability:* The object of consideration here is a workstation which determines the output rate and has sequence dependent setup times. Processing the orders in the sequence of

the planned start date of operations would increase setup times, while a setup time minimizing processing would increase the variance of the throughput times and thus the output lateness. Various possibilities for dealing with the competition between these logistic objectives are discussed in Sect. 25.3.1.

2. *Competition between Schedule Reliability and Service Level:* This case arises in mixed productions in which custom orders and stock orders compete for resources. In order to coordinate orders, stock orders can be continually re-scheduled (see Sect. 10.1). The planned start date or end date of the operations are then suitable criteria for sequencing rules, or, as an alternative the priority numbers for make-to-stock orders described in Sect. 25.2 (Eqs. 25.2 and/or 25.3). They can be directly compared to the critical ratio for custom orders (see Eq. 25.4 as well as generally Sect. 25.2).

## 29.5 Selecting a Capacity Control Method

In comparison to the remaining manufacturing control tasks, the capacity control is determined to a particular degree by the operational conditions. This applies to both the reaction time with which the capacity adjustment can be carried out in as well as to the measures that are available for the adjustment. Thus, the selection of a capacity control method described here should also be supplemented with an enterprise specific selection of measures.

The fundamental question in selecting a capacity control is regarding the available flexibility of the capacities. Three cases can be differentiated:

- capacity flexibility available
- no capacity flexibility
- temporary capacity flexibility

Especially when there are limited capacities it can sometimes be difficult to decide if there is any capacity flexibility. In this case, the following questions may be helpful:

Do the capacities or the market demand limit the enterprise's sales? In the case where the capacities limit the sales, there is not sufficient flexibility of the capacities, even when the work content or capacities could still be adjusted on the non-bottleneck workstations of the production.

Are there measures with which the working hours or capacities can be quickly adjusted? If this is not the case, the requirements for conducting a capacity control are not met and thus also not for a high level of logistic target attainment. It is thus a priority for the enterprise to create capacity flexibility.

Independent of the question about the capacity flexibility of the entire production, the enterprise should correct deviations of the actual capacities from the planned capacities as quickly as possible with a Plan Oriented Capacity Control (Sect. 28.1). Implementing additional methods for controlling the capacities should then be oriented on the available capacity flexibility and the type of order trigger (e.g., stock or customer orders).

## *Available Capacity Flexibility*

If there is sufficient capacity flexibility, generally Backlog Control (Chap. 27), Inventory based Capacity Control (Sect. 28.4) or Due Date Oriented Capacity Control (Sect. 28.2) come into consideration.

The most important criterion in selecting a method is the type of order trigger: INCAP is generally only suitable when a large portion of the production volume is make-to-stock. In comparison, Due Date Oriented Capacity Control is conceived for a make-to-order production. Backlog Control, however, can be implemented in both make-to-stock and make-to-order productions.

Thus, simplified the following recommendations apply:

- With a make-to-stock production with comparable unit processing times for the variants INCAP is frequently suitable. Advantages of the method include its simplicity, the close coupling of the capacity and the customer demand or store consumption and the reaction to the impact of disruptions on the production. With a manageable number of variants it also seems reasonable to combine it with Kanban.
- In the remaining cases implementing a Backlog Control seems to be an obvious choice. Basic advantages include that it can be universally implemented and has a positive impact on the schedule reliability. When the schedule reliability is very significant, the enterprise can supplement the Backlog Control with a Due Date Oriented Capacity Control. This applies in particular to the last operations in the material flow: The advantages of a Due Date Oriented Capacity Control have an especially strong influence on these.

## *No Capacity Flexibility*

If the capacities limit the sales of an enterprise, attaining the logistic targets is endangered. In a make-to-order production the enterprise can no longer legitimately promise the desired delivery times and frequently the delivery reliability also suffers due to capacity bottlenecks. This is particularly true, when enterprises do not take capacity bottlenecks into consideration when planning. In a make-to-stock production the inventory sinks and consequently the service level as well. For conditions such as this the Output Rate Maximizing Capacity Control (see Sect. 28.3) is well suited despite being unable to solve the problem of the lacking capacity flexibility, mentioned above. In the long run therefore, designing the capacities in this way is recommendable only in limited exceptional cases (e.g., with very high fixed cost ratios and price sensitive demands).

### *Temporary Capacity Flexibility*

To some extent an enterprise will go through different phases i.e., there will be phases where there is greater demand and the capacities are exhausted and those where the demand is lower and there is available capacity flexibility. This means enterprises need to be able to adjust their planning and control logic to these conditions. For the capacity control it means changing between an Output Rate Maximizing Capacity Control and a Backlog Control, for example.

## 29.6 Case Studies

Let us now turn our attentions to possible configurations of the manufacturing control based on three exemplary case studies which to a large extent depict typical types of productions in the field.

- Section 29.6.1 describes the configuration of the manufacturing control in a *make-to-order production*. The industry sector selected for this case study is the circuit board industry. Here, particular attention is dedicated to the question of how an enterprise can react to pronounced business cycles.
- Section 29.6.2 examines the configuration of the manufacturing control in an enterprise that produces a large number of different variants of precision work tools for the metal processing industry. The enterprise considered is a *mixed production* in which some of the variants are produced according to the make-to-stock principle and others according to the make-to-order principle.
- Section 29.6.3 depicts the manufacturing control in a supply chain within the automobile industry. The focus of our considerations is a pure *make-to-stock production* of valves, as serial product for the engine plant of a car maker.

All three cases studies are based on actual enterprises. However, the described manufacturing control is not oriented on the actual specification of the manufacturing control. Instead, it follows the discussed procedure for configuring the manufacturing control. Accordingly, the results deviate considerably from the actual existing manufacturing control.

### *29.6.1 Manufacturing Control in a Make-to-Stock Production*

Circuit Boards Incorporated manufactures customized circuit boards according to the make-to-order principle. The business is distinctly cyclical. During peak periods many enterprises in the industry sector struggle with increased WIP levels, long throughput times and poor delivery reliability. Even with diminishing demand however, only a minority of enterprises manage to attain high delivery reliability.

In this next section, we will examine the manufacturing control tasks and a possible configuration of manufacturing control methods.

**Logistic Targets** In peak demand periods the priorities are an as high as possible utilization rate and delivery reliability. The high utilization is required for targeting the most sales possible. High delivery reliability prevents costs for express expeditions and conventional penalties. Moreover, it is an excellent means for the enterprise to develop long term business relationships with customers since during peak demand periods the delivery reliability clearly sinks on average in the industry sector.

During good business cycles most customers accept longer delivery times, however, they are also willing to pay more for shorter delivery times. Therefore, it should at least be possible to promise shorter delivery times for a portion of the orders. The WIP is less significant. As a make-to-order manufacturer, Circuit Board Incorporated does not maintain any finished goods stores. Since the circuit boards are manufactured from standardized base materials, the capital tied-up in the raw materials store is also comparably minimal. Due to quality reasons and in order to keep the tied-up capital to a minimal, the WIP in the production should nevertheless be kept as low as possible.

In low demand periods a high utilization of resources cannot be targeted due to a lack of demand. Accordingly, the logistic objectives ‘delivery time’ and ‘delivery reliability’ become all the more important, whereas the WIP is also here not so significant.

**Manufacturing Environment** The production is in part organized as a job shop production e.g., for the drilling, milling and printing. In other areas it is organized according to the flow principle.

The size of the production runs for the series varies considerably; the mean lot-size is approximately 100 pieces. Due to numerous changes in the layout of the circuit boards the repetition frequency of the variants is minimal. Most variants are only reproduced once or twice.

Most parts of the production are characterized by lot-wise transport. The enterprise has however already implemented a one-piece-flow-principle for a smaller part of the operations and wants to extend this section in the future.

Due to the products being custom-made the number of variants is very high and practically unlimited. Despite the diversity of variants the complexity of the material flow is relatively low. The production of the circuit boards generally follows the same sequence for the basic processing steps. However, some processing steps are needed only for special types of PCBs and the sequence of these may vary.

Trade in the circuit board industry is strongly cyclical. In peak periods Circuit Board Inc. is generally not able to accept all of the customer orders. Nevertheless, during low demand periods, operating the plant with three shifts is to some extent not sustainable.

The load flexibility – also due to the broad customer basis – is comparably high. With the exception of orders from important regular customers, the enterprise can

**Table 29.2** Control relevant manufacturing characteristics of Circuit Board Inc.

| Criterion | Character | | | | |
|---|---|---|---|---|---|
| Manufacturing Principle | Workbench | On-Site | **Job Shop** | Cell | **Flow** |
| Type of Production | Non-Repetitive | **Single and Small Series** | Series | Mass Production | |
| Part Flow | Bulk | **Lot-Wise** | Overlapped | **One-Piece-Flow** | |
| Number of Variants | Very Low | Low | Medium | High | **Very High** |
| Material Flow Complexity | Very Low | **Low** | Medium | High | Very High |
| Fluctuations of Capacity Requirements | Very Low | **Low** | Medium | High | **Very High** |
| Capacity Flexibility | Very Low | **Low** | Medium | High | Very High |
| Load Flexibility | Very Low | **Low** | Medium | High | Very High |

afford to reject customer orders. Moreover, the throughput time for a large percentage of the orders is considerably shorter than the delivery time.

In comparison, the capacity flexibility is low. It generally consists of the possibility to integrate weekend shifts. Occasionally, operations or orders can also be processed externally. Nevertheless, the cost of doing so is quite high especially in peak periods. The low capacity flexibility has to be considered when determining the delivery date.

Table 29.2 summarizes the manufacturing characteristics that are relevant for the manufacturing control.

**Production Planning** The determination of the delivery date is the essential interface between the production planning and manufacturing control for attaining logistic goals. It takes into consideration the load and production progress on all workstations with critical capacities. Even during high demand periods the capacities are not completely planned in order to maintain reserves for rush orders and to be able to ensure a high rate of schedule compliance. A WIP control ensures though that the capacities are fully utilized as long as there is a sufficient supply of work. In order to facilitate this, orders are also released early (see the following discussion on order release).

**Order Generation** Circuit Board Inc. generally converts customer orders directly into production orders, whereby they to some degree also take into consideration expected rejects and increase the order accordingly. The enterprise divides extremely large orders into sub-orders thus reducing the variance of the work contents. As a result the trends of the Logistic Operating Curves for the production are more favorable and the enterprise can then plan with lower throughput times and WIP/stock levels (see Sect. 3.2.2). Usually, the task of generating orders in the circuit board production is comparably simple (see flow of information for the order generation in Fig. 29.4).

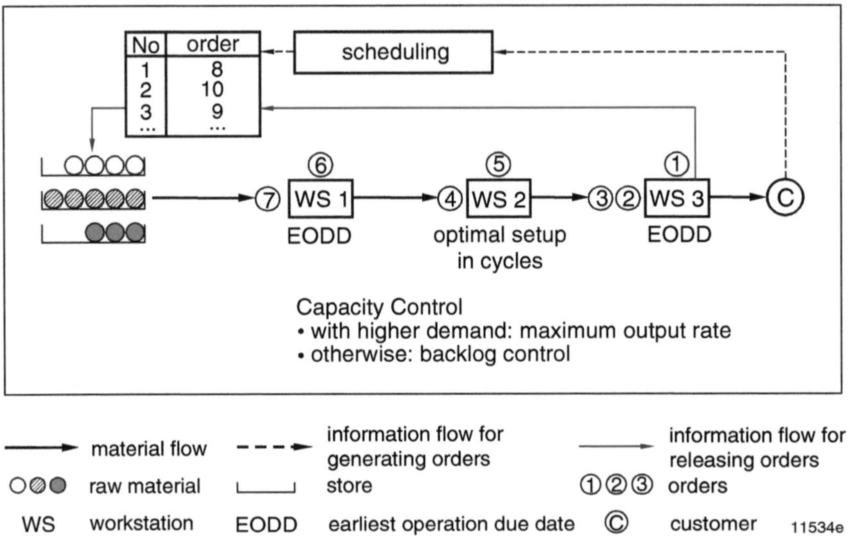


**Fig. 29.4** Configuration of the manufacturing control of Circuit Board Inc

**Order Release** Seeing the enterprise already considers the load of the capacities when determining the delivery date it would not be very reasonable for Circuit Board Inc. to conduct an order release with a workstation specific load balance. Moreover, the complexity of the material flow tends to be minimal. Sequence interchanges in releasing the orders would therefore only cause unnecessary turbulence.

In comparison, a *WIP control* is reasonable and critical: There is considerable load flexibility and although the utilization of the capacities is significant, the due date and capacity planning does not ensure the utilization of the capacities.

There is a range of possibilities with regards to whether a centralized or decentralized WIP control should be conducted. Due to the simplicity of it, Circuit Board Inc. decides to implement a centralized WIP control. Since there is no clear bottleneck, applying a Bottleneck Control is not practical; CONWIP remains as an alternative (see flow of information for order release in Fig. 29.4).

During periods where there is a high order input the order release takes on particular significance in the manufacturing control concept. It then ensures the utilization of the capacities. CONWIP maintains the WIP (in number of panels[1]) as a constant. If the WIP in the production sinks below the limit due to the completion of an order, then an order from the list of urgent jobs is released for production. The

[1] The term "panel" refers to the base material of the circuit boards in standardized sizes. Generally a number of circuits are on each panel. In one of the last operations the panels are then divided into circuit boards that contain only one circuit.

release list contains known orders, organized according to planned start date. If the actual output rate exceeds the planned output rate, space for accepting new orders on short notice is created. Since the production planning usually leaves a capacity reserve, this is generally the case. Short delivery times can then be promised for newly acquired orders. During peak demand times it is therefore usually possible to sell these capacities at a good price. Even when it is not possible to utilize the capacities directly with new orders, there is rarely a loss of utilization since CONWIP then releases orders from the pool of existing ones. As a result it is possible to attain a higher utilization rate than with a Due Date Based Order Release.

During low demand periods the WIP control is less significant since completely utilizing the resources capacities is no longer possible. Circuit Board Inc. usually then stops the WIP control or at least reduces the advance release window. By doing so it prevents orders from being released too soon and thus causing a WIP build-up. Compliance with the delivery times is then ensured via a Backlog Control.

**Sequencing** The most important logistic objective both in peak demand periods as well as in low demand periods is the delivery reliability. The majority of the workstations therefore process orders in the sequence of the earliest operation due date (see sequencing specifications amongst the workstations in Fig. 29.4). In this way the delay of orders can frequently be compensated for.

A small number of workstations have to be re-set when the format of the base material (the panels) is changed e.g., for the exposure process. Constantly changing the format would give rise to high costs and reduce the production's output. These workstations have (variable) production cycles. Within a sub-cycle only circuit boards with the same format are produced. The sequence within a sub-cycle – as with the other workstations – is formed according to the earliest planned end date. Whereas standard circuit boards are manufactured at least once a day, the more unusual formats are produced in a 2-day rhythm. This sequence thus forms a compromise between the logistic objectives "output rate" and "schedule reliability".

**Capacity Control** The capacity control differs greatly between peak demand periods and low demand periods. If the possible demand exceeds the available capacities, the goal is to produce as much as possible (Output Rate Maximizing Capacity Control). At least workstations with critical capacities are then typically operated in three-shifts and on weekends. Staggering breaks allows these workstations to be operated (almost) 24-h a day. It is then the responsibility of the remaining workstations to ensure that the workstations with these critical capacities are steadily supplied.

If the capacities are sufficient for coping with the demand, then Circuit Board Inc. initiates a Backlog Control. The backlog is calculated by comparing the planned output with the actual output both for the individual workstations as well as the entire production. Due to the relative small work content and the minimal effort required for feedback the enterprise uses a discrete backlog measurement; the work already completed on orders is thus not taken into consideration.

Overtime and weekend shifts define the enterprise's capacity flexibility. On workstations with large capacity reserves the enterprise uses overtime almost exclusively. This can be managed as required and without prior notice when the operator is in agreement. If the backlog exceeds half of a shift, the enterprise arranges overtime for the next days until the backlog is eliminated.

On workstations with critical capacities weekend shifts are to some extent unavoidable; these need to be arranged with employees at the latest by Tuesday of the respective week. The decision to implement weekend shifts is made when the backlog attains or exceeds 10 standard hours.

Circuit Board Inc. coordinates the measures for the capacity control together with the production workers. Since many of the workers are cross-trained usually more workers are able to work overtime than required. The enterprise tries to take into consideration the personal interests of the employees and to evenly distribute the workload. Due to the additional wages however, usually more employees apply for overtime than required.

## *29.6.2 Manufacturing Control in a Mixed Production*

The Mechanical Engineering Co. produces a very large array of precision tool variants for the metal processing industry. A number of the variants are produced for stock, whereas a number of others are customer orders. In the manufacturing control, The Mechanical Engineering Co. adopts the basic concepts of Capacity Oriented Materials Management (see Chap. 10).

**Logistic Targets** In the make-to-order production the delivery reliability is the most critical objective: The automobile industry, including both suppliers and manufacturers, are important customers of The Mechanical Engineering Co., and late deliveries to these customers can cause standstills in production lines. This in turn can give rise to extreme costs related to delays. The delivery time in comparison, is frequently less decisive. The demand for tools can be reliably determined and often considerably in advance.

In the make-to-stock production the service level is the decisive logistic objective. Due to the competition within the industry sector and the interchangeability of standardized end products, customer orders which cannot be supplied directly from stock are generally lost to a competitor. Moreover, the price competition is intensive. The Mechanical Engineering Co. is thus forced to limit the capital tied-up in stores of finished goods. This reduces among other factors the considerable inventory risk that is created due to product innovation and ever shortening product lifecycles. Excess goods either have to be sold at considerable discounts or scrapped.

Moreover, the price competitions only allow minimal over capacities. The Mechanical Engineering Co. therefore requires a high utilization rate in order to make profits.

**Manufacturing Environment** The production is organized as a job shop production. The arrangement of the job shops is oriented on the material flow. It is a repetitive production with varying lot-sizes, whereby the average lot-size is 500 pieces. The orders are transported in complete lots from workstation to workstation (lot-wise transport). There are a huge number of variants: The Mechanical Engineering Co. manufactures approximately 2,000 variants as stock and almost exactly the same amount of variants are made-to-order. The material flow is not especially complex, however, the bottleneck varies and not all product families flow through all of the workstations with critical capacities. The customer demand for each variant fluctuates considerably. This impacts the capacity requirements only minimally though. The increased demand for a few variants is usually accompanied by a reduced demand for other variants so that the impact of the demand fluctuations on the capacity requirements, at least partially balance themselves out. However, due to demand cycles the capacity requirements vary to the degree typical in the industry sector. The large majority of time the capacities are just sufficient enough to satisfy the customer demand.

The capacity flexibility of The Mechanical Engineering Co. is relatively minimal. The workstations with critical capacities are operated in three shifts. Outsourcing orders is not possible due to the special machinery. Nevertheless, special shifts on weekends can be arranged with the work council on short notice.

In comparison, the load flexibility is relatively high. The delivery times for many of the make-to-order jobs exceed the production throughput time while make-to-stock orders can be generated early for the purpose of balancing the load. Furthermore, by taking into consideration the demand during the replenishment time there is often additional leeway during the order throughput time.

Table 29.3 summarizes the essential manufacturing characteristics relevant for the manufacturing control of The Mechanical Engineering Co.

**Table 29.3** Control relevant manufacturing characteristics of The Mechanical Engineering Co.

| Criterion | Character | | | | |
|---|---|---|---|---|---|
| Manufacturing Principle | Work-bench | On-Site | **Job Shop** | Cell | Flow |
| Type of Production | Non-Repetitive | **Single and Small Series** | Series | Mass Production | |
| Part Flow | Bulk | **Lot-Wise** | Overlapped | One-Piece-Flow | |
| Number of Variants | Very Low | Low | Medium | High | **Very High** |
| Material Flow Complexity | Very Low | Low | **Medium** | High | Very High |
| Fluctuations of Capacity Requirements | Very Low | **Low** | **Medium** | High | Very High |
| Capacity Flexibility | Very Low | **Low** | Medium | High | Very High |
| Load Flexibility | Very Low | Low | Medium | **High** | Very High |

**Production Planning** Due to the make-to-stock production, the multitude of variants and the broad customer base, the production planning is marked by a high degree of uncertainty. Prognoses with a sufficient amount of precision can at best be made on the level of product families. The enterprise uses these prognoses to make investment decisions. The capacity requirements for the workstation in the first part of the material flow cannot even be precisely predicted a few days in advance since custom orders in particular (but also stock orders) that are especially urgent can be generated at any time. Furthermore, the time at which make-to-stock orders actually have to be completed is dependent on the customer demand during the replenishment time (see Chap. 10 on probability scheduling).

**Order Generation** Generating customer orders is simple. These are directly converted into production orders (production lot-size = order lot-size). At most the enterprise divides very large lots so that the variance of the work content in the production is minimized. For the variants that are produced for stock a simple Order Point System is implemented. As a result of the very broad customer basis and the frequently not so pronounced customer loyalty (the products are almost exclusively C parts for the customer), no other method can up to now be justified. In contrast, there is a trend towards electronically processing orders. This streamlines the order handling without changing the method itself.

In applying the Order Point System though, The Mechanical Engineering Co. positively stands-out from the competition. On the one hand, it sets the safety stock levels based on a Logistic Positioning with Storage Operating Curves. On the other hand, it generates the replenishment orders before reaching the order point in order to allow a load balancing. For this purpose, it calculates the 'look ahead time' within which the stock of the variants will probably fall below the order point (see Chap. 10).

The information flow of the order generation as well as the entire manufacturing control is summarized in Fig. 29.5.

**Order Release** The firm conducts a workstation specific load balance with the order release (see information flow for the order release in Fig. 29.5). This is reasonable for a number of reasons (see Sect. 29.3.1):

1. The steady utilization of capacities is an important logistic goal.
2. Feedback about the completion of operations is quickly and reliably registered via an electronic operational data compilation system.
3. With all stock orders and a portion of the custom orders, the jobs are known before the planned start date. There is thus a relatively high load flexibility.
4. The due date and capacity planning is marked by uncertainty. Therefore a finite scheduling of the orders would be quite laborious and require constant re-scheduling.
5. The orders flow through different workstations that have critical capacities. The complexity of the material flow is therefore high enough to enable a load balance.

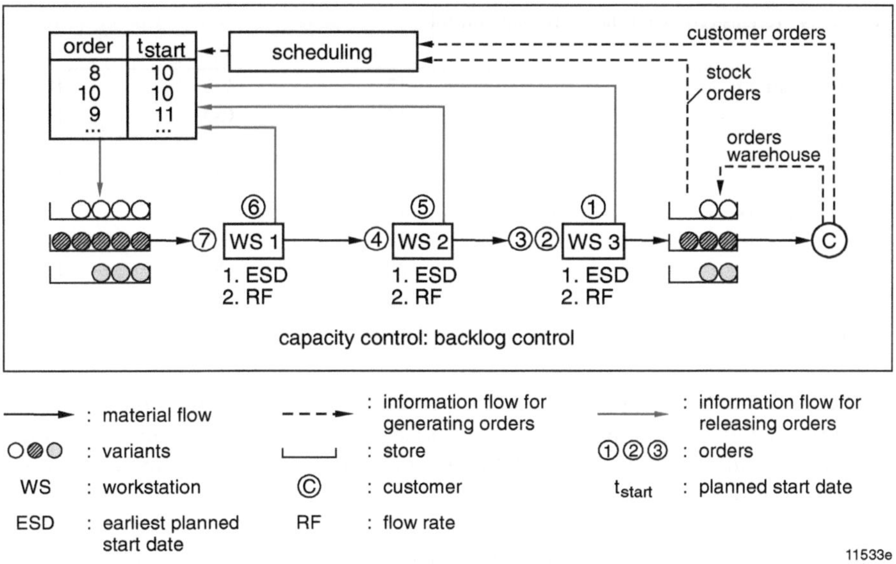


**Fig. 29.5** Configuration of the manufacturing control for The Mechanical Engineering Co

The Mechanical Engineering Co. applies Workload Control to release orders oriented on events. This is unproblematic because the positions of the workstations do not vary greatly in the order flow; otherwise the enterprise would have chosen to apply Load Oriented Order Release. The Order Release with Linear Programming was not suitable due to the complexity of it and because the enterprise wanted to determine the sequence of the order release to a large degree themselves.

The release list is sorted according to the following criteria:

- First, customer orders whose planned start date has been reached or passed are released in the sequence of their planned start date.
- Following that the remaining orders are sorted according to their planned start date. In the case where orders have the same planned start date, customer orders are prioritized before stock orders.

Table 29.4 depicts an excerpt from a list of urgent orders for Shop Calendar Day 11.

**Sequencing** As with the order release The Mechanical Engineering Co. sets a two stage priority system for the sequencing (see sequencing tasks in Fig. 29.5).

- Customer orders, whose planned start date on a workstation is already reached or passed, are processed according to the sequence of their planned start date.
- For the remaining orders the ratio of the throughput time remaining until their (expected) planned completion date and the operation time is calculated (Eqs. 25.2 and 25.4). The smaller this ratio is the greater the priority is.

**Table 29.4** Release list with the workload control

| | No. | Order | Type | $T_{start}$ |
|---|---|---|---|---|
| Urgent customer orders | 1 | 12 | CO | 10 |
| | 2 | 14 | CO | 11 |
| | 3 | 15 | CO | 11 |
| | 4 | 13 | CO | 11 |
| Remaining orders | 5 | 17 | SO | 10 |
| | 6 | 18 | SO | 11 |
| | 7 | 20 | CO | 12 |
| | 8 | 22 | SO | 12 |

*CO* customer order, *SO* stock order, $T_{start}$ planned start date

The aim of assigning priorities is to promote the delivery reliability and attain a high service level. In doing so, the enterprise weights customer orders more than stock orders in the sequencing. This is not necessarily reasonable for other enterprises or industries.

**Capacity Control** The Mechanical Engineering Co. conducts a periodic Backlog Control. It measures the backlog for all workstations and for the entire production daily and in standard hours. In order to do so, the enterprise schedules all stock orders backwards using Flow Rate Oriented Scheduling. The result is planned end dates for all of the stock orders' operations. These dates already exist for customer orders as a result of the due date and capacity planning. Based on the planned end date of all the orders, The Mechanical Engineering Co. can calculate the planned output of all workstations. The backlog is then determined by comparing these with the actual output.

For the decision about which measure to use to increase the capacities, the backlog on a workstation (and the entire production) is divided by the output rate, resulting in the (calculated) expected lateness (see Eq. 3.22). If the calculated lateness exceeds a specific limit, the enterprise arranges overtime or weekend shifts (or both) with the employees.

*Overtime:* For the arrangement of overtime, the limit for the expected lateness is 0.2 SCD. Due to the different utilization of capacities the manufacturing control aligns the capacity increases on minimally utilized workstations with the possible capacity increase on workstations with critical capacities. If the capacities on these workstations cannot be increased by overtime then the capacities on the remaining workstations remain the same.[2] Otherwise, on the workstations with critical capacities the capacity increase results from the backlog or the possible capacity increase. The remaining workstations implement this capacity increase to the same extent i.e. when the capacity on the workstations with critical capacities is increased by 10%, the remaining workstations also increase their capacities by 10%. If the

[2] This requires that the bottleneck workstations are supplied with a sufficient amount of WIP.

output can be attained on these workstations to some degree by increasing the utilization then the capacities are not increased as much.

*Weekend Shifts:* One weekend shift has to be announced at the latest by Tuesday of the same week, whereas two or three weekend shifts have to be announced by the Tuesday of the week before. The quicker reaction time for the one weekend shift is due to the availability of workers: It is easier to quickly find operators for one shift than it is to find them for two or three shifts.

The decision about a weekend shift is dependent on a number of factors: the current backlog, the future load of the production, the normal capacity and the arranged overtime. In addition, the enterprise conducts a capacity planning – as depicted in Fig. 27.3a. It verifies whether or not the planned output might also be at least approximately attained without a weekend shift. If it could be, then a weekend shift can be refrained from. If the lateness is expected to exceed a limit of 0.5 SCD then a weekend shift is arranged with the employees.

The enterprise has a similar procedure for making decisions about multiple weekend shifts. The planning horizon in this case is longer and higher limits of 0.8 SCD (for two weekend shifts) and 1.2 SCD (for three weekend shifts) are arranged with the employees.

The limits for the calculated lateness should be coordinated with the delivery time buffer for the customer orders: At The Mechanical Engineering Co. the delivery time buffer is 2 days. Furthermore, with the weekend shifts the lowest possible capacity that can be supplied needs to be considered. With a very small backlog the planned output would be (to some degree considerably) exceeded. Since the calculated lateness is quite small, implementing a weekend shift in this case can barely be justified.

The Mechanical Engineering Co. programmed an interface between the PPC system and a spreadsheet program which allows the required data to be exported from the PPC system to a previously configured spreadsheet. The decision with regards to overtime or weekend shifts is thus simplified via a simple software tool.

### *29.6.3 Manufacturing Control in a High Volume Make-to-Stock Production*

The automobile industry was selected as an example for the manufacturing control in a high volume make-to-stock production. More specifically we will consider a supply chain consisting of a valve manufacturer, an engine building plant and an assembly plant. Although our main focus is the valve manufacturer, Valve Technology Inc., we will also take a brief look at how the enterprise is integrated into the supply chain of the different car manufacturers.

**Logistic Targets** The dominating logistic objective for Valve Technology Inc. is a service level of 100%. A line standstill in the engine building plant of a customer would be extremely costly and at least with repeated occurrences would seriously endanger the supply relationship with the customer.

The second most important logistic objective is the WIP which is required to maintain the delivery certainty. Due to the customers' high demand rate and the comparable high value of the products a large amount of capital is tied-up even with a small range.

The utilization of resources has the least priority of all the objectives. It is essentially determined by the demand of the car manufacturers which can in turn be barely influenced. Valve Technology Inc. can plan the capacities of the resources quite reliably, because the customer's capacities are also limited.

**Manufacturing Environment** Essential parts of the production are organized according to the one-piece-flow principle, either in the form of manufacturing cells or in synchronized production lines. The one-piece-flow principle however is not consistently implemented in all processes. Above all, the earlier production processes are characterized by a production with lot-wise transports which are occasionally necessary due to technology dependent minimum inter-operation times. These arise primarily as a result of cooling times after processes that strongly heat the work pieces.

Furthermore, the production lot-sizes vary due to the different setup costs in the material flow. As a result there is a build-up of semi-finished stock in the production. The material flow is relatively simple. Valve Technology Inc. manufactures approximately 200 variants for 20 customers. The majority of sales are generated by approximately 20 variants from three to four key customers.

The predictability of the customer demand varies from customer to customer. Some customers constantly change delivery dates and quantities to a large degree and without warning, while other customers very reliably predict their expected deliveries in the next 2 weeks. Order quantities and frequencies are also influenced by the geographical distance to the customers' engine building plants.

The cycles of the automobile trade impact the order input of Valve Technology Inc. Since the enterprise can implement overtime and weekend shifts to increase capacities, their capacity flexibility is sufficient though for counteracting the load fluctuations. During normal demand periods the enterprise operates two shifts. The second shift then begins 2 h after the end of the first shift. This makes it possible to extend the first shift by 2 h without influencing the start of the second shift. The load flexibility is in comparison minimal. For key customers it is practically impossible to reject orders or shift the delivery date. Table 29.5 summarizes the essential manufacturing characteristics relevant for the manufacturing control of Valve Technology Inc.

**Order Generation** The delivery terms and requirements vary considerably from customer to customer. Valve Technology Inc. tries to integrate the order generation via Cumulative Production Figures. The enterprise decided to implement Cumulative Production Figures to control production for the following reasons (see Sect. 29.2):

**Table 29.5** Control relevant manufacturing characteristics of Valve Technology Inc.

| Criterion | Character | | | | |
|---|---|---|---|---|---|
| Manufacturing Principle | Workbench | On-Site | **Job Shop** | Cell | **Flow** |
| Type of Production | Non-Repetitive | Single and Small Series | **Series** | Mass Production | |
| Part Flow | Bulk | **Lot-Wise** | Overlapped | **One-Piece-Flow** | |
| Number of Variants | Very Low | **Low** | **Medium** | High | Very High |
| Material Flow Complexity | Very Low | **Low** | Medium | High | Very High |
| Fluctuations of Capacity Requirements | Very Low | Low | **Medium** | High | Very High |
| Capacity Flexibility | Very Low | Low | Medium | **High** | Very High |
| Load Flexibility | **Very Low** | Low | Medium | High | Very High |

- It is a make-to-stock production.
- Requirements can be generated over a number of stages.
- The demand fluctuations can (within limits) be forecasted (good predictability).
- The capital tied-up in WIP is considerable (high planning necessity).
- Due to the different customer requirements the order generation has to be able to be handled flexibly.
- The system should – within limits – allow leeway for decentralized decisions.

The configuration of the Cumulative Production Figures differs from variant to variant and from customer to customer:

- One of the customers also applies the Cumulative Production Figures principle in order to coordinate the production of its engine building plant with the vehicle assembly. With this customer the Cumulative Production Figures are directly integrated via an interface in the EDP system (see information flow of order generation from car manufacturer 2 in Fig. 29.6). If this customer changes the Cumulative Production Figures for a specific type of motor, this change is directly transferred to the Cumulative Production Figures for the corresponding type of valve. The assembly plant arranged limits for changes that should not be exceeded with the engine building plant and the valve manufacturer. Applying Cumulative Production Figures across a number of enterprises already fulfils fundamental requirements of Supply Chain Management i.e., the car manufacturer provides clear requirements, nevertheless, without coordinating these in detail with the supplier. Due to the number of parts and suppliers a coordination process such as this would be extremely complicated. Here, it is more important that the supplier receives comprehensive information about the requirements early enough to be able to fulfill them. The supplier is informed about the WIP levels of the engine building plant as well as about the requirements that are to be expected in the future.

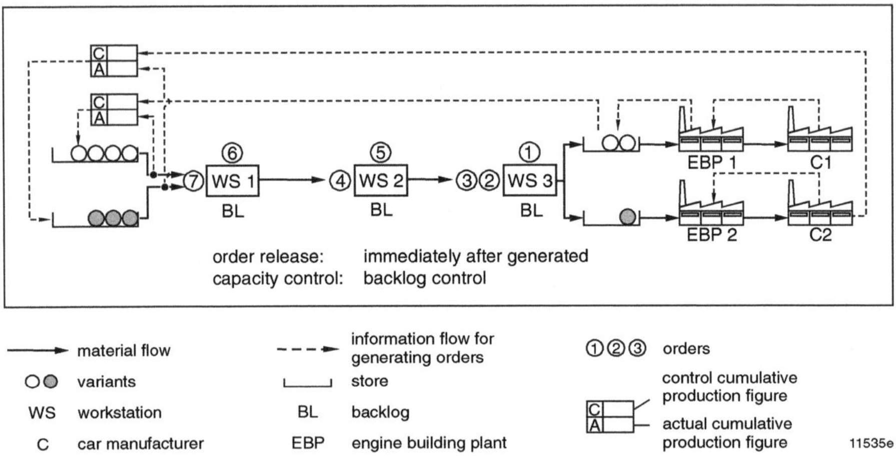


**Fig. 29.6** Configuration of the manufacturing control for Valve Technology Inc

- With another customer whose engine building plant is nearby, the supply is arranged via Kanban (see information flow of order generation from Engine Building Plant 1 in Fig. 29.6). Store withdrawals are directly converted into cumulative production and planning figures for the components. Valve Technology Inc. receives the expected requirements for each variant 2 weeks in advance. Upper and lower limits for the deviation from the predicted demands are also arranged.
- The remaining customers place each order individually with the exact delivery quantity and exact delivery date. The entering of these orders into the Cumulative Production Figures is either manual or – for most – partially automated.

Furthermore, Valve Technology Inc. calculates actual and planned cumulative production figures for all of the workstations using a WIP offset. They serve to coordinate the sequencing and to calculate the backlog. For reasons of simplification it is assumed here that the WIP offset results in a realistic and leveled production plan. This is realistic since the variants are manufactured in large volumes and supplied in high frequencies. If the delivery frequency is reduced, the corresponding variants should be planned so that the resulting capacity requirements are as steady as possible. This is also reasonable with higher capacity flexibility and prevents unnecessary fluctuations in the employees work.

**Order Release** Valve Technology Inc. opted not to conduct a workstation specific load balance or a WIP control (see Sect. 29.2):

- The steady utilization of the production is not a primary logistic target for Valve Technology Inc.
- There is minimal load flexibility.
- The capacity flexibility is sufficient to compensate for load fluctuations.

Valve Technology Inc. thus releases orders immediately after they are generated.

**Sequencing** The service level is the dominating logistic objective for sequencing. The workstations consistently reproduce the variants with the highest backlog first (see sequencing tasks in Fig. 29.6):

$$BL_i = CPF_{plan,i} - CPF_{act,i} \tag{29.1}$$

where

| | |
|---|---|
| $BL_i$ | backlog Variant *i* [units], |
| $CPF_{plan,i}$ | planned Cumulative Production Figure for Variant *i* [units], |
| $CPF_{act,i}$ | actual Cumulative Production Figure for Variant *i* [units]. |

Valve Technology Inc. ascertains both the actual as well as the planned Cumulative Production Figures in real time. Changes in the customers' desires are thus directly conveyed to all of the workstations and can be immediately taken into consideration with the sequencing. The enterprise can thus react very quickly to changes in the customers' demand.

**Capacity Control** Due to the available capacity flexibility and the significance of the schedule reliability Valve Technology Inc. implements a Backlog Control. The backlog is determined from the planned and actual cumulative production figures at the end of every shift (see Eqs. 27.5 and 27.6). As a result of the – to some extent – large lot-sizes the enterprise takes into consideration the already completed portion of the orders being processed (continuous backlog measurement). Backlogs that arise are reduced as quickly as possible after the shift using (unplanned) overtime. Fluctuations in the capacity requirements are taken into consideration by the due date and capacity planning in (planned) overtime and weekend shifts.

## 29.7 Implementing a Configuration

The best configuration remains useless when an enterprise cannot transfer it to the shop floor. Enterprises therefore have to ensure:

- that orders are actually generated as specified by the order generation method (strong order generation compliance),
- that orders are actually released as specified by the release method (strong release compliance),
- that orders are actually processed in the given sequence (strong sequencing compliance), and
- that capacities are actually adjusted as specified by the capacity control (strong capacity control compliance).

To conclude our discussion on configuring and implementing the manufacturing control let us direct our attention to a few suggestions regarding critical aspects of the implementation. These can be divided into two parts. First, five general guidelines for implementing instructions are formulated. Following that a number of stumbling blocks, which based on experience, cause enterprises particular problems and which are specific to PPC, are presented. Finally, we conclude with a brief summary.

**General Guidelines for Implementing Instructions** The probability that an operator implements an instruction increases when the following conditions are fulfilled.

- *The instruction is easy to implement:* Enterprises should provide employees with suitable tools for implementing the decision. In the order generation and order release departments, software may be required. With sequencing and backlog controls it is probably sufficient enough to clearly identify the order priority and the production's progress.
- *The employees are convinced the instructions are correct:* Enterprises should communicate and explain the objectives of the instructions, how each instruction is a part of an overall concept and which advantages result from that. Employee trainings may be required.
- *Management monitors and ensures that instructions are followed:* Otherwise, instructions remain non-binding.
- *Following instructions is connected to positive incentives:* Enterprises can connect monetary or non-monetary incentives with strongly complying to a sequence or release method.
- *Not following instructions is connected with negative consequences:* Penalties are generally neither required nor desired in order to implement a manufacturing control configuration. It can however not be tolerated that individual workers consciously endanger the agreed upon logistic targets of a production.

**PPC Stumbling Stones** Wiendahl, Begemann and Nickel formulate seven stumbling blocks for the production planning and control which prevent logistic targets from being frequently attained [Wien-03b]. H.-H. Wiendahl takes up these stumbling blocks and supplements them [Wien-03c].

***1. Business Environment***
The most important causes of trouble are unreliable suppliers or unpredictable customers who change the delivery quantities and dates regularly and with little advance notice. Even when both areas lay outside of the enterprise, a closer cooperation offers the possibility to influence the supplier and customer behavior. This is a fundamental objective of Supply Chain Management (see Chap. 6).

***2. Model***
Many enterprises fail to use a model or at least a suitable model for systematically analyzing why logistic targets are not satisfactorily attained and for estimating potential improvement. Consequently, how the manufacturing control influences the attainment of logistic targets is also not systematically understood. It is

recommendable to develop a logistic production monitoring based on the funnel model (see Chaps. 2 and 3).

***3. Method and Parameterization***
In many instances, enterprises select the wrong method for a PPC task or at least do not properly set the parameters. In the worst cases, enterprises totally fail to perform a task. How manufacturing control methods can be systematically selected is the subject of the first sections of this chapter. The sections on the method description in each of the chapters are dedicated to the parameterization of the respective methods.

***4. Data Quality***
Successful PPC requires the input data for the planning to be correct. This data can be sub-divided into basic order-independent data consisting of master and structure data as well as in mostly order oriented transaction data [Wien-03b]. Analyses conducted by the Institute of Production Systems and Logistics demonstrate a lack of quality particularly in the following areas:

*Routings and Standard Times:* To some extent neither the standard times nor the order throughput correspond with reality due to changes in technology. Consequently, both the calculations of the required capacities as well as the planned throughput times become inapplicable.

*Planning and Control Parameters:* Many enterprises do not systematically set the PPC parameters such as the throughput times or safety stock in a Logistic Positioning, but rather at best do so based on historical data. Moreover, they fail to update these parameters or fail to do so frequently enough (see also the 3rd stumbling block).

*Feedback Data:* Current feedback data is a key input variable for a logistic analysis of the production. Moreover, manufacturing control methods – such as a WIP Regulating Order Release or a Backlog Control – are frequently dependent on current feedback data. Business analyses, however, show that many employees provide late feedback data for orders and operations or fail to provide it at all. Furthermore, industrial data compilation systems are to some degree not integrated into the PPC software [Beck-02].

***5. Process Disruptions***
Process disruptions can impact machinery, employees or materials [Trac-97]. Here only machine disruptions are considered. If a machine breaks down the planned output can then not be attained even when all other necessary prerequisites are given. Just because of the fear of a machine breaking down, a number of enterprises maintain very high safety stock levels so that the service level can be ensured even when there is a longer disruption. Machine disruptions can however also be guarded against with regular and preventative maintenance or by using proven and reliable technology. Furthermore, emergency plans can be developed in order to minimize the impact of process disruptions.

***6. Employee Qualifications***
Frequently, both personnel working on the shop floor as well as those in production planning and control do not have the basic knowledge of production logistics that is required. On the shop floor this leads to poor sequencing compliance and to delayed feedback data (see 4th stumbling block), whereas, in the production planning and control, the incorrect parameterization of methods (see 3rd stumbling block) and poor compliance in the order release (manual overriding) and capacity control result. It is therefore critical to uncover lacking qualifications and to close knowledge gaps.

***7. Diverging Enterprise and Employee Interests***
The enterprise's and employees' interests do not always coincide. Differences can be present both objectively as well as – often due to a lack of employee qualifications (see 6th stumbling block) – only subjectively felt. An example of the latter would be an employee, who after reducing the WIP works slower because he fears that there will not be sufficient work.

More difficult cases are those in which there is objectively a (considerable) difference between the interests of the employees and the enterprise. An example here is so-called 'progress chasers'. When the PPC functions properly their work is redundant, because the important orders are then completed on-time without their progress being chased [Wien-03b]. In this case and in other comparable situations an enterprise should try to employ the worker who is impacted for other tasks. An example of this would be e.g., the retraining of 'progress chasers' to experts in reducing setup times, for statistical process controls or of course for production planning.

To some degree diverging interests also arise due to unsuitable payment systems. If an employee's wages are calculated alone through the yielded production quantity, incentive arises for shifting setup operations or for producing more than the market demand.

At any rate the enterprise should therefore thoroughly analyze the employees' interests in order to recognize right from the start possible resistances to a new configuration of the manufacturing control.

## 29.8 Conclusion

The theoretical comprehension of the correlations between the manufacturing control tasks, the logistic objectives and the logistic actuating and control variables has increased considerably in the last years. It is now possible to pointedly control the logistic objectives with an acceptable amount of effort. Moreover, enterprises can identify the causes for insufficiently attaining logistic targets comparably easily and derive suitable measures for improvement.

The lack of a strong logistic performance can thus be traced to external influences less and less often. Well functioning production logistics are therefore an increasingly important prerequisite for maintaining competitiveness.

Conversely, there are substantial opportunities for enterprises that quickly transfer existing theoretical knowledge of production logistics to the industrial practice. The key advantages are reduced tied-up capital due to smaller inventories and a greater customer loyalty due to shorter delivery times and higher delivery reliability. From there, chances to accelerate the growth of sales and increase profits develop.

## References


[Bonv-97*] Bonvik, A.M., Couch, C.E., Gershwin, B.S.: A comparison of production-line control mechanisms. Int. J. Prod. Res. **35**(3), 789–804 (1997)

[Beck-02] Beckert, B., Hudetz, W.: Stand und Potenzial produktionsnaher Datenverarbeitung. PPS-Management **7**(2), 35–39 (2002)

[Chan-94*] Chang, T.M., Yih, Y.: Generic kanban systems for dynamic environments. Int. J. Prod. Res. **32**(4), 889–902 (1994)

[Ever-99] Eversheim, W. et al.: Ausprägungen der Produktionsplanung und -steuerung. In: Eversheim, W., Schuh, G. (eds.): Produktion und Management, vol. 4, pp. 14–60 - 14–83. Springer, Berlin (1999)

[Gläß-95] Gläßner, J.: Modellgestütztes Controlling der beschaffungslogistischen Prozeßkette. VDI Progress Reports, Series 2, No. 337, Düsseldorf (1995)

[Hopp-96*] Hopp, W.J., Spearman, M.L.: Factory Physics. Irwin, Chicago (1996)

[Trac-97] Tracht, T.: Auditierung der Produktionsplanung und -steuerung. VDI Progress Reports, Series 2, No. 430, Düsseldorf (1997)

[Wien-02] Wiendahl, H.-H.: Situative Konfiguration des Auftragsmanagements im turbulenten Umfeld. Jost-Jetter Verlag, Heimsheim (2002)

[Wien-03b] Wiendahl, H.-P., Begemann, C., Nickel, R.: Die klassischen Stolpersteine der PPS und der Lösungsansatz 3-Sigma-PPS. In: Baumgarten, H., Wiendahl, H.-P., Zentes, J. (eds.) Springer Experten System Logistik-Management. Strategien, Konzepte, Praxisbeispiele, vol. 2, part 7, Beitrag 7/03/01/02, pp. 1–28

[Wien-03c] Wiendahl, H.-H.: Marktanforderungen verstehen, Stolpersteine erkennen. In: Flexible atmende Produktion: Auftragsschwankungen in den Griff bekommen. Management Circle Seminar, Munich, (2003)

# Index



H. Lödding, *Handbook of Manufacturing Control*,
DOI 10.1007/978-3-642-24458-2, © Springer-Verlag Berlin Heidelberg 2013 

MIX
Papier aus verantwortungsvollen Quellen
Paper from responsible sources
FSC® C105338

If you have any concerns about our products,
you can contact us on
**ProductSafety@springernature.com**

In case Publisher is established outside the EU,
the EU authorized representative is:
**Springer Nature Customer Service Center GmbH**
**Europaplatz 3, 69115 Heidelberg, Germany**

Printed by Libri Plureos GmbH
in Hamburg, Germany